Molecule and Ion Transport through Polymer Membranes

Yong Soo Kang

WILEY VCH

Author

Professor Yong Soo Kang
Hanyang University
Seoul 04763
Republic of Korea

Cover Design: Wiley
Cover Image: © Jimin Ha

Typesetting: Lumina Datamatics, Chennai, India

Library of Congress Card No.: applied for

British Library Cataloguing-in-Publication Data A catalogue record for this book is available from the British Library.

Bibliographic information published by the Deutsche Nationalbibliothek
The Deutsche Nationalbibliothek lists this publication in the Deutsche Nationalbibliografie; detailed bibliographic data are available on the Internet at <http://dnb.d-nb.de>.

Print ISBN: 9783527353835
ePDF ISBN: 9783527846993
ePub ISBN: 9783527846986
oBook ISBN: 9783527847006

Printed and bound by
CPI Group (UK) Ltd, Croydon, CR0 4YY

C9783527353835_171225

Contents

Preface

In an era of significant advancements in energy storage, environmental conservation, and materials science, the need for efficient separation technologies has increased. Membrane technology, particularly focused on molecule and ion transport through polymer membranes, is critical to many of these developments. This book, *Molecule and Ion Transport through Polymer Membranes*, provides a comprehensive yet accessible overview of the fundamental principles governing transport phenomena in polymeric materials. It is intended for students, newcomers to the field, and professionals seeking a structured reference.

The book is divided into two parts: Part I covers the transport of neutral molecules, such as gases and vapors, explaining concepts such as like diffusivity, solubility, and permeability. Through detailed examination of the solution–diffusion model, Fick's law, and various permeation mechanisms, this section establishes a solid framework for understanding the transport properties of uncharged species. Part II focuses on the behavior of ions in polymeric membranes, introducing foundational electrochemical concepts essential for understanding ion transport, alongside an in-depth discussion on polymer electrolytes and ion exchange membranes. By paralleling the study of molecular and ionic transport, this book highlights both their unique behaviors and shared principles, such as the influence of polymer structure on permeation.

This work stands out by combining molecular and ionic transport within a single volume. It shows how free volume theory, typically applied to gas permeation, also helps explain ionic mobility. Such insights underscore the versatility and relevance of polymer membranes in various applications, from fuel cells to water water-splitting systems. The text book also examines emerging concepts, such as facilitated transport, which can enhance permeability and selectivity, offering possibilities for future research.

Ultimately, *Molecule and Ion Transport through Membranes* aims to be a valuable resource, encouraging readers to explore and innovate within the field of polymeric membrane science. Whether for foundational knowledge or practical insights, this book serves as both a guide and an inspiration for advancing materials designed to meet modern challenges.

Acknowledgement

The book cover, designed by Jimin Ha, illustrates stepping stones that help a child cross a stream, analogous to carriers facilitating molecule and ion transport in solid polymer membranes.

1

Overview of Molecule and Ion Transport Through Polymer Membranes

Energy and environmental issues have become increasingly important on a global scale, as they are often interconnected. For example, hydrogen fuel cells, which generate electricity with minimal CO_2 emissions, offer an energy-efficient solution that addresses both environmental and energy concerns. To tackle these challenges, industries such as electrochemical energy conversion and storage as well as petrochemicals are focusing on more energy-efficient processes to generate fuels such as hydrogen. Among the key technologies supporting these efforts is membrane-based separation, which plays a growing key role in both academic research and industrial applications.

Membrane technologies enable the selective transport of molecules and ions through different mechanisms based on the pore size of the polymeric membrane, as shown in Figure 1.1 [1]. When pores are large, the transport of molecules, for instance, follows Newtonian flow, while smaller pores lead to Knudsen flow or molecular sieving. In nonporous membranes, the solution–diffusion mechanism dominates (Figure 1.1d). This book will focus specifically on diffusional transport of small molecules – gases, vapors, and liquids – and ions, through nonporous polymer membranes based on the solution-diffusion mechanism.

A solid understanding of molecule and ion transport through polymer membranes is essential for designing efficient materials and optimizing their performance. The

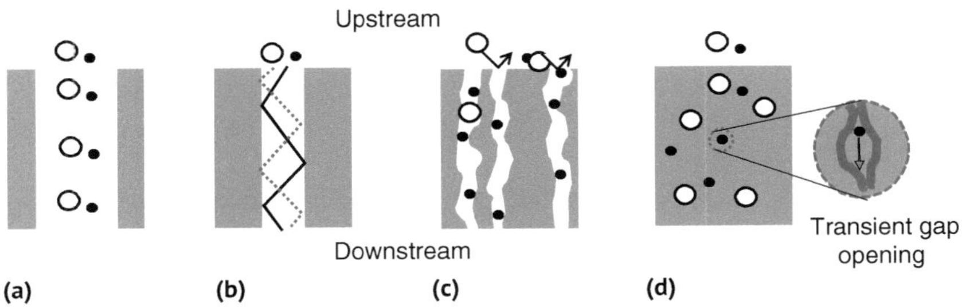

Figure 1.1 Schematic representation of main mechanisms of membrane-based separations: (a) Newtonian flow; (b) Knudsen flow with pore 2–50 nm; (c) sieving mechanism with pore <0.7 nm; and (d) solution–diffusion mechanism without pore. *Source*: Adapted from [1].

Molecule and Ion Transport through Polymer Membranes, First Edition. Yong Soo Kang.

solution–diffusion model is represented by Fick's law for molecule transport and additional Ohm's law for ion transport, both of which describe how diffusional flux is driven by concentration and electric potential gradients, respectively. Since polymer membranes are widely used in separation processes, understanding these transport mechanisms is crucial for improving membrane efficiency in various applications. This book starts with the "Introduction to Polymeric Materials" and then continues with two parts containing "Molecule Transport Through Polymer Membranes" and "Ion Transport Through Polymer Membranes." Thus, in this chapter, similarities and differences between molecule and ion transport will be briefly addressed first, followed by overviews of the two parts.

1.1 Molecule Transport

Diffusional transport of molecules can be readily visualized by dropping a black ink droplet into water without any convectional flow. Black-colored ink spreads out through water three-dimensionally by diffusion process (Figure 1.2a). Mass transport through polymeric materials is also evidenced by air transport, both oxygen and nitrogen, through a balloon wall made of rubbery polymers. When a balloon is blown to expand its size with a higher pressure (p_{in}) than the outer atmospheric pressure (p_{out}) at time zero, the balloon after a certain time (hours) will be shrunk into a smaller or crushed shape because of the mass (oxygen and nitrogen) transport across the balloon wall (Figure 1.2b).

Adolph Fick in 1855 wrote that "The transfer of salt and water occurring in a unit of time, between two elements of space filled with differently concentrated solutions of the same salt, must be, *caeteris paribus*, directly proportional to the difference of concentration, and inversely proportional to the distance of the elements from one another." In mathematical language, this may be thus expressed as [2]:

$$J_j = -D_j \frac{dc_j}{dx} \tag{1.1}$$

where one-dimensional molar flux J_j (mol/cm^2 sec) for x (cm) direction is proportional to the concentration gradient dc_j/dx (mol/cm^3 cm) of species j, and the proportionality constant D_j (cm^2/sec) is called diffusion coefficient [2, 3]. Equation (1.1), commonly called as the Fick's first law, has been conveniently used to describe the diffusional mass transport.

In the case of gases, the pressure gradient dp_j/dx is frequently used rather than the concentration one, mostly for convenience:

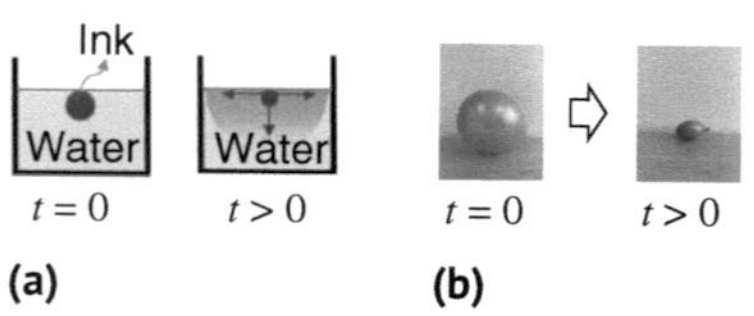

Figure 1.2 Schematics for (a) diffusion of black-colored ink in water and (b) photographs for air transport across the balloon wall with time because $p_{in} > p_{out}$. *Source*: Y.S. Kang.

$$J_j = -P_j \frac{dp_j}{dx} \tag{1.2}$$

Here, P_j is called the permeability or permeation coefficient, defined as a product of the diffusion coefficient D_j and the solubility coefficient S_j: $P_j = D_j \times S_j$. The more detailed explanation is available in Section 3.1 of Chapter 3.

However, mass flux is more rigorously expressed by the chemical potential gradient, $d\mu_j/dx$, rather than the concentration gradient, dc_j/dx. The chemical potential, μ_j, is given by $\mu_j = \mu_j^0 + RT \ln a_j$, where a_j is the chemical activity, defined as $a_j = \gamma_j c_j$ with γ_j being the activity coefficient and μ_j^0 is the standard chemical potential at $a_j = 1$.

Under ideal conditions with the activity coefficient $\gamma_j = 1$, $d\mu_j/dx$ is directly related with dc_j/dx as:

$$\frac{d\mu_j}{dx} = \frac{RT}{c_j}\frac{dc_j}{dx} \tag{1.3}$$

Substituting Equation (1.3) into (1.1) gives:

$$J_j = -\left(\frac{D_j c_j}{RT}\right)\frac{d\mu_j}{dx} \tag{1.4}$$

Therefore, Equation (1.4) is another expression of Fick's law in terms of chemical potential gradient.

1.2 Ion Transport

Ion possesses electrical charge in addition to its mass. Therefore, driving forces for ion transport involve two different modes: one for mass and the other for charge. The former is chemical potential gradient, as described previously, and the latter is the electric potential gradient. Therefore, the total driving force can be expressed as electrochemical potential gradient, where it is summation of chemical potential gradient and electrical potential gradient.

Electrochemical potential $\bar{\mu}_j$ is the total energy required to move a charged particle (like an ion or electron) from one point to another, combining both its chemical potential μ_j (J/mol) (related to concentration) and the electrical potential φ (V = J/C) (related to charge), which is mathematically expressed as:

$$\bar{\mu}_j = \mu_j + z_j F\varphi \tag{1.5}$$

where the charge number z_j is positive for cation and negative for anion, and F (C/mol) is the Faraday constant.

Then, the electrochemical potential gradient $\bar{\mu}_j/dx$ can be readily obtained from Equations (1.3) and (1.5) as:

$$\frac{d\bar{\mu}_j}{dx} = RT\left(\frac{1}{c_j}\frac{dc_j}{dx} + \frac{z_j F}{RT}\frac{d\varphi}{dx}\right) \tag{1.6}$$

Replacing the chemical potential gradient $d\mu_j/dx$ in Equation (1.4) with electrochemical potential gradient $\bar{\mu}_j/dx$ of Equation (1.6) yields:

$$J_j = -D_j c_j \left(\frac{1}{c_j} \frac{dc_j}{dx} + \frac{z_j F}{RT} \frac{d\varphi}{dx} \right) \tag{1.7}$$

Equation (1.7) is called the Nernst–Planck electrodiffusion equation contributed from two different transport mechanisms (the first term for diffusion and the second for migration), which will be explained in more detail in Section 8.5 of Chapter 8.

Electron transport across an electronic conductor can be conveniently described by Ohm's law. It says that the electric current I (A) is linearly proportional to applied voltage V (V) and inversely proportional to resistance R (Ω): $I = V/R$, which can be rearranged to:

$$j = -\sigma \frac{d\varphi}{dx} \tag{1.8}$$

where j (A/cm^2 = C/cm^{2}sec) is the current density, $d\varphi/dx$ is the electric potential gradient and σ (S/cm = 1/Ω cm) is the conductivity.

Equation (1.8) can be applied to the ion transport associated with the electrochemical potential gradient because the ion contains charges in addition to mass. In real electrolyte solutions containing salts dissolved in a solvent, mass transport always accompanies the charge transport because every ion in the electrolyte has its own charge. Therefore, there will be two different ion transport modes: diffusion due to the concentration gradient and migration or drift due to the electric potential gradient. Such differences must be always recognized as long as you deal with the ion transport.

1.3 Similarities and Differences Between Molecule and Ion Transport

Free volume theory and facilitated transport phenomena help explain how both molecules and ions transport through solid-state polymers, although their driving forces are different. For ion transport, overall electroneutrality should be macroscopically met in any electrochemical devices. This section briefly introduces the key concepts to transport in polymers, including free volume theory, facilitated transport, diffusion, migration, transport parameters, and electroneutrality.

1.3.1 Free Volume Theory

When a molecule dissolved in a polymer finds an empty space or free volume in a polymer matrix, it can move from one position to another by diffusional jump due to its thermal motion at a given temperature. In this case, diffusion process can be determined by the diffusional jump distance δ and also by the jump frequency ϕ, which are mostly determined by the amount of the free volume and its size, as shown in Figure 1.3a.

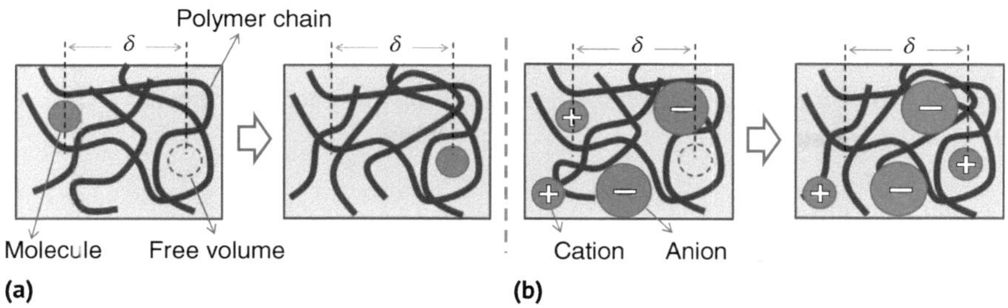

Figure 1.3 Schematic drawings of diffusional jump with distance δ of (a) molecule and (b) ion through free volume or empty space among polymer chains. *Source*: Y.S. Kang.

The direct relationship between the diffusion coefficient and the amount of the free volume in a given polymer has been observed. One of the most popular models is the free volume theory, suggesting that the diffusion coefficient D exponentially increases with the amount of free volume or the size of the free volume:

$$D = D_0 \exp\left(-\frac{B_d}{f}\right) \tag{1.9}$$

where f is the fractional free volume (FFV) defined as $f = V_F/V$ with V_F and V being the total specific free volume and the specific volume, respectively. B_d represents the minimum volume size required for diffusional jump, and D_0 is the pre-exponential factor. Since no extra energy is assumed to be necessary for redistribution of the free volume in a liquid-like matrix, it works well with most rubbery polymers because of their high chain mobility. Surprisingly, it also works well even common glassy polymers where free volume redistribution would hardly occur because of their practically frozen chain mobility in glassy states.

The free volume theory can also be applied for ion transport through solid polymer electrolytes (SPEs) and ion exchange membranes (IEMs). This is because the basic concepts of the diffusional process for both molecule and ion transport are nearly equivalent to each other as schematically shown in Figure 1.3a and b. More detailed descriptions for diffusion of molecules and ions will be covered in Chapters 3 and 8, respectively, in this book. All these results suggest that the fundamental concepts of transport of (small and neutral) molecules and ions are nearly equivalent to each other according to the free volume theory.

1.3.2 Facilitated Transport in the Solid State

Another similarity between molecule and ion transport is the facilitated transport in the solid state in addition to the free volume theory. Facilitated transport was originally observed in liquid blood for oxygen transport from heart to organs in animals. However, facilitated transport in the solid state has been observed and intensively investigated recently, which will be described in detail in Chapter 4 of this book. The facilitated transport is the sum of the two different transport modes of the Fickian transport and carrier-mediated transport. The Fickian transport, governed by the solution–diffusion mechanism, is a mass transport caused by the concentration gradient. The carrier-mediated transport occurs through the reversible interactions between a solute A and carrier C, where the carrier C is defined as any kind of chemical compound, which *selectively* and *reversibly* interacts the specific solute A from its mixture as:

$$A + C \underset{k_2}{\overset{k_1}{\rightleftarrows}} [A \cdot\cdot C] \qquad K = k_1/k_2 \tag{1.10}$$

where k_1 and k_2 are forward and backward reaction rate constants with equilibrium constant K. One characteristic feature of facilitated transport in the solid state is to simultaneously increase the flux of A, represented by its permeability, and the selectivity of A over the others, which are commonly in conflict among most polymers: high permeability with low selectivity and *vice versa*. Therefore, facilitated transport in the solid state has huge advantage to improve the separation performance markedly.

The first example of the facilitated molecule transport in the solid state is O_2 transport through a membrane containing Co(II) or Fe(II) coordinated in polymers, as shown in Figure 1.4a, resulting in the high permeability as well as high O_2/N_2 selectivity. Transport of olefins like ethylene and propylene is markedly increased by the olefin carrier action of silver ions Ag^+ dissolved in polymer to make polymer electrolytes (Figure 1.4b). Here, Ag^+ has selective and reversible interactions with only olefins, but not with paraffins like ethane and propane. As a result, the propylene permeability increases more than 100-fold with almost 500-fold increase in the propylene selectivity over propane through silver polymer electrolytes. These are representative examples of facilitated molecule transport in the solid state.

In the case of ion transport through SPEs and IEMs facilitated ion transport in the solid state has been frequently observed, which is not commonly termed as facilitated transport anyway. Here representative examples of facilitated ion transports in the solid state will be introduced. For instance, lithium salts like $LiCF_3SO_3$ are dissolved in a polymer matrix containing ligands of O, N, or S atoms to form polymer electrolyte. Then we can imagine the specific and reversible interactions between Li^+ and oxygen atoms in polymers like poly(ethylene oxide) for facilitated Li^+ transport in SPEs (Figure 1.5a). In IEMs like proton exchange membranes (PEMs) with microphase-separated structures containing ion channels with fixed charge group of sulfonate anions —SO_3^- dispersed in hydrophobic polymer matrix, specific and reversible interactions between H^+ and —SO_3^- at the interface of ion channels (Figure 1.5b), which facilitate the proton transport through the interfacial layer of the ion channels.

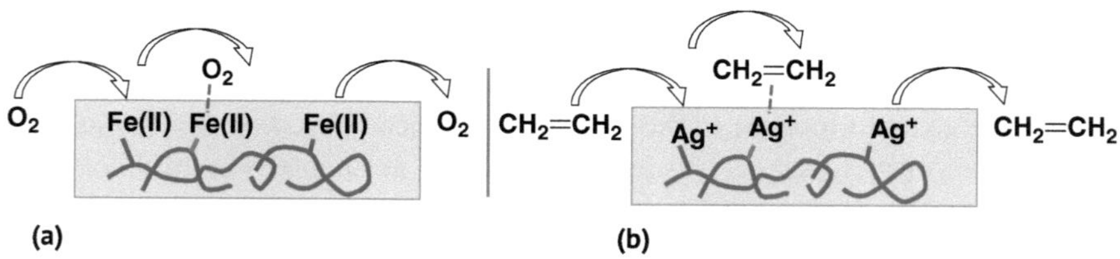

Figure 1.4 Schematic illustrations of facilitated transport in the solid state for (a) oxygen and (b) ethylene. (--) represents coordinative interaction. *Source*: Y.S. Kang.

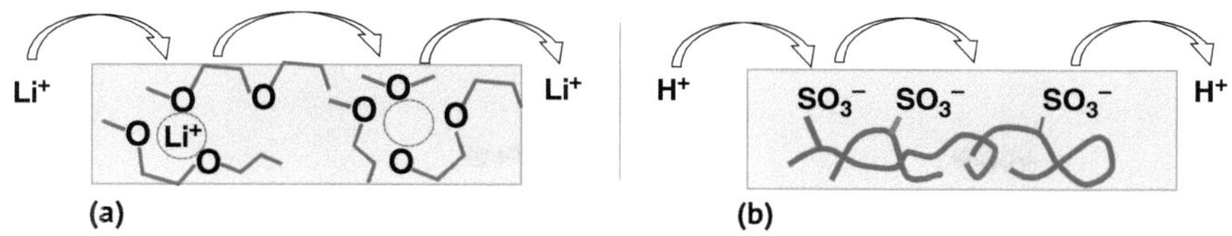

Figure 1.5 Schematic illustrations of facilitated transport in the solid state for (a) Li^+ in polymer electrolyte and (b) H^+ in cation exchange membrane. *Source*: Y.S. Kang.

Table 1.1 Characteristics and Differences of molecule transport and ion transport.

	Molecule transport	**Ion transport**
Transporting species	Molecules with mass only	Ions with mass and charge
Transport mechanism	Diffusion	Diffusion and migration (or drift)
Driving force	Chemical potential gradient (concentration gradient)	Electrochemical potential gradient (concentration and electric potential gradient)
Governing law	Fick's law	Fick's law and Ohm's law
Transport model	Free volume theory	Free volume theory
Facilitated transport	With carriers	With carriers
Constraint	No	Electroneutrality

Therefore, as a short summary, facilitated transport similarly prevails in both molecule and ion transport in the solid state because of the selective and reversible interactions of molecules and ions with their corresponding carriers.

1.3.3 Diffusion and Migration

Even though the transport mechanisms of molecules and ions are commonly described by the free volume theory, the fundamental difference between molecule and ion transports is due to the fact that ion has electric charge whereas molecule does not. Therefore, ion has two different transport modes or fluxes: diffusional flux j_d due to the concentration gradient, described by the Fick's law, and migrational flux j_m due to the electric potential gradient, described by Ohm's law. Therefore, it is anticipated that the directions of the diffusion and the migration are the same or opposite, depending on the voltage applied, as shown in Figure 1.6a and b. It is important to note here that the total current density j (C/cm^2 sec) is the vectorial summation of diffusion and migration: $j = j_d + j_m$, suggesting that j becomes either small or big depending on the directions of j_d and j_m. A summary of the characteristics and differences of molecule and ion transports is provided in Table 1.1.

1.3.4 Transport Parameters and Electroneutrality

Transport properties of gas in polymeric membranes are mostly characterized by diffusion coefficient D_j, solubility coefficient S_j, permeability P_j of molecular species j,

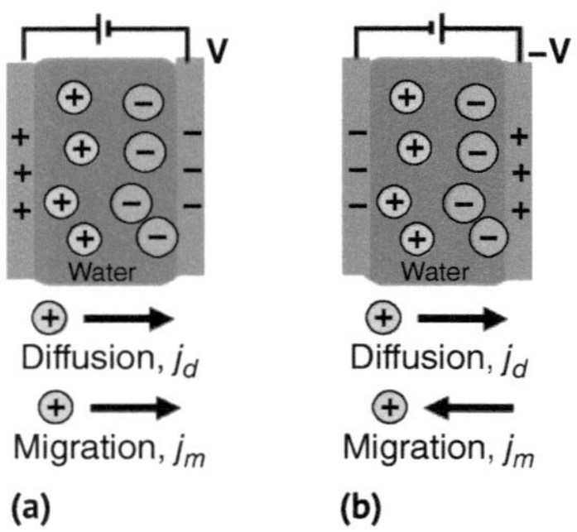

Figure 1.6 Schematic transport of cation under applied potential difference *V*: (a) co-current and (b) counter-current between diffusion and migration. *Source*: Y.S. Kang.

Table 1.2 Transport parameters for molecule transport and ion transport.

Molecule transport	Ion transport
Mass flux, J_j	Current density (charge flux), j_j
Permeability, P_j	Conductivity, σ
Diffusion coefficient, D_j	Diffusion coefficient, D_j
Mobility, u_j	Mobility, u_j
Selectivity, α_{ij}	Permselectivity, P_j^m
—	Transport number, T_j
—	Charge transfer resistance, R_{CT}
—	Ion exchange capacity, IEC

and their permeability ratio α_{ij} of molecular species i and j as a selectivity, whereas those of ion by conductivity σ, mobility u_j, permselectivity P_j^m, and transport number T_j. In addition, charge transport resistance R_{CT} and ion exchange capacity (IEC) also play important roles in determining charge transport in most SPEs and IEMs, respectively. Characteristics of molecule and ion transports are listed in Table 1.2.

1.4 Overview of Molecule and Ion Transport Through Polymer Membranes

This book is divided into two main parts: Part I focuses on molecule transport, while Part II explores ion transport through polymer membranes. In this context, "molecule" refers to uncharged, small-sized species such as gases or organic solvents. Since all materials discussed in this book are polymer-based, Chapter 2 briefly introduces the fundamentals of polymer materials.

1.4.1 Fundamentals on Polymeric Materials

A comprehensive understanding of polymeric materials forms the foundation for studying molecule and ion transports. Chapter 2 explores the interactions of polymer with small molecular solvents, including polymer solution and phase separation behavior. The former will be applied for the sorption behavior of organic vapors and liquid sorbents, while the latter for the formation of microphase-separated structures for asymmetrically microporous support of composite membranes.

The chapter then introduces the concepts of free volume and free volume theory, where the free volume refers to the unoccupied spaces within a polymer matrix that allow for molecules and ions to diffuse. The theory provides a framework for understanding how molecules and ions transport through these spaces and therefore offers insights into the relationship between polymer structure and transport properties.

The nonequilibrium features of glassy polymers and physical aging are also highlighted. Glassy polymers, being in a nonequilibrium state, undergo physical aging at sub-glass transition temperatures, leading to changes in free volume and transport properties and their relationship over time. Lastly, sub-glass transitions, such as the γ-relaxation, involve localized segmental motions that can influence diffusion processes.

Finally, the chapter briefly touches on the crystal structure, crystallization kinetics, and the viscoelastic and mechanical properties of polymers, rounding out the foundational knowledge needed for the study of molecule and ion transport.

1.4.2 Part 1: Molecule Transport Through Polymer Membranes

Understanding the fundamental principles governing molecule transport is essential for designing membranes with tailored properties. Part 1 (Chapters 3–7) of this book discusses both the theoretical and practical aspects of molecular diffusion, sorption, and permeation in polymer systems.

Chapter 3 introduces Fick's Law, a widely accepted solution–diffusion model essential for understanding how molecules permeate polymeric membranes. This model provides the basis for determining key parameters such as the diffusion coefficient, solubility coefficient, permeability, and selectivity. A detailed discussion of the free volume theory, including models proposed by Cohen–Turnbull, Fujita, and Vrentas–Duda, illustrates how molecular transport is influenced by the available free volume within polymer matrices. This chapter also emphasizes mathematical models for transient sorption and permeation through glassy polymers and composite membranes, offering a strong analytical basis for further study.

Chapter 4 builds on the facilitated transport phenomena in solid-state polymer membranes. It highlights various mathematical models, such as the dual transport model, effective diffusion model, limited chain mobility model, and concentration fluctuation model, which provide insights into how molecular transport can be facilitated by carrier-mediated processes. Special attention is given to oxygen, carbon dioxide, and olefin transport, where carriers like metallic ions, surface-activated nanoparticles, and ionic liquids play a pivotal role in improving transport efficiency.

Chapter 5 transitions into selective transport membranes, where the concepts of permeability and selectivity, mathematically defined here, play crucial roles in understanding separation performance. This part investigates the relationship between polymer structure and transport properties, categorized into polymers with flexible chains, stiff chains, and extremely stiff chains with high free volume. The role of physical aging and plasticization in glassy polymers is discussed, providing insights into how membrane performance can change over time.

Chapter 6 covers the measurement of key transport parameters, such as diffusion coefficient, solubility coefficient, and permeability. It provides detailed theoretical backgrounds and experimental methods, including manometric, gravimetric, and time-lag techniques, for assessing these transport properties. This chapter is critical for readers interested in practical methodologies for evaluating polymer membranes.

Chapter 7 introduces practical applications, including separations of hydrogen, oxygen/nitrogen, carbon dioxide, and olefin/paraffin, including propylene/propane and ethylene/ethane mixtures, offering real-world relevance for the theories and models covered in earlier chapters.

The chapters collectively provide a comprehensive overview of small neutral molecular transport through polymer membranes to offer a clear understanding of the challenges and opportunities in this field.

1.4.3 Part 2: Ion Transport Through Polymer Membranes

Part 2 (Chapters 8–12) of the book, titled *Ion Transport Through Polymer Membranes*, shifts the focus from small neutral molecules to charged species, examining the unique challenges and mechanisms involved in ion transport. Chapters 8–12 discuss the electrochemical fundamentals, polymer electrolytes, and IEMs that are pivotal for energy conversion and storage devices involving ion transport, such as secondary batteries, solar cells, fuel cells, and water-splitting applications.

Chapter 8 provides the necessary background in electrochemistry, focusing on key concepts of electrochemical potential and its influences on electrochemical reaction kinetics and ultimately ion transport. The ion transport mechanism, including diffusion and migration or drift due to the concentration gradient and electric potential gradient, respectively, is also introduced to provide insight into ion conductivity and transport number. This chapter also explores the electric double layer and how it impacts charge transfer and mass transport of ions, providing a theoretical basis for the subsequent chapters on polymer electrolytes and IEMs.

Chapter 9 delves into polymer electrolytes, which are essential for devices like secondary batteries and sensitized solar cells. It covers the formation of polymer electrolytes, the generation of free charge carriers, and the various mathematical models of ionic conduction, such as the free volume model, conformational entropy model, dynamic bond percolation (DBP) model, and Anderson and Stuart model. Additionally, it explores the role of temperature and glass transition in influencing polymer chain mobility, which directly affects ionic transport. Emerging polymer electrolytes, including polymer-in-salts (PISs) and nanocomposite structures, are discussed as promising materials for future energy storage applications.

Chapter 10 focuses on IEMs, including cation and anion exchange membranes (CEMs and AEMs), as well as bipolar membranes (BPMs). This chapter highlights the structure–property relationships that dictate conductivity and selectivity in IEMs. The concept of permselectivity is introduced to explain how these membranes differentiate between ions of the same charge. It also examines the formation of the space charge region at the bipolar junction in BPMs and the generation of huge electric fields, which can be utilized for applications like water electrolysis for energy-saving processes.

Chapter 11 outlines the experimental methods used to measure ionic transport properties and transport numbers, including direct current and alternating current techniques. The chapter provides practical insights into theoretical bases and experimental methods like linear sweep voltammetry (LSV), cyclic voltammetry (CV), potentiostatic intermittent titration technique (PITT), and electrochemical impedance spectroscopy (EIS), which are key to characterizing SPEs and IEMs.

Finally, Chapter 12 explores the diverse applications of polymer electrolytes and IEMs in real-world technologies. From secondary batteries and sensitized solar cells to fuel cells and electrochemical water splitting, this chapter provides a comprehensive overview of how the theoretical principles discussed in previous chapters are applied in cutting-edge technologies.

Overall, this part of the book provides an in-depth understanding of ion transport through polymer membranes, emphasizing both theoretical models and their practical applications in energy-related devices.

References

1 W.J. Koros, G. Fleming, Journal of Membrane Science 83(1) (1993) 1–80.
2 A. Fick, Journal of Membrane Science 100(1) (1995) 33–38.
3 A. Fick, The London, Edinburgh, and Dublin Philosophical Magazine and Journal of Science 10(63) (1855) 30–39.

2

Introduction to Polymeric Materials

Polymers are macromolecules made up of long chains of repeating units, or monomers, connected by covalent bonds. Polymer chains can vary significantly in molecular weight from several thousand to millions, with a distribution of molecular weights being a common characteristic. The high molecular weight of polymers, combined with their entangled chain structures, plays a crucial role in determining their physicochemical properties. For instance, these features largely influence mechanical strength and the transport properties of small molecules such as gases or organic solvents and ions. Polymers, despite their size, have similar chemical properties to their smaller counterparts.

This chapter provides a foundational understanding of polymer materials to support the exploration of molecule and ion transport through polymer membranes. By examining the relationship between the structure of polymer materials and their transport properties, this discussion aims to offer valuable insights into how molecular architecture influences key behaviors in these systems.

For further in-depth information and as primary references throughout this chapter, I would like to acknowledge the following two books by: (1) *Polymer Physics* [1] and (2) *Polymer Materials: Macroscopic Properties and Molecular Interpretations* [2]. These references provide extensive discussions on the macroscopic properties of polymer materials and their molecular interpretations, serving as essential resources for readers seeking a deeper understanding of the subject.

2.1 Interactions of Polymer with Solvent and Polymer Solution

The interaction between a polymer and a solvent largely determines the behavior of a polymer solution and the quality of the solvent. Unlike small molecules, polymers have significantly higher molecular weights, requiring modifications to traditional solution theories. In 1942, Flory and Huggins independently developed mathematical equations to describe polymer solutions based on lattice model, leading to the Flory–Huggins theory. This theory remains a fundamental framework

Molecule and Ion Transport through Polymer Membranes, First Edition. Yong Soo Kang.

for understanding polymer solution properties, including phase behavior and phase separation.

When mixing two components 1 and 2, the free energy change of mixing ΔG_m is:

$$\Delta G_m = \Delta H_m - T\Delta S_m \tag{2.1}$$

For ideal solutions $\Delta G_m < 0$ because $\Delta H_m = 0$ and $\Delta S_m > 0$, and therefore, the ideal solution is always miscible. In polymer solutions, however, deviations from ideality are common and play a significant role in determining miscibility and phase separation.

2.1.1 Flory–Huggins Lattice Theory

The Flory–Huggins theory of polymer solutions was developed by assuming (1) a mean field approximation, (2) no volume change on mixing, and (3) dilute solution. It was treated that each repeating unit of a polymer, called a *segment*, occupies one position in a lattice and so does each solvent molecule, as shown in Figure (2.1).

Thus, Flory and Huggins obtained as:

$$\frac{\Delta S_m}{N} = -R\left(v_1 \ln v_1 + \frac{v_2}{x} \ln v_2\right) \tag{2.2}$$

where N is the mole number, v_1 and v_2 are the volume fraction of the solvent 1 and the polymer 2 in the mixture, respectively, defined as $v_1 = n_1/(n_1 + n_2x)$ and $v_2 = n_2 x/(n_1 + n_2x)$ in the lattice model as shown in Figure 2.1. n_i is the number of molecules of component i and x is the average number of segments of a polymeric chain. Thus, total number n of lattice sites becomes $n = n_1 + n_2x$ [3].

The enthalpy of mixing, ΔH_m, for polymer solution is obtained as:

$$\frac{\Delta H_m}{N} = z\Delta\omega_{12} v_1 v_2 \tag{2.3}$$

where $N = N_1 + xN_2$ with N_i being the mole number of components i. $\Delta\omega_{12} = \omega_{12} - (\omega_{11} + \omega_{22})/2$ with ω_{ij} representing the interaction energy between i and j and thus having a negative value. When the 1–2 attractions are stronger than the 1–1 and 2–2 attractions, both $\Delta\omega_{12}$ and ΔH_m become negative, and the mixing process is

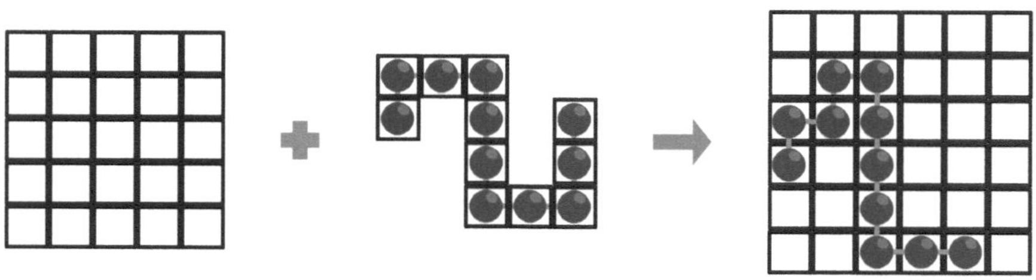

Figure 2.1 Mixing of one polymer chain ($n_2 = 1$ and $x = 11$, the average number of segments of a polymeric chain) with 25 solvent molecules ($n_1 = 25$) in lattice model. The solvent and the segment of the polymer chain are assumed to have the same size as the lattice. *Source*: Redrawn from [1].

exothermic [1]. Note that the strong attraction forces mean highly negative ω value. Equation (2.3) can be rewritten by Flory and Huggins as:

$$\frac{\Delta H_m}{N} = RT\chi_{12}v_1v_2 \tag{2.4}$$

where the Flory–Huggins interaction parameter, χ_{12}, is defined as $\chi_{12} = z\Delta\omega_{12}/RT$. Equation (2.4) suggests that a good solvent having strong interactions with polymer shows a large negative value of both $\Delta\omega_{12}$ and χ_{12}.

The combination of the entropy and enthalpy terms (Equations (2.2) and (2.4)) gives the changes in Gibbs free energy ΔG_m of mixing:

$$\frac{\Delta G_m}{N} = RT\left(v_1 \ln v_1 + \frac{v_2}{x}\ln v_2 + \chi_{12}v_1v_2\right) \tag{2.5}$$

The unit of the F–H interaction parameter χ_{12} is lattice volume. Therefore, the lattice site volume V_0 must be specified whenever χ_{12} is discussed. Typical χ_{12} values for polymer and solvent pairs are listed in ref. [4].

The chemical potential, μ_1, of component 1 representing how ΔG_{mix} changes as one mole of component 1 is added to the polymer solution at constant temperature, pressure and amount of component j except 1 ($\mu_i = (\mathrm{d}\Delta G_m/\mathrm{d}n_i)_{T,p,nj}$) may be mathematically obtained as [1]:

$$\frac{\mu_1 - \mu_1^0}{RT} = \ln a_1 = \ln v_1 + \left(1 - \frac{1}{x}\right)v_2 + \chi_{12}v_2^2 \tag{2.6}$$

$$\frac{\mu_2 - \mu_2^0}{RT} = \ln a_2 = \ln v_2 + (1 - x)v_1 + x\chi_{12}v_1^2 \tag{2.7}$$

where a_i is the chemical activity of component i.

In the case of gas mixture, $a_i = p_i/p$ with p_i and p being partial and total pressures, respectively, and v_i represents the amount of solvent in a polymer–solvent mixture, i.e. the solubility of component i. Therefore, Equation (2.6) represents the relationship between the gas solubility, represented by v_1, and applied pressure, represented by a_1.

2.1.2 Solubility Parameter Approach

Another approach to obtain the enthalpy of mixing is using the solubility parameter, δ. According to Flory, $\Delta H_m = n_1 z v_2 \Delta\omega_{12}$, where $\Delta\omega_{12} = \omega_{12} - (\omega_{11} + \omega_{22})/2$ as described previously. In the case where the attraction forces are nonspecific, ω_{12} is approximated to be the geometric mean value of ω_{11} and ω_{22}.

$$\omega_{12} = \sqrt{\omega_{11}}\sqrt{\omega_{22}} \tag{2.8}$$

Thus,

$$\Delta H_m = \frac{1}{2} z n_1 v_2 \left\{\sqrt{\omega_{11}} - \sqrt{\omega_{22}}\right\}^2 \tag{2.9}$$

The interaction energy per lattice volume $z\omega_{ii}/2V_0$ in the pure state is related to the cohesive energy density (CED_i), which is defined as $CED_i = \Delta E_{vi}/V_i$ (J/cm^3 or

MPa). Here, ΔE_{vi} (J/mol) is the energy of vaporization ($\Delta E_{vi} > 0$) representing the energy required to overcome molecular interactions in the pure state of component i, and V_i (cm^3/mol) is the molar volume of molecule i.

CED_i represents the interaction energy per unit volume among the molecules in the pure substance. It quantifies the total energy needed to separate all neighboring molecules [5]. Therefore, we can write:

$$CED_1 = \frac{\Delta E_{v1}}{V_1} = -\frac{z\omega_{11}}{2V_0} = \delta_1^2 \quad \text{and} \quad CED_2 = \frac{\Delta E_{v2}}{V_2} = -\frac{z\omega_{22}}{2V_0} = \delta_2^2 \tag{2.10}$$

where δ_i ($(J/cm^3)^{1/2}$, or $MPa^{1/2}$) is the Hildebrand solubility parameter of component i. And for the interaction between different molecules:

$$-\frac{z\omega_{12}}{2V_0} = \delta_1\delta_2 \tag{2.11}$$

Substitution of the relationship between ω_{ii} and δ_i (Equations (2.10) and (2.11)) into Equation (2.9) gives:

$$\frac{\Delta H_m}{N} = V_0 v_1 v_2 (\delta_1 - \delta_2)^2 \tag{2.12}$$

This equation shows that the enthalpy of mixing depends on the difference in solubility parameters between two components. The relation between χ_{12} and the solubility parameter can be thus obtained as:

$$\chi_{12} = \frac{V_0}{RT}(\delta_1 - \delta_2)^2 \tag{2.13}$$

Since χ_{12} is related to the square of the difference in solubility parameters (Equation (2.13)), it is clear why the Flory interaction parameter χ_{12} is usually positive. The above approach works reasonably well for nonpolar interactions, which have only van der Waals forces between species, and does not work in mixtures with strong polar or specific interactions, such as hydrogen bonds [5]. The solubility parameter of a good solvent, δ_1, should be similar to that of the polymer, δ_2, to have the least ΔH_m for better mixing of the polymer and the solvent.

The Hildebrand solubility parameter has taken only for dispersion forces between structural units into account. For many liquids and polymers, however, interactions such as polar–polar and hydrogen bonding interactions are present and consequently affect the cohesive energy and solubility parameter. The effects of the interactions, such as both polar–polar and hydrogen bonding interactions, are conveniently accounted by Hansen's solubility parameter comprising three parameters:

$$\delta^2 = \delta_D^2 + \delta_P^2 + \delta_H^2 \tag{2.14}$$

where δ_D, δ_P, and δ_H are for dispersive, polar, and hydrogen bonding interactions. The Hansen's solubility parameter represents the molecular properties of common polymers much more accurately. The solubility parameters are useful in practice because their values of common polymers and solvents are readily available independently in literature and also estimated by the group contribution method [4].

2.1.3 Limitations and Modifications of Flory–Huggins Theory

Basic assumptions Flory and Huggins made were (1) the mean field approximation, (2) no volume change on mixing, and (3) dilute solution, which can be relaxed for applications to real polymer solutions. For instance, experimental results suggest that χ_{12} should consider additional effects for specific interaction, concentration, and temperature. In this section, the effects of only specific interactions will be introduced, but not for concentration and temperature effects. When you are interested in those area, you may refer to reference including the experimental data for concentration effects [6] and for temperature effects [1].

The Flory–Huggins theory was originally derived without considering any interactions among the polymeric segments and the solvent, which may be referred to as mean field approximation. Many solvents and polymers, however, have polar and hydrogen bonding interactions [1]. Such interactions lead to reorientation of the molecules, differing from the pure components, which violates the approximations of the mean field and no volume change on mixing. It is thus postulated that such interactions contribute to the entropy of mixing [1]. However, the effects of the interactions on the entropy by local packing (or reorientation) and the volume changes on mixing in practice are not fully understood yet. Nonetheless, the magnitude of the entropy term should be proportional to the number of paired contacts by analogy with the enthalpy contribution.

Therefore, in practice, the interaction parameter χ_{12} should regard as a free energy term, i.e., contributions from both enthalpy χ_{h12} and entropy χ_{s12}:

$$\chi_{12} = \chi_{h12} + \chi_{s12} \tag{2.15}$$

The enthalpic contribution, χ_{h12}, depends on the size and shape of solvent molecules and also on those of polymers, whereas entropy contribution, χ_{s12}, accounts for the specific interactions [1]. The above equation is rewritten as:

$$\chi_{12} = \frac{V_1}{RT}(\delta_1 - \delta_2)^2 + \chi_{s12} \tag{2.16}$$

where V_1 is the molar volume of the solvent, being replaced by V_0. χ_{s12} for nonpolar/nonpolar systems is known to be about 0.34. In the case of poly(methyl methacrylate) (PMMA) in toluene, for example, $\chi_{h12} = 0.03$ and $\chi_{s12} = 0.42$, suggesting, even in this case, $\chi_{h12} \ll \chi_{s12}$ [1]. The parameters χ_{h12} and χ_{s12} have been tabulated for many polymer solutions and polymer blends, indicating the large contribution of parameter χ_{s12} [5].

The Flory–Huggins theory predicts only upper critical solution temperature (UCST), but not lower critical solution temperature (LCST) in phase diagrams. The concept of phantom chains was also adopted in deriving the Flory–Huggins theory, which is hardly observable in real systems.

Despite the limitations of the Flory–Huggins theory, the lattice theory gives us, in a relatively simple and instructive way, semiquantitative information on the thermodynamics of polymer solutions, including phase separations [6].

2.2 Phase Diagram and Phase Separation

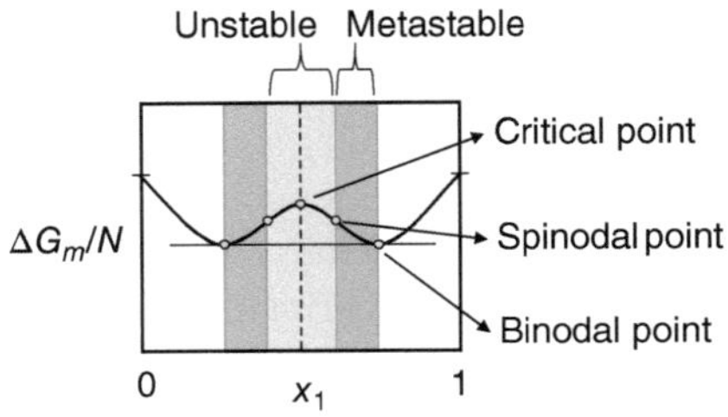

Figure 2.2 ΔG_m vs mole fraction x_1 plot for illustrating binodal, spinodal, and critical points at a given temperature. *Source*: Y.S. Kang.

When two components are mixed, their phase behavior and miscibility can be characterized by ΔG_m as a function of composition and temperature.

When the ΔG_m curve against the mole fraction x_1 at a given temperature is perturbed by the change in, for example, composition or temperature, two minima and two inflection points can be observed, as shown in Figure 2.2.

Two inflection points, frequently referred to as spinodal points, are obtained at [1]:

$$\frac{d^2(\Delta G_m/N)}{dx_1^2} = 0 \Rightarrow \frac{d\mu_1}{dx_1} = 0 \tag{2.17}$$

Region between two spinodal points where $d^2(\Delta G_m/N)/dx_1^2 < 0$ or $d\mu_1/dx_1 < 0$ is called the spinodal region. In the spinodal region, the system becomes unstable due to the decrease in the chemical potential μ_1 with respect to any fluctuation in composition x_1, and therefore the phase separation occurs spontaneously [1].

A common tangent to the plot of ΔG_m against x_1 gives two binodal points where the chemical potentials both components (1 and 2) are equal, indicating phase-separation into regions of A and B :

$$\mu_1^A - \mu_1^0 = \mu_1^B - \mu_1^0 \quad \text{and} \quad \mu_2^A - \mu_2^0 = \mu_2^B - \mu_2^0 \tag{2.18}$$

The critical point where the binodal and spinodal curves meet is associated with [1]:

$$\frac{d^3(\Delta G_{mix}/N)}{dx_1^3} = 0 \Rightarrow \frac{d^2\mu_1}{dx_1^2} = 0 \tag{2.19}$$

Thus,

$$\frac{d\mu_1}{dx_1} = 0 \Rightarrow -\frac{1}{v_1} + \left(1 - \frac{1}{x}\right) + 2\chi_{12}v_2 = 0 \tag{2.20}$$

and

$$\frac{d^2\mu_1}{dx_1^2} = 0 \Rightarrow -\frac{1}{v_1^2} + 2\chi_{12} = 0 \tag{2.21}$$

The solutions for the above conditions, then, gives [1]:

$$v_{2,c} = \frac{1}{1 + \sqrt{x}} \tag{2.22}$$

$$\chi_{12,c} = \frac{1}{2} + \frac{1}{2x} + \frac{1}{\sqrt{x}} \tag{2.23}$$

where $v_{2,c}$ and $\chi_{12,c}$ are the polymer concentration and χ_{12} at the critical point, respectively. For a regular solution ($x = 1$), $v_{2,c} = 1/2$. For a polymer with $x = 100$, $v_{2,c} = 1/11{\sim}0.1$. Therefore, the maximum point is shifted to the left. For high molar mass polymer ($x{\sim}\infty$), the critical value for the interaction parameter becomes:

$$\lim_{x\to\infty} \chi_{12,c} = \frac{1}{2} \tag{2.24}$$

Therefore, when $\chi_{12} > 1/2$, the system becomes immiscible.

The different phase behavior between polymer solution and regular solution is mainly originated from the entropy term. The entropy term for a polymer solution of (v_2/x) ln (v_2) is much smaller than that for a regular solution of x_2 ln (x_2). Therefore, ΔG_m is more unlikely to be negative for polymer solutions, compared to regular solutions. In other words, solvent for polymers is more restrictive than that for regular low molecular components. It is also important to note from Figure 2.2 that $\Delta G_m < 0$ is a necessary condition for miscibility but not a sufficient one.

2.2.1 Phase Diagram

When ΔG_m curve shows two minima with two binodal and two inflection points at T_1, as shown in Figure 2.2, the corresponding points in the phase diagram are positioned. Repeating the procedure by varying the temperature, the data points at different temperatures make the binary phase diagram as shown in Figure 2.3.

2.2.2 Phase Separation

Phase diagram shows that the system is unstable in the range of $v_2^{\mathrm{I}} < v_2 < v_2^{\mathrm{II}}$, metastable in the range of $v_2^A < v_2 < v_2^{\mathrm{I}}$ and $v_2^{\mathrm{II}} < v_2 < v_2^B$, and stable in the other concentration ranges at temperature T_1. The phase stability refers to as the system stability or miscibility with respect to change in composition or temperature.

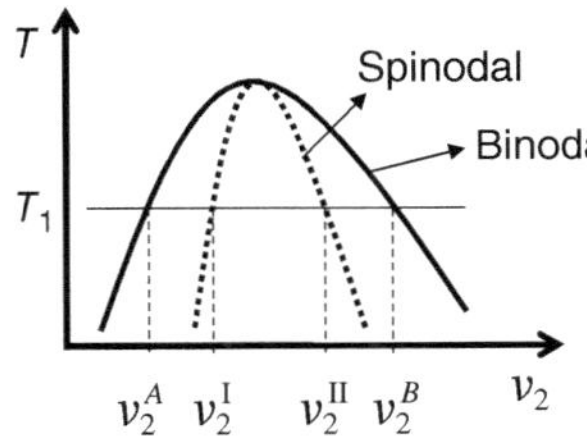

Figure 2.3 A phase diagram, binodal and spinodal curves for two-component system. *Source*: Y.S. Kang.

Phase separation into a polymer-rich phase and a polymer-lean phase takes place via two different mechanisms: the spinodal decomposition (SD) for the unstable region and the nucleation and growth (NG) mechanism for the meta-stable region. In the range of $v_2^{\mathrm{I}} < v_2 < v_2^{\mathrm{II}}$, the system becomes $(\partial\mu_i/\partial v_i) < 0$ and unstable. Thus, SD occurs spontaneously.

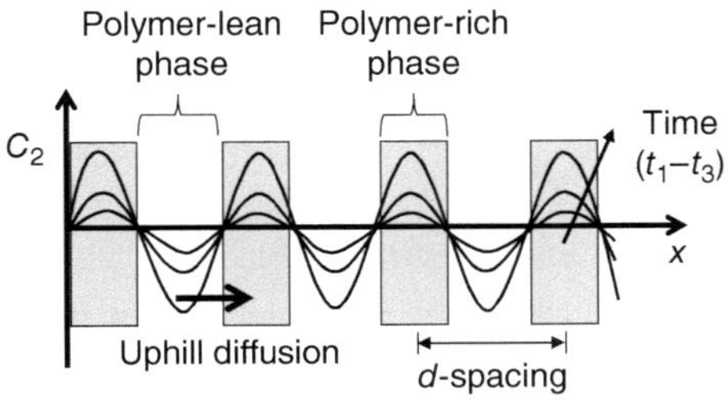

Figure 2.4 Schematic concentration change with time ($t_1 < t_2 < t_3$) in initial stage by the spinodal decomposition. *Source*: Y.S. Kang.

During SD, polymer molecules diffuse against their concentration gradient, i.e. uphill diffusion, as shown in Figure 2.4. Therefore, the concentration of the polymer in

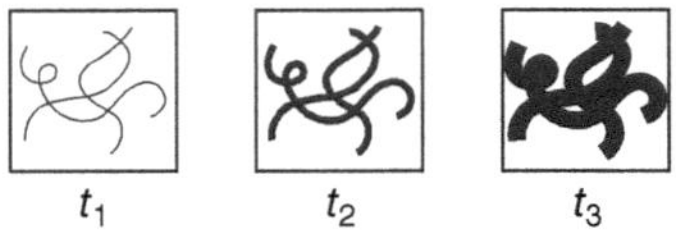

Figure 2.5 Phase-separated structures with time ($t_1 < t_2 < t_3$) by the spinodal decomposition. *Source*: Y.S. Kang.

the polymer-rich phase becomes richer whereas that in the polymer-lean phase it becomes leaner. Furthermore, the initial structure, for example, the *d*-spacing in Figure 2.4, maintains unchanged during the early stage. However, at the late stage, the *d*-spacing changes with time. Finally, the phase-separated structure becomes normally interconnected, lacy structures (Figure 2.5 at $t = t_3$).

The nucleation and growth (NG) mechanism is prevailing in the metastable region of $v_2^A < v_2 < v_2^{\mathrm{I}}$ and $v_2^{\mathrm{II}} < v_2 < v_2^B$. The solution becomes unstable toward small fluctuations in composition. Demixing can commence only when stable nuclei have been formed (Figure 2.6). The nuclei are formed randomly and grow in spherical shape. As a result, spherical particles with various sizes are formed, as shown in Figure 2.7.

Figure 2.6 Concentration change of polymer by the nucleation and growth mechanism. *Source*: Y.S. Kang.

The phase diagram of Figure 2.3 is for a binary system containing polymer and solvent where phase separation is commonly induced by varying the temperature of the system.

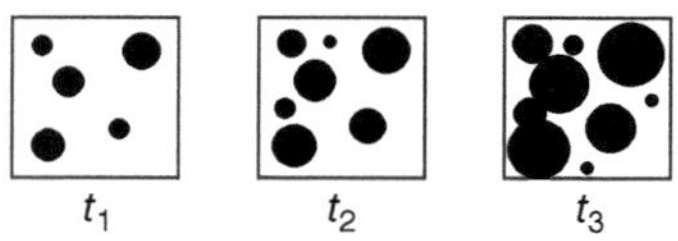

Figure 2.7 Phase-separated structures with time by the nucleation and growth mechanism. *Source*: Y.S. Kang.

The solution behavior of a binary system can also be differently controlled by incorporating a new component to make a ternary system. In particular, when the third component is a nonsolvent (NS), phase separation behavior can be readily controlled by the amount of NS at a given temperature. Typical ternary phase diagram of polymer, P, solvent, S, and NS, at a given temperature is illustrated in Figure 2.8.

Therefore, the morphology of final structure can be controlled by understanding phase behavior including phase separation mechanisms.

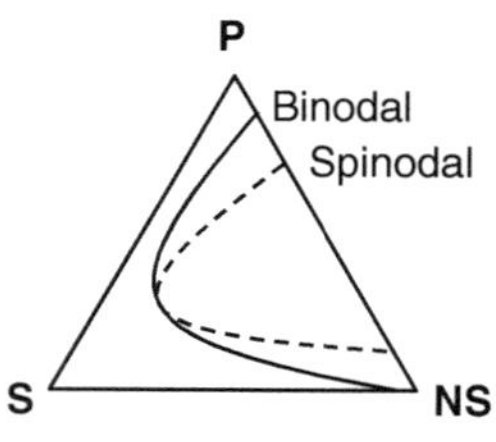

Figure 2.8 Ternary phase diagram of polymer/solvent/nonsolvent system at a given temperature T_1. *Source*: Y.S. Kang.

2.3 Rubbery and Glassy Polymers

Polymers can exhibit either rubbery or glassy behavior depending on their temperature relative to their glass transition temperature, T_g. Below T_g, polymers exist in the glassy state, where their molecular chains are practically frozen in place, resulting in a rigid and brittle material. In contrast, at temperatures above T_g, polymers enter the rubbery state, where increased thermal energy allows the polymer chains to move more freely, imparting elasticity and flexibility.

2.3.1 Rubbery vs Glassy Polymers

Rubbery polymers exist above their glass transition temperature ($T > T_g$), where polymer chains have significant mobility and are in an equilibrium state. They are flexible, elastic, and soft, allowing them to undergo substantial deformation and quickly recover their shape in response to applied stress, a characteristic behavior of elastomers.

In contrast, glassy polymers exist below their glass transition temperature ($T < T_g$), where chain mobility is practically frozen. They are typically rigid, strong, and dimensionally stable but remain in a nonequilibrium state, leading to time-dependent changes in their properties.

A key distinction is that rubbery polymers, being in equilibrium, can instantaneously absorb and dissipate energy due to their high mobility and elasticity. In contrast, glassy polymers, in a nonequilibrium state, experience gradual changes in their physicochemical properties over time, i.e. time-dependent, due to their restricted chain mobility.

2.3.2 Glass Transition Temperature

When a hard glassy polymer is heated slowly, its modulus, E, defined as $E = \sigma/\varepsilon$, where σ and ε are the stress and strain, respectively, may drop several orders of the magnitude from 10^9 to 10^6 Pa at a temperature to become rubbery, which is called glass transition temperature, T_g [7], as shown in Figure 2.9a. At T_g, the polymer's-specific volume also undergoes a noticeable change in its temperature-dependent slope. This change is characterized by the thermal expansion coefficients α_r and α_g, corresponding to the rubbery and glassy states, respectively, as illustrated in Figure 2.9b.

The glass transition temperature, T_g, marks the point at which a hard, glassy polymer begins to soften as increasing temperature. This transition occurs due to the onset of long-range cooperative molecular motion, involving coordinated movement of polymer chain segments comprising 30–50 carbon atoms as the material is heated.

The glass transition temperature, T_g, of polymers can vary widely with their chemical structure. In general, polymers with flexible backbones and small substituent groups (e.g. polyethylene (−82 to −78 °C) and polydimethyl siloxane (−123 °C)) have

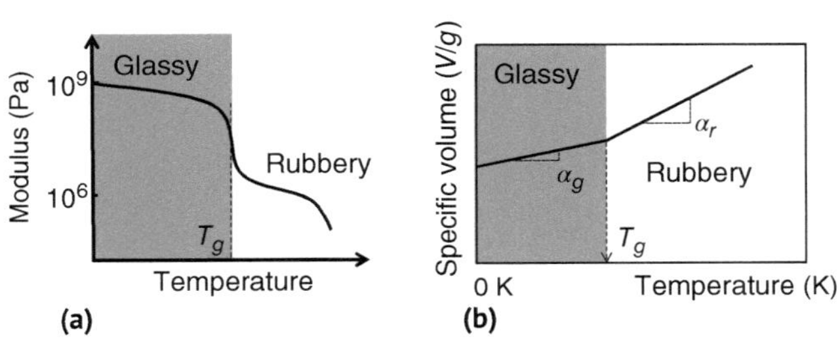

Figure 2.9 Schematic plots of (a) modulus E of amorphous polymer as a function of temperature; and (b) specific volume change by varying temperature. *Source*: Y.S. Kang.

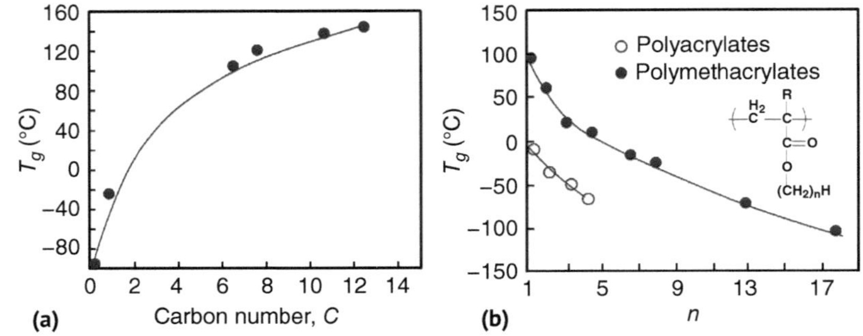

Figure 2.10 Glass transition temperature, T_g, of (a) different vinyl polymers $-(-CH_2-CHX-)_m-$ showing the influence of the size of the inflexible pendant group. Here, the carbon number C = 0 for X = —H, C = 1 for $-CH_3$, C = 6 for -phenyl, C = 7 for -*o*-methyl-phenyl, *C* = 10 for -1-naphthalene, *C* = 12 for -*p*-biphenyl. *Source*: Adapted from [8]; (b) T_g of polyacrylates and polymethacrylates as a function of the number of methylene unit, *n*, in the flexible alkyl group $(-(CH_2)_n-H)$. *Source*: Adapted from [1, 9, 10].

low T_g, while those with rigid backbones, such as polymers containing main-chain aromatic groups (e.g. polysulfone (190 °C) and poly(2,6,-dimethyl-1,4-phenylene oxide, PPO) (220 °C)), have high T_g. Therefore, T_g represents the chain mobility: The more flexible polymer chains have lower T_g, and *vice versa*.

In the cases of vinyl polymers of $-(-CH_2-CHX-)_m-$, the glass transition temperature strongly depends on the flexibility of the pendant X.

When the pendant X is inflexible, it poses restrictions on torsion about a bond in the backbone chain, i.e. the backbone chain becomes less flexible, and this causes the increase in T_g, as shown in Figure 2.10a. However, when X is relatively flexible like alkyl group in polyacrylates or polymethacrylates, T_g decreases with the length of the pendant alkyl group, *n* in $-(CH_2)_n-H$, as shown in Figure 2.10b. The flexible side chains induce the increase in the main-chain mobility and thereby the decrease in T_g.

The glass transition temperature T_g strongly depends on the chain mobility, which can also be influenced by both the CED, δ^2, and the internal mobility, *m*, of the groups in a single chain as [1]:

$$T_g = \frac{2\delta^2}{mR} + C_1 \tag{2.25}$$

where *R* is a gas constant and C_1 is a constant. Equation (2.25) indicates that T_g is mostly determined by both the intermolecular force represented by the cohesive force and also the intrinsic chain mobility represented by *m*.

2.3.3 Theories for Glass Transition Temperatures

Several phenomenological models have been proposed to provide an understanding of the glass transition phenomena [11]. One is that the glass transition marks an *iso-viscous* state. This means that, as a polymer is cooled from its melt state, viscosity increases rapidly to a common (maximum) value, ~ 10^{12} Pa s (10^{13} poise) at T_g, for all glassy materials of both low molecular weight and polymer.

A second view is that the glass transition represents a state of *iso-free volume.* The free volume theory states that the material at T_g takes a universal fractional free volume, denoted $f_g = 0.025$, where $f = V_f/V$, with V_f and V being total specific free volume and specific volume, respectively. The glass transition is characterized by an *iso-free volume* state, i.e. it considers that T_g is the temperature at which the polymers have a certain universal free volume fraction [1, 7, 11].

A third view of the glass transition represents an *iso-entropic* state. Gibbs and DiMazio suggested an equilibrium glass transition temperature, T_2, at which the conformational entropy, S_c (a measure of the total number of ways of arranging a polymer molecule), goes to zero. It can be shown that T_2 lies approximately 52 °C below the experimentally measured T_g ($T_2 = T_g - 52$). Polymeric chains can exist in a large number of possible spatial conformations in solution or in the melt state. Each of these conformations corresponds to a different energy state. If the melt is cooled at an infinitely slow rate to assure the attainment of equilibrium, eventually a temperature will be reached (i.e. T_2) at which only the lowest-energy conformation is available. At this point, the conformation entropy will be zero [12–14].

Recently, machine learning technologies have been utilized to predict glass transition temperature, T_g, from effective mobility value. A strong general correlation was observed between the experimentally measured T_g and the effective mobility value, ζ, for conjugated polymers with different backbones and alkyl side chains. An effective mobility value, ζ, is calculated using an assigned atomic mobility value within each repeat unit. Solid line in Figure 2.11 represents the best linear fit through all data. Group 1 (squares) is composed of thiophene-rich backbones, while Group 2 (circles) consists of phenyl-rich backbones. Comparison between experimental T_g and prediction by $T_g = 979\text{–}1,102\ \zeta$ [°C] for all 32 alkylated conjugated polymers yields good agreement with a root-mean-square error of 13 °C [15].

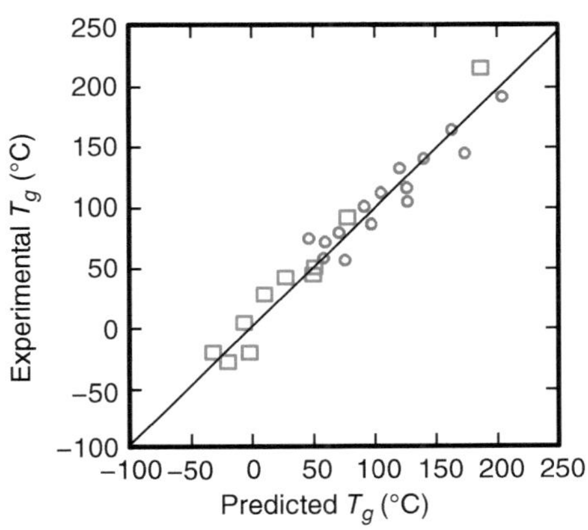

Figure 2.11 Predicting glass transition temperature, T_g, of alkylated conjugated polymers from effective mobility value, ζ with a root-mean-square error of 13 °C. *Source*: Ref. [15]/Springer Nature/CC BY 4.0.

2.4 Nonequilibrium Features of Glassy Polymers

Below the glass transition temperature, glassy polymers are rigid and hard, with polymer chains practically frozen in a disordered state. Despite this rigidity, some molecular mobility still exists, allowing continuous localized motion of polymer segments and consequently the decrease in the specific volume with time, which indicate a nonequilibrium state.

2.4.1 Nonequilibrium Features of Glassy Polymers and Physical Aging

Nonequilibrium state can be evidenced by the change in the experimental glass transition temperature depending on the cooling rate from points I to II in Figure 2.12: T_g increases with faster cooling rates, often called kinetic glass transition. Experimentally, the measured T_g changes by about 3 K for a 10-fold change in cooling rate. If the cooling rate is infinitely slow, the glass transition reaches an equilibrium temperature, T_2 or T_0, which is typically 52 K lower than the measured T_g.

When a polymer is isothermally maintained at point III, a few degrees below II, the specific volume extremely slowly decreases over time, transitioning to an equilibrium at point IV. This process is called "physical aging" or "isothermal volume recovery" due to the nonequilibrium feature [1]. Thus, the physical aging in glassy polymers refers to the changes in their properties over time at temperatures below T_g [16]. The key characteristics of physical aging include the decrease in free volume, changes in mechanical properties, and decrease in entropy.

Polymers exhibit delayed responses upon changes in temperature, mechanical stress, or chemical environments due to their large molecular weight and entangled chains. Relaxation processes are generally observed when the polymer is taken below or near its T_g [17, 18]. This time-dependent response process, known as relaxation, leads to an equilibrium state over time. Relaxation processes in polymers are categorized into two types based on the scale of motion: primary relaxation and secondary relaxation.

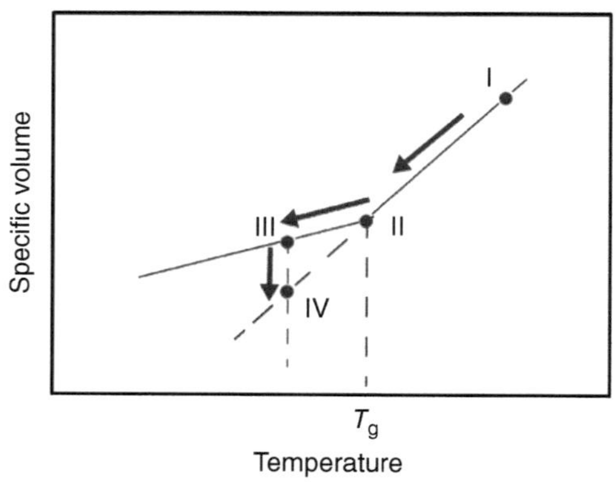

Figure 2.12 Schematic illustration of physical aging from states III to IV at a given temperature. *Source:* Y.S. Kang.

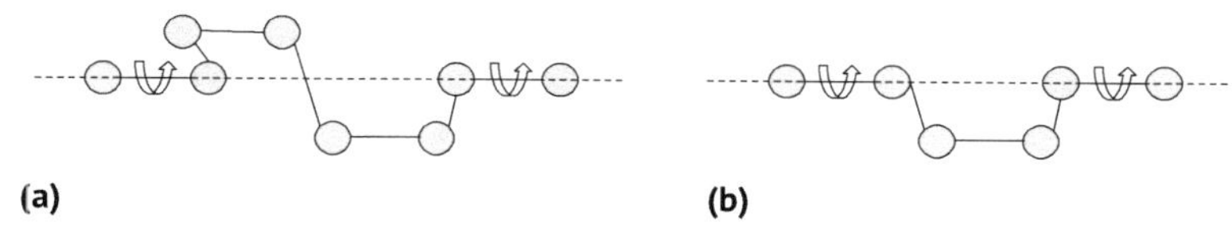

Figure 2.13 (a) Schatzki and (b) Bovey models for crank-shaft motion of a carbon–carbon backbone. The dashed line represents the virtual axis around which bonds rotate. *Source*: Adapted from [19].

Primary relaxation in glassy polymers refers to the process, where segments of the polymer chains undergo localized motion, without the entire chain moving, often occurring at the glass transition. That is, the glass transition involves the motion of polymer chains at the segmental level, where segments of the polymer chains undergo localized motion without the entire chain moving.

Even in the glassy state, polymer chains exhibit microscale motions, called secondary or sub-T_g relaxations. These include rotations or flips of side groups and short main-chain segments, such as aromatic ring rotations in polycarbonate. The crank-shaft rotations are also examples for a main-chain secondary relaxation [19]. According to this model, five contiguous bonds join in rotation around the main chain consisting of C—C bonds, referred to as Schatzki model (Figure 2.13a). A similar crank-shaft model proposed by Boyer involves the rotation of three contiguous bonds (Figure 2.13b). Such limited or noncooperative motions may occur at very low temperatures (e.g. ~−120 °C).

In addition to these secondary main-chain motions, substituent groups can rotate or wag at extremely low temperatures, as low as 70 K. All these motions can occur in the glassy state, below T_g, as a precursor to the onset of long-range, main-chain cooperative-motions that mark the glass transition [16].

The secondary-relaxation processes can have an influence on glassy state properties. In particular, they affect the molecular mobility and free volume of the polymer matrix, which in turn influences the diffusion and permeation of gases through the material.

2.4.2 Characterization of Sub-glass Transitions by Dynamic Mechanical Analysis

Sub-T_g relaxations can be studied and characterized using techniques like dynamic mechanical analysis (DMA), dynamic dielectric spectroscopy (DES), nuclear magnetic resonance (NMR), and differential scanning calorimetry (DSC). Sub-glass transition temperatures represent temperature ranges below T_g of glassy polymers where secondary-relaxation processes occur.

DMA is a powerful tool to characterize the secondary-relaxation phenomena. When dynamic, i.e. time-dependent stress or strain is applied in a glassy polymer, its response is measured as a function of time t to characterize the material properties (Figure 2.14). For instance, when sinusoidal strain, $\varepsilon^* = \varepsilon_0 \exp(i\omega t)$, with the angular frequency, ω, is applied to a body, the resulting complex stress, σ^* can be observed with a phase difference, δ, as ($\sigma^* = \sigma_0 \exp\left[i(\omega t + \delta)\right]$. In the case of

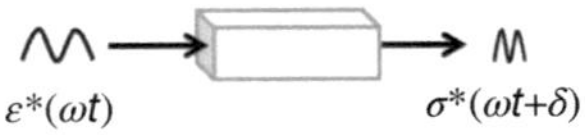

Figure 2.14 Schematic drawing of dynamic mechanical experiment with input strain $\varepsilon^*(\omega t)$ and output stress $\sigma^*(\omega t + \delta)$. *Source*: Y.S. Kang.

a Hookean solid such as an ideal rubber, the phase difference δ becomes zero, whereas a Newtonian fluid shows $\delta = \pi/2$. Thus, common polymers show the δ value in the range of $0 \leq \delta \leq \pi/2$.

It follows that a complex modulus, E^*, can be defined as the ratio of complex stress to complex strain: $E^* = \sigma^*/\varepsilon^* = E' + iE''$ where E' and E'' are storage and loss moduli, respectively, and expressed $E' = (\sigma_0/\varepsilon_0)\ \cos(\delta)$ and $E'' = (\sigma_0/\varepsilon_0)\ \sin(\delta)$. The ratio of loss to storage moduli defines another useful parameter called $\tan(\delta) = E''/E' = 1/\omega\tau$.

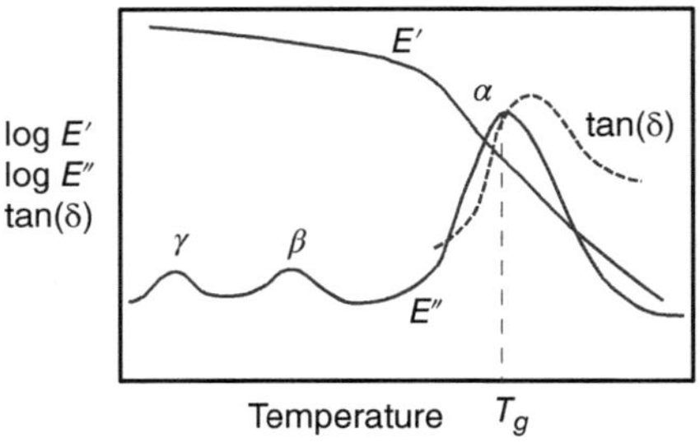

Figure 2.15 Dynamic mechanical analysis (DMA) spectra showing α-, β-, and γ-peaks as a function of temperature. *Source*: Y.S. Kang.

Dynamic mechanical behavior of a glassy polymer is illustrated in Figure 2.15 (log modulus vs. temperature). The storage modulus E' generally follows the behavior of Young's modulus, whereas E'' and $\tan(\delta)$ display several maxima at sub-T_g temperatures. In particular, several $\tan(\delta)$ peaks are commonly observed and these peaks correspond to different secondary-relaxation processes including α-, β-, and γ-relaxations in order of decreasing temperature from the glass transition temperature. Thus, the α-peak is assigned to the glass transition temperature, T_g.

The α-peak in DMA is associated with the glass transition temperature, T_g, of the polymer as described above. The β-peak for a secondary-relaxation process is typically associated with localized segmental motion or side group motions within the polymer chains, and also unrelaxed stress built during sample preparation. The γ-peak is yet another secondary-relaxation process at lower temperatures than the β-peak. Like the β-relaxation, the γ-relaxation is related to specific molecular motions within the polymer, such as substituent wagging motions, local chain rotations, or other hindered motions and therefore strongly associated with the gas transport properties.

2.5 Free Volume Theory and Applications

The concept of free volume has significant importance for several subjects in polymer science, including glass transition, transport properties, time-temperature superposition of viscoelastic properties, melt viscosity [7, 20]. The free volume in a polymer medium is the space not occupied by polymer chains, as shown in Figure 2.16a [21]. The presence of such free volume can be schematically redrawn in Figure 2.16b for a lattice model.

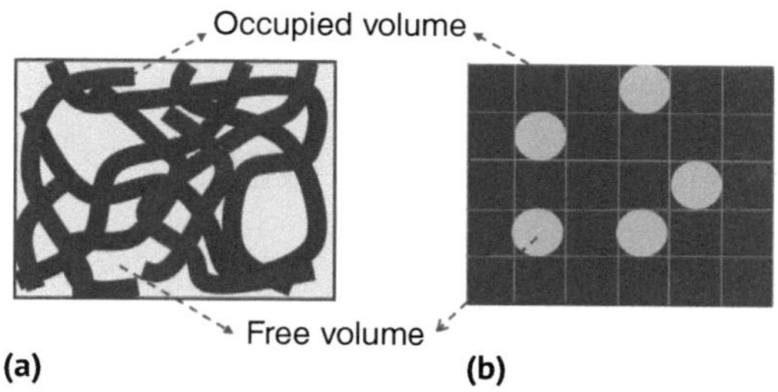

Figure 2.16 Free volume and occupied volume in (a) polymers and (b) lattice model. *Source*: Adapted from [21].

The measured specific volume, V (cm^3/g), comprises the occupied volume, V_O (cm^3/g), and the free volume, V_f (cm^3/g), as shown in Figure 2.17.

Therefore,

$$V_f = V - V_O \quad \text{and} \quad f = \frac{V_f}{V} \tag{2.26}$$

where f is the fractional free volume (FFV).

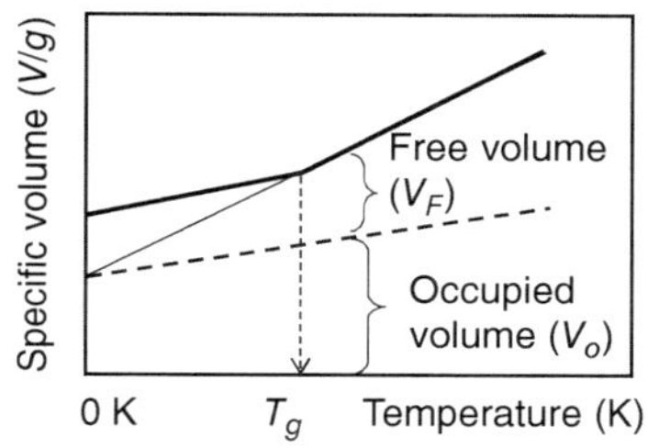

Figure 2.17 Specific volume change as a function of temperature. *Source*: Y.S. Kang.

2.5.1 Free Volume Theory

According to the empirical Doolittle equation, as the free volume, V_f, increases the viscosity, η, of polymers decreases as:

$$\eta = A \exp\left(B \frac{V_O}{V_f} \right) \tag{2.27}$$

Here, A and B are constants. Since $f = V_f/(V_o + V_f) \sim (V_f/V_o)$ (because $V_o \gg V_f$), Equation (2.27) can be rewritten as follows:

$$\eta = A \exp\left(\frac{B}{f} \right) \tag{2.28}$$

Following the Stokes–Einstein–Sutherland equation for diffusion of spherical particles with a radius r through a liquid as:

$$D = \frac{k_B T}{6\pi\eta r} \tag{2.29}$$

where diffusion coefficient, D, is inversely proportional to viscosity, η. Therefore, Equation (2.28) can be rearranged again for diffusion coefficient D:

$$D = A_d \exp\left(-\frac{B_d}{f} \right) \tag{2.30}$$

where A_d and B_d are constants for diffusion. Details on the free volume theory for diffusion coefficient will be described at Section 3.5 of Chapter 3.

2.5.2 Applications of Free Volume Theory

Since the free volume theory is conceptually and mathematically simple, it can be applied for many different fields. In the following, diverse application areas will be introduced.

2.5.2.1 Temperature-dependent Diffusion and Permeation

The f(=FFV) is primarily determined by chemical structure of polymer and also affected by the amount of additives inside and temperature. For instance, it has been found that f increases linearly with temperature T at rubbery state as follows:

$$f = f_g + \alpha_f(T - T_g) \tag{2.31}$$

where $\alpha_f = \alpha_r - \alpha_g$ is the thermal expansion coefficient of the free volume or the difference in the thermal expansion coefficients in rubbery, α_r, and glassy, α_g, states, which has been found roughly constant at $\alpha_f = 4.8 \times 10^{-4}$/K by Simha et al., and f_g is f at the glass transition temperature, which has been known to be 0.025 for many polymers according to the universal *iso-free volume theory* [1].

According to Equation (2.31), f linearly increases with temperature, which can be applied to explain the increases in both the diffusion and permeation coefficients as increasing the temperature.

2.5.2.2 Molecular Weight Effects on T_g

The fraction free volume, f, increases with the concentration of chain ends, which are proportional to M^{-1} with M being the molecular weight of the polymer. Thus, the free volume is related to the molecular weight of polymers and T [1] as:

$$f = (C_1 + C_2T) + \frac{C_3}{M} \quad \text{and consequently} \quad T_g = C_4 - \frac{C_5}{M} \tag{2.32}$$

where C_1–C_5 are all constants.

In the case of cross-linked polymers, T_g is expressed: $T_g = C_6 + C_7/M_c$, where M_c is the average molecular weight between two cross-links.

2.5.2.3 T_g for Polymer with Diluent

When diluents such as solvent and plasticizers are mixed with a polymer, the free volume, and consequently, the glass transition temperature depend on their composition. In the case of a mixture of a polymer, 2, and a diluent, 1, the free volume fraction depends on both the diluent concentration and temperature. It is thus treated that f linearly increases with concentration or commonly with the volume fraction of diluent 1, v_1:

$$f(v_1, T) = f(0, T) + \beta v_1 \tag{2.33}$$

where β is the proportionality constant.

Another expression for f for a mixture of a polymer, 2, and diluent, 1, system at constant temperature and pressure, is given as [19]:

$$f = f_g + \alpha_1(T - T_{g,1})v_1 + \alpha_2(T - T_{g,2})v_2 \tag{2.34}$$

where v_2 is the volume fraction of component 2 ($v_1 + v_2 = 1$). α_i and $T_{g,i}$ are the thermal expansion coefficient for free volume and the glass transition temperatures of component i, respectively. At T_g, the equation reduces to:

$$\alpha_1(T_g - T_{g,1})v_1 + \alpha_2(T_g - T_{g,2})v_2 = 0 \tag{2.35}$$

Rearrangement of Equation (2.35) gives the widely known Kelly–Bueche equation [19, 22]:

$$T_g = \frac{\alpha_1 v_1 T_{g,1} + \alpha_2 v_2 T_{g,2}}{\alpha_1 v_1 + \alpha_2 v_2} \tag{2.36}$$

If $\alpha_1 = \alpha_2$, commonly accepted due to the universal value of the thermal expansion coefficient of free volume, the above equation reduces to:

$$T_g = v_1 T_{g,1} + v_2 T_{g,2} \tag{2.37}$$

This suggests that T_g of polymer alloy depends on the volume fraction of each component, which is different from the Fox equation (Equation (2.46)).

2.5.2.4 Williams–Landel–Ferry Equation

Temperature dependence of the relaxation processes of polymers, characterized by the relaxation time τ, is commonly expressed by the William–Landel–Ferry (WLF) equation for near glass transition, particularly in the range of $T_g < T < T_g + 100$ K as [1]:

$$\log a_T = \log\left(\frac{\tau_T}{\tau_{T_g}}\right) = \frac{-17.44\,(T - T_g)}{51.6 + T - T_g} \tag{2.38}$$

where a_T is the shift factor at temperature T, defined as the relaxation time ratio ($a_T = \tau_T/\tau_{Tg}$) by setting a reference temperature at T_g. Here, τ_T is the relaxation time at T, defined as $\tau_T = \eta_T/E_T$ with η_T and E_T being the viscosity and elastic modulus, respectively, of viscoelastic materials like polymers at T.

By assuming that $(\tau_T/\tau_{Tg}) = (\eta_T/\eta_{Tg})$, a_T can be expressed in terms of the viscosity, η, as:

$$a_T \cong \frac{\eta_T}{\eta_{T_g}} \tag{2.39}$$

The WLF equation (Equation (2.38)) will be directly derived by using the Doolittle equation (Equation (2.27) or (2.28)), where it demonstrates the relationship between η and f. This provides the theoretical backgrounds for the empirical WLF equation (Equation (2.38)).

The substitution of the Doolittle equation (Equation (2.28)) into Equation (2.39) yields

$$\ln a_T = B\left(\frac{1}{f} - \frac{1}{f_g}\right) \tag{2.40}$$

Substituting Equation (2.31) into Equation (2.40) results in:

$$\ln a_T = \frac{-(B/f_g)(T - T_g)}{(f_g/\alpha_f) + T - T_g} \tag{2.41}$$

Comparing Equation (2.41) with the WLF equation (Equation (2.38)) gives $f_g = 0.025$ and $\alpha_f = 4.8 \times 10^{-4}$(/K), when B is set to 1, which are universal constants [1, 23].

The outcome is that the empirical WLF equation is consistent with the free volume theory, giving the equilibrium glass transition temperature $T_2 = T_g - 52$(K), universal α_f and f_g values at T_g.

2.5.3 Measurement of Free Volume

The total amount of the free volume, V_f, can be either directly measured by means of several experimental techniques such as positron annihilation lifetime spectroscopy (PALS), inverse gas chromatography, ^{129}Xe NMR spectroscopy, wide-angle X-ray scattering (WAXS), photochromic and fluorescence techniques and spin probe measurements, or estimated indirectly from V and V_O by $V_f = V - V_O$.

Since the PALS is highly rigorous to measure size and amount of free volume, it has become a valuable tool in material characterization [24]. The PALS technique is based on the measurement of the lifetimes, τ, of positrons in a material. A positron can exist either as a free positron (e^+) or as positronium (Ps), a metastable hydrogen-like bound state of an electron (e^-) and positron (e^+), which can exist in two spin states. The triplet state, known as ortho-positronium (*o*-Ps), has lifetime of 142 ns in vacuum. However, when *o*-Ps interacts with the electrons from the walls of the free volume elements, its lifetime is to ~ 1.5–4.0 ns.

The two shortest lifetimes of τ_1(~ 0.2 ns) and τ_2(0.3–0.5 ns) are due to the annihilation of singlet state Ps (*p*-Ps) and the free positrons (e^+), respectively. The long-lived lifetime(s) τ_i (i = 3 or 4) are due to *o*-Ps localized in free volume elements and depend on the size of the free volume elements; the larger the free volume elements, the longer the lifetimes. Thus, it can be used to characterize the free volume. The collected data, lifetimes (τ_i), and corresponding intensities (I_i) can be processed using computer programs to obtain information on the size and the amount of the free volume elements [24].

The indirect method is to measure or estimate V and V_O first and then to calculate V_f. Here, V is the inverse of the density and V_O can be estimated by several different methods: (1) van der Waals radii of the atoms; (2) crystalline volume at 0 K; and (3) the extrapolated volume of liquid at 0 K. The van der Waals volume, V_w, has been the most employed value, which can be readily estimated from the group contribution method, such as Bondi's group contribution method [4]. It is worth to note that the occupied volume includes the thermal expansion due to anharmonicity of vibrations. Therefore, V_O is commonly related with V_w as [4]:

$$V_O \approx 1.3 \times V_w \tag{2.42}$$

Therefore, f can be obtained indirectly as:

$$f = \frac{V_f}{V} \cong \frac{V - 1.3V_w}{V} \tag{2.43}$$

2.6 Crystalline Polymers

Polymers such as polyethylene and polyamide are semicrystalline, consisting of both crystalline and amorphous domains, whereas poly(methyl methacrylate) is totally amorphous. A semicrystalline polymer is typically referred to as a crystalline polymer in polymer studies just for convenience. Crystallites are generally nonpermeable to common gases and ions, and therefore, strongly influence transport properties. Thus, structures of single crystals, lamellae, and spherulites in crystalline polymers are highlighted first in this section. Then, crystallization kinetics will be addressed, which affect the size and number of crystallites and, consequently, various physicochemical properties including molecule and ion transport and mechanical properties.

2.6.1 Crystal Structures

Polymeric single crystals have been experimentally observed from very dilute solutions (0.01% of polymer). For example, linear polyamide 6 in a solvent at 70 °C gives hollow pyramid-type single crystal with a thickness of 6 nm [1]. In the crystals, the polymer chains were found to be folded upon themselves to form lamellae from dilute polymer solutions. The lamella structure is characterized by three folded-chain models of (1) nonadjacent, (2) regular adjacent, and (3) irregular adjacent reentry models, depending on how polymer chains fold back, as illustrated in Figure 2.18.

When polymers are crystallized from the melt or from concentrated solutions, crystallites or lamella can be organized into larger spherical structures called spherulites, which comprise several individual lamellar plates and amorphous region among lamellae. X-ray and electron diffraction experiments indicated that the *c* axis of the lamellar crystals was normal to the radial growth direction of the spherulites, as shown in Figure 2.19.

The chain arrangements in unit cells exhibit the lowest energy state of the all-*trans* or *gauche* conformation of the mainchains. The former gives zigzag conformations, whereas the latter results in helical structure, which can be interpreted in terms of steric hindrance between atoms that forbids the planar zigzag conformation [25]. Polyethylene is an example of zigzag chains, while polyamide-6,6 and polyamide-6

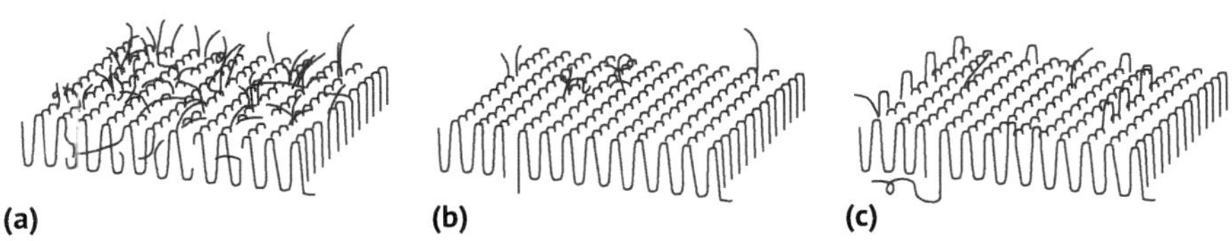

Figure 2.18 Three idealized models for chain folding in polymer crystals: (a) nonadjacent reentry; (b) regular adjacent reentry; and (c) irregular adjacent reentry. *Source:* Adapted from [1].

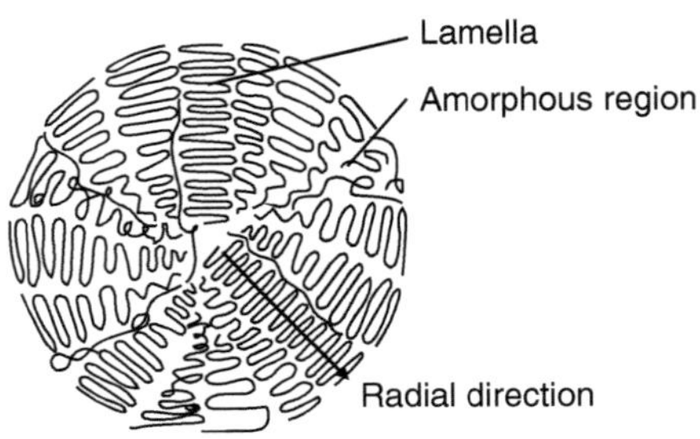

Figure 2.19 Schematic model of a spherulite. Black arrows indicate the direction of molecular alignment. *Source:* Adapted from spherulite (polymer physics) in Wikipedia.

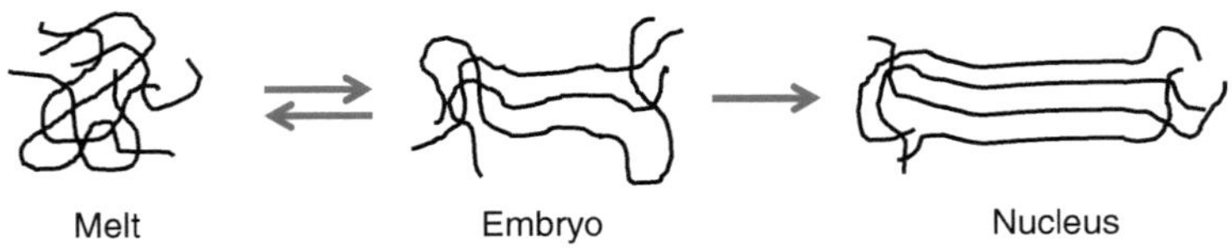

Figure 2.20 Schematic illustration of nucleation steps. *Source*: Y.S. Kang.

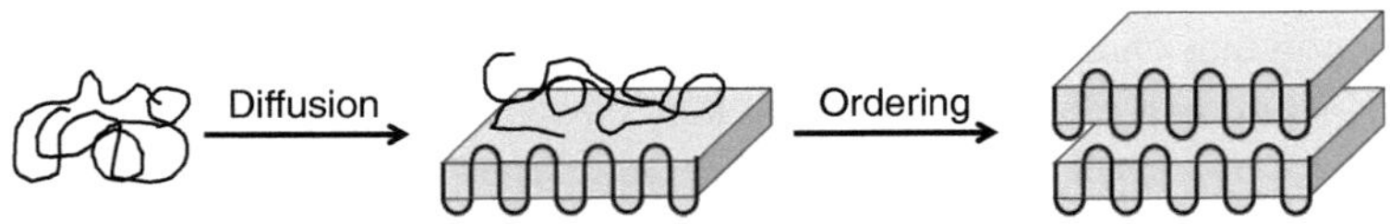

Figure 2.21 Schematic crystallization process: diffusion of the polymeric chain on the crystal front and ordering to make the crystal bigger. *Source*: Adapted from [1].

are planar zigzag type because of hydrogen bonds. The helical chain conformations are frequently observed in vinyl polymers and poly(ethylene oxide).

2.6.2 Crystallization Kinetics and the Degree of Crystallinity

As in any crystallizing materials, crystallization of polymeric chains starts from melt by a nucleation step. Thermal fluctuations first form reversible embryos, small particles with an enhanced inner order, in the melt (Figure 2.20). In order for the embryo to be stable, it has to exceed a critical size; it is then called a nucleus.

Once stable nuclei are formed, crystallization proceeds from the nuclei to form spherulites by two successive steps [2]. First polymeric chains in melts diffuse to the crystal surface or front, and subsequently, the polymeric chains order to make the crystal size bigger. In short, crystallization involves both diffusion of the crystallizable units to the crystal front and growth of crystals or spherulites [1], which is schematically drawn as (Figure 2.21).

It is worth noting that the nucleation process described above is a homogeneous nucleation, occurring in a homogeneous pure melt. However, this is not the usual situation; under practical conditions, nucleation mostly starts on the surface of low molecular weight particles, which come into the sample either uncontrollably or deliberately as nucleating agents; in this case, the nucleation is heterogeneous.

The addition of nucleating agents is of considerable industrial importance because it decreases the supercooling necessary for the onset of crystallization [2].

The isothermal radial growth of the spherulites is usually observed to be linear with time, i.e. $r = gt$, where r is the radius of spherulites at crystallization time t and g is the linear radial growth rate. When g is plotted as a function of crystallization temperature, T_c, a maximum is commonly observed, as shown in Figure 2.22. The g value increases due to the increase in the driving force or the degree of supercooling $\Delta T\ (= T_m^0 - T_c)$ as the T_c is lowed from the melting temperature, T_m, or equilibrium melting temperature, T_m^0. As T_c is lowered still further, molecular motion becomes sluggish as the glass transition, T_g, is approached, and g decreases again. Thus, g becomes theoretically zero at both T_m and T_g. Below T_g, molecular motion is so slow that the rate of crystallization effectively becomes zero.

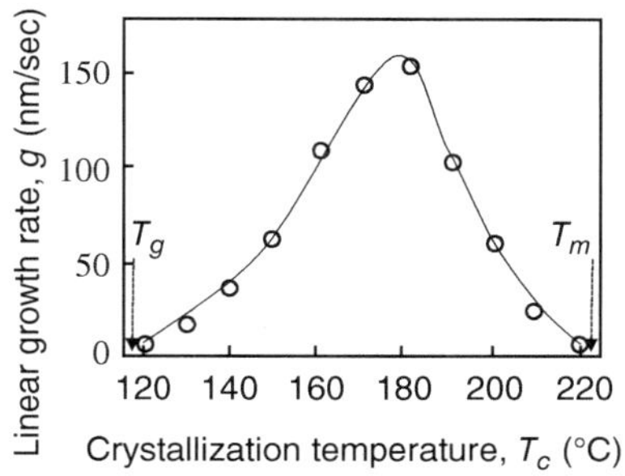

Figure 2.22 Plot of linear growth rate, *g*, versus crystallization temperature, T_c, of poly(ethylene terephthalate). *Source*: Adapted from [26].

Crystallinity can be determined from density data and calorimetric measurements. If we know the densities of the crystalline and amorphous phases, ρ_c and ρ_a, respectively, along with the sample density ρ, the volume crystallinity, X_c, can be obtained from:

$$X_c = \frac{\rho - \rho_c}{\rho_c - \rho_a} \tag{2.44}$$

Alternatively, differential scanning calorimetry (DSC) provides the enthalpy of melting for the sample, ΔH_m, and for a fully crystalline reference, ΔH_m^0, from which X_c can also be estimated from:

$$X_c = \frac{\Delta H_m}{\Delta H_m^0} \tag{2.45}$$

2.7 Polymer Blends

When two polymers are blended together, their structure becomes either homogeneously mixed or phase-separated, depending on their compatibility or miscibility, which is characterized often by measuring glass transition temperature, T_g.

Homogeneously miscible polymer blends show only one glass transition temperature, whereas phase-separated polymer blends show multiple glass transition temperatures, corresponding to their constitutions. For example, acrylonitrile-1,3-butadiene-styrene (ABS) resin, a polymer blend of incompatible poly(1,3-butadiene) (BR) and poly(styrene-co-acrylonitrile) (SAN), shows two glass transition temperatures, corresponding to SAN and BR. However, they have

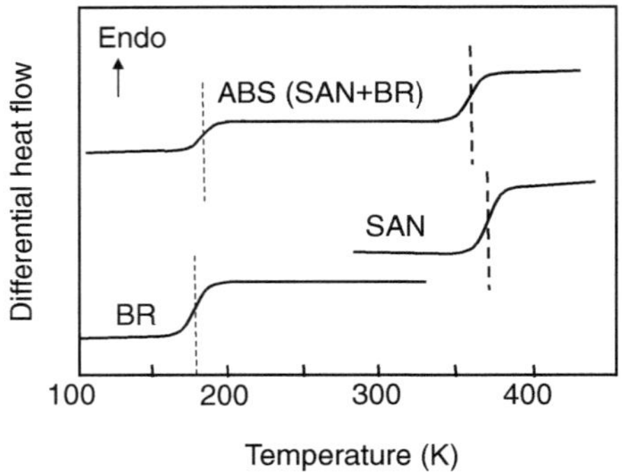

Figure 2.23 Schematic differential scanning calorimetry (DSC) thermograms of acrylonitrile-1,3-butadiene-styrene (ABS) and its constituent poly(1,3-butadiene) (BR) and poly(styrene-co-acrylonitrile) (SAN). *Source*: Adapted from [1].

a mutual, limited solubility or partial miscibility indicated by a shift in the two T_gs accompanying a change in the phase composition of the blend (Figure 2.23).

When miscible, the glass transition temperature commonly follows the Fox equation, depending on the individual compositions:

$$\frac{1}{T_g} = \frac{w_1}{T_{g1}} + \frac{w_2}{T_{g2}} \tag{2.46}$$

where w_i and T_{gi} are the weight fraction and the glass transition temperature of component i, respectively. In addition, Equation (2.37) based on free volume theory, can also be used to characterize the glass transition temperatures of miscible polymer blends.

2.8 Viscoelastic and Mechanical Properties

The mechanical properties of polymers change over time due to their viscoelastic nature, which is characterized by relaxation time. This relaxation time varies with temperature according to the WLF equation, which also provides a theoretical basis for the time-temperature superposition principle. Accordingly, this section first discusses the viscoelastic properties of polymers. Then, key mechanical properties such as modulus, yield stress, elongation at break, and toughness are briefly described based on stress-strain behavior. Understanding these properties allows polymers to be engineered for a wide range of applications, from flexible films to structural materials.

2.8.1 Viscoelastic Properties

The two extreme relationships between stress, σ, and strain, ε, for viscous fluids and elastomers are mathematically expressed as:

$$\sigma = \eta \frac{d\varepsilon}{dt} \tag{2.47}$$

$$\sigma = E\varepsilon \tag{2.48}$$

Figure 2.24 Schematic deformation of an original sample ($A_0 \times L_0$) into $A \times L$ by applying stress σ. *Source*: Y.S. Kang.

where η and E are viscosity and elastic modulus, respectively. Here, strain $\varepsilon = (L - L_0)/L_0$ with L and L_0 being the length of the extended specimen and its original length, respectively, as schematically shown in Figure 2.24.

These two extreme properties combine together in series or parallel to model mechanical properties and a relaxation time τ, defined as $\tau = \eta/E$, representing the time necessary to relax polymer materials into their equilibrium state. Mechanical properties of polymer materials cannot be generally described by simple expressions of a single relaxation time, τ, but by several relaxation times, called relaxation time spectrum.

Temperature dependence of the relaxation time, τ, is commonly expressed by the Arrhenius equation for sub-glass relaxation processes as:

$$\tau = \tau_0 \exp\left(+\frac{\Delta E}{RT}\right) \tag{2.49}$$

where ΔE is the activation energy of the relaxation process. ΔE for secondary relaxation is typically 10–20 kcal/mol.

At temperatures near T_g, particularly in the range of $T_g < T < T_g + 100$ K, the relaxation time may be expressed by the empirical WLF equation as:

$$\log \frac{\tau_T}{\tau_{T_g}} = \frac{-17.44\ (T - T_g)}{51.6 + (T - T_g)} \tag{2.50}$$

which predicts the relaxation time as a function of temperature. The WLF equation can be applied to the time-temperature superposition principle by defining a shift factor $a_T = \tau_T/\tau_{Tg}$ where τ_T and τ_{Tg} are the relaxation times at temperature T and T_g, respectively.

2.8.2 Stress–Strain Behavior

Polymers exhibit a wide range of stress–strain behavior, as shown in Figure 2.25, which also depends on temperature and rate of deformation. At normal use temperatures, brittle polymers (e.g. atactic polystyrene, poly(methyl methacrylate)) are represented by Curve 1, showing a small strain at failure, ε_f. The stress at failure is called the *ultimate stress* or *stress-at-break*.

Ductile polymers, including many engineering thermoplastics and toughened plastics, exhibit stress–strain behavior represented by Curves 2 and 3. Their Young's modulus, given by the initial slope of the stress–strain curve, is somewhat lower than that of the brittle, glassy polymer. As shown, the stress reaches a maximum value, which is called its *yield stress* at a certain strain. As strain is further increased,

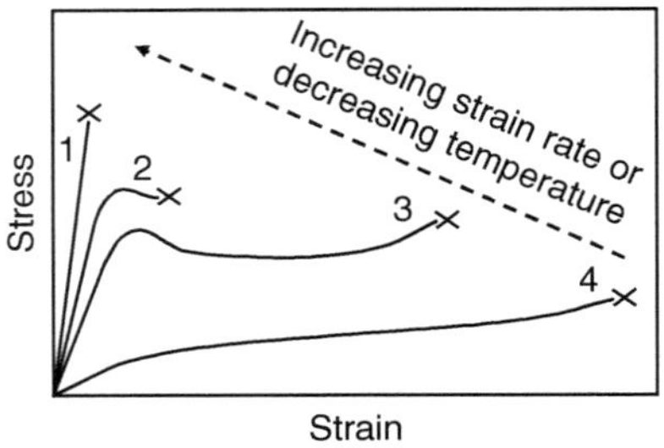

Figure 2.25 Typical stress–strain curves exhibiting brittle failure (Curve 1), ductile failure with neck formation (Curve 2), ductile failure with neck propagation and orientational hardening (Curve 3), and rubbery behavior with evidence of strain-induced crystallization (Curve 4). Point of failure is indicated by the symbol x. *Source:* Adapted from [19].

stress at first decreases, followed by either failure (Curve 2) or extensive elongation at almost constant stress (Curve 3). This process is called *plastic flow region* or *strain softening*, by *neck propagation*, which usually occurs at strains between 5 and 50%. At further increases in strain, the sample experiences *orientational hardening* before failure (Curve 3). During orientational hardening, polymer chains are stretched locally in the tensile direction. Chain extension causes a resistance to further deformation; stress is, therefore, observed to increase [19].

Rubbery polymers follow stress–strain behavior similar to that of Curve 4. Modulus is low, but ultimate extension can be very large, on the order of several hundred percent. Before failure, the rubber may experience an increase in stress as a consequence of strain-induced crystallization caused by molecular orientation in the stretched direction.

Additionally, toughness can be estimated quantitatively from the stress–strain curve. It represents the total mechanical energy per unit volume that a material can absorb before fracture, at a strain ε_f. Because it reflects both strength and ductility, toughness is a key indicator of a material's resistance to mechanical failure.

Toughness T ($\text{Nm/m}^3 = \text{J/m}^3$) can be mathematically derived from the mechanical work $\text{d}w$ required to deform a material by a small deformation $\text{d}L$ under an applied force f: $\text{d}w = f\text{d}L$:

$$T = \frac{w}{A_0 L_0} = \int_0^{\varepsilon_f} \sigma \text{d}\varepsilon \tag{2.51}$$

where A_0 and L_0 are initial surface area and length of the sample, respectively. Equation (2.51) represents that toughness is the area under the stress–strain curve up to fracture.

References

1 U. Gedde, Polymer Physics, Springer Science & Business Media, 1995.
2 J.L. Halary, F. Lauprêtre, L. Monnerie, Polymer Materials: Macroscopic Properties and Molecular Interpretations, John Wiley & Sons, 2011.
3 P.J. Flory, Principles of Polymer Chemistry, Cornell University Press, 1953.
4 D.W. van Krevelen, K. te Nijenhuis, Properties of Polymers: Their Correlation with Chemical Structure; Their Numerical Estimation and Prediction from Additive Group Contributions, Elsevier Science, 2009.

5 M. Rubinstein, R.H. Colby, Polymer Physics, OUP Oxford, 2003.
6 H.R. Allcock, F.W. Lampe, J.E. Mark, Contemporary Polymer Chemistry, Pearson/Prentice Hall, 2003.
7 R.P. White, J.E.G. Lipson, Macromolecules 49(11) (2016) 3987–4007.
8 A. Eisenberg, Physical Properties of Polymers, American Chemical Society, 1984.
9 S. Rogers, L. Mandelkern, The Journal of Physical Chemistry 61(7) (1957) 985–991.
10 J.A. Shetter, Journal of Polymer Science Part B: Polymer Letters 1(5) (1963) 209–213.
11 S.F. Edwards, Polymer 35(18) (1994) 3827–3830.
12 T.S. Chow, Macromolecules 13(2) (1980) 362–364.
13 E.A. Dimarzio, J.H. Gibbs, Journal of Polymer Science Part A: General Papers 1(4) (1963) 1417–1428.
14 J.H. Gibbs, E.A. DiMarzio, The Journal of Chemical Physics 28(3) (1958) 373–383.
15 R. Xie, A.R. Weisen, Y. Lee, M.A. Aplan, A.M. Fenton, A.E. Masucci, F. Kempe, M. Sommer, C.W. Pester, R.H. Colby, Nature Communications 11(1) (2020) 893.
16 L.C.E. Struik, Physical Aging in Amorphous Polymers and Other Materials, Elsevier Scientific Publishing Company, 1978.
17 G.B. McKenna, S.L. Simon, Macromolecules 50(17) (2017) 6333–6361.
18 L.P. Singh, Polymer Journal 55(2) (2023) 141–151.
19 J.R. Fried, Polymer Science and Technology, Prentice Hall, 2014.
20 P.M. Budd, N.B. McKeown, D. Fritsch, Journal of Materials Chemistry 15(20) (2005) 1977–1986.
21 P.M. Budd, N.B. McKeown, Polymer Chemistry 1(1) (2010) 63–68.
22 F.N. Kelley, F. Bueche, Journal of Polymer Science 50(154) (1961) 549–556.
23 L.H. Sperling, Introduction to Physical Polymer Science, Wiley, 2006.
24 Z.-X. Low, P.M. Budd, N.B. McKeown, D.A. Patterson, Chemical Reviews 118(12) (2018) 5871–5911.
25 J.L. Halary, F. Laupretre, L. Monnerie, Polymer Materials: Macroscopic Properties and Molecular Interpretations, John Wiley & Sons, 2008.
26 L.H. Palys, P.J. Phillips, Journal of Polymer Science: Polymer Physics Edition 18(4) (1980) 829–852.

Part I

Molecule Transport Through Polymer Membranes

3

Molecule Transport Through Polymer Membranes

Mass transport through polymers involves sorption and diffusion of small neutral molecules in polymers, which have been practically applied for diverse fields, such as separation membranes, drug delivery systems, packaging materials, etc.

Fundamentals on the Fick's law for diffusion of gases, vapors and liquids through polymer membranes will be first addressed, followed by the sorption and permeation features. The free volume theory, the most common mathematical model for diffusion phenomenon in polymer membranes, will also be emphasized along with mathematical models for sorption and permeation. Finally, mass transport will be briefly overviewed in terms of the Fickian and non-Fickian behavior. It is also imperative to acknowledge the three major references for the fundamentals on mass transport through polymers: (1) *Diffusion in Polymers* [1]; (2) *The Mathematics of Diffusion* [2]; and (3) *Materials Science of Membranes for Gas and Vapor Separation* [3].

3.1 Fick's Law and Solution-diffusion Mechanism

Consider a polymeric slab, referred to as a membrane, with a thickness of L that separates two sides: an upstream and a downstream, each containing either gas or liquid. When a constant pressure p_0 (or bulk concentration C_0^B) is applied and maintained in both sides, the concentration throughout the membrane reaches equilibrium at C_0 (where $C(x,0) = C_0$, initial concentration) after a certain period, as illustrated in Figure 3.1a.

Changing the pressures (or concentrations) in upstream and downstream sides to p_1 (or C_1^B) and p_2 (or C_2^B), respectively, new surface concentrations C_1 (where $C(0,t) = C_1$) and C_2 (where $C(L,t) = C_2$) at $x = 0$ and $x = L$ will be established immediately, as shown in Figure 3.1b. Given $C_1 > C_2$, a concentration gradient $\mathrm{d}C(x,t)/\mathrm{d}x$ forms across the membrane, driving diffusional mass transport from $x = 0$ to $x = L$. At steady state, a linear concentration profile inside the membrane, depicted in Figure 3.1b, is achieved only if the diffusion coefficient D remains constant. On the downstream side at $x = L$, desorption or evaporation occurs, completing the mass transport process across the membrane.

Molecule and Ion Transport through Polymer Membranes, First Edition. Yong Soo Kang.

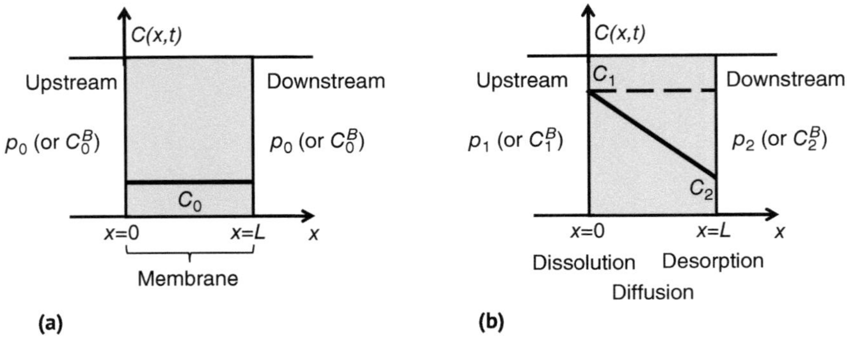

Figure 3.1 (a) Initial concentration C_0 for a semi-infinite slab with thickness of L; and (b) solution–diffusion mechanism of mass transport and linear concentration profile with a constant diffusion coefficient D at steady state. *Source*: Y.S. Kang.

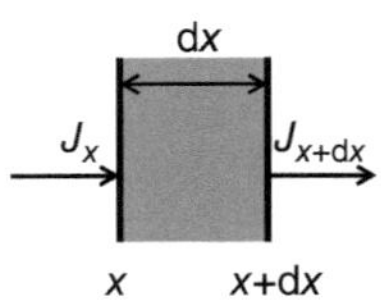

Figure 3.2 Unidirectional mass transport across dx between x and x + dx. *Source*: Y.S. Kang.

Thus, mass transport involves sorption or dissolution at the front interface ($x = 0$), diffusion through the membrane ($0 \leq x \leq L$), and desorption at the back interface ($x = L$). This mass transport mechanism through a polymeric membrane is described by the solution–diffusion or sorption–diffusion mechanism.

Fick's first law in the x-direction describes that mass flux, J (g/cm^{2}sec), is proportional to the concentration gradient, dC/dx (g/cm^{3}cm), and its proportionality constant is called the diffusion coefficient D (cm^2/sec) as:

$$J = -D\frac{\mathrm{d}C}{\mathrm{d}x} \tag{3.1}$$

Considering a slab membrane with a thickness of dx between x and x + dx, as shown in Figure 3.2, the increase in the concentration dC over the time interval dt can be expressed as:

$$\frac{\mathrm{d}C}{\mathrm{d}t} = -\frac{\mathrm{d}J}{\mathrm{d}x} = \frac{\mathrm{d}}{\mathrm{d}x}\left(D\frac{\mathrm{d}C}{\mathrm{d}x}\right) \tag{3.2}$$

This is the equation for Fick's second law of diffusion, also called the continuity equation, describing the time-dependent mass transport behavior. When the diffusion coefficient is constant, or not a function of concentration or position, it reduces to:

$$\frac{\mathrm{d}C}{\mathrm{d}t} = D\frac{\mathrm{d}^2C}{\mathrm{d}x^2} \tag{3.3}$$

This equation is second-order, linear, and homogeneous. Therefore, the solution must be made to fit one initial condition and two boundary conditions.

At steady state, Equation (3.1) can be converted to:

$$J = D\frac{\Delta C}{L} \tag{3.4}$$

where $\Delta C = C_1 - C_2$ the concentration difference across a membrane with a thickness of L. For convenience, both the surface concentrations C_1 and C_2 are assumed to be linearly proportional to the applied pressure p_1 and p_2, respectively: $C_1 = S_1 p_1$ and $C_2 = S_2 p_2$, here S_1 and S_2 are the solubility coefficients, which will be treated in more detail in the Section 3.2. In the case of liquid or solid phase, $C_1 = K_1 C_1^B$ and $C_2 = K_2 C_2^B$, instead, where C_1^B and C_2^B are the concentrations of the bulk phases with corresponding partition coefficient K_1 and K_2, respectively, as shown in Figure 3.1b.

When $\Delta p = p_1 - p_2$ or $\Delta C^B = C_1^B - C_2^B$ is not so large, both S_1 and S_2 will be assumed to be equivalent to each other: $S_1 = S_2 = S$ or equivalently $K_1 = K_2 = K$, respectively. Then, equation (3.4) is readily converted into:

$$\begin{aligned} J &= DS\frac{\Delta p}{L} = P\frac{\Delta p}{L} \quad \text{for gas phase} \\ J &= DK\frac{\Delta C^B}{L} = P\frac{\Delta C^B}{L} \quad \text{for liquid or solid phase} \end{aligned} \tag{3.5}$$

Here, we define a new parameter, P, the permeability or permeation coefficient, which is defined as a product of the diffusion coefficient D and the solubility coefficient S or the partition coefficient K:

$$\begin{aligned} P &= DS \text{ for gas phase} \\ P &= DK \text{ for liquid or solid phase} \end{aligned} \tag{3.6}$$

The advantage of Equation (3.5) is that P is the parameter depending only on the material properties used, whereas Δp (or ΔC^B) and L are determined by the process and the membrane thickness, respectively. In this regard, P is a very important parameter, characterizing the transport properties of materials along with D and S. One important advantage of using P rather than D and/or S is the easiness of the flux estimation because the surface concentration is rather difficult or cumbersome to obtain, whereas the pressure for gas phases and the bulk concentration for the liquid phase are much easier to attain.

It is important to note that units of S and K are different: S is expressed in (cm^3(STP)/cm^3 cmHg) or (mol/cm^3 cmHg) while K is dimensionless (−). Consequently, the permeability P for gases is measured in (cm^3(STP)cm/cm^2sec cmHg or Barrer), where 1 Barrer = 1.0×10^{-10} cm^3(STP)cm/cm^2sec cmHg. For liquids, P shares the same units as the diffusion coefficient (cm^2/sec).

3.2 Sorption and Permeation Features in a Slab

The transport properties of small molecules through polymeric materials are commonly characterized by D, S, and P, which can be obtained from two different methods: Transient sorption and permeation. Both D and S are obtained from transient sorption, whereas D and P are obtained from transient permeation. Detailed experimental procedures for both methods will be discussed in Chapter 6 of this book.

3.2.1 Transient Sorption

One practically important solution of Equation (3.3) is for a semi-infinite slab of a thickness of $2L$, in which the same surface concentrations are held to be constant at both surfaces (C_1 at $x = -L$ and $x = L$) and the initial concentration is also constant (C_0). In this case, mass transport occurs inwardly symmetrically from both interfaces, called sorption process as shown in Figure 3.3a. Refer to the book by Crank [2] for detailed mathematical descriptions of concentration and the total amount sorbed M_t as a function of time t for diverse geometries.

The penetrant concentration C depends on both position x and time t, and thus will be expressed as $C(x, t)$. The initial and boundary conditions are mathematically stated as:

$C(x, 0) = C_0$ for initial condition where $-L \leq x \leq L$ and $t < 0$
$C(-L, t) = C(L, t) = C_1$ or $\mathrm{d}C(0, t)/\mathrm{d}x = 0$ for boundary conditions with $t \geq 0$

$C(x, t)$ can be mathematically obtained from the integration of Equation (3.3) subject to the initial and boundary conditions [2]:

$$\frac{C - C_0}{C_1 - C_0} = 1 - \frac{4}{\pi}\sum_{n=0}^{\infty} \frac{(-1)^n}{2n+1} \exp\left\{-D(2n+1)^2\pi^2 t/4L^2\right\} \cos\frac{(2n+1)\pi x}{2L} \tag{3.7}$$

The schematic concentration change with respect to time is illustrated in Figure 3.3a. The total amount of penetrant M_t (mole or gram) sorbed in the slab at time t is calculated by:

$$M_t = A\int_{-L}^{L} C(x, t)\,\mathrm{d}x = -2A\int_0^t D\frac{\mathrm{d}C(L, t)}{\mathrm{d}x}\mathrm{d}t \tag{3.8}$$

Thus, Equations (3.7) and (3.8) give [2]:

$$\frac{M_t}{M_\infty} = 1 - \sum_{n=0}^{\infty} \frac{8}{(2n+1)^2\pi^2} \exp\{-D(2n+1)^2\pi^2 t/4L^2\} \tag{3.9}$$

In the M_t vs $\sqrt{t}$ curve for sorption experiment, M_t initially increases linearly, followed by concave to the square-root of time axis to reach an equilibrium, M_∞

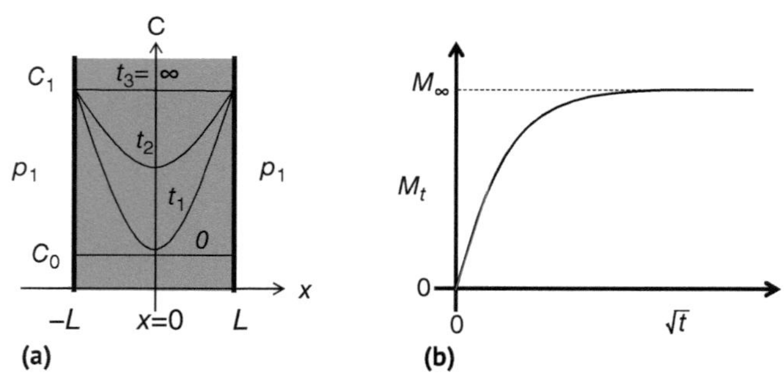

Figure 3.3 Schematic changes for sorption in (a) the concentration $C(x, t)$; and (b) the amount M_t of species i sorbed at time t as a function of square-root of time t during sorption experiment. *Source*: Y.S. Kang.

(Figure (3.3b)). Both D and S can be readily obtained from the M_t vs $\sqrt{t}$ curve. Detailed procedures to get those parameters will be given in Chapter 6 of this book.

3.2.2 Transient Permeation

When the surface concentrations are different ($C_1 > C_2$) in a semi-infinite slab of a thickness of L, the mass transport occurs from high concentration side to low concentration side, resulting in permeation (Figure 3.4a).

The initial and boundary conditions are mathematically stated as:

$C(x,0) = C_0$ for the initial condition where $0 \leq x \leq L$ and $t < 0$
$C(0,t) = C_1$ and $C(L,t) = C_2$ for boundary conditions with $t \geq 0$

$C(x, t)$ can be mathematically obtained from the integration of Equation (3.3) subject to the initial and boundary conditions [2]:

$$C = C_1 + (C_2 - C_1)\frac{x}{L} + \frac{2}{\pi}\sum_{n=1}^{\infty}\frac{C_2\cos n\pi - C_1}{n}\sin\frac{n\pi x}{L}\exp\{-Dn^2\pi^2 t/L^2\} + \frac{4C_0}{\pi}\sum_{m=0}^{\infty}\frac{1}{2m+1}\sin\frac{(2m+1)\pi x}{L}\exp\{-D(2m+1)^2\pi^2 t/L^2\} \tag{3.10}$$

Therefore, the total amount Q_t (mole or g/cm^2) of the species permeated through the unit area of slab increases with time for constant p_1 and p_2 (constant C_{1b} and C_{2b} for liquid) and increases linearly at steady state after long experimental time (Figure 3.4b). Q_t can be obtained [2]:

$$Q_t = -\int_0^t D\frac{\mathrm{d}C(L,t)}{\mathrm{d}x}\,\mathrm{d}t \tag{3.11}$$

Thus, Equations (3.10) and (3.11) give [2]:

$$Q_t = D(C_1 - C_2)\frac{t}{L} + \frac{2L}{\pi^2}\sum_{n=1}^{\infty}\frac{C_1\cos n\pi - C_2}{n^2}\{1 - \exp(-Dn^2\pi^2 t/L^2)\} + \frac{4C_0 L}{\pi^2}\sum_{m=0}^{\infty}\frac{1}{(2m+1)^2}\{1 - \exp\{-D(2m+1)^2\pi^2 t/L^2\}\} \tag{3.12}$$

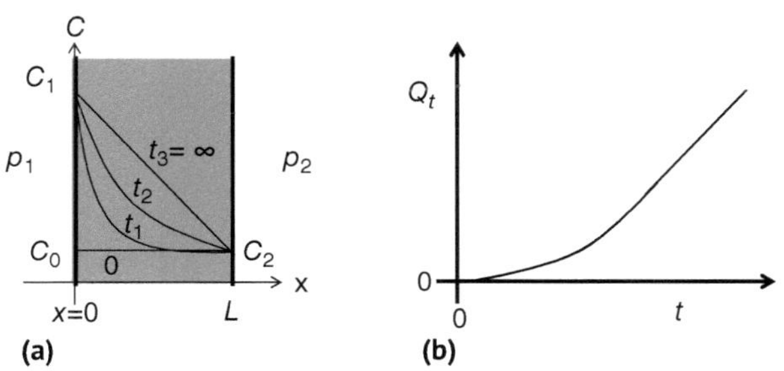

Figure 3.4 Schematic changes in (a) the concentration $C(x,t)$ as a function of position x and time t; and (b) the total amount Q_t of the species permeated through a slab membrane obtained by the pressure measurement in the downstream side p_2 with time t during permeation experiment. *Source:* Y.S. Kang.

Both *D* and *P* can be obtained from the permeation experiment. Detailed procedure to get those parameters will be given in Chapter 6 of this book.

3.3 Diffusion Through Polymers

A molecule diffuses through a continuous polymeric medium by cooperative motion of polymeric chains. Such cooperative motion is possible because common polymeric chains are flexible at least to some extent. Thus, diffusional displacement is strongly associated with cooperative motion of common polymeric chains, whereas molecular sieving could be dominant through exceptionally "rigid" or frozen chain polymers like carbon molecular sieve membranes. Therefore, the diffusion coefficient of a penetrant or diffusant through polymers is dependent on its size and concentration in some occasions, and also strongly on the mobility of the polymeric medium. Since the diffusant concentration affects the chain mobility, in particular at high concentrations like organic vapor, the diffusion coefficient may consequently depend on the concentration of the diffusant dissolved. However, the diffusion coefficient in common polymers is normally accepted to be constant in low soluble gases such as permanent gas.

3.3.1 Constant Diffusion Coefficient

Since the diffusion coefficient is strongly influenced by the mobility of polymer chains, highly mobile polymers with low glass transition temperatures, such as natural rubbers (T_g~ −70 °C), exhibit higher gas diffusion coefficients compared to rigid-chain polymers like poly(vinyl chloride) (PVC, T_g~82 °C), a typical glassy-state polymer as demonstrated in Figure 3.5. Here, the infinite dilution diffusion coefficient is measured to eliminate the effect of the concentration dependency of diffusant.

In fact, the diffusion coefficient in the natural rubber is more than 1,000 times higher than that in the glassy PVC (Figure 3.5) at room temperature. Such a large difference in the diffusion coefficient can be interpreted by the substantial difference

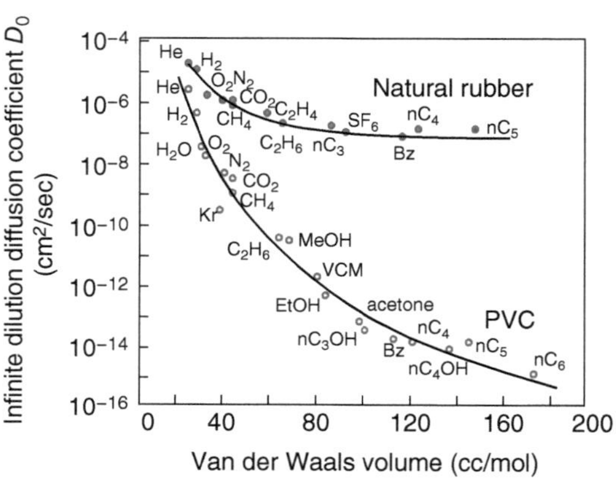

Figure 3.5 Infinite dilution diffusion coefficient through polymers as a function of van der Waals volume. *Source:* Adapted from [4].

in the segmental mobility of polymeric chains, which can be often characterized by the glass transition temperature. The low glass transition temperature represents the high segmental mobility as in rubbery state, whereas the high one represents that the macroscopic mobility is practically frozen but only minor, secondary motions are effective, including the crank-shaft motion, flipping motion of phenylene groups, and substituent motion. Such secondary motions are frequently referred to as the β- and γ-transitions, undergoing in glassy state. The glass transition temperatures of natural rubber and PVC are $-70\,°C$ and $82\,°C$, respectively. Therefore, it suggests that the much higher diffusion coefficient in the natural rubber is primarily associated with the much higher chain mobility. In other words, the mobility of polymer segments and the diffusion coefficient may be strongly correlated with each other: the higher the mobility of polymeric chain segments and the higher the diffusion coefficient is.

Figure 3.5 also illustrates that the diffusion coefficient strongly depends on the size of the diffusant represented by the van der Waals volume. The diffusion coefficient through the natural rubber decreases with the diffusant volume from 1.0×10^{-5} to 1.0×10^{-7} cm^2/sec, whereas that through PVC, decreases much more sharply from 1.0×10^{-6} to 1.0×10^{-15} cm^2/sec with the given volume change.

3.3.2 Concentration-dependent Diffusion Coefficient

When diffusant is in particular highly soluble in polymeric medium, the concentration-dependent diffusion coefficient is frequently observed. Since methanol, acetone, isopropyl amine, and propyl chloride are highly soluble in poly(vinyl acetate), the diffusion coefficient D is not a constant, instead, it increases exponentially with the solvent concentration inside, called the concentration-dependent diffusion coefficient (Figure 3.6). One common empirical relationship between the diffusion coefficient, D, and concentration of the diffusant, C, can be readily obtained as:

$$D = D_0 \exp(\alpha C) \tag{3.13}$$

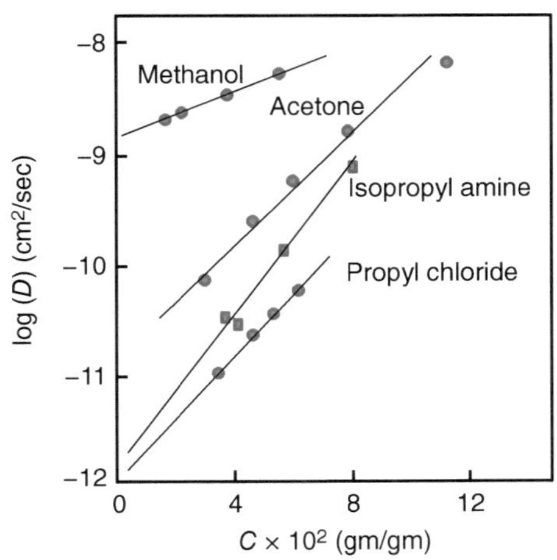

Figure 3.6 Diffusion coefficient D against the diffusant concentration C(g/g) in poly(vinyl acetate) at 40 °C. *Source*: Adapted from [5].

where D_0 is the infinite dilution diffusion coefficient and α is constant depending on both polymer matrix and penetrant. A linear relation between D and C is also suggested as $D = \alpha C + \beta$ with α and β being constants.

3.3.3 Temperature-dependent Diffusion and Theories for Activation Energy

Figure 3.7 shows the diffusion coefficients of various gases in copolymer of acrylonitrile and methyl acrylate as a function of the inverse temperature, demonstrating a typical feature of the Arrhenius behavior. Interestingly, a marked break in the Arrhenius behavior is observed in many cases when the measurement temperature passes through T_g.

Since the diffusion of small molecules in polymers is an activated process, its temperature dependency can be described by the Arrhenius type equation [7]:

$$D = D_0 \exp\left(-\frac{E_d}{RT}\right) \tag{3.14}$$

where D_0 is a front factor and E_d is the activation energy of diffusion. It is observed that the front factor is correlated with the activation energy according to the transition state theory simply as [8]:

$$\ln D_0 = a\frac{E_d}{RT} - b \tag{3.15}$$

where a is independent of both gas and polymer types and has a universal value of 0.68 with an equivalent dimension of D_0 (cm^2/sec) at 298 K, and b is independent of gas type and has a value of 9.2 for rubbery polymers and 11.5 for glassy polymers [9]. The linear correlation between $\ln D_0$ and E_d is very strong for rubbery polymers

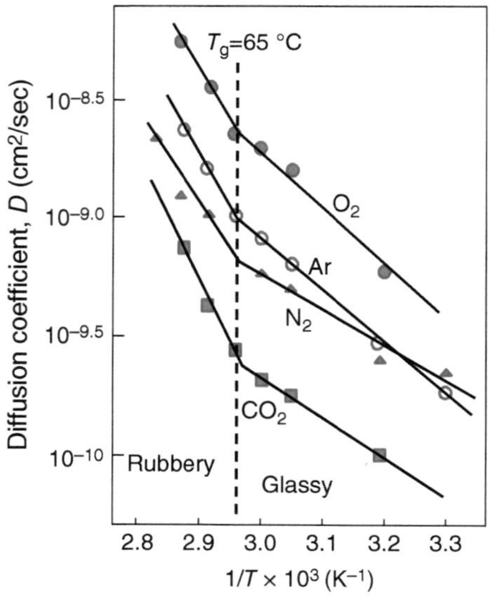

Figure 3.7 Gas diffusion coefficient in copolymer of acrylonitrile and methyl acrylate as a function of temperature. T_g of the copolymer is 65 °C. *Source*: Adapted from [6].

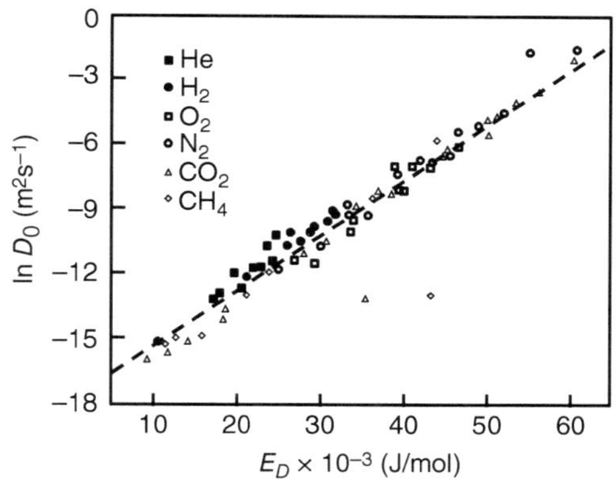

Figure 3.8 Correlation of the natural logarithm of the pre-exponential factor (ln D_0) and activation energy of diffusion (E_D) for six gases in amorphous rubbery poly(phenylmethylsiloxane). *Source*: Adapted from [10].

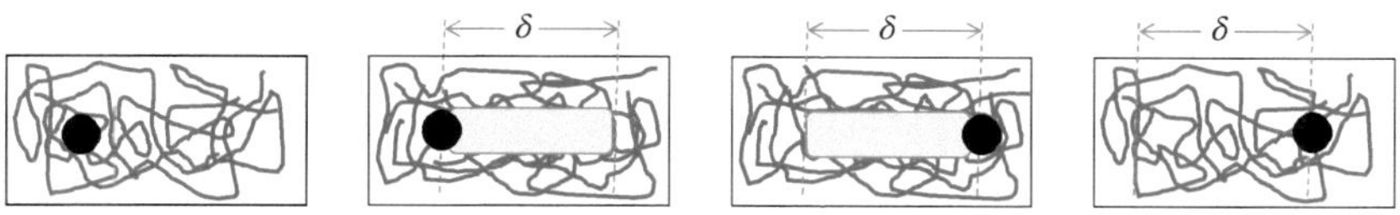

Figure 3.9 Schematic activation process for diffusion by generating free volume ($\pi r_A^2\,\delta$). *Source*: Y.S. Kang.

compared to that for glassy polymers. The above relation is often referred to as a "linear energy theory." Their correlation was demonstrated in Figure 3.8 [10].

Mears developed a quantitative microscopic explanation based on the notion that the activation energy is proportional to the volume of the activated state (Figure 3.9), which is proportional to the product of penetrant diameter squared and penetrant jump distance [11]. Mears considered that diffusion takes place along a cylindrical volume equal to $\pi r_A^2\,\delta$, where r_A is the radius of the diffusant and δ is the length of a diffusional step or the diffusional jump distance. He has equated the activation energy for diffusion to the energy necessary to expand the cylindrical volume required for the passageway of component A. Thus, the activation energy can be mathematically expressed as the product of the polymer cohesive energy density (CED) and the volume of the passageway thus:

$$E_d = \pi r_A^2\,\delta(CED) \tag{3.16}$$

Brandt's model also supports for the notion that the activation energy depends on the penetrant diameter squared for stiff chain glassy polymers [12]. Therefore, the effect of penetrant size on activation energy is modeled as follows:

$$E_d = cd_A^2 - f \tag{3.17}$$

where c and f are constants, which depend on the polymer. van Krevelen reports the c values, representing chain stiffness, from 250 cal/mol A^2 for extremely flexible poly(dimethyl siloxane) to approximately 1,100 cal/mol A^2 for stiff, glassy PVC. The value of f, representing interchain separation, ranges from 0 cal/mol for rubbery polymers and low-performance glassy polymers to approximately 14,000 cal/mol for the polyimide. van Krevelen also reports from the summary of many experimental data available that the activation energy of diffusion linearly correlates best with the square of penetrant collision diameter for a wide range of polymers [9].

3.3.4 Diffusion in Crystalline Polymers

Many polymeric materials such as polyethylene are indeed semicrystalline, despite that they are often called crystalline for convenience, indicating the coexistence of both crystalline and amorphous phases. Crystallites act as an impermeable barrier to molecular diffusion, primarily attributed from the higher packing density of polymeric chain segments within crystallites.

A relationship between the apparent diffusion coefficient, D, of crystalline polymer and the diffusion coefficient D_a of the totally amorphous phase has been proposed by [13]:

$$D = \alpha_a D_a = \frac{D_a}{\tau\beta} \tag{3.18}$$

where α_a is the volume fraction of the amorphous phase, τ is the tortuosity factor due to the presence of impermeable crystals, and β is the chain immobilization factor.

Figure 3.10a illustrates how impermeable crystallites create tortuous diffusion pathways. The tortuosity factor τ is often related to α_a using a power–law relationship:

$$\tau = \alpha_a^{-n} \tag{3.19}$$

where n typically ranges from 1 to 2 for unoriented polymers [14].

In addition, the diffusion coefficient is influenced by the restricted motion of polymer segments near the amorphous–crystalline interface. For instance, polymer chains in the amorphous phase may be partly involved in crystallization to make the polymer segments less mobile or immobilized (Figure 3.10b). This effect is quantitatively represented by the chain immobilization factor β.

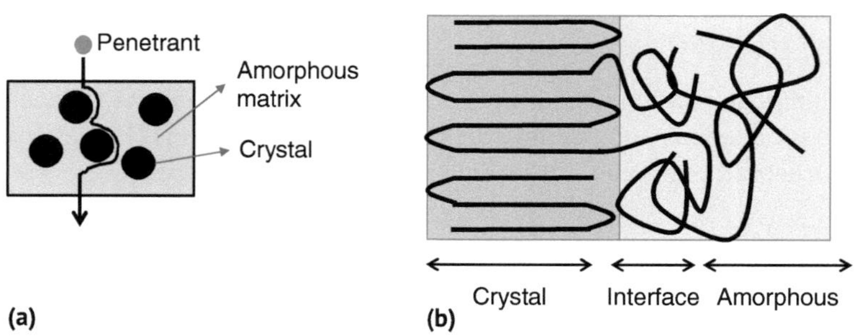

Figure 3.10 (a) Tortuous diffusion through crystalline polymer with crystals denoted as black circles; and (b) chain immobilization at interface between crystal and amorphous phases. *Source*: Y.S. Kang.

3.4 Statistical View of Diffusion Coefficient

Diffusion is a process by which a mass transports from one position of a medium to another as a result of random thermal molecular motion. The probability $p(x)$ that a molecule will one-dimensionally diffuse to a region between x and $x + \mathrm{d}x$ at time t is simply the number of the molecules $C(x)$ between x and $x + \mathrm{d}x$ with area A divided by the total number in the original source N_0 ($= n_0 N_A$ with n_0 and N_A being the mole number and the Avogadro number, respectively) [15].

$$p(x)\,\mathrm{d}x = \frac{AC(x)N_A}{N_0}\,\mathrm{d}x = \frac{1}{2(\pi Dt)^{1/2}}\exp\left(-\frac{x^2}{2Dt}\right)\mathrm{d}x \tag{3.20}$$

The mean square distance, $\langle x^2 \rangle$, diffused by a molecule in time t, is:

$$\langle x^2 \rangle = \int x^2 p(x)\,\mathrm{d}x = 2Dt \tag{3.21}$$

and

$$\langle x^2 \rangle^{1/2} = (2Dt)^{1/2} \tag{3.22}$$

which demonstrates that the average distance $\langle x^2 \rangle^{1/2}$ traveled by diffusing particle is linearly proportional to the square root of time t.

The diffusion coefficient can also be rewritten from Equation (3.22) as:

$$D \equiv \frac{\langle x^2 \rangle}{2t} \tag{3.23}$$

The calculation method of $\langle x^2 \rangle$ is equivalent to the same mathematical problem as finding the mean square distance from the beginning at (0, 0, 0) to the chain end at (x, y, z), or the end-to-end distance, r, of a freely jointed chain of N bonds with a bond length of l, commonly referred to as a random flight model as shown in Figure 3.11.

Hence,

$$\langle r^2 \rangle = Nl^2 \tag{3.24}$$

Here, N is the number of bonds from the beginning to the end or the steps of random walk and can be expressed in terms of diffusional jump frequency ϕ (i.e. $N = \phi t$), and l is replaced with the diffusional jump distance, δ, respectively. Thus,

$$3\langle x^2 \rangle = (\phi t)\delta^2 \tag{3.25}$$

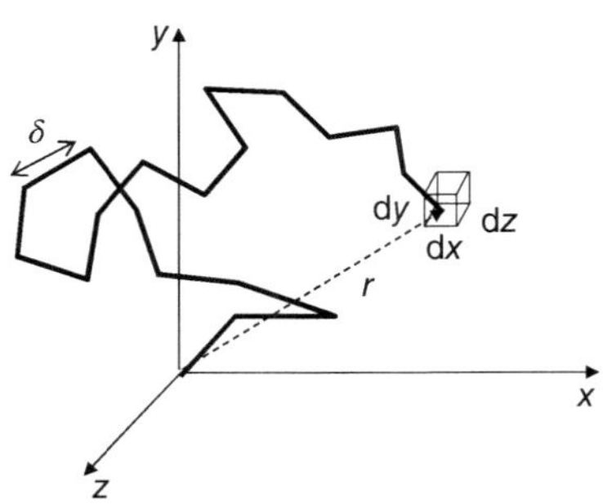

Figure 3.11 Statistical view of diffusion process in a three-dimensional space. Movement of a penetrant from initial position (0, 0, 0) to another (x, y, z) after time t with a jump distance δ. The average displacement is expressed as $\langle r^2 \rangle^{1/2}$. *Source*: Y.S. Kang.

The factor of 3 in Equation (3.25) arises from the fact that $\langle r^2 \rangle = \langle x^2 \rangle + \langle y^2 \rangle + \langle z^2 \rangle = 3\langle x^2 \rangle$. When Equation (3.25) is substituted in Equation (3.23), it gives:

$$D = \frac{\phi \delta^2}{6} \tag{3.26}$$

Since the diffusional jump distance δ will not vary greatly from material to material (~1 nm), the diffusion constant will vary dominantly as a result of changes in the jump frequency, ϕ. In other words, the diffusion coefficient, D, is linearly proportional to the jump frequency, ϕ, from one site to the other with a jump distance, δ.

3.5 Free Volume Theory for Diffusion

Diffusion is closely linked to the mobility of a system, which is primarily influenced by the amount of free volume. Free volume can be visualized as unoccupied or vacant spaces within a lattice model, as shown in Figure 3.12.

In the lattice model, all sites in a box are initially occupied by molecules. When one molecule is removed, creating a vacant site or free volume, the mobility of the remaining molecules increases. This makes it easier for the molecules to rearrange. If additional molecules are removed to generate 2 or 3 free volumes, the mobility increases significantly. This enhanced mobility facilitates the movement of a molecule from one position in the box to another, a process referred to as diffusion.

In essence, diffusion is largely governed by the amount of free volume in the system, as greater free volume provides more opportunities for molecular motion and rearrangement, resulting in faster and easier diffusion.

When the lattice model is applied to polymers, the specific volume, V (cm^3/g), of polymers consists of the occupied volume by polymeric chains, V_o (cm^3/g), and the free volume, V_f (cm^3/g). Thus,

$$V = V_o + V_f \tag{3.27}$$

In characterizing the amount of the free volume, the fractional free volume FFV or f has been frequently used:

$$f = \text{FFV} = \frac{V_f}{V} \tag{3.28}$$

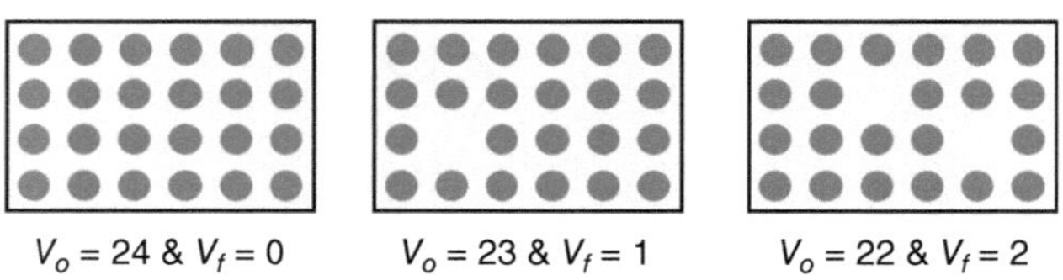

Figure 3.12 Free volumes in lattice model where V_o and V_f are the numbers of occupied sites and unoccupied, free sites, respectively. *Source*: Y.S. Kang.

3.5.1 Free Volume in Polymers

Specific volume V of a polymer medium increases with temperature T, and this increase includes contributions from both V_o and V_f, where $V = V_o + V_f$ as shown in Figure 3.13. The occupied volume, V_o, accounts for the intrinsic volume of the polymer molecules, including their van der Waals volume and the volume associated with molecular vibrations [16]. The free volume, V_f, on the other hand, refers to the excess volume beyond V_o, which is directly associated with the mobility of polymer chains.

As illustrated in Figure 3.13, V increases linearly with T up to a critical temperature, known as the glass transition temperature T_g. Beyond T_g, both V and V_f increase more sharply, reflecting a significant rise in the mobility of the polymer chains. This behavior also implies that f, defined by Equation (3.28), increases linearly with T.

In polymer–diluent systems, such as those containing a diffusant, the diluent concentration, ϕ_1, representing the fractional volume of the diluent, also affects f. Consequently, f generally depends on both T and ϕ_1, expressed as $f(\phi_1,T)$.

It is generally argued that the temperature dependence of fraction free volume $f(0,T)$ of a neat polymer above T_g may be expressed as:

$$f(0,T) = f_g(0,T_g) + \alpha_f (T - T_g) \tag{3.29}$$

where α_f is the thermal expansion coefficient of the free volume, expressed as $\alpha_f = \alpha_r - \alpha_g$, where α_r and α_g are thermal expansion coefficients of rubbery and glassy states, respectively. According to iso-free volume theory for glass transition, $f_g(0, T_g)$ is set at 0.025 for many polymers and α_f has been found roughly constant at 3.2×10^{-4} /K by Simha et al. [17]. Bueche has pointed out that the use of Equation (3.29) is valid only when the temperature T is above T_g [18].

In the case of a polymer–diluent mixture, it is generally known that the free volume increases with the amount of diluent. For small ϕ_1, f increases linearly with ϕ_1 at a given T, and may be written in the form:

$$f(\phi_1, T) = f(0, T) + \beta(T)\phi_1 \tag{3.30}$$

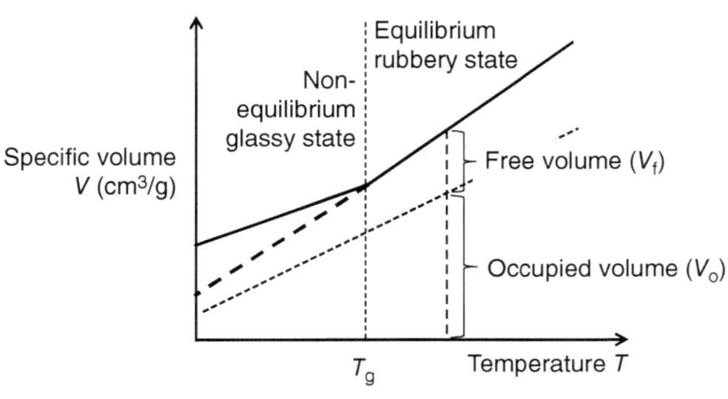

Figure 3.13 Specific volume increase as a function of temperature and the definition of the occupied and free volumes. *Source*: Y.S. Kang.

where $f(0,T)$ is the fractional free volume of neat polymer at a given T, and $\beta(T)$ is a function of T, which represents the effectiveness of the diffusant for the increase of free volume.

In characterizing the size and the amount of free volume, one of the three parameters including (1) the average free volume per molecule $\langle v_f \rangle$, originally defined by Cohen and Turnbull [19], (2) the average size of each free volume $\langle v_h \rangle$, and (3) the fractional free volume f (or FFV), has been frequently used. They are interrelated together as:

$$\langle v_f \rangle = \langle v_h \rangle (N_f / N) \quad \text{and} \quad f = \langle v_h \rangle (N_f / V) \tag{3.31}$$

where N_f is the number of the free volume per 1 g of polymer, N is the number of the molecule or the repeating unit per 1 g of polymer, and V is the specific volume. N_f may vary from polymer to polymer. Equation (3.31) is obtained since $V_f = \langle v_h \rangle N_f = \langle v_f \rangle N = f \cdot V$ and $N = N_A / M_{\text{rep}}$ with M_{rep} being the MW of the repeating unit of polymer.

3.5.2 Theory of Cohen and Turnbull

The free volume approach is one of the simplest ways to understand polymer dynamics, including diffusion and glass transition behavior. For instance, Cohen and Turnbull treated diffusion as transport of hard spheres in a liquid [1, 19]. In their view, diffusion occurs not as a result of an activation in the ordinary sense, but rather as a result of the redistribution of the free volume within the liquid (Figure 3.14).

Cohen and Turnbull derived for diffusion coefficient as [1, 19]:

$$D = D_0 \exp\left(-\frac{\gamma v*}{\langle v_f \rangle} \right) \tag{3.32}$$

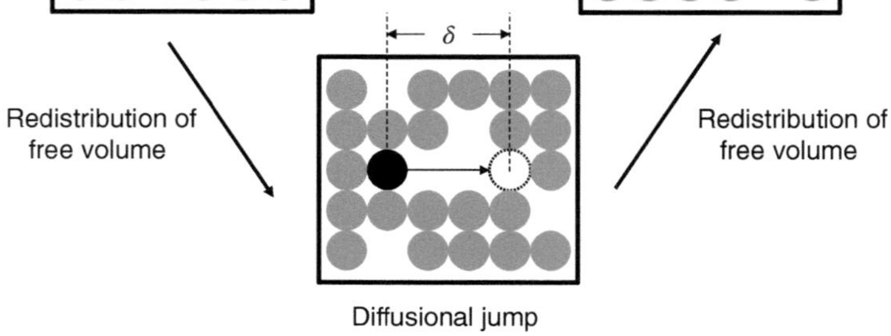

Figure 3.14 Schematic view of diffusion process according to the free volume theory by Cohen and Turnbull [19], where the diffusional jump distance δ is in the range of 0.4–0.8 nm for permanent gases. *Source*: Adapted from [19].

where γ is a constant that allows for overlap of the free volume ($0.5 \leq \gamma \leq 1.0$), v^* is the minimum size of a free volume required for transport. Equation (3.32) indicates that D is directly determined by both v^* and $\langle v_f \rangle$.

Since the average free volume size $\langle v_h \rangle$ is frequently measured, Equation (3.32) is just for convenience rearranged into:

$$D = D_0 \exp\left(-\frac{B}{\langle v_h \rangle}\right) \tag{3.33}$$

The effects of $\langle v_h \rangle$ on D can be experimentally verified by measuring D of various gas molecules through polyimides and their copolymers with varying $\langle v_h \rangle$ as shown in Figure 3.15. Note that $\langle v_h \rangle = \langle v_f \rangle (N/N_f)$ as introduced previously.

Figure 3.15 shows a good agreement between the theoretical predictions and diverse experimental data, supporting the validity of the Cohen and Turnbull theory. It also illustrates that, for a given hole size $\langle v_h \rangle$, smaller gas molecules exhibit higher diffusion coefficients due to their greater ability to access the available free volume. In terms of the free volume approach, the size dependence is attributed to a critical volume fluctuation large enough for the diffusant to accomplish an elementary diffusion jump [20]. The pre-factor D_0 is the smallest for largest CH_4 in these experimental range, which can be interpreted as being indicative of the largest activation energy while accepting $D_0 = D_0^0 \exp(-E_d/kT)$ [20, 21]. The activation energy E_d is generally attributed to the critical energy of a penetrant molecule that must be obtained to overcome the attractive forces holding it to its neighbors.

The value of B for Equation (3.33) obtained from the slope of Figure 3.15 is observed to be linearly proportional to the molecular size ($4\pi r^3/3$) of diffusant with the radius r as shown in Figure 3.16. The B values were estimated to be in the range of 0.4–0.8 nm^2 for gases (He, H_2, O_2, N_2, CO_2, and CH_4), calculated from the kinetic diameter (filled circles in Figure 3.16) [20]. The minimum volume v^* of the critical free volume or hole was additionally found to be near the van der Waals volume V_w or, more specifically, $0.66\ V_w < v^* < 0.86\ V_w$ with $\gamma = 1$ and $1.32\ V_w < v^* < 1.7\ V_w$ with $\gamma = 1/2$ [19, 22].

Vrentas and Duda [21] proposed a linear mathematical relationship between B and the molecular size of the diffusant as:

$$B = \frac{\gamma V_1(0)}{N_f M_j} \tag{3.34}$$

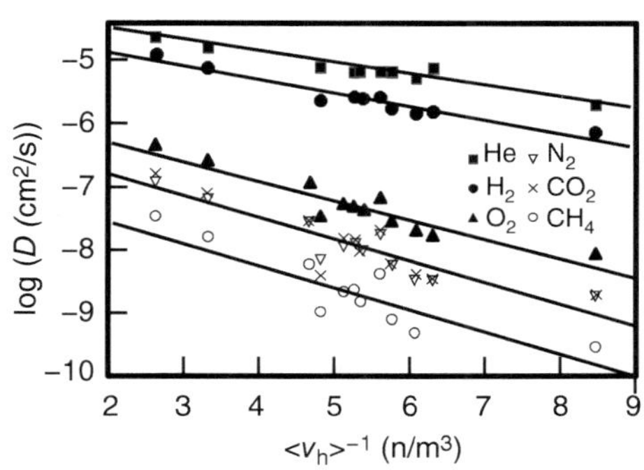

Figure 3.15 Plot of the logarithm of the diffusion coefficients D of various gas molecules vs reciprocal average hole size $\langle v_h \rangle$ of polyimides and their copolymers from positron annihilation lifetime spectroscopy. Straight lines are linear regressions based on Equation (3.33). *Source*: Ref. [20]/American Chemical Society.

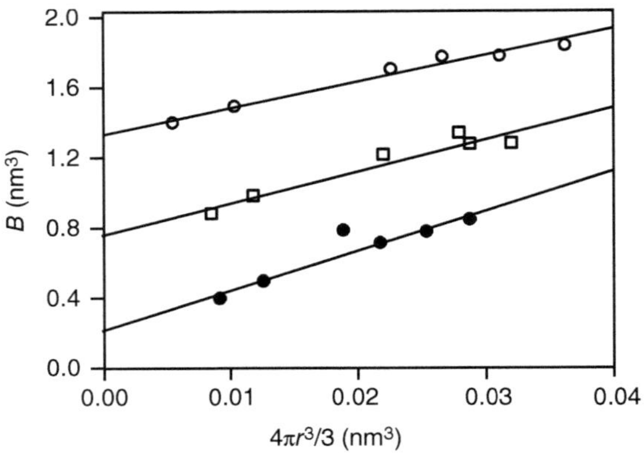

Figure 3.16 Free volume parameter B in Equation (3.33) plotted against the molecular volume ($4\pi r^3/3$) of the gas molecules where r is the kinetic radius. Radii were calculated from zeolite sieving (solid circles), Lennard-Jones collision data (open squares), and viscosity measurements (open circles). *Source*: Adapted from [20].

where γ is a free volume overlap factor of the order of 0.5, $V_1(0)$ is the molecular gas volume at 0 K, which can be estimated from the kinetic diameter, and N_f is the number of free volumes or holes, which can be measured from positron annihilation lifetime spectroscopy (PALS). The most uncertain parameter in Equation (3.34) is the magnitude of the polymer jump unit M_j. However, many glassy polymers showed $M_j \approx 1.5\ M_{\text{mono}}$, where M_{mono} is the molar mass of the monomer unit [23].

3.5.3 Theory of Fujita

An elegant treatment of diffusion based on the free volume theory in a polymer–diluent system was given by Fujita [24]. Both Fujita and Mears showed that the free volume approach provides a fairly reasonable explanation of the principal features of the concentration- and temperature-dependence of the diffusion coefficient of organic vapors in amorphous rubbery polymers. The basic idea of these works is that the mobilities of both the polymer segment and the diffusant in a polymer–diluent system are primarily determined by the amount of free volume. Similar to the Doolittle theory for viscosity, it was assumed that the free volume directly relates with the mobility, m_d, of the diffusant relative to the polymer component:

$$m_d = A_0 \exp\left(-\frac{B_d}{f}\right) \tag{3.35}$$

where f is the fractional free volume, defined as $f = V_f/V$, and the parameters A_0 and B_d depend only on the type of the diffusant, but independent of diffusant concentration and temperature [25]. B_d represents a constant critical value of free volume for a diffusion process.

Thermodynamic diffusion coefficient, D_T, of the diffusant relative to the polymer component is defined as [24]:

$$D_T = RTm_d \tag{3.36}$$

Therefore, D_T may be expressed as:

$$D_T = A_0 RT \exp\left(-\frac{B_d}{f}\right) \tag{3.37}$$

The mutual diffusion coefficient D can be related to the thermodynamic diffusion coefficient D_T through the thermodynamic factor Φ, which accounts for the non-ideal interactions between particles in a mixture. This relationship can be expressed as [1, 24]:

$$D = D_T \cdot \Phi \tag{3.38}$$

where Φ is given in a binary system by:

$$\Phi = \frac{d\ln(a_1)}{d\ln(C_1)} = 1 + \frac{d\ln(\gamma_1)}{d\ln(C_1)} \tag{3.39}$$

Here, a_1 and C_1 are the activity and the concentration of species 1 or diluent and they are related as $a_1 = \gamma_1 C_1$ where γ_1 is the activity coefficient. Thus, the activity accounts for the effective concentration of species, considering nonideal interactions. If the system is in ideal state ($\gamma_1 = 1$), $\Phi = 1$ and consequently $D = D_T$.

Therefore, D can be expressed as:

$$D = A_0 RT\exp\left(-\frac{B_d}{f}\right)\left(1 + \frac{d\ln(\gamma_1)}{d\ln(C_1)}\right) \tag{3.40}$$

and more conveniently

$$D = A_d\exp\left(-\frac{B_d}{f}\right) \tag{3.41}$$

where $A_d = A_0 RT[1 + d\ln(\gamma_1)/d\ln(C_1)]$.

Since f is a function of T and C_1 (*or* ϕ_1) as in Equations (3.29) and (3.30), expressed as $f(\phi_1, T)$ for convenience, Equation (3.41) is frequently used to analyze the concentration- and temperature-dependent diffusion coefficients, which appears to be a huge advantage.

Figure 3.17a shows a strong correlation between log(D) and $1/f$, aligning well with Equation (3.41). The B_d values are closely correlated with the square of the Lennard-Jones diameter (d^2) and mostly fall within the range of 0.9–1.1 (Figure 3.17b),

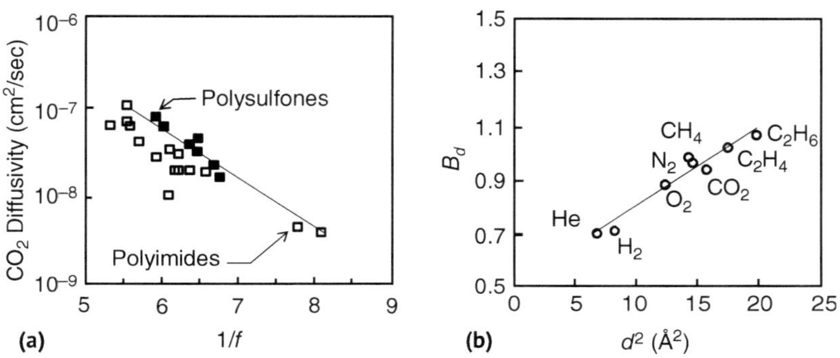

Figures 3.17 (a) Correlation of CO_2 diffusivity D with fractional free volume f in glassy aromatic polymers of polysulfones and polyimides [3, 26]; and (b) dependence of B_d in Equation (3.41) on square of penetrant Lennard-Jones diameter d^2 in crosslinked poly(ethylene glycol diacrylate). *Source*: Adapted from [27].

which is thus commonly approximated as unity. It is interesting to note that Figure 3.16 shows a strong correlation between B and d^3, representing the size of the penetrant [20], while Figure 3.17b shows a good correlation between B_d and d^2, rather than d^3.

3.5.4 Theory of Miyamoto and Shibayama, and Vrentas and Duda

The original free volume theory, proposed by Cohen and Turnbull [19], has been re-examined to improve the correlation between the diffusion coefficient and the free volume. One interesting approach is a modified free volume model for glassy polymers by Miyamoto and Shibayama for ion conduction [28], and Vrentas and Duda [21, 29]. Taking into account the critical energy, E^*, a penetrant molecule must obtain to overcome the attractive forces holding it to its neighbors, Equation (3.42) has been developed for diffusion coefficient in the limit of zero penetrant concentration $D(0)$:

$$D(0) = D_0 \exp\left(-\frac{\gamma v^*}{\langle v_f \rangle}\right) \exp\left(-\frac{E^*}{RT}\right) \tag{3.42}$$

The formulation of Equation (3.42) was based upon the assumption that diffusion in liquids is governed by (1) the presence of an adjacent free volume of certain size into which a molecule can jump and (2) the acquisition of sufficient energy by the diffusing molecule to escape from the force field of its neighbors [30].

3.5.5 Temperature- and Concentration-dependent Diffusion Coefficients

Temperature- and concentration-dependent diffusion coefficients are readily explained by the free volume theory, because the free volume changes with temperature and concentration of diffusant. These relationships are mathematically expressed in Equations (3.29) and (3.30).

For example, substituting Equation (3.29) into Equation (3.41) produces either the Williams–Landel–Ferry (WLF) equation or the Vogel–Tammann–Fulcher (VTF) equation, which describe how the diffusion coefficient varies with temperature. The WLF equation (Equation (2.38) in Section 2.5.2 and Equation (9.28) in Section 9.6.2) characterizes the temperature dependence of polymer relaxation processes near the glass transition, represented by the relaxation time τ. On the other hand, the VTF model explains the temperature dependence of ionic conductivity in electrolytes, Equation (9.26) in Section 9.6.1.

For a polymer–diluent mixture containing a polymer 2, and a diluent 1, substituting Equation (3.30) into Equation (3.41) yields:

$$D = A_d \exp\left(-\frac{B_d}{f(0,T) + \beta\phi_1}\right) \tag{3.43}$$

Equation (3.43) demonstrates that the diffusion coefficient varies with the diluent fractional volume ϕ_1. Thus, the free volume theory provides a framework for understanding both temperature- and concentration-dependent diffusion behavior.

3.5.6 Advantages and Limitations of Free Volume Theory

The free volume theory is conceptually and mathematically simple, making it easy to learn and apply across various fields. The theory's basic assumption is that when the free volume is large, system mobility increases, leading to decreased viscosity, and increased diffusion coefficient.

In polymers, the mobility of polymer chains and the diffusion coefficient of small molecules are directly influenced by the amount of free volume. Cohen and Turnbull initially assumed that the redistribution of free volumes for diffusion did not require additional energy input or change [19, 31]. This assumption is widely accepted for highly flexible, rubbery polymers, allowing the free volume model to be commonly used to analyze mass transport phenomena in these materials without difficulty.

However, the free volume theory can also be applied to mass transport through glassy polymers by disregarding the original assumption. For glassy polymers, redistribution of free volume requires additional energy due to their limited chain mobility. Figure 3.17 demonstrates the validity of the free volume theory even in glassy polymers. Table 3.1 summarizes key differences in diffusion phenomena between rubbery and glassy states. Nonetheless, the free volume theory has been successfully applied across various fields, including mass transport in glassy polymers, as listed in Table 3.2.

3.5.6.1 Complexities in Glassy Polymers

Applying the free volume theory to glassy polymers presents several issues and complexities. Those include (1) applicability (the theory must be adapted to account for the low mobility of chains in glassy polymers, which affects diffusion mechanisms); (2) definition of free volume (variations in how free volume is defined can lead to inconsistencies when comparing models or interpreting data); (3) estimation and characterization of free volume; (4) terminology; and (5) time-dependent changes in free volume (physical aging and secondary relaxations introduce variability in the size and amount of free volume over time, complicating predictions).

For example, Lipson and White critically reviewed the concept of polymer free volume and its relationship to the glass transition temperature [32], highlighting how variations in free volume definitions influence our understanding of polymer

Table 3.1 Gas transport through rubbery and glassy polymers.

	Rubbery	Glassy
Material state	Equilibrium	Non-equilibrium
Transport mechanism	Fickian	Non-Fickian or flow-like
Chain flexibility	Flexible	Practically frozen
Free volume redistribution	Easy	Practically none
Main factor in determining D	Total free volume	Size and amount of free volume

Table 3.2 Diverse applications of free volume theory.

Field	Equation	Remark
Chain mobility	$m_d = A_0 \exp\left(-\frac{B_d}{f}\right)$	(3.35)
Doolittle equation	$\eta = A \exp\left(+\frac{B}{f}\right)$	(2.28)
Molecular diffusion	$D = D_0 \exp\left(-\frac{\gamma v*}{\langle v_f \rangle}\right)$	(3.32)
Molecular diffusion	$D = A_d \exp\left(-\frac{B_d}{f}\right)$	(3.41)
Ion diffusion	$D = A_d \exp\left(-\frac{B_d}{f}\right)$	(3.41)
Permeation	$P = A_p \exp\left(-\frac{B_d}{f}\right)$	(3.57)
WLF equation	$\log(a_T) = \frac{-17.44(T - T_g)}{(51.6 - (T - T_g))}$	(2.38)
VTF equation	$\sigma(T) = AT^{-\frac{1}{2}} \exp\left(\frac{-B}{(T - T_0)}\right)$	(9.26)
T_g of polymer	$f = 0.025$ at $T = T_g$	
T_{g12} of miscible blends	$T_{g12} = v_1\ T_{g1} + v_2\ T_{g2}$	(2.37)

behavior during phase transitions. Historically, the term "free volume" has referred to both the total free volume, V_f, and the interstitial or excess free volume, $V_{f,\text{int}}$.

The free volume theory assumes that the redistribution of free volume occurs without requiring additional energy. However, in glassy polymers, the chain mobility is extremely low, and their chain conformations are practically frozen, with only minor secondary relaxations. Despite these challenges, the free volume theory offers useful semi-predictive insights for gas transport through glassy polymers, as illustrated in Figure 3.17 [3].

3.5.6.2 Definition of Free Volume

In rubbery polymers, free volume is defined as $V = V_o + V_f$ as illustrated Figure 3.13, where V_o and V_f increase with temperature. For glassy polymers, the relationship $V = V_o + V_f$ still holds as shown in Figure 3.18, but V_f is defined differently as:

$$V_f = V_{if} + V_{mv} \quad \text{for glassy state (Figure 3.18)} \tag{3.44}$$

$$V_f = V_{if} \quad \text{for rubbery state (Figure 3.13)} \tag{3.45}$$

where V_{if} and V_{mv} are the interstitial free volume in the equilibrium matrix and the microvoid volume, respectively. Microvoids, as schematically illustrated in Figure 3.24b, explain the unique gas sorption behavior of glassy polymers and are often treated as a type of large free volume.

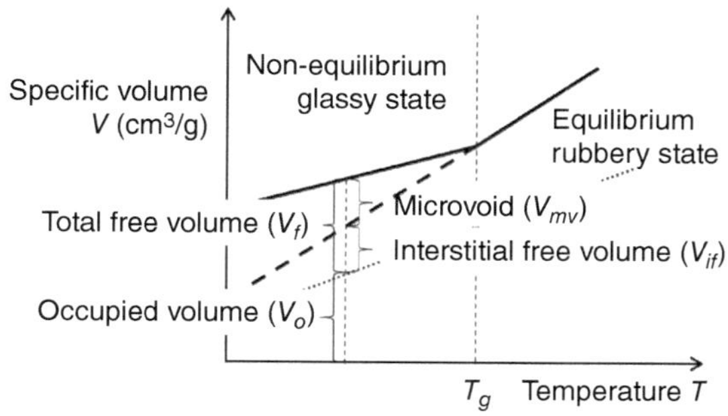

Figure 3.18 Schematic free volume in glassy polymer: occupied volume (V_o), microvoid (V_{mv}), interstitial free volume (V_{if}) in equilibrium matrix, and total free volume (V_f). *Source*: Y.S. Kang.

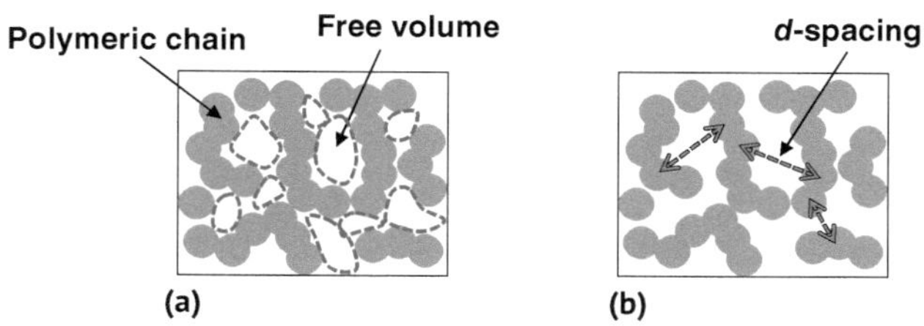

Figure 3.19 Schematic (a) free volume and (b) *d*-spacing between polymeric segments. *Source*: Y.S. Kang.

3.5.6.3 Time-dependent Free Volume

Free volume changes over time due to physical aging or secondary relaxation in the nonequilibrium state of glassy polymers. Advanced theories accounting for these changes could involve models that incorporate the dynamic nature of free volume, such as its size and redistribution over time, as well as the energy barriers associated with these changes. These models could address limitations by better predicting gas transport and diffusion behaviors in glassy polymers under varying conditions.

3.5.6.4 Summary

The free volume theory, while conceptually straightforward, requires adaptation for glassy polymers. This includes incorporating models that address the dynamic nature of free volume, such as changes in its size and redistribution over time, energy barriers associated with these changes, and physical aging effects. These advanced approaches can provide more precise predictions of diffusion phenomena, including gas transport and sorption behaviors, even under varying conditions.

3.5.7 Measurement and Estimation of Free Volume and *d*-spacing

The free volume and the *d*-spacing are schematically visualized in Figure 3.19, where the *d*-spacing is the distance between the nearest polymeric segments. Thus, the size of both the free volume and the *d*-spacing are interrelated and distributed statistically.

The free volume in polymers can be measured directly and indirectly. Direct experimental methods include PALS, inverse gas chromatography, ^{129}Xe NMR spectroscopy, wide-angle X-ray scattering (WAXS), photochromic and fluorescence techniques and spin probe measurements as listed in Table 3.3. In particular, PALS technique has shown a good correlation between the lifetime of *ortho*-positronium

Table 3.3 Experimental methods used for probing free volume in polymers.

Method	Probe	Size (nm)	Information
PALS[a]	*o*-Ps	0.106	Size, its distribution, concentration, and their T and p effects
Inverse gas chromatography	Organic vapor	>0.5 >C_3	Temperature-averaged mean size
^{129}Xe NMR	^{129}Xe	~0.4	Size and its temperature-dependence
Spin probe method	TEMPO, stable nitroxyl radicals	>10	Size distribution of large free volume
Photochromic probe method	Stilbene and azobenzene	13.60	
Electrochromic probe method	Azo-dyes	>80	

[a]Positron annihilation lifetime spectroscopy [3].
Source: Adapted from [3].

Table 3.4 Fractional free volume and glass transition temperatures for representative polymer materials at ambient temperature.

Polymer	T_g(°C)	f
Polysulfone	186	0.18
PIM-1	300	0.25
TR-1	>250	0.2 [35]
Poly(1-trimethylsilyl-1-propyne)	>250	0.29

Source: Adapted from [34].

(*o*-PS) atoms and the free volume for both simple liquids and polymers and provides a powerful tool to obtain information on the size, number and its distribution of free volumes. Interestingly, IGC and PALS estimates of free volume element size distributions are consistent [33]. The measured fractional free volumes, f, of representative polymers are summarized in Table 3.4.

An indirect method is to first measure or estimate both the specific volume, V, and the occupied volume, V_o, and then calculate the free volume, V_f where $V_f = V - V_o$. V is simply the reciprocal of the density ρ, i.e. $V = 1/\rho$, where ρ can be readily measured conveniently and accurately. One reliable way to measure ρ is to utilize a density gradient column.

The occupied volume V_0 is commonly estimated differently from: (1) van der Waals radii of the atoms; (2) crystalline volume at 0 K; and (3) the extrapolated volume of liquid at 0 K. V_0 can also be estimated from the molecular volume in the absence of thermal vibrations at 0 K [36]. It is worth to note that V_0 commonly includes the thermal expansion due to anharmonicity of vibrations. In many cases, it can also be readily estimated semi-empirically using the group contribution method, such as Bondi's group contribution method [9, 37]. Bondi proposed calculating $V_o = 1.3\,V_w$ where V_w is the van der Waals volume. Some values of V_w are tabulated in the book by van Krevelen [9].

Hence:

$$V_f = V - 1.3V_w \tag{3.46}$$

The factor of "1.3," called the Bondi's factor, has been reinterpreted as the reciprocal of the packing density of a polymer crystal at 0 K [38]. However, a new method particularly for a mixed system was proposed to obtain the V_o value [38]. Free volume in polycarbonate and polysulfone is in the range of 0.02–0.07 nm^3 in size and 0.3–0.45 nm in diameter. Note that the bond lengths of C—C and C—H are 0.15 and 0.11 nm, respectively.

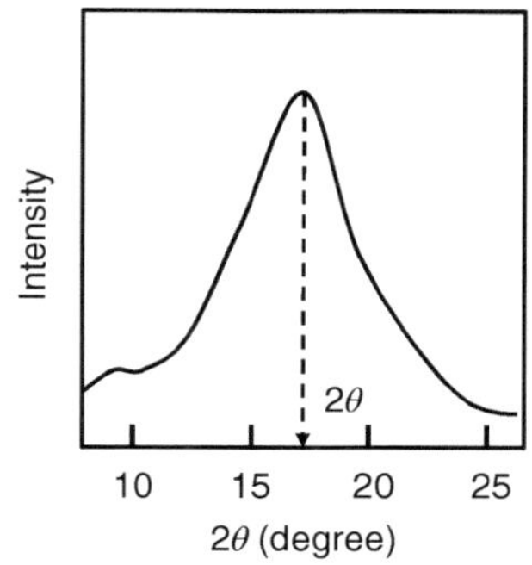

Figure 3.20 Schematic wide angle X-ray diffraction spectrogram. *Source*: Y.S. Kang.

Intersegmental d-spacing can also represent the approximate size of free volume, which can be readily obtained from wide-angle X-ray diffraction (WAXS) by means of the Bragg equation, in the form [39]:

$$d = \frac{n\lambda}{2\sin\theta} \tag{3.47}$$

where d is the d-spacing, n is the diffraction order ($n = 1$ is first order, $n = 2$ is second order), λ is the wavelength (0.154 nm for Cu Kα radiation) and θ is the scattering angle corresponding to the maximum of the principal peak in a plot of intensity versus the scattering angle as shown in Figure 3.20.

3.6 Sorption in Polymers

Solubility of gases and vapors plays a critical role in determining the mass transport properties through polymers in addition to the diffusion coefficient as described previously. More specifically, the amount C (mole/cm^3 or cm^3 (STP)/cm^3 polymer) of a penetrant sorbed in a polymeric slab significantly influences mass flux J (mole/cm^2 sec or cm^3(STP)/cm^2 sec) according to the Fick's law, where C is termed as the concentration or the solubility. As a first approximation, the concentration C of a gas in a polymer is linearly proportional to the applied pressure p (atm or cmHg). The proportionality constant S or k_D (mole/cm^3 cmHg or cm^3(STP)/cm^3 atm) is called as the solubility coefficient of the gas penetrant expressed as:

$$S = C/p \quad \text{or} \quad k_D = C/p \tag{3.48}$$

When S is not a function of concentration but a constant, S is commonly expressed as k_D, also termed as the Henry's law constant.

For a liquid penetrant, the above equation is replaced by:

$$K = C/C^B \tag{3.49}$$

where K (dimensionless) and C^B (mole/cm^3) are the partition coefficient and the bulk concentration, respectively. K is typically considered to be concentration-dependent because of larger solubility of common liquid solvent in polymers. Its features are very close to S, and thus will not be treated here further.

3.6.1 Thermodynamic View of Sorption

Gas or vapor sorption in polymers at a given temperature is regarded as a two-step thermodynamic process: (1) condensation of the gaseous penetrant to a liquid-like condensed state; and (2) mixing of the liquid-state penetrant with the polymer segment (Figure 3.21).

Thus, the enthalpy of sorption, ΔH_s, can be expressed as:

$$\Delta H_s = \Delta H_c + \Delta H_m \tag{3.50}$$

where ΔH_c is the enthalpy of condensation of a pure gaseous penetrant into a liquid state, representing the condensability, and ΔH_m is the molar enthalpy of mixing the condensed penetrant with polymer segments, representing polymer–penetrant interactions. The enthalpy of condensation of the pure penetrant is directly related to its enthalpy of vaporization: $\Delta H_v = -\Delta H_c$ [3].

For light, permanent gases, such as H_2, O_2, and N_2 in polymers, ΔH_m is small due to their weak interactions. Thus, ΔH_s is primarily determined by ΔH_c. As the size or molecular weight of the penetrant molecules increases, ΔH_c decreases while ΔH_m increases. This is because larger penetrant molecules are more condensable, resulting in a smaller ΔH_c, while their interactions, including van der Waals forces, polar interactions, and hydrogen bonding, become stronger, leading to a larger ΔH_m. Consequently, ΔH_m, which measures these interactions, plays a crucial role in determining ΔH_s and solubility. In this regard, this section will be divided into two subsections: (1) permanent gas sorption in polymers; and (2) condensable gas and vapor sorption in polymers.

3.6.2 Sorption of Permanent Gases in Polymers

For light permanent gas penetrants having low penetrant boiling point T_b and critical temperature T_c that do not undergo specific interactions with the polymer, ΔH_s

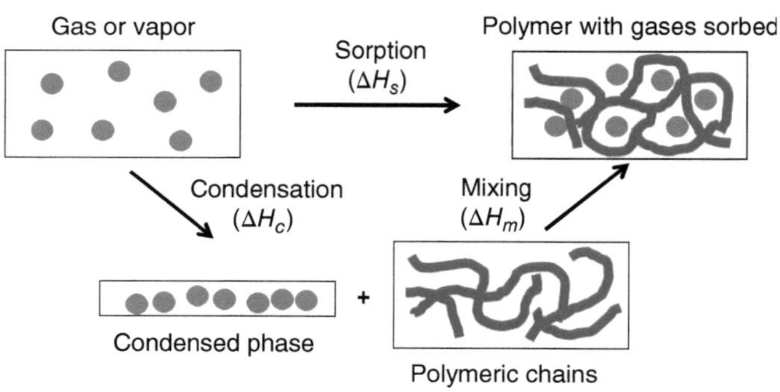

Figure 3.21 Schematic of sorption of gaseous penetrant into a polymeric medium by two step process: condensation and mixing. *Source:* Y.S. Kang.

is dominantly determined by ΔH_c rather than ΔH_m. As a result, the penetrant solubility in polymers typically scales with a convenient measure of penetrant condensability, such as T_b, T_c, or Lennard–Jones temperature, (ε/k), with k and ε being the Boltzmann's constant and the potential well depth parameter in the Lennard–Jones potential curve, respectively [40]. A strong experimental correlation between ΔH_c and T_c or (ε/k) was obtained as [3]:

$$\Delta H_c = a_c R T_c = a_e R\left(\frac{\varepsilon}{k}\right) \tag{3.51}$$

where a_c and a_e are all adjustable parameters. For most penetrants and polymers, the values of a_c are quite similar. Typically, a_c is between 13 and 19 [41].

Accepting the van't Hoff equation (Equation (3.55) in Section 3.6.5) with the case of $\Delta H_c \gg \Delta H_m$, which is often the case for permanent gases in polymers, then [3]:

$$\ln S = \ln S_0 + a_c T_c = \ln S_0 + a_e\left(\frac{\varepsilon}{k}\right) \tag{3.52}$$

This suggests that the solubility of light permanent gas having no specific interactions with polymer segments depends mostly on its own condensability, frequently measured by T_c and ε/k. Table 3.5 lists T_b and T_c of gases and vapors to give qualitative information on condensability of gases.

The experimental data in Figure 3.22 show a good linear relationship between log (S) and T_c for both rubbery poly(dimethyl siloxane) (PDMS) and glassy polysulfone (PSF), demonstrating the validity of Equation (3.52). However, a small difference in

Table 3.5 Boiling T_b and critical T_c temperatures of representative gases and vapors.[a]

	He	H_2	N_2	O_2	CO_2	CH_4	VCM	*n*-Hexane
T_b(K)	4.22	20.27	77.36	90.19	194.7	111.7	259.7	341.4
T_c(K)	5.19	33.20	126.2	154.6	304.2	190.8	—	—

[a] Data from Wikipedia: critical point (thermodynamics).

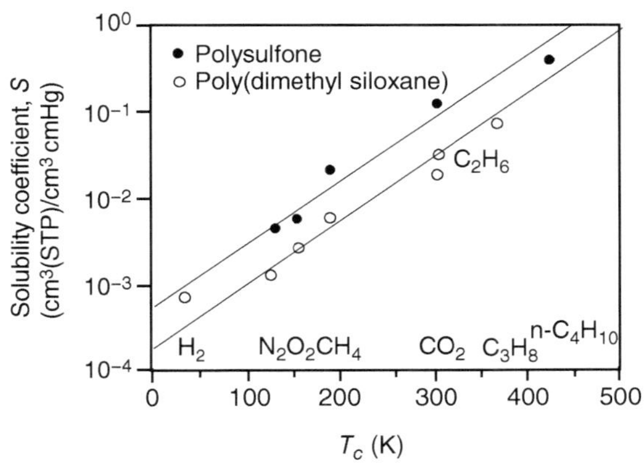

Figure 3.22 Solubility coefficient, S, of gases as a function of their critical temperature, T_c. *Source:* Adapted from [42].

the solubility coefficient was observed depending on the polymers used, suggesting the effect of the small interaction energy difference or ΔH_m between penetrant and polymer segments on the solubility [42].

3.6.3 Sorption of Condensable Gas and Vapor in Rubbery Polymers

For condensable gases and organic vapors in rubbery polymers, a linear relationship between solubility C and applied pressure p is generally observed at low pressures at a given T, frequently referred to as Henry's law. However, as the applied pressure p_i or precisely the chemical activity a_i increases, Henry's law may no longer be valid, but the Flory–Huggins equation (Equation (3.53)) becomes applicable instead. In other words, experimental data follow Henry's law at low chemical activities but conform the Flory–Huggins equation at high chemical activities as schematically illustrated in Figure 3.23a.

$$\ln a_1 = \ln \phi_1 + \left(1 - \frac{1}{x}\right)\phi_2 + \chi_{12}\phi_2^2 \tag{3.53}$$

where ϕ_1 and ϕ_2 are the volume fractions of penetrant 1 and polymer 2, respectively, x is the degree of polymerization, and χ_{12} is the Flory–Huggins interaction parameter between the penetrant and polymer pair.

In the case of higher chemical activities of gases or vapors, the solubility becomes higher and starts to deviate from the linear relationship between C and p, indicating the concentration-dependent solubility coefficient $S(C)$. For instance, the solubility of n-hexane vapor in *cis*-polybutadiene rubber, in which intermolecular interactions are principally of the nonspecific van der Waals type, linearly increases with the activity initially, followed by deviation from the linearity at higher activities (Figure 3.23b). Here, the penetrant chemical activity $a_i = p_i/p_i^0$, where p_i and p_i^0 are the applied vapor pressure and the saturation vapor pressure at a given temperature, respectively.

Figure 3.23b demonstrates the agreement between experimental data with the theoretical results from the Flory–Huggins equation (Equation (3.53)) with $\chi_{12} = 0.58$. Thus, the solubility coefficient, S, in rubbery polymers can be expressed

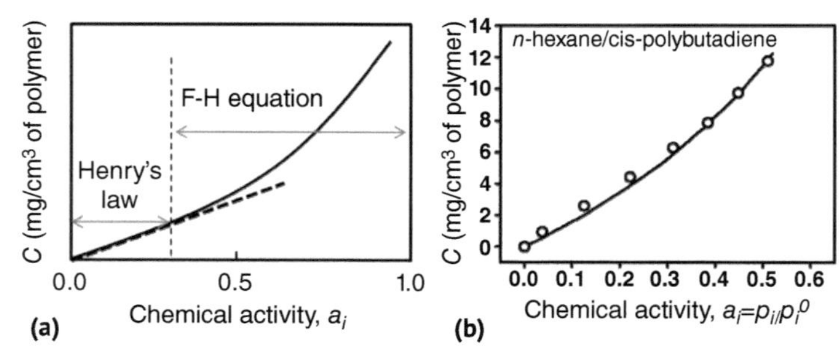

Figure 3.23 Equilibrium solubility of (a) condensable gas or vapor in rubbery polymer as function of the chemical activity a_i; and (b) n-hexane vapor in rubbery *cis*-polybutadiene as a function of the penetrant activity a_i. The solid line in (b) is obtained from the Flory–Huggins theory (Equation (3.53)) with $\chi = 0.58$. *Source*: Adapted from [43].

by either the Henry's law or the Flory–Huggins equation, depending on the chemical activity a_i of the penetrant: Henry's law for low a_i and the Flory–Huggins theory for high a_i. The equation of state theory is also adapted to explain the equilibrium sorption data when the volume change is significant on mixing of organic vapor with polymers [44].

3.6.4 Sorption of Condensable Gas and Vapor in Glassy Polymers: Dual Sorption Model

Sorption features of condensable gases and vapors are totally different between rubbery and glassy polymers. Such a difference is mostly associated with the difference of the phase: the equilibrium state in rubbery polymers and the nonequilibrium state in glassy polymers.

A specific volume of a given polymer at rubbery state decreases with decreasing temperature linearly and its slope changes at a glass transition temperature T_g. The rubbery state is known to be in the equilibrium state, representing that a new equilibrium can be immediately achieved whenever exposed to a new environment and consequently its specific volume does not change with time. However, glassy polymers are not in the equilibrium but in the nonequilibrium, consequently relaxing polymeric segments into the equilibrium state isothermally, which is termed as physical aging or secondary relaxation process. Such relaxation process results in the decrease in the specific volume with time extremely slowly at a given temperature. Therefore, it can be treated that glassy polymers have extra-volume, associated with the nonequilibrium state, in addition to the equilibrium volume as shown in Figure 3.24a. Such extra volume, commonly called microvoids or holes, is also sketched in Figure 3.24b.

One simple way to visualize the structure of glassy polymers is to have a distribution of microvoids or holes frozen into an equilibrium matrix as shown in Figure 3.24b, a dual phase model. Thus, it can be anticipated that microvoids or holes decrease during relaxation processes.

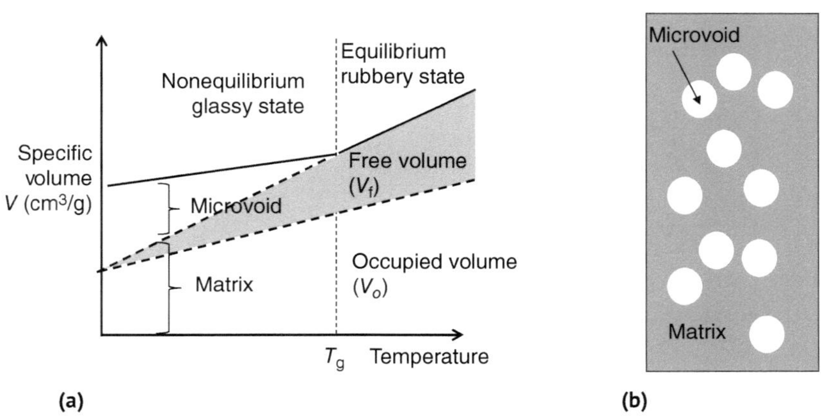

Figure 3.24 (a) The specific volume change as a function of temperature; and (b) a schematic dual phase structure of a glassy polymer. *Source:* Y.S. Kang.

Accepting the dual phase model, gases can be dissolved into both the equilibrium matrix, C_D, and the holes, C_H, suggesting a dual mode sorption. C_D represents sorption of normally diffusible species in the equilibrium matrix while C_H represents the amount of penetrant adsorbed in monolayer on surface of microvoids or holes. Therefore, the total solubility C is expressed as [44]:

$$C = C_D + C_H = k_D p + C'_H \frac{bp}{1 + bp} \tag{3.54}$$

where k_D is the solubility coefficient, C'_H is hole saturation constant, and b is hole affinity constant related to the ratio of rate constants of sorption and desorption of penetrant in holes.

Condensable gas such as CO_2 and CH_4 in glassy polycarbonate ($T_g \sim 147\,°C$) showed typical dual sorption mode, concave to the pressure axis, as shown in Figure 3.25a, whereas less condensable gas such as N_2 or He followed apparently linear Henry's model.

Solubility of a more condensable gas such as VCM (vinyl chloride monomer) in PVC, $T_g = 85\,°C$) at 90 °C, higher than T_g, follows the Flory–Huggins equation with $\chi = 0.98$, as shown in Figure 3.25b, because it is in an equilibrium rubbery state. However, as the experimental temperature decreased from 90 to 30 °C, the solubility deviation from the Flory–Huggins equation (Equation (3.53)) becomes larger, which seems to be associated with the contribution of the holes. Therefore, in this case, the total solubility C is equal to the sum of C_D and C_H (Equation (3.54)), where C_D can be obtained from the Flory–Huggins equation (Equation (3.53)) and C_H from Langmuirian sorption $C_H = C'_H bp/(1 + bp)$ (Equation (3.54)), demonstrating a dual sorption mode [46].

3.6.5 Temperature-dependent Sorption

Temperature-dependence of the solubility coefficient, S (cm^3(STP)/(cm^3 polymer) · atm), according to the van't Hoff equation, is expressed as:

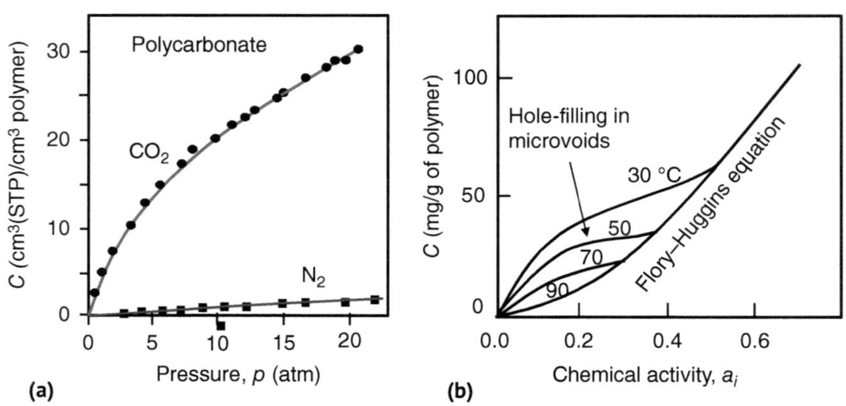

Figure 3.25 Sorption isotherms of (a) CO_2 and N_2 in glassy polycarbonate at 35 °C. *Source*: Adapted from [45]. (b) Vinyl chloride monomer in poly(vinyl chloride) by varying temperature ($T_g = 85\,°C$). *Source*: Adapted from [46].

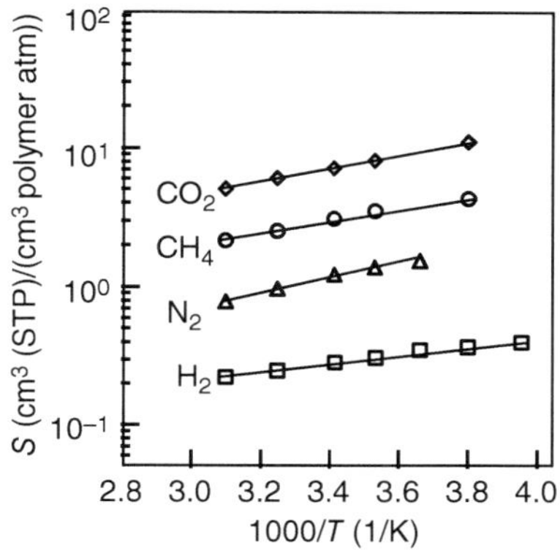

Figure 3.26 Solubility coefficients, S, for CO_2, CH_4, N_2, and H_2 for 1 hour at 10 atm as a function of temperature for the thermally rearranged polymer, TR-400. *Source*: Ref. [47]/ with permission of Elsevier.

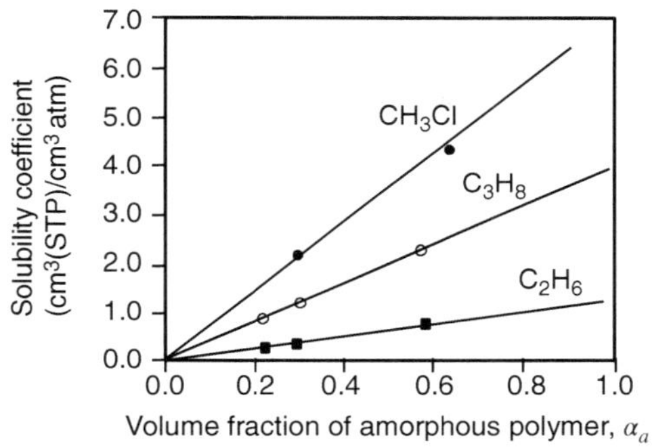

Figure 3.27 Effect of the volume fraction of amorphous phase on gas solubility coefficients in polyethylene. *Source*: Adapted from [42].

$$S = S_0 \exp\left(\frac{-\Delta H_s}{RT}\right) \tag{3.55}$$

where ΔH_s (kJ/mol) is the enthalpy of sorption and S_0 is the pre-exponential factor. Figure 3.26 shows the solubility coefficients for CO_2, CH_4, N_2, and H_2 for 1 hour at 10 atm as a function of temperature for the thermally rearranged polymer (TR400) [47]. The enthalpy of sorption, obtained from the slope of the curve at 10 atm, has negative values, suggesting endothermic behavior.

3.6.6 Sorption in Crystalline Polymers

Following a two-phase model in crystalline polymers for diffusion (Figure 3.10), the Henry's law solubility coefficient, S *or* k_D, is proportional to the solubility coefficient of the completely amorphous phase, S_a, as [39]:

$$S = \alpha_a S_a \tag{3.56}$$

where the proportionality constant, α_a, represents the amorphous volume fraction of the polymer [42]. Equation (3.56) indicates that the solubility coefficient is linearly proportional to the volume fraction α_a of amorphous phase as demonstrated in Figure 3.27, and also suggests that the crystals are non-permeable to common gases, consistent with the effective diffusion coefficient through crystalline polymers [42].

3.7 Permeation Through Polymers

The permeability P of gases or vapors through a polymer membrane is defined as the product of the diffusion coefficient D and the solubility coefficient S : $P = D \times S$ (Equation (3.6)). D is nearly proportional to the mobility of polymeric chains in rubbery states, while it is mostly governed by size and amount of free volume in glassy polymers. S is primarily determined by either the condensability of permanent gases or the interactions with condensable gases, like highly soluble gases or organic vapors. Figure 3.28 presents representative values of D, S, and P for gases in rubbery poly(dimethyl siloxane) (PDMS) and glassy polysulfone (PSF) as a function of the critical volume V_c of gas penetrants, where V_c serves as a measure of molecular size [3].

In Figure 3.28a and b, D decreases slightly in rubbery PDMS ($T_g = -150\,°C$) as V_c increases, while D decreases sharply in glassy and tough PSF ($T_g = 188\,°C$). On the other hand, S increases with V_c in both PDMS and PSF though the correlation

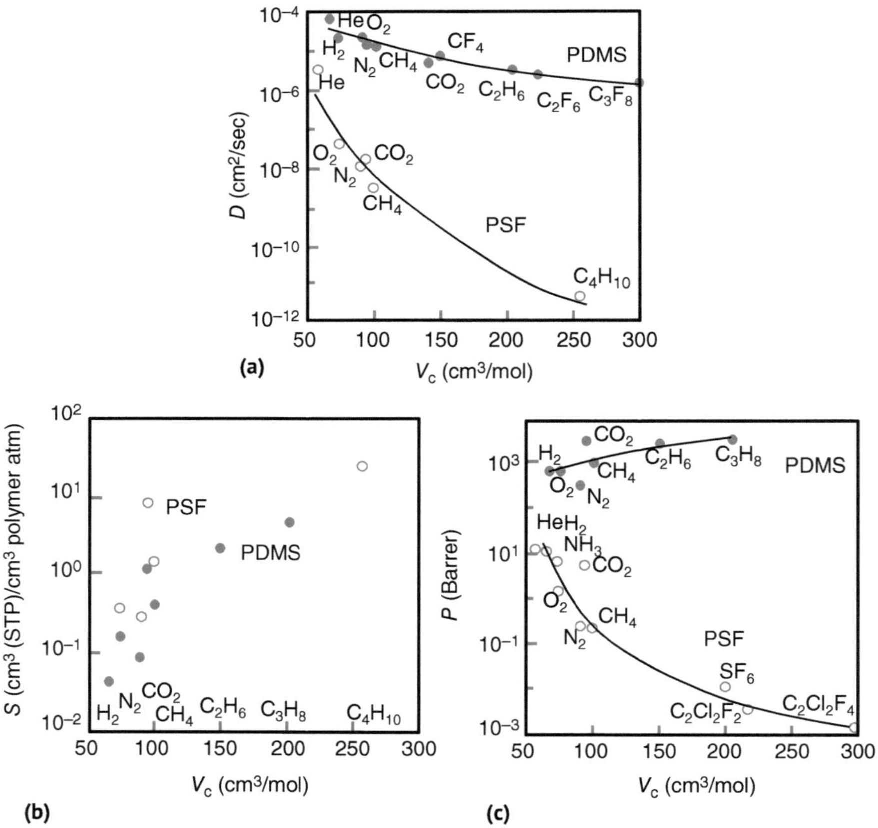

Figure 3.28 Diffusion coefficients D, solubility coefficients S, and permeabilities P of gases as a function of penetrant critical volume V_c through poly(dimethyl siloxane) (PDMS) at 35 °C and polysulfone (PSF) at 23 °C. *Source*: Adapted from [3].

is weak. Interestingly, P in the rubbery PDMS increases slightly as V_c increases, whereas, in PSF, it decreases sharply with V_c.

These observations suggest that flexible rubbery polymers like PDMS, which possess large free volumes, can readily redistribute these volumes without significant energy barriers, as proposed by Cohen and Turnbull [19]. This redistribution allows large penetrants to diffuse, resulting in a slight decrease in D with increasing V_c. Concurrently, S increases significantly with V_c, causing P in rubbery PDMS to increase slightly.

In contrast, for glassy polymers like PSF, D decreases sharply with V_c from approximately 1×10^{-6} for He to 1×10^{-12} cm^2/sec for n-butane (Figure 3.28a). S in PSF increases with V_c from about 10^{-2} to 10^1 cm^3(STP)/cm^{3}polymer · atm (Figure 3.28b). Consequently, P in PSF is significantly lower than in PDMS, and P decreases sharply with V_c (Figure 3.28c) primarily due to the steep decline in D. This behavior indicates that the redistribution of the free volumes is limited in the glassy state primarily due to restricted conformational changes of polymer chains. As a result, glassy polymers like PSF exhibit limited chain mobility, significantly hindering the diffusion of large penetrants while allowing smaller ones, leading to a pronounced size-sieving effect or steep slope in Figure 3.28c.

3.7.1 Structure–Properties Relationships for Gas Permeation

The studies on the transport of gases and vapors in polymers indicate that chemical structure of polymers dictates the permeation properties. The gas permeability P of glassy polymers is much more sensitive to their chemical structure than those of rubbers. Such sensitivity can be observed by the difference in P for different polymers: $P_{CO_2} = 0.002$ Barrer for polyacrylonitrile and $P_{CO_2} = 27{,}000$ Barrer in poly(trimethylsilyl propyne) [48]. Therefore, fundamentals on the chemical structure–permeability relationships will be discussed hereafter for rubbery and glassy polymers [49].

The free volume theory, applicable for both rubbery and glassy polymers, suggest that free volume should be enlarged to improve the gas permeability. Strategies to increase the fractional free volume vary depending on the state of polymers: rubbery and glassy polymers.

For rubbery polymers, main chain mobility, as characterized by glass transition temperature T_g, determines the total free volume and consequently gas permeability. Then, it can be anticipated that low T_g rubbers may have high gas permeabilities. Figure 3.29 shows that the nitrogen permeabilities in rubbers decrease with T_g as expected. While a general trend in this figure can be seen, there is also quite a bit of scatter, suggesting factors other than T_g contribute in an important way to P [3].

For glassy polymers, free volumes are increased by increasing polymeric segmental rigidity, which is opposite to the strategy for rubbery polymers. They can be enhanced by employing rigid backbone chains and/or by incorporating bulky and flexible substituents to backbone chains for prohibiting compact packing.

The chemical structures of the main backbone chain in aromatic polymers, including polysulfones, polycarbonates, poly(phenylene oxide), and polyimides,

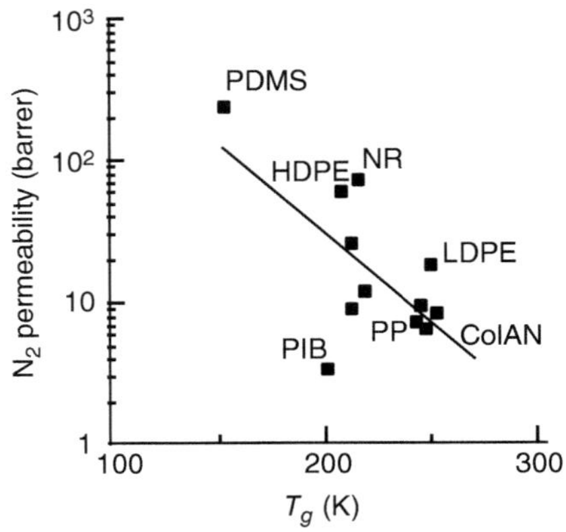

Figure 3.29 Permeability of nitrogen in rubbers with different glass transition temperatures: poly(dimethyl siloxane) (PDMS); high density polyethylene (HDPE); low density polyethylene (LDPE); natural rubber (NR); polyisobutylene (PIB); polypropylene (PP); poly(isoprene-co-acrylonitrile) (CoIAN). *Source*: Adapted from [3].

primarily strongly influence the permeation properties. For example, bulkier connector group X between two aromatic groups in polyimides can stiffen the chain backbones, inhibit their dense packing, and thus increase free volume and permeability [50]. When X is changed from —O—, —(CH_2)—, —C(CH_3)$_2$— to —C(CF_3)$_2$—, P_{CO_2} changes from 23, 19.3, 30, and 63.9 Barrer, respectively. These effects are also revealed in the transport properties of polycondensation materials [51].

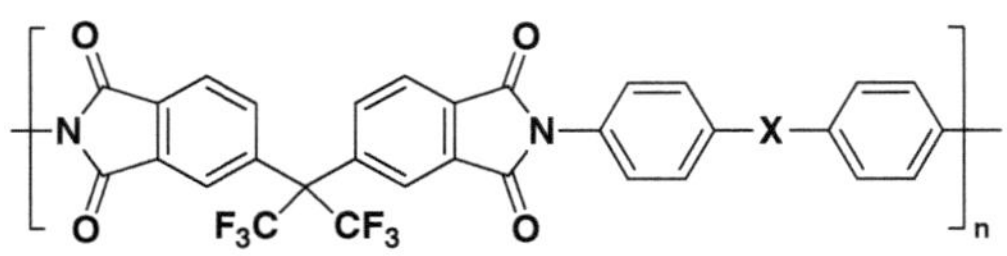

In addition, substituent groups on the mainchain backbones can also influence the permeation properties depending on their chemical structure, size, mobility, substitution-position, and the degree of substitution. In polymers with aromatic backbone polymers, a marked increase in permeability can be achieved by symmetrical or asymmetrical substitution at phenylene rings [51]. For example, R_1 in polycarbonates is systematically changed from H, CH_3, Cl, and Br. Table 3.6 lists physical, thermal, and transport properties of substituted polycarbonates [3, 45]. The methyl substitution increases both T_g and f (or FFV), resulting in the increase in P_{CO_2}. The substitution of halogens (i.e., Cl, Br) introduces additional polar groups into the polymers, which causes T_g to increase additionally and significantly, indicating a reduction in chain mobility. The halogen substitution also reduces f. The more efficient chain packing and loss of chain mobility in the halogen tetrasubstituted polycarbonates lower CO_2 and CH_4 permeability.

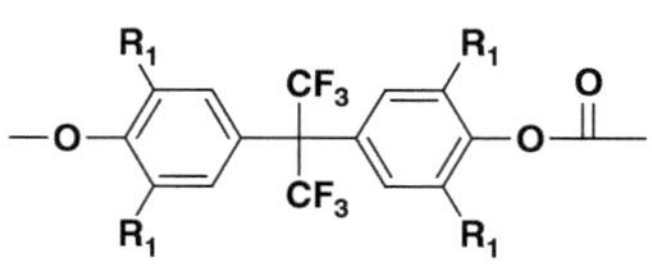

Bulky side groups or substituents, including —Si(CH_3)$_3$, —Ge(CH_3)$_3$, and —CH(CH_3)$_2$, are popular in increasing P, among which —Si(CH_3)$_3$ is most frequently used. For example, when one group of —Si(CH_3)$_3$ is introduced in the every repeating unit of polysulfone P_{CO_2} increased from 12.6 to 190 Barrer, higher than

Table 3.6 Physical, thermal, and transport properties of substituted polycarbonates [3, 45].

R_1	T_g	P_{CO_2}	α_{CO_2/CH_4}	f
H	150	6.0	23.3	0.164
CH_3	193	17.6	22.0	0.180
Cl	230	6.7	29.7	0.179
Br	263	4.2	33.6	0.133

15 fold [52, 53]. When two groups are introduced in PSF, P_{CO_2} increased from 4.92 to 15.8 Barrer [54].

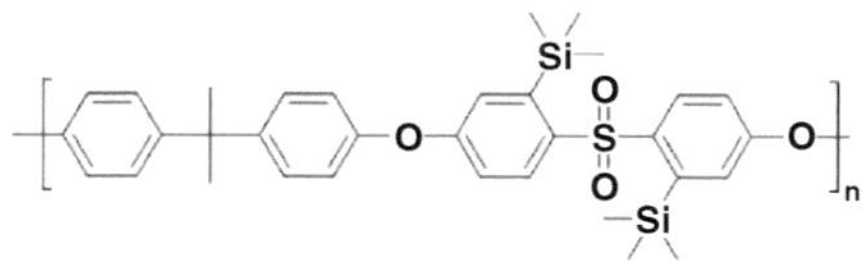

The most detailed information on such effects was obtained for polyacetylenes [55]. Examination of numerous Si-substituted glassy polymers allows a formulation of the rules for stronger enhancement of permeability: (1) symmetry of side groups plays the strongest effects, for example, the stronger effect of $SiMe_3$ group than that of $SiMe_2$ Et; (2) a bulky group should be attached directly to the main chain and not *via* a spacer; and (3) long linear substituents (e.g. $SiMe_2$-n-C_6H_{13}) show significant decreases in permeability.

So far, only the effects of nonpolar groups have been considered. Meanwhile, the introduction of substituents, capable to dipole-dipole interactions or to form hydrogen bonds, can strongly influence the transport properties due to increasing the interchain interactions or interactions with some penetrants. For example, in carboxylated poly(vinyl trimethylsilane) with the —COOH group content of 20 wt %, the permeability of hydrocarbon gases decreased two-fold, whereas that of water vapors increased by a factor of 4.5 [52]. The introduction of the —COOH group in the *m*-phenylenediamine moiety of 6FDA polyimide also resulted in a decrease in P_{O_2} from 3 to 1 Barrer [56].

Side-chain mobility can also affect the T_g value and therefore the transport properties in addition to its size [57, 58]. However, attempts to find a general correlation between T_g and P failed [59]. This could be mostly due to the several experimental findings to display the inverse relationship: high T_g polymers can display a very large permeability. For example, PTMSP, PIMs, and TR polymers, whose chemical structures are illustrated in Figure 3.30, have extremely high gas permeabilities while their glass transition temperatures are unmeasurably high (higher than 250 °C) [55].

3.7.2 Gas Permeation in Polymers with Extremely Stiff Chains and High Free Volume

When designing new membrane materials with high gas permeability, one might initially consider materials having high free volume with flexible chains and facile redistribution of free volumes, based on the original free volume theory by Cohen and Turnbull [19]. However, such high free volume materials with flexible chains

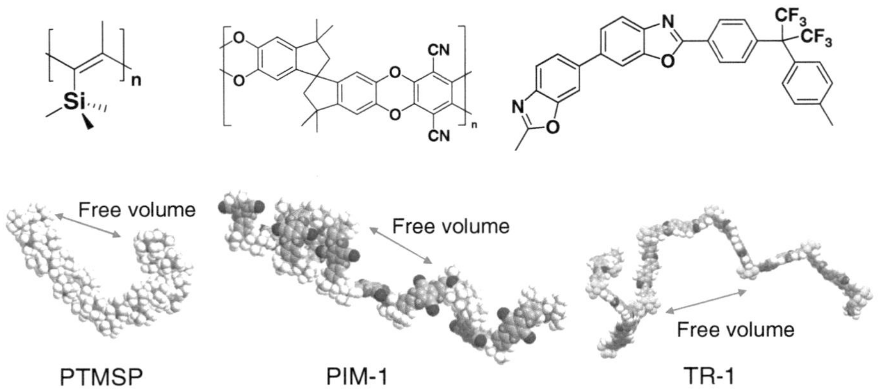

Figure 3.30 Chemical structures of poly(1-trimethylsilyl-1-propyne) (PTMSP), polymers of intrinsic microporosity (PIM-1), and thermally rearranged polymer (TR-1). *Source:* Y.S. Kang.

tend to be soft and mobile, similar to liquids, which is impractical for membrane applications. Another approach to solve this issue could be to design materials having high free volume with highly rigid backbone chains. In such materials, free volume redistribution is minimal, which appears to be not compatible with the original free volume theory, while expecting high gas permeability. This concept was realized with the synthesis of poly(1-trimethylsilyl-1-propyne) (PTMSP) [60].

Since then, highly microporous polymers with rigid backbone chains and high glass transition temperatures T_g have been developed, exhibiting markedly high permeability for common gases such as H_2, O_2, N_2, CO_2, and CH_4. Notable examples include polymers of intrinsic microporosity (PIM) by Budd and McKeown [61, 62], and thermally rearranged polymers (TR polymers) by Park and Lee [35].

3.7.2.1 Poly(1-trimethylsilyl-1-propyne)

PTMSP is known as the most permeable of all polymers to gases. It has a backbone of alternating single and double bonds, like other polyacetylenes, but the double bonds are not significantly conjugated. The bulky trimethylsilyl and methyl groups cause the polymer chain to be twisted. Furthermore, there is a very high barrier to rotation about the single bonds 170 kJ/mol, so that at normal temperatures it cannot easily change its conformation. These twisted, inflexible polymer chains contain a large amount of excess free volume. Sorption measurements suggest that PTMSP is essentially porous, with a fractional free volume f of about 0.25. Such a large f value leads to the suggestions that free volume elements are connected to result in the extremely high gas permeability, up to $P_{O_2} = 7{,}700$ Barrer. However, although the gas permeability is the highest among polymeric membranes, their separation performance is still limited for practical applications because of the low selectivity, like $P_{O_2}/P_{N_2} = 1.4$. In addition, such high permeability depends on the sample history and dramatically decreases over time, which has been attributed mostly to changes in the amount and distribution of free volume [63, 64].

PALS studies of the ultrapermeable PTMSPs indicate a bimodal size distribution of free volume element with small microvoids of diameter in the range 0.5–0.8 nm and larger ones of diameter in the range 1.0–1.5 nm. On moving to less and less permeable polymers, there is a loss of the larger free volume element, and the bimodal distribution converges ultimately on a unimodal distribution [63].

3.7.2.2 Polymers of Intrinsic Microporosity

Polymers of intrinsic microporosity (PIMs) are defined as solids that contain interconnected free volumes or pores of less than 2 nm in size, and consequently they possess large and accessible surface areas, typically 300–2,000 m^2/g as measured by gas adsorption. They exhibit analogous behavior to that of conventional microporous materials such as activated carbons [61, 62].

PIMs are designed with a highly contorted and rigid backbone, leading to inefficient packing and substantial microvoids or free volume. In particular, the lack of rotational freedom along the polymer backbone ensures that the polymeric chain conformations cannot rearrange to collapse the open structure. The fractional free volume f of PIMs is typically very high, often in the range of 0.22–0.26 due to these structural characteristics with diameter, often in the range of 0.3–1.0 nm. This high f contributes to their excellent gas permeabilities. The interconnected pores within PIMs allow for extremely high gas permeability, making them several orders of magnitude more efficient than traditional gas separation polymers [63, 65].

3.7.2.3 TR Polymers

Thermally rearranged polymers (TR polymers) are another kind of polymer with extremely stiff chains and high free volume. Their free volume sizes depend on the chemical structure of specific precursors as well as the thermal treatment applied for converting diverse precursor polymers to TR polymers [35]. Generally, the free volume elements in TR polymers are also in the nanometer range, with pore sizes between 0.2 and 0.8 nm. It is interesting to note that, upon thermal rearrangement at a temperature around 350–450 °C as illustrated in Figure 3.31, the free volume increases significantly from $f = 0.15$ for precursor to 0.20 for TR polymers, and the free volume distribution is also optimized to favor permeability. As a result, the gas permeability is additionally improved to show $P(O_2) = 747$ Barrer [35, 65, 66].

Permeabilities of various gases in diverse TR polymers correlate fairly well with the inverse of the average free volume size $\langle v_h \rangle$ as shown in Figure 3.32, demonstrating the validity of the free volume theory again. The highest permeability was obtained from the TR polymer made with 2,2-bis(3-amino-4-hydroxyphenyl)-hexafluoropropane (APAF) and 4,4′-(hexafluoroisopropylidene) diphthalic anhydride (6FDA) [35, 65, 66].

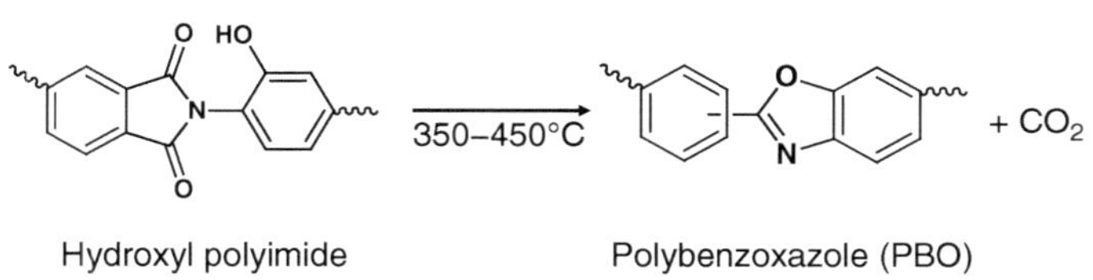

Figure 3.31 Thermal rearrangement of *o*-hydroxy polyimide to polybenzoxazole (PBO, thermally rearranged-1 [TR-1] polymer) by heat treatment at 350–450 °C. *Source*: Adapted from [35].

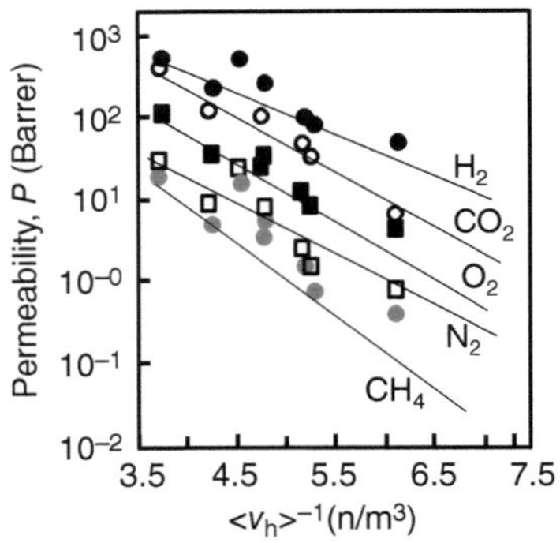

Figure 3.32 Comparison of gas permeability P (Barrer) in TR polymers as a function of the inverse of the average free volume element size $\langle v_h \rangle$ (nm^3) determined by PALS. *Source*: Adapted from [66].

3.7.2.4 PIMs and TR Polymers

Both PIMs and TR polymers exhibit exceptionally high gas permeability compared to traditional polymeric membranes. Gas permeabilities of TR polymers are often two orders of magnitude higher than those of the common glassy polymers, but are still lower than those of PTMSP, the most permeable polymer. However, selectivities (e.g. O_2/N_2 and CO_2/CH_4) are much higher than those in PTMSP. For example, P_{O_2} and P_{CO_2} of the TR-1 polymer are 700 and 1,700 Barrer, respectively [35], significantly higher than the common glassy polymers, for example $P_{O_2} = 1\text{–}10$ Barrer and $P_{CO_2} = 10\text{–}20$ Barrer for PSF. Consequently, various types of PIMs and TR polymers have been developed recently to enhance permeability further while minimally sacrificing selectivity for practical applications. This high permeability is primarily due to the increased free volume and its optimized size and distribution, facilitated by highly rigid and inflexible main chain backbones, which favor gas separation performance. Additionally, it is noteworthy that there is a rather good correlation between permeability and free volume, despite some scatters in the data for both PIMs and TR polymers [66].

3.7.3 Effects of Free Volume or *d*-spacing on Permeation

The permeability P is often directly correlated with the free volume:

$$P = A_P \exp\left(-\frac{B_d}{f}\right) \tag{3.57}$$

where $A_p = A_d S$. This is because gas solubility coefficients in polymers vary over a much narrower range than diffusion coefficient, and the effect of the free volume on solubility is generally much weaker than that of diffusivity. Figure 3.33 illustrates the good correlation between the CO_2 permeability P and $1/f$.

The d-spacing of polymer segments, illustrated schematically in Figure 3.19b, serves as a measure of the free volume size, which can be readily adjusted by modifying the polymer's chemical structure, either through changes to the main backbone chain or by altering side group substituents. This value, along with its distribution, can be easily determined using wide-angle X-ray spectroscopy (WAXS) and the Bragg equation. This method offers a significant advantage over measuring free volume experimentally.

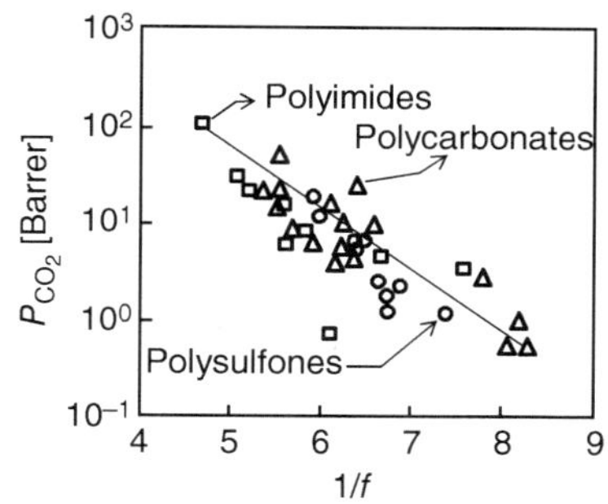

Figure 3.33 Correlation between CO_2 permeability coefficients *P* with fractional free volume *f* in glassy aromatic polyimides, polysulfones and polycarbonates. *Source*: Adapted from [3].

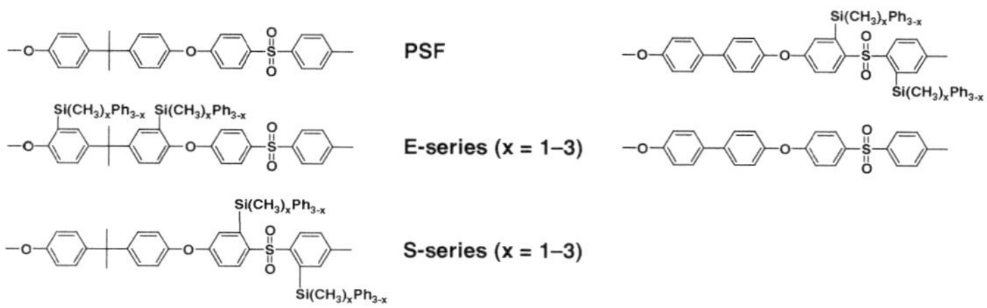

Figure 3.34 Chemical structures of polysulfone (PSF) and poly(phenyl sulfone) (PPSF) and their derivatives. *Source*: Adapted from [68].

The *d*-spacing was systematically changed by varying the size and mobility of silyl substituents in polysulfones (PSFs) and poly(phenyl sulfone) (PPSF) to investigate how these factors affect gas permeability. The substitution site was also varied from *ortho* to either the ether (E series) or sulfone (S series) linkage of PSFs, and the sulfone linkage (R series) of PPSF chains, as drawn in Figure 3.34. The size of the substituents was increased from trimethylsilyl (EM3, SM3, RM3), to dimethylphenylsilyl (EM2, SM2, RM2), and to methyldiphenylsilyl (EM1, SM1, RM1), and the degree of substitution was controlled to be nearly 1.7–1.9 [67, 68].

Figure 3.35 illustrates that both P_{O_2} and *d*-spacing unexpectedly increases from base materials (PSF and PPSF) to M1 series, M2 series and finally to M3 series. For instance, EM3 with the spatially smallest trimethylsilyl group markedly increased P_{O_2} from 1.1 to 7.1 Barrer compared to that of unmodified neat PSF by increasing *d*-spacing from 0.495 to 0.553 nm [68]. Unexpectedly, the oxygen permeability for PSFs with bulkier $SiMe_2$ Ph and $SiMePh_2$ substituents remained almost unchanged.

The improved P_{O_2} appear to be associated with the size as well as mobility of the silyl substituent primarily due to the lengthen *d*-spacing, also evidenced from the increase in the specific volume [68]. Thus, the mobility of pendant substituents was characterized by the subglass transition temperature such as γ-peak from dynamic mechanical analysis (DMA). From the DMA results, it was concluded that the mobility of the substituents decreases in the order of $SiMe_3 > SiMe_2$ Ph $> SiMePh_2$ as expected. The decreased mobility with phenyl group could be mostly due to its planar structure in addition to the molecular weight effect, which may thus be readily associated with other phenylene groups in both main chains and substituents.

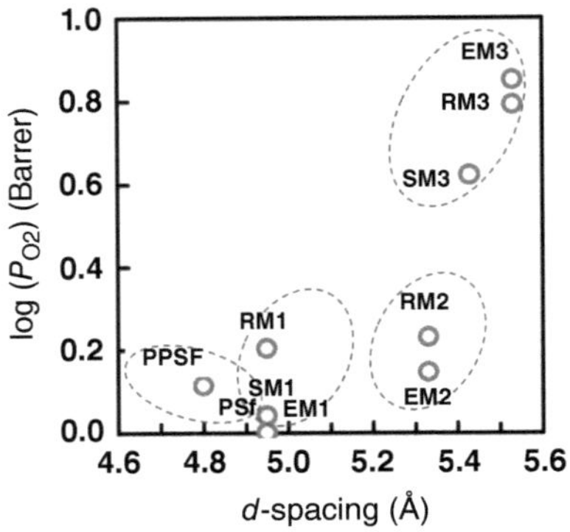

Figure 3.35 Log(P_{O_2}) vs d-spacing of modified aromatic polysulfones (PSF) and poly(phenylene sulfone)s (PPSF). *Source*: Adapted from [68].

Thus, the decreased mobility of the substituents leads to the smaller d-spacing, lower specific volume, and consequently lower gas permeability. This leads us to conclude that the substituent mobility rather than its size is more critical in determining d-spacing and fractional free volume and consequently transport properties [67, 68]. The importance of the local mobility of substituent rather than its size has been also demonstrated by the difference in meta/para substituent effects [69].

The substitution at the *ortho* position to the ether linkage (E-series) was found to be more effective than that near the sulfone linkage (S-series) in enhancing transport properties between EM3 and SM3. The improved gas separation performance appears to be primarily linked to increased free volume, as indicated by larger d-spacing and higher specific volume [67, 68].

3.7.4 Temperature-dependent Permeation

Temperature-dependence of diffusion and solubility coefficients is typically described by Arrhenius and van't Hoff type equations: $D = D_0\exp(-E_d/RT)$ (Equation (3.14)) and $S = S_0\exp(-\Delta H_s/RT)$ (Equation (3.55)). Therefore, temperature-dependent permeation coefficient P can be readily expressed as:

$$P = P_0 \exp\left(-\frac{E_P}{RT}\right) \tag{3.58}$$

simply by adapting the relation $P = D \times S$ as described previously. Thus,

$$E_P = E_d + \Delta H_s \tag{3.59}$$

where E_d is always positive and ΔH_s depends on penetrant and polymeric matrix, but mostly negative. Thus, the sign of E_p depends on the relative magnitude of E_d and ΔH_s.

For strong size-sieving polymers with rigid chains, $E_d > |\Delta H_s|$, E_p becomes positive, and P increases with T as shown in Figure 3.36a. However, E_p can be negative in gas permeation through rubbery PDMS and glassy PTMSP, where ΔH_s exhibits large absolute values or E_d is very small, as shown in Figure 3.36b [70].

Figure 3.36a shows the changes of slopes of Arrhenius plots at T_g, which were observed with some but not all gases [6]. For instance, the slope change was observed

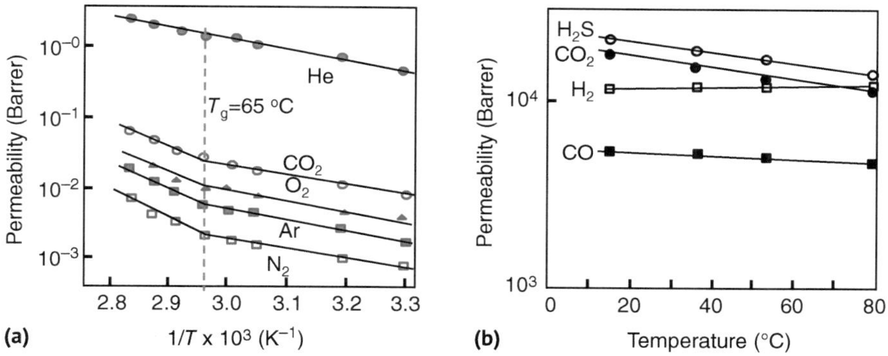

Figure 3.36 Gas permeabilities through (a) copolymer of acrylonitrile and methyl acrylate; and (b) PTMSP as a function of temperature. *Source*: Adapted from [6] and [70].

with poly(viny1 acetate) [11] and copolymer of acrylonitrile and methyl acrylate [6], whereas no inflection was observed with poly(ethyl methacrylate) [71] and vinyl chloride-vinyl acetate copolymer [72].

3.7.5 Gas Permeation Through Crystalline Polymers

Crystallites in crystalline polymers have been known to be mostly non-permeable to common gases or vapors. Thus, the apparent diffusion D and solubility S coefficients can be expressed approximately as expressed previously:

$$D = \alpha_a D_a = \frac{D_a}{\tau\beta} \quad \text{and} \quad S = \alpha_a S_a \tag{3.60}$$

where α_a is the volume fraction of the amorphous phase, D_a and S_a are the diffusion and the solubility coefficients of the penetrant in the amorphous phase, respectively. τ and β are the tortuosity and chain immobilization factors, respectively. τ is related to α_a as written in Equation (3.18) and β represents the immobilization factor for polymer chains at the interface between crystallites and amorphous phase as illustrated in Figure 3.10b.

Thus, the permeability coefficient is:

$$P = \alpha_a^2 D_a S_a \tag{3.61}$$

3.8 Mathematical Models for Transient Sorption and Permeation Through Glassy Polymers and Composite Membranes

The transport of gases, vapors, and liquids through polymers can be classified as either Fickian or non-Fickian, depending on whether their behavior follows the Fick's law. Fickian behavior typically prevails in rubbery polymers, while non-Fickian behavior is more common in glassy polymers. For sorption and permeation for a thin slab, mathematical solutions of Fick's second law with a constant diffusion coefficient are Equations (3.9) and (3.12), respectively, as described in Section 3.2. Other mathematical solutions for Fick's second law in different geometries, such as spheres, cylinders, and cubes, are well established and summarized in "*Mathematics of Diffusion*" [2].

However, mass transport in glassy polymers does not generally follow Fick's law and is often referred to as non-Fickian. Numerous studies have aimed to develop rigorous mathematical models for this behavior. One such model adopts the basic concept of dual-mode sorption, which involves both equilibrium matrix absorption and microvoid adsorption, as illustrated in Figure 3.24b [73].

In the dual sorption model, when a glassy polymer sample is exposed to a constant pressure p, a fraction of the dissolved gas C_1 was assumed to be *irreversibly* immobilized or adsorbed in the microvoids C_2 of the glassy polymer. However, such *irreversible* immobilization assumption was relaxed in advanced models. For instance, it is assumed instead that the immobilization process is *reversible* with the equilibrium constant $K = k_1/k_2$, following a first-order reaction with forward and backward rate constants k_1 and k_2, respectively, described by:

$$C_1 \underset{k_2}{\overset{k_1}{\rightleftarrows}} C_2 \quad \text{and} \quad K = k_1/k_2$$

3.8.1 Transient Sorption and Permeation for a Glassy Slab

Accepting the concept of *reversible* immobilization between C_1 and C_2 as described above in glassy polymers, we can analyze transient sorption and permeation in a slab of thickness $2L$ and L, respectively, as depicted in Figures 3.3a and 3.4a. The local material balances for the mobile species C_1 and the reversibly immobilized species C_2 are expressed as follows as [74, 75]:

$$\frac{dC_1}{dt} = D\frac{\partial^2 C_1}{\partial x^2} - \frac{dC_2}{dt} \tag{3.62}$$

$$\frac{dC_2}{dt} = k_1 C_1 - k_2 C_2 \tag{3.63}$$

3.8.1.1 Transient Sorption

For transient sorption, Equations (3.62) and (3.63) are solved subject to as [74, 75]:

$$\begin{aligned} &C_1(x,0) = C_2(x,0) = 0 && \text{at} \quad -L \le x \le L \,\text{and}\, t < 0 \\ &C_1(-L,t) = C_1(L,t) = C_1^L && \text{at} \quad x = -L \,\text{and}\, x = L \,\text{and}\, t \ge 0 \,\text{or} \\ &dC_1(0,t)/dx = 0 && \text{at} \quad x = 0,\, t \ge 0\,(\text{symmetric}) \end{aligned}$$

Since M_t/M_∞ is expressed as:

$$\frac{M_t}{M_\infty} = \frac{1}{(1+K)C_1^L L}\int_0^L (C_1 + C_2)\,dx \tag{3.64}$$

Integration of Equations (3.62) and (3.63) over dx using Laplace transformation technique and associated convolution theorem yields the fractional uptake M_t/M_∞ as a function of time t as:

$$\frac{M_t}{M_\infty}=\frac{1}{(1+K)}\left\{1+2\sum_{n=0}^{\infty}\frac{e^{p_n\tau}}{p_n(d\phi/dp)_n}\right\}+\frac{K}{(1+K)}\left\{(1-e^{-\varphi\tau})+2\varphi\sum_{n=0}^{\infty}\frac{e^{p_n\tau}-e^{-\varphi\tau}}{p_n(p_n+\varphi)(d\phi/dp)_n}\right\} \tag{3.65}$$

where $\tau=Dt/L^2, \varphi=k_2L^2/D, \phi(p)=p+Kp\varphi/(p+\varphi)$, and p_n are the nonzero roots of $\phi(p_n)=-((n+1/2)\pi)^2$. Equation (3.65) indicates that sorption behavior depends only upon φ (i.e. the inverse of the Deborah number) and K.

3.8.1.2 Transient Permeation

The permeation process through a slab of thickness L (Figure 3.4a) is governed by Equations (3.62) and (3.63) again, albeit with different boundary conditions as [74, 75]:

$$C_1(x,0)=C_2(x,0)=0 \quad \text{at} \quad 0\le x\le L,\ t<0$$
$$C_1(0,t)=C_1^0=Sp \quad \text{at} \quad x=0,\ t\ge 0$$
$$C_1(L,t)=C_1^L=0 \quad \text{at} \quad x=L,\ t\ge 0$$

Accumulation Q_t per unit area can be defined as:

$$Q_t=-\int_0^t D_1\frac{dC_1(L,t)}{dx}\,dt \tag{3.66}$$

Dimensionless accumulation q per unit area, defined as $q=Q_t/LC^0$, can be obtained from:

$$q=\frac{1}{1+K}\left[\tau-2\pi^2\sum_1^{\infty}(-1)^n\frac{n^2}{p_n^2(d\phi/dp)_n}(e^{p_n\tau}-1)\right] \tag{3.67}$$

For slow reactions, p_n has two roots: $-(n\pi)^2$ and $-\varphi$. Equation (3.67) reduces to:

$$q=\tau-\frac{1+K}{6}-\frac{2}{\pi^2}\sum_1^{\infty}\frac{(-1)^n}{n^2}\left[e^{-(n\pi)^2\tau}+Ke^{-\varphi\tau/K}\right] \tag{3.68}$$

Equations (3.67) and (3.68) demonstrate the two apparent time lags of θ_1 and θ_2 from the plot of q as a function of $\tau: \tau_1=1/6$, extrapolated from q in the region $\tau\ll 1$; and a later one $\tau_2=(1+K)/6$ from in the period of $\tau\gg 1/\pi^2$, where $\theta=L^2/D$. These two time-lags yield D and K, while the permeability coefficient, P independently obtained from the slope of either asymptote.

3.8.2 Transient Sorption for a Glassy Sphere

For transient sorption in a glassy polymeric sphere with diameter $r=R$, the local material balances for mobile species C_1 and the reversibly immobilized species C_2 are expressed, respectively, [80, 82] as:

$$\frac{dC_1}{dt}=\frac{1}{r^2}\frac{d}{dr}\left(r^2D\frac{dC_1}{dr}\right)-\frac{dC_2}{dt} \tag{3.69}$$

$$\frac{dC_2}{dt} = k_2(KC_1 - C_2) \tag{3.70}$$

Initial and boundary conditions for sorption are:

$$\begin{aligned} &C_1(r,0) = C_2(r,0) = 0 && \text{at} \quad 0 \le r \le R, t < 0 \\ &C_1(R,t) = C_1^R && \text{at} \quad r = R, t \ge 0 \\ &dC_1(0,t)/dr = 0 && \text{at} \quad r = 0, t \ge 0 \end{aligned}$$

Integration of Equations (3.69) and (3.70) over space using Laplace transformation technique and associated convolution theorem yields the fractional uptake M_t/M_∞ as a function of time t:

$$\begin{aligned} \frac{M_t}{M_\infty} = &\frac{1}{(1+K)}\left\{1 + 6\sum_{n=1}^{\infty}\frac{e^{p_n\tau}}{p_n(d\phi/dp)_n}\right\} \\ &+\frac{K}{(1+K)}\left\{(1-e^{-\varphi\tau}) + 6\varphi\sum_{n=1}^{\infty}\frac{e^{p_n\tau} - e^{-\varphi\tau}}{p_n(p_n+\varphi)(d\phi/dp)_n}\right\} \end{aligned} \tag{3.71}$$

where $\tau = Dt/R^2, \varphi = k_2 R^2/D, \phi(p) = p + Kp\varphi/(p+\varphi)$ and p_n are the nonzero roots of $\phi(p_n) = -(n\pi p)^2$. Equation (3.71) indicates that sorption behavior depends only upon φ (i.e. the inverse of the Deborah number) and K.

When diffusion is much faster than immobilization or conversion i.e. $\varphi \ll \pi^2/(1+K)$, Equation (3.71) reduces to Equation (3.72), illustrating a two-stage sorption:

$$\frac{M_t}{M_\infty} = \frac{1}{(1+K)}\left\{1 - 6\sum_{n=1}^{\infty}\frac{e^{-n^2\pi^2\tau}}{n^2\pi^2}\right\} + \frac{K}{(1+K)}\{1 - e^{-\varphi\tau}\} \tag{3.72}$$

On the other hand, when φ is large, Equation (3.71) reduces to:

$$\frac{M_t}{M_\infty} = 1 - 6\sum_{n=1}^{\infty}\frac{1}{n^2\pi^2}\exp\left(\frac{-n^2\pi^2\tau}{1+K}\right) \tag{3.73}$$

Equation (3.73) results in very similar to equation for Fickian transport with the reduced diffusion coefficient by $1/(1+K)$ [2].

In addition, the penetrant surface concentration with time t can be obtained by integration of Equation (3.70) with $C_1(R,t) = C_1^R$ as:

$$C_s = C_1^R\left[1 + K(1 - \exp(-k_2 t))\right] \tag{3.74}$$

which demonstrates the time-dependent surface concentration.

3.8.3 Time-dependent Surface Concentration

A straightforward approach involves using time-dependent surface concentration to analyze various non-Fickian sorption features, including pseudo-Fickian, sigmoidal, and two-stage sorption. The following mathematical expressions are frequently used to describe the time-dependent surface concentration [2]:

$$C^s = C^0\left[1 - \exp(-\beta t)\right] \quad \text{or} \quad C^s = \gamma t \tag{3.75}$$

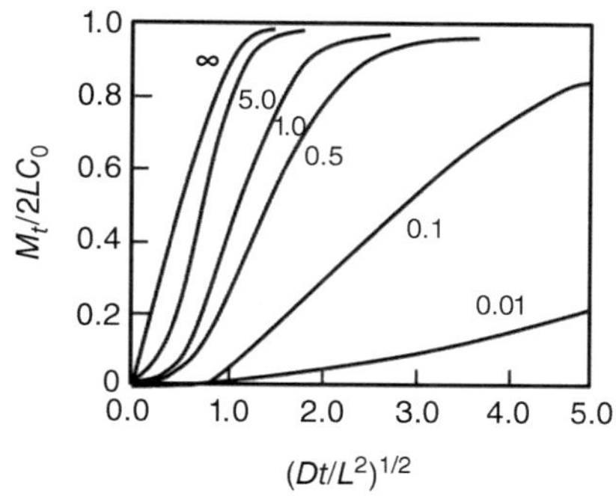

Figure 3.37 Sorption curves for variable surface concentrations of $C^s = C^0[1 - \exp(-\beta t)]$ by varying β. *Source*: Ref. [2]/Oxford University Press.

$$C^s = C_1^0 + C_2^0[1 - \exp(-\beta t)] \tag{3.76}$$

where C^0 in Equation (3.75) represents the equilibrium surface concentration. C_1^0 and C_2^0 for Equation (3.76) denote the equilibrium surface concentrations for mobile species and the reversibly immobilized species, respectively, where C_1^0 is instantaneously obtained by the applied pressure $p\left(C_1^0 = Sp\right)$. The constants β and γ are specific to the system. The dimensionless sorption–time curves are obtained from the analytical solutions as illustrated in Figure 3.37. These results indicate that various non-Fickian behaviors, including pseudo-Fickian, sigmoidal, and two stage sorptions, can be effectively analyzed utilizing the simple concept of time-dependent surface concentration. The mathematical solutions and corresponding sorption curves are detailed in the book of Crank [2, 76].

3.8.4 Transient Sorption for a Composite Film

A simple mathematical model is developed that simulates transient sorption and permeation in a composite film composed of slowly permeable, identically sized microspheres, 2, with a radius of $r = R$ that are dispersed uniformly within a continuous, highly permeable matrix, 1, with a thickness of $2L$ (Figure 3.38). Note that the volume fraction of the microspheres $\phi_2 = 4\pi R^3 N/3$, where r is defined separately for each sphere with the radius $r = R$ and the number density of microspheres N; and the effective diffusion cross-section in the continuous phase is proportional to $(1 - \phi_2)$. The concentrations in the continuous and spherical phases are expressed for convenience as $C_1(x, t)$ and $C_2(x; r, t)$, respectively.

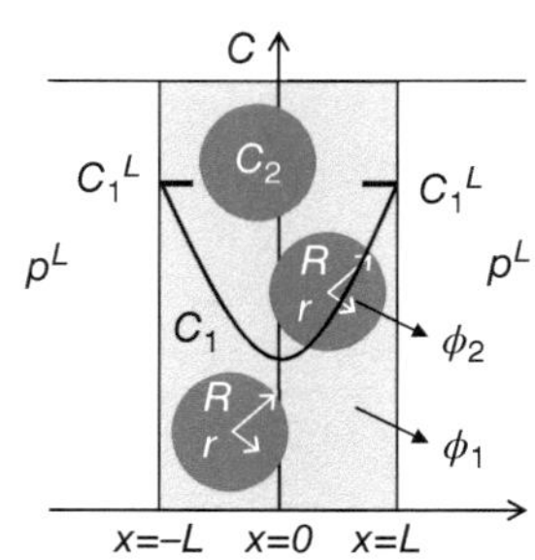

Figure 3.38 Schematic structure of composite membrane containing microspheres, 2, with radius $r = R$, which are dispersed in a polymeric matrix, 1, for sorption experiments. *Source*: Y.S. Kang.

As zero order approximations, it is assumed that: (1) diffusion in the continuous phase is effectively unidirectional and mediated by local penetrant consumption by the spheres; (2) the dispersed phase surface concentration $C_2(x; R, t)$ at a given distance into the slab depends upon the (time-varying) concentration $C_1(x, t)$ in the continuous phase at that position, but not upon location on the spherical surface; and (3) no diffuse interphase between phases 1 and 2 is assumed, but a simple equilibrium relationship, $C_2(x; R, t) = K\, C_1(x, t)$ has been applied. The equations to be solved are then [74]:

$$\frac{\partial C_1}{\partial t} = \frac{\partial}{\partial x}\left[D_1 \frac{\partial C_1}{\partial x}\right] - \frac{4\pi R^2 N}{1-\phi_2}\left[D_2 \frac{\partial C_2}{\partial r}\right]_{r=R} \tag{3.77}$$

$$\frac{\partial C_2}{\partial t} = \frac{2}{r}\left[D_2 \frac{\partial C_2}{\partial r}\right] + \frac{\partial}{\partial r}\left[D_2 \frac{\partial C_2}{\partial r}\right] \tag{3.78}$$

where subscripts 1 and 2 denote the continuous and dispersed spherical phases, respectively.
in the case of sorption, subject to:

1. $C_1(x, 0) = C_2(x; r, 0) = 0$ at $t < 0$
2. $C_1(\pm L, t) = C_1^L = Sp$ at $x = \pm L$ and $t \geq 0$ or
3. $dC_1(0, t)/dx = 0$ at $x = 0$ and $t \geq 0$
4. $C_2(x; R, t) = K\, C_1(x, t)$ at $t \geq 0$
5. $dC_2(x; 0, t)/dr = 0$ at $r = 0$ and $t \geq 0$

and, in the case of permeation, subject to:

6. $C_1(x, 0) = C_2(x; r, 0) = 0$ at $t < 0$
7. $C_1(-L, t) = C_1^L = Sp$ at $x = -L$ and $t \geq 0$
8. $C_1(+L, t) = 0$ at $x = +L$ and $t \geq 0$
9. $C_2(x; R, t) = K\, C_1(x, t)$ at $t \geq 0$
10. $dC_2(x; 0, t)/dr = 0$ at $r = 0$ and $t \geq 0$

Equations (3.77) and (3.78) allow for concentration-dependent diffusion coefficients. For simplicity we have assumed constant diffusion coefficients and numerically solved by the finite difference method [74].

Figure 3.39 depicts a family of curves of M_t/M_∞ versus $\tau^{1/2}$ with ϕ_2 fixed at 0.5 and K at 1, as η ($\eta = (L^2D_2/R^2D_1)$) varies over seven orders of magnitude. The results span the range of behavior from near-Fickian ($\eta = 1$) to distinctly two-stage ($\eta \ll 1$). Although it is not apparent in the figure, *all* curves asymptotically approach an ordinate of unity.

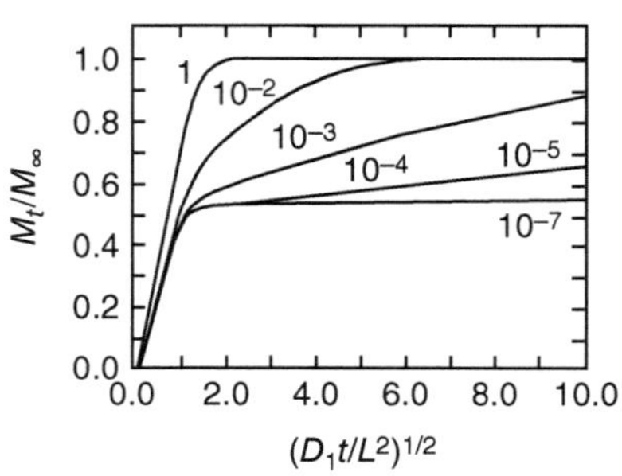

Figure 3.39 Theoretical sorption curves in a composite membrane as a function of $\tau^{1/2}$ ($= (Dt/R^2)^{1/2}$) by varying η where $\eta = (L^2D_2/R^2D_1)$ with $K = 1$ and $\phi_2 = 0.5$. *Source*: Ref. [74]/John Wiley & Sons (Figure 5).

3.9 Transient Sorption of Organic Vapor in Polymers

Since organic vapors usually have higher solubility and lower permeability, leading to longer experimental times, transient sorption experiments are more frequently performed than transient permeation experiments for practical reasons. This section will present and compare experimental transient sorption results of *n*-hexane in a rubbery *cis*-polybutadiene slab, glassy polystyrene spheres, and their composite membranes with theoretical data derived from new mathematical models introduced in previous Section 3.8. These comparisons aim to deepen the understanding of the physical structure and organic vapor transport properties of glassy polymers and composite membranes composed of a glassy matrix dispersed in a rubbery matrix.

3.9.1 Transient Sorption of Organic Vapor in Rubbery Polymers

The experimental transient sorption data of *n*-hexane in a *cis*-polybutadiene, PBD, slab with thickness $L = 0.26$ mm at $p_1/p_1^0 = 0.4$ and 25 °C, were plotted in the reduced sorption curve (M_t/M_∞ vs $t^{1/2}/L$) together with the theoretical predictions (solid line) calculated from the Fickian solution (Equation (3.9)) with a constant diffusion coefficient, $D = 5.7 \times 10^{-5}$ cm^2/sec as shown in Figure 3.40a. Here, M_t and M_∞ are the amount of *n*-hexane sorbed at time t and the equilibrium solubility, respectively. The results demonstrate the close fit between the experimental and theoretical data and therefore the Fickian [74].

The diffusion coefficients of *n*-hexane as a function of its volume fraction, ϕ_1, were plotted in Figure 3.40b. Based on the free volume theory for diffusion coefficient and the Flory–Huggins theory, the diffusion coefficient, D, can be expressed as a function of the volume fraction ϕ_1 of *n*-hexane as [43]:

$$D = A_d RT \exp\left[\frac{-B_d}{f(\phi_1, T)}\right][(1-\phi_1) - 2\chi\phi_1(1-\phi_1)](1-\phi_1) \tag{3.79}$$

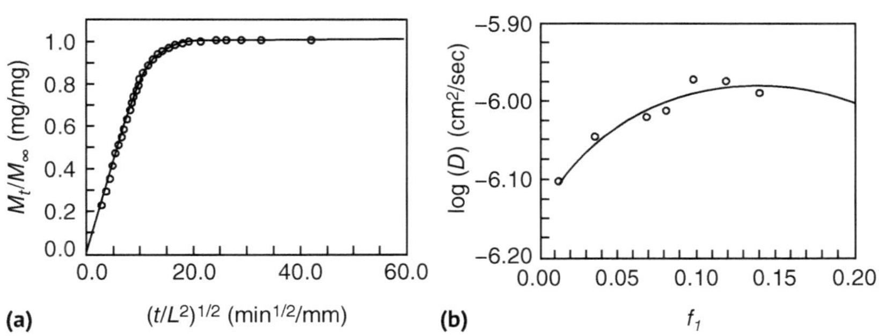

Figure 3.40 (a) Fit of the Fickian sorption solution (solid line) of Equation (3.9) with a constant diffusivity coefficient $D = 5.7 \times 10^{-5}$ cm^2/sec to the experimental data for *n*-hexane in a *cis*-polybutadiene slab with thickness L = 0.26 mm at $p_1/p_1^0 = 0.4$ and 25 °C. *Source*: Adapted from [74]; and (b) the experimental diffusion coefficient, *D*, as a function of the volume fraction ϕ_1 of *n*-hexane at 25 °C with theoretical predictions with Equation (3.79). *Source*: Adapted from [43].

where A_d and B_d are constants for the free volume theory with χ being the Flory–Huggins interaction parameter. The fractional free volume of polymer with organic vapor $f(\phi_1,T)$ is expressed as $f(\phi_1,T) = f(0,T) + \beta(T)\,\phi_1$ with the fractional free volume $f(0,T)$ of neat PBD and $\beta(T)$ being estimated to be 0.088 and 0.44, respectively, from independent experimental data. The experimental diffusion coefficients, D, were fitted to Equation (3.79) to get the values of A_d and B_d: $A_d = 1.08 \times 10^{-16}$ mole sec/g and $B_d = 0.11$. Both experimental and theoretical results show a maximum diffusion coefficient at the n-hexane volume fraction of $\phi_1 = 0.12$ [43].

It has been well accepted that the transient sorption and permeation in rubbery polymers follow the Fick's law, called the Fickian transport behavior, mostly because rubbery polymers are in equilibrium state.

3.9.2 Transient Sorption of Organic Vapor in Glassy Polymers

The transient experimental results for the sorption of n-hexane in glassy polystyrene (T_g~100 °C) microspheres (radius, $R = 0.06\,\mu$m) at $p/p^0 = 0.1$ and 25 °C were plotted in Figure 3.41a. The theoretical predictions (the solid line in Figure 3.41a based on the Fick's second law with a constant diffusion coefficient of $D = 5.53 \times 10^{-14}$ cm^2/sec, calculated from the initial slope, were compared against the experimental data, illustrating the strong disagreement between them and consequently the non-Fickian sorption behavior.

Figure 3.41b shows sorption curves of n-hexane in polystyrene microspheres of various sizes, represented by M_t/M_∞ as a function of $(t/R^2)^{1/2}$. The smallest microspheres with $R = 0.04\,\mu$m, in this experimental range, show the very slow second-stage sorption, while the largest one with $R = 0.31\,\mu$m display nearly Fickian, indicating the thickness-dependent, non-Fickian sorption behavior.

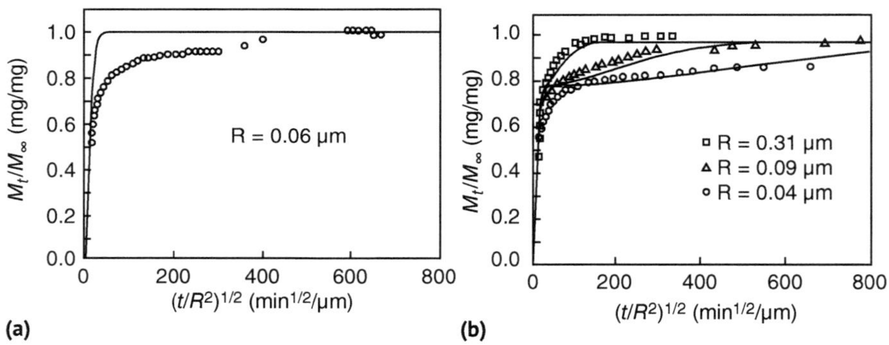

Figure 3.41 Reduced sorption curves for n-hexane in polystyrene microspheres at $T = 25$ °C (a) with a fixed size of $R = 0.06\,\mu$m at $p/p^0 = 0.1$ and the theoretical Fick's second law curve (solid line) with a constant diffusion coefficient of 5.53×10^{-14} cm^2/sec. *Source*: Adapted from [74] (Figure 3); (b) with varying radii ($R = 0.31, 0.09$, and $0.04\,\mu$m) at $p/p^0 = 0.4$. Solid curves calculated from Equation (3.71) with $D = 5.62 \times 10^{-14}$ cm^2/sec, $k_2 = 2.7 \times 10^{-5}$ /sec, and $K = 0.25$. *Source*: Adapted from [77] (Figure 1).

These experimental results are analyzed using the mathematical model (solid line by Equation (3.71)). As can be seen in Figure 3.41b, the agreement between the experimental data and the theoretical results based on Equations (3.69) and (3.70) and their analytical solution of Equation (3.71) is markedly improved, compared to the curve in Figure 3.41a. These findings suggest that the mathematical model described by Equations (3.69) and (3.70) provides a markedly improved explanation of the transient sorption behavior of organic vapor in glassy polymers.

3.9.3 Transient Sorption of Organic Vapor in Composite Membranes

The transient sorption of a composite membrane containing glassy microspheres, homogeneously dispersed in a rubbery matrix offers valuable insights into the transport features of more complexed systems, commonly used for various applications. A composite membrane containing polystyrene (T_g~100 °C) microspheres 2 dispersed in a *cis*-polybutadiene (T_g~−60 °C) matrix 1, having a thickness of 2*L*, was prepared to measure the transient sorption curves of *n*-hexane represented by M_t/M_∞ as a function of $(t/L^2)^{1/2}$, as shown in Figure 3.42a and its extended figure in Figure 3.42b.

The radius, *R*, of polystyrene microspheres varies from 0.06, 0.09, and 0.31 μm while maintaining both their volume fraction $\phi_2 = 0.2$ and the slab thickness 2L = 0.15 mm kept constant. These variations allow for the examination of how the size of the glassy microspheres influences the sorption behavior in the composite membrane.

Equations (3.77) and (3.78) are numerically solved using the finite difference method. Their theoretical predictions are compared against the experimental results displayed in Figure 3.42a to investigate the size effects of polystyrene microspheres. The results for polystyrene microspheres with $R = 0.06$ μm, the smallest in this experimental range, polystyrene beads show nearly Fickian, while the curve for the biggest microspheres with $R = 0.31$ μm clearly shows the two-stage behavior. Such behavior is even more pronounced in the extended version, Figure 3.42b. With

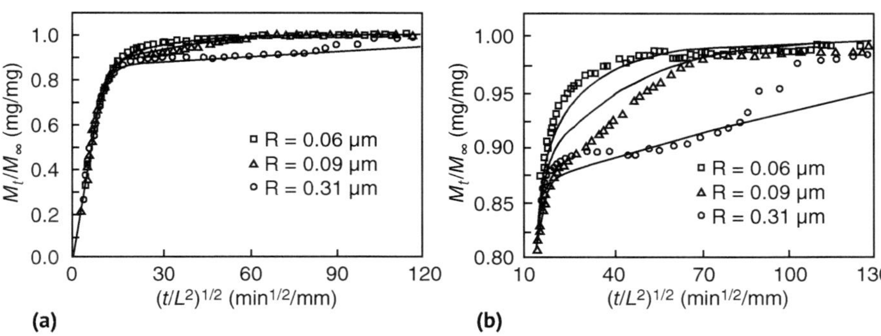

Figure 3.42 (a) *n*-Hexane sorption in a polystyrene/*cis*-polybutadiene composite slab with thickness $L = 0.15$ mm and volume fraction of polystyrene $\phi_2 = 0.2$. Polystyrene bead radiuses of 0.06, 0.09, and 0.31 μm; and (b) the extended figure of (a). *Source:* Adapted from [74] (Figure 4).

intermediate-sized 0.09 μm spheres, the results fall between these two extremes. Interestingly, this behavior is somewhat analogous to the sorption in glassy polymers, though the size effects are apparently inverse [74].

These theoretical results derived from the mathematical model of sorption in the composite indicate that the mere presence of two phases with different diffusion time constants ($\eta \neq 1$) is a sufficient condition for apparently non-Fickian overall behavior. Consequently. definitive morphological inferences cannot be drawn from such observations. In addition, results of modeling the corresponding permeation processes are analogous to those obtained in the study of permeation with reversibly immobilizing reaction between C_1 and C_2 in glassy polymers [74]. When this model is compared to the experimental results as illustrated in Figure 3.42, the fairly good agreement is clearly demonstrated even though the many assumptions made for mathematical derivation.

3.10 Mass Transport Overview: Fickian vs non-Fickian Behavior

The diffusion behavior in rubbery polymers can be mostly described adequately by a concentration-dependent form of Fick's law with constant boundary conditions, whereas the diffusion behavior in glassy polymers cannot be described adequately by a concentration-dependent form of the Fick's law with constant boundary conditions, termed non-Fickian, mostly due to their nonequilibrium state. When a penetrant causes an extensive swelling of the polymer, non-Fickian behavior is also commonly observed.

The essential distinction is that polymer chains in the rubbery state respond nearly instantaneously to changes in their environments. For example, a change in temperature causes an almost immediate change to a new equilibrium volume. The properties of glassy polymers, however, tend to be time-dependent; for example, the stress developed in a glassy polymer may slowly decay with time after tensile deformation, representing time-dependent and nonequilibrium properties. Deviations from Fickian behavior are considered to be associated with the finite rates at which the polymer structure may change in response to sorption or desorption of penetrant molecules [1]. Diffusion and sorption below the glass transition were discussed, followed by a review and classification of the various theoretical models [73].

The following are different types of transient sorption behavior in Figure 3.43 [2]:

1. Fickian (also called Case I): The sorption curve (M_t vs $t^{1/2}$) is linear in the initial stage up to 60% or more of M_∞. For $D(C)$ increasing with C, the absorption curve is linear almost up to the final sorption equilibrium. The sorption curves of slabs of different thickness are superimposable in reduced sorption plot of M_t/M_∞ vs $t^{1/2}/L$.
2. The pseudo-Fickian has been used to describe sorption curves of the same general shape with Fickian, but for which the initial linear portion persists for shorter times (less than 60% equilibrium sorption).

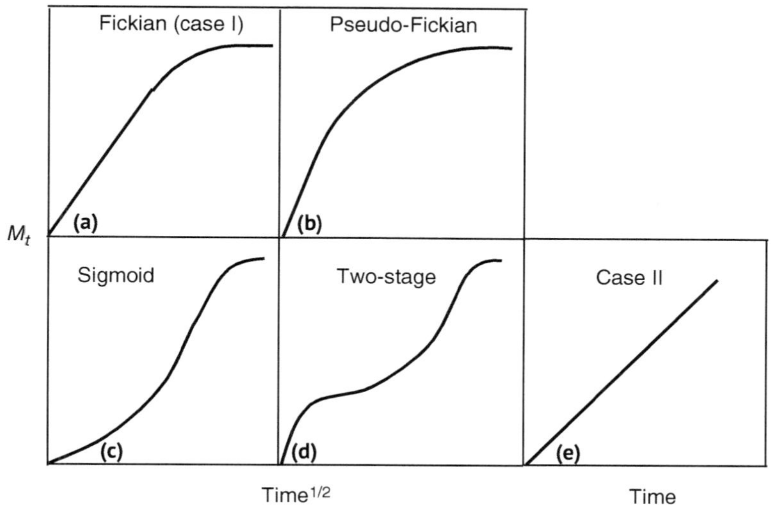

Figure 3.43 Fickian and non-Fickian sorption curves. *Source*: Adapted from [2].

3. Sigmoid: The sorption curves are sigmoid in shape with a single inflection point often at about 50 percent equilibrium sorption.
4. Two-stage sorption: An initial rapid uptake, a linear function of (time)$^{1/2}$, leads to a quasi-equilibrium uptake, followed by a slow approach to a final true equilibrium. A simple explanation is that a quasi-equilibrium is reached rapidly first at the polymer surface and then by simple diffusion through the polymer sheet. The second stage sorption is associated with an increase in surface concentration, which occurs slowly compared with the diffusion process and is the rate-determining factor for sorption: time-dependent surface concentration.
5. Case II: When a liquid penetrant diffuses into a polymer sheet, sharp boundary can often be seen under a microscope, where the boundary advances linearly proportional to (time)^1 rather than (time)$^{1/2}$.

References

1 J. Crank, G.S. Park, Diffusion in Polymers, Academic Press, London, 1968.
2 J. Crank, The Mathematics of Diffusion, 2 ed., Clarendon Press, Oxford, 1979.
3 Y. Yampolskii, I. Pinnau, B.D. Freeman, Materials Science of Membranes for Gas and Vapor Separation, John Wiley & Sons, West Sussex, England, 2006.
4 V.T. Stannet, R.T. Chern, W.J. Koros, H.B. Hopfenberg, Materials Science of Synthetic Membranes, American Chemical Society, Washington, D.C., 1985, 2.
5 R. Kokes, F. Long, Journal of the American Chemical Society 75(24) (1953) 6142–6146.
6 H. Yasuda, T. Hirotsu, Journal of Applied Polymer Science 21(1) (1977) 105–112.
7 R.M. Barrer, Transactions of the Faraday Society 35(1) (1939) 0644–0655.
8 R. Barrer, G. Skirrow, Journal of Polymer Science 3(4) (1948) 549–563.

9 D.W. van Krevelen, K. te Nijenhuis, Properties of Polymers, 4 ed., Elsevier B.V, 2009.
10 J.M. Zheng, J. Qiu, L.M. Madeira, A. Mendes, Journal of Physical Chemistry B 111(11) (2007) 2828–2835.
11 P. Meares, Journal of the American Chemical Society 76(13) (1954) 3415–3422.
12 W.W. Brandt, Journal of Physical Chemistry 63(7) (1959) 1080–1084.
13 A.S. Michaels, R.B. Parker, Journal of Polymer Science 41(138) (1959) 53–71.
14 W.J. Koros, Barrier Polymers and Structures, American Chemical Society, Washington, D.C., 1989.
15 P.W. Atkins, Physical Chemistry, 4 ed., Oxford University Press, Oxford, 1990.
16 S. Matsuoka, Polymer Engineering & Science 21(14) (1981) 907–921.
17 R. Simha, R.F. Boyer, The Journal of Chemical Physics 37(5) (1962) 1003-1007.
18 F. Bueche, Physical properties of polymers, Wiley-Interscience, New York, 1962.
19 M.H. Cohen, D. Turnbull, The Journal of Chemical Physics 31(5) (1959) 1164–1169.
20 C. Nagel, K. Günther-Schade, D. Fritsch, T. Strunskus, F. Faupel, Macromolecules 35(6) (2002) 2071–2077.
21 J.S. Vrentas, J.L. Duda, Journal of Polymer Science: Polymer Physics Edition 15(3) (1977) 403–416.
22 C.A. Kumins, T.K. Kwei, Free volume and other theories, In: Diffusion in Polymers, Academic Press, London, 1968.
23 J.M. Zielinski, J.L. Duda, AIChE Journal 38(3) (1992) 405–415.
24 H. Fujita, Organic vapors above the glass transition temperature, In: Diffusion in Polymers, London, 1968.
25 K. Haraya, S.T. Hwang, Journal of Membrane Science 71(1-2) (1992) 13–27.
26 M. Pixton, D. Paul, Polymeric Gas Separation Membranes, CRC Press, 2018, pp. 83–153.
27 H.Q. Lin, B.D. Freeman, Macromolecules 38(20) (2005) 8394–8407.
28 T. Miyamoto, K. Shibayama, Journal of Applied Physics 44(12) (1973) 5372–5376.
29 A. Thran, G. Kroll, F. Faupel, Journal of Polymer Science Part B:Polymer Physics 37(23) (1999) 3344–3358.
30 H.S. Chung, Journal of Chemical Physics 44 (1966) 1362–1364.
31 D. Turnbull, M.H. Cohen, Journal of Chemical Physics 34(1) (1961) 120–125.
32 R.P. White, J.E.G. Lipson, Macromolecules 49(11) (2016) 3987–4007.
33 A.Y. Alentiev, V.P. Shantarovich, T.C. Merkel, V.I. Bondar, B.D. Freeman, Y.P. Yampolskii, Macromolecules 35(25) (2002) 9513–9522.
34 R.W. Baker, Membrane Technology and Applications, Wiley, 2012.
35 H.B. Park, C.H. Jung, Y.M. Lee, A.J. Hill, S.J. Pas, S.T. Mudie, E. Van Wagner, B.D. Freeman, D.J. Cookson, Science 318(5848) (2007) 254–258.
36 M.R. Pixton, D.R. Paul, In: D.R. Paul, Y.P. Yampolsky (Eds.), Polymeric Gas Separation Membranes, CRC Press, 1994, 3, p. 92.
37 J.Y. Park, D.R. Paul, Journal of Membrane Science 125(1) (1997) 23–39.
38 N.R. Horn, Journal of Membrane Science 518 (2016) 289–294.
39 J.R. Fried, In: Y. Yampolski, I. Pinnau, B.D. Freeman (Eds.), Materials Science of Membranes for Gas and Vapor Separation, Wiley, NJ, USA, 2006, 3, p. 101.

40 J.H. Petropoulous, In: D.R. Paul, Y.P. Yampolski (Eds.), Polymeric Gas Separation Membranes, CRC Press, Boca Raton, FL., 1994, pp. 17–81.
41 R.S. Prabhakar, T.C. Merkel, B.D. Freeman, T. Imizu, A. Higuchi, Macromolecules 38(5) (2005) 1899–1910.
42 A.S. Michaels, H.J. Bixler, Journal of Polymer Science 50(154) (1961) 393-412.
43 Y.S. Kang, Polymer (Korea) 14 (1990) 234–239.
44 D.S. Pope, I.C. Sanchez, W.J. Koros, G.K. Fleming, Macromolecules 24(8) (1991) 1779–1783.
45 N. Muruganandam, W. Koros, D. Paul, Journal of Polymer Science Part B: Polymer Physics 25(9) (1987) 1999–2026.
46 A.R. Berens, Die Angewandte Makromolekulare Chemie 47(1) (1975) 97–110.
47 K.A. Stevens, Z.P. Smith, K.L. Gleason, M. Galizia, D.R. Paul, B.D. Freeman, Journal of Membrane Science 533 (2017) 75–83.
48 T.C. Merkel, V. Bondar, K. Nagai, B.D. Freeman, Journal of Polymer Science Part B: Polymer Physics 38(2) (2000) 273–296.
49 Y. Yampolskii, Macromolecules 45(8) (2012) 3298–3311.
50 M.R. Coleman, W.J. Koros, Journal of Membrane Science 50(3) (1990) 285–297.
51 M.R. Pixton, D.R. Paul, Polymeric Gas Separation Membranes, CRC Press, Boca Raton, FL, 1994, 3, pp. 83–154.
52 N.A. Plate, Y.P. Yampolskii, Polymeric Gas Separation Membranes, CRC Press, Boca Raton, FL, 1994, 4, pp. 155–207.
53 A.S. Michaels, H.J. Bixler, Journal of Polymer Science 50(154) (1961) 413–439.
54 H.J. Kim, S.I. Hong, Journal of Applied Polymer Science 76(3) (2000) 391–400.
55 K. Nagai, T. Masuda, T. Nakagawa, B.D. Freeman, I. Pinnau, Progress in Polymer Science 26(5) (2001) 721–798.
56 K. Tanaka, M. Okano, H. Toshino, H. Kita, K.I. Okamoto, Journal of Polymer Science Part B: Polymer Physics 30(8) (1992) 907–914.
57 Y. Kawakami, H. Toda, M. Higashino, Y. Yamashita, Polymer Journal 20(4) (1988) 285–292.
58 Y. Kawakami, H. Karasawa, T. Aoki, Y. Yamamura, H. Hisada, Y. Yamashita, Polymer Journal 17(11) (1985) 1159–1172.
59 Y. Hirayama, T. Yoshinaga, Y. Kusuki, K. Ninomiya, T. Sakakibara, T. Tamari, Journal of Membrane Science 111(2) (1996) 169–182.
60 T. Masuda, E. Isobe, T. Higashimura, K. Takada, Journal of the American Chemical Society 105(25) (1983) 7473–7474.
61 P.M. Budd, E.S. Elabas, B.S. Ghanem, S. Makhseed, N.B. McKeown, K.J. Msayib, C.E. Tattershall, D. Wang, Advanced Materials 16(5) (2004) 456–459.
62 P.M. Budd, B.S. Ghanem, S. Makhseed, N.B. McKeown, K.J. Msayib, C.E. Tattershall, Chemical Communications (2) (2004) 230–231.
63 P.M. Budd, N.B. McKeown, D. Fritsch, J Mater Chem 15(20) (2005) 1977–1986.
64 Y.P. Yampolskii, V.P. Shantorovich, F.P. Chernyakovskii, A.I. Kornilov, N.A. Plate, Journal of Applied Polymer Science 47(1) (1993) 85–92.
65 S. Kim, Y.M. Lee, Progress in Polymer Science 43 (2015) 1–32.
66 Z.P. Smith, G. Hernández, K.L. Gleason, A. Anand, C.M. Doherty, K. Konstas, C. Alvarez, A.J. Hill, A.E. Lozano, D.R. Paul, B.D. Freeman, Journal of Membrane Science 493 (2015) 766–781.

67 K.J. Lee, J.Y. Jao, Y.S. Kang, Y. Dai, G.P. Robertson, M.D. Guiver, J. Won, Journal of Membrane Science 212(1-2) (2003) 147–155.
68 I.W. Kim, K.J. Lee, J.Y. Jho, H.C. Park, J. Won, Y.S. Kang, M.D. Guiver, G.P. Robertson, Y. Dai, Macromolecules 34(9) (2001) 2908–2913.
69 C. Wright, D. Paul, Polymer 38(8) (1997) 1871–1878.
70 T. Merkel, R. Gupta, B. Turk, B. Freeman, Journal of Membrane Science 191 (1–2) (2001) 85–94.
71 V. Stannett, J.L. Williams, Journal of Polymer Science Part C: Polymer Symposia 10(1) (1965) 45–59.
72 C.A. Kumins, J. Roteman, Journal of Polymer Science Part A: Polymer Chemistry 1(1) (1963) 527-540.
73 H. Frisch, Polymer Engineering & Science 20(1) (1980) 2–13.
74 Y.S. Kang, J.H. Meldon, N.H. Sung, Polymer Engineering & Science 26(15) (1986) 1045–1049.
75 J.H. Meldon, Y.S. Kang, N.H. Sung, Industrial & Engineering Chemistry Fundamentals 24(1) (1985) 61–64.
76 X.Y. Wang, F.T. Willmore, R.D. Raharjo, X.C. Wang, B.D. Freeman, A.J. Hill, I.C. Sanchez, Journal of Physical Chemistry B 110(33) (2006) 16685–16693.
77 Y.S. Kang, J.H. Meldon, N. Sung, Journal of Polymer Science Part B: Polymer Physics 28(7) (1990) 1093–1103.

4

Facilitated Transport Phenomena in the Solid State

The fundamental feature of facilitated transport is its capability to enhance permeability of a given species while unaffecting permeabilities of the others in a mixture, thereby increasing the selectivity of the given species to others. Therefore, permeability and selectivity of a given species increase simultaneously, a challenge often encountered for conventional polymeric materials due to their conflicting nature. In addition, permeability increases as decreasing the applied pressure, which is a huge advantage to reducing the operation cost in practical applications.

Facilitated transport can be described as the sum of the Fickian transport and carrier-mediated transport as illustrated in Figure 4.1, with the latter playing a pivotal role in determining the overall mass transport. Here, the carrier must selectively and reversibly react or interact with the target component.

The origins of facilitated transport are rooted in biological systems, particularly the human bloodstream, where specialized carriers like heme in hemoglobins enable the efficient transport of oxygen from the lungs to various organs. The adaptation of these principles to solid-state systems, particularly within polymeric materials, has opened new avenues for research and application. These developments have spurred interest in utilizing facilitated transport for practical purposes in advanced material systems.

Considering these developments, this chapter aims to provide a comprehensive introduction to the fundamentals of facilitated transport phenomena in the solid state. It begins with foundational principles, carrier characteristics focusing their interactions with target species. The mechanisms underlying transport and the mathematical models used to describe these processes are discussed in depth. Additionally, the chapter highlights key applications, such as oxygen separation from air, CO_2 capture from flue gas, and olefin/paraffin separation, i.e. ethylene/ethane

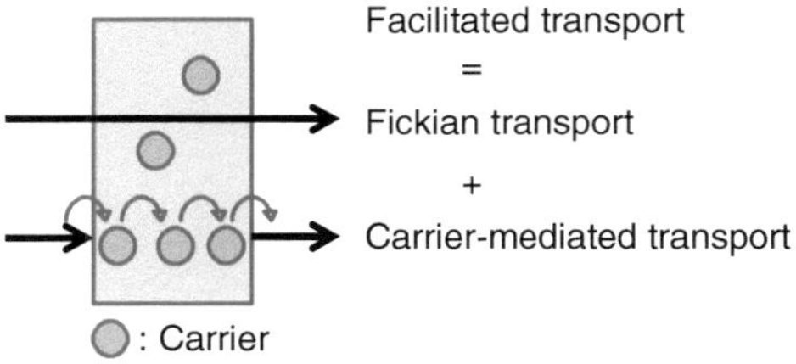

Figure 4.1 Concept of facilitated transport in the solid state as the sum of the Fickian transport and carrier-mediated transport.
Source: Y.S. Kang.

Molecule and Ion Transport through Polymer Membranes, First Edition. Yong Soo Kang.

and propylene/propane mixtures, illustrating the broad utility and future potential of facilitated transport systems in addressing critical industrial and environmental challenges.

4.1 Facilitated Transport in the Liquid State and the Solid State

The facilitated oxygen transport in the human body starts by forming an oxygen–heme (O_2:heme) complex in lungs, which flow through blood vessels to reach organs with the help of heart pumping. Once the complex reaches the target organ, oxygen is liberated by decomplexation, resulting in the regeneration of neat heme, which flows back to the lungs (Figure 4.2a). The heme group contains an iron ion (Fe^{2+}) coordinated by a porphyrin (Figure 4.2b) and can interact with oxygen *specifically* and *reversibly*. Therefore, Fe^{2+} coordinated by porphyrin plays as an *oxygen carrier* for facilitated O_2 transport.

Fundamentals on facilitated transport behavior in the liquid state have been thus investigated to improve the separation performance of O_2 from air along with long-term operation stability. Since then, the concept of facilitated transport has been expanded to the application for separation of CO_2 from flue gas. For practical purposes, liquid membranes containing carriers are frequently immobilized in a microporous polymeric support to make supported or immobilized liquid membranes (SLMs or ILMs). In this case, liquid solvent will evaporate during operation, resulting in a serious hurdle for long-term operational stability.

Notably, in the 1980s, Professors Nishide and Tsuchida first demonstrated the facilitated oxygen transport through *solid-state* polymer membranes containing either Co(II)- or Fe(II)-porphyrin or Schiff's base complexes as an analog of the heme structure [1–3]. Here, Co(II) or Fe(II) interacts specifically with oxygen reversibly, thereby acts as a *fixed-site* oxygen carrier for facilitated oxygen transport

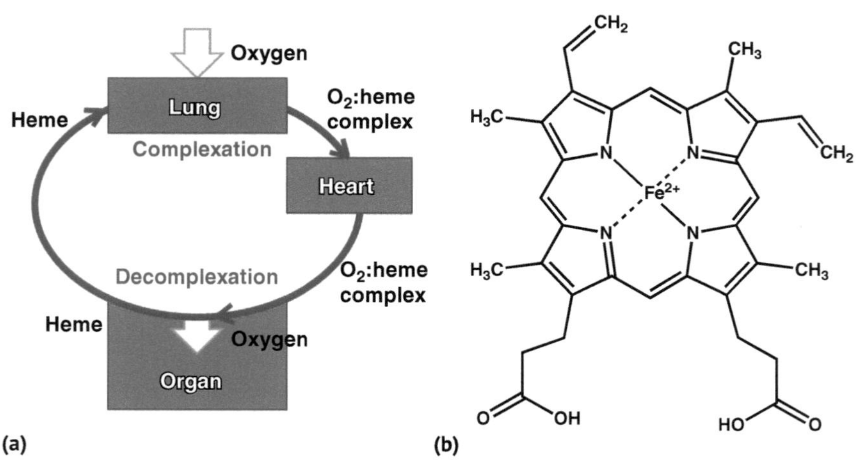

Figure 4.2 (a) Schematic facilitated oxygen transport in a human body by complexing and decomplexing with heme of hemoglobin in blood. *Source*: Y.S. Kang. (b) Structure of Fe(II)-porphyrin subunit of heme B. Redrawn from [20].

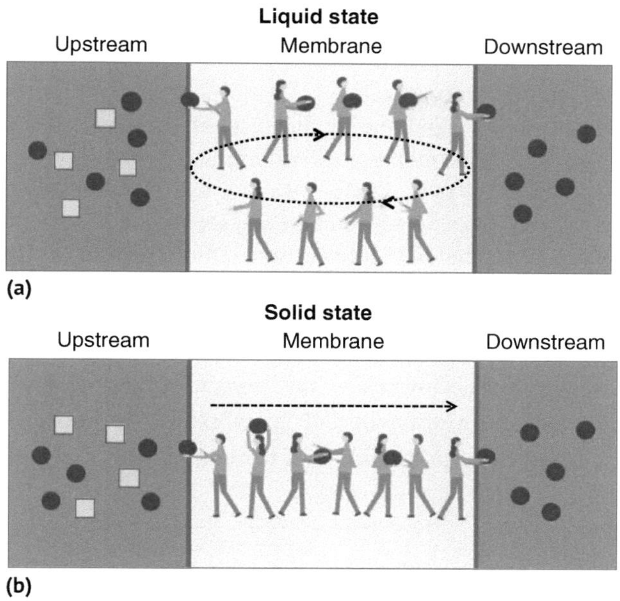

Figure 4.3 Difference in the facilitated transport mechanisms between (a) the liquid state and (b) the solid state. *Source*: Y.S. Kang.

in the solid state. Hereafter, a fixed-site carrier in the solid state will be referred to as a simply "carrier" just for convenience.

Figure 4.3 schematically illustrates the difference in the facilitated transport mechanism in (a) the liquid state and (b) the solid state. The reversible interaction kinetics could be important in determining the oxygen transport for both liquid and solid states. However, kinetics is more critical in the solid state than those in the liquid state because the multiple complexing and decomplexing steps are required for the solid state while only a single step of complexing and decomplexing is required for the liquid state.

Facilitated transport membranes in the solid state comprise fixed-site carriers incorporated into a solid polymeric matrix. Here, the carrier can be any chemical, which *reversibly* interacts with a *specific* solute as described previously. Thus, its extent and kinetics of interactions with the specific solute are essential in determining transport properties. For instance, consider a carrier C that reversibly interacts with a specific component A but not with B in a mixture of A and B, which can be expressed chemically:

$$A + C \underset{k_2}{\overset{k_1}{\rightleftarrows}} [A \cdot\cdot C] \qquad K = k_1/k_2 \tag{4.1}$$

where k_1 and k_2 are forward and backward interaction or reaction rate constants, respectively, to form solute–carrier complex $[A \cdot\cdot C]$ and K is equilibrium constant.

According to the concentration fluctuation model [4, 5], a mathematical model introduced in Section 4.2.4, two key parameters primarily influence the transport properties: the backward interaction rate constant, k_2, and the equilibrium constant, K, in addition to the carrier concentration, C_C. Among these, k_2 is the most influential because a faster k_2 ensures that carrier C is readily available to interact with the solute A. This makes k_2 more influential than k_1 and K. A more detailed analysis of these factors is provided in Section 4.2.

4.2 Mathematical Models for Facilitated Transport in the Solid State

We begin by highlighting the distinct characteristics and advantages of facilitated transport membranes in the solid state. For instance, CoP membranes containing Co(II)P_4, Co(II)P_2B_2, and Co(III)P_4 linked to poly(octyl methacrylate-co-vinylimidazole) exhibited unique separation characteristics of oxygen over nitrogen as illustrated in Figure 4.4, where Co(II)P_4 and Co(II)B_2P_2 are (meso-α, α,α-tris(*o*-methacrylamidophenyl)-β-(*o*-methacrylamidophenyl)-porphinato)cobalt(II) and (meso-α, α-bis(*o*-methacrylamidophenyl)-α, α-bis-(*o*-pivalamidophenyl)porphinato) cobalt(II), respectively [3].

Figure 4.4a shows how the oxygen permeability, P_{O_2}, changes with applied pressure, p, in different CoP membranes. We notice that P_{O_2} sharply increases as p decreases for membranes containing active oxygen carriers of Co(II)P_4 and

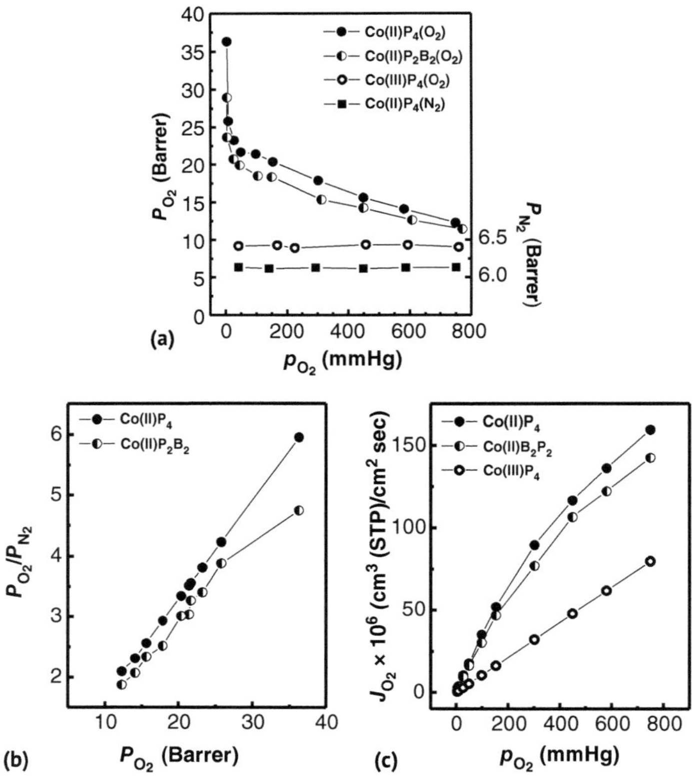

Figure 4.4 (a) Effects of upstream gas pressure p (mmHg) on the permeability coefficient P (Barrer) of oxygen and nitrogen through the CoP polymer membranes at 45 °C: the oxygen permeability P_{O_2} through Co(II)P_4 (●), Co(II)B_2P_2 (◐) and Co(III)P_4 (○), and the nitrogen permeability P_{N_2} (■) through Co(II)P_4; (b) the relationship between P_{O_2} and P_{O_2}/P_{N_2} from the data of (a); and (c) the relationship between the oxygen fluxes J_{O_2} and the applied oxygen pressure (P_{O_2}), where the flux data from Figure 4.3a. *Source*: Adapted from [3].

$Co(II)P_2B_2$, compared to the reference membrane with the inactive carrier of $Co(III)P_4$, while maintaining the nitrogen permeability P_{N_2} remains unaffected. It is important to note that Co(II) is capable of interacting with oxygen and thus serves as an oxygen carrier, whereas Co(III) lacks this ability. Figure 4.4b illustrates the simultaneous increases in both P_{O_2} and P_{O_2}/P_{N_2}, a huge advantage of facilitated transport membranes, as they often conflict with conventional membrane materials. Additionally, the mass fluxes in $Co(II)P_4$ and $Co(II)P_2B_2$ membranes far exceed those in the Fickian transport membrane of $Co(III)P_4$, further highlighting their attractiveness (Figure 4.4c).

In this section, we will provide a brief overview of the mathematical models developed, including dual sorption, effective diffusion, limited mobility of carriers, and concentration fluctuation models. These models will enhance our understanding of facilitated transport in solid-state membranes and underscore their advantages over traditional Fickian transport membranes.

4.2.1 Dual-mode Transport Model

The dual-mode transport model, originally developed to interpret the sorption behavior of gases or vapors in glassy polymers, has been employed to explain facilitated transport properties because the dual sorption behavior is conceptually analogous to the mass transport in a facilitated transport membrane in solid state. The simple dual-mode transport model was developed by assuming that only two independent diffusional pathways of the ordinary Fickian diffusion represented by D_D and the direct diffusion between carriers represented by D_C (Figure 4.5a).

Mathematically, the permeability, P, a product of the diffusion and solubility coefficients, through facilitated transport membranes is obtained by simple production of the corresponding diffusion and the solubility coefficients obtained from the dual sorption model as [6]:

$$P = k_D D_D + C'_C \frac{K}{1 + Kp} D_C \tag{4.2}$$

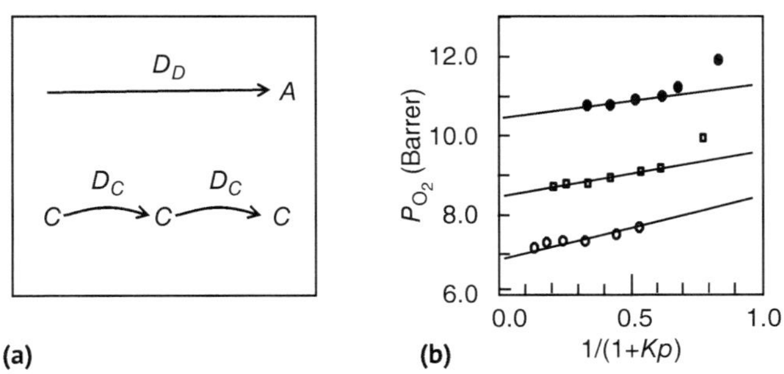

Figure 4.5 (a) Schematic of two transport modes; and (b) oxygen permeability in the CoPIm/PBMA membrane plotted according to Equation 4.2 (at (○) 20 °C, (□) 25 °C, and (●) 30 °C). *Source:* Adapted from [2].

where k_D is the solubility coefficient of Henry's law, C'_C is a saturated amount of solute A reversibly bound to fixed-site carriers and K is the equilibrium constant for the reversible reaction between the solute A and carrier C (Equation (4.1)). A fairly linear relationship between P and $(1 + Kp)^{-1}$ was obtained from the experimental oxygen permeabilities of CoPIm/PBMA membranes, where CoPIm is the complex of Co(II)porphyrin with 1-methylimidazole and PBMA is poly(butyl methacrylate), demonstrating the validity of the dual-mode transport model (Figure 4.5b) [2].

In this model, only two independent transport pathways, the ordinary Fickian diffusion, D_D, and the direct diffusion between carriers, D_C, were assumed without interactions, which may not be always valid in real system. Thus, this assumption was relaxed to adopt four diffusional pathways with four corresponding diffusion coefficients: (1) diffusion between dissolved solutes D_{DD}, (2) diffusion from dissolved solutes to carriers D_{DC}, (3) diffusion from carriers to dissolved solutes D_{CD}, and (4) diffusion between fixed carriers D_{CC} [3, 7, 8].

$$P = k_D D_{DD} + D_{CC}\frac{C'_C K}{1 + Kp} + \left[\frac{D_{CD}C'_C K - k_D D_{DC}}{1 + Kp} + \frac{2k_D D_{DC}}{Kp}\ln(1 + Kp)\right] \quad (4.3)$$

In the case of the CoP_4 membrane, D_{DD}, D_{DC}, D_{CD}, and D_{CC} were estimated to be 2.2×10^{-6}, 3.2×10^{-7}, 2.2×10^{-7}, and 1.4×10^{-7} cm^2/sec, respectively, with $k_D = 7.2 \times 10^{-4}$ $cm^3(STP)/cm^3cmHg$, and $C'_C = 0.2$ $cm^3(STP)/cm^3$. The CoP_4 concentration in the membranes was 1.3 wt% [3].

The dual-mode transport model has been occasionally adopted to analyze the permeability data because it is conceptually and mathematically simple and predicts the dependency of permeability on upstream pressure qualitatively. However, it does not predict facilitated transport without direct diffusion between carriers or when $D_{CC} = 0$ for original dual-mode transport model (Equation (4.2)). Furthermore, it neglects the importance of the complexation kinetics between solute and carrier on facilitated transport, which is found to be very important in determining facilitated transport.

4.2.2 Effective Diffusion Coefficient Model

A more rigorous analysis of the facilitated transport in the solid state was presented by introducing a concept of "the effective diffusion coefficient" between fixed-site carriers (Figure 4.6a) [9]. Conceptually, this model was modified by introducing the kinetic effects of the complexation between A and C into the dual sorption model.

In the presence of excess carriers, the concentration of the unreacted carrier is constant, which leads to [9, 10]:

$$\frac{J}{J_0} = \left[1 + \frac{\alpha K}{1 + \alpha K}\right] / \left[1 + \frac{\alpha K}{1 + \alpha K}\left(\frac{\tanh\varphi}{\varphi}\right)\right] \quad (4.4)$$

where $\varphi = \frac{1}{2}\left[\frac{1 + (\alpha + 1)K}{\varepsilon(1 + K)}\right]^{1/2}$ and $\alpha = \frac{D_{AC}C_C}{D_A C_{A0}}$ with $D_{AC} = h^2\left(k_j^+ + k_{j+1}^-\right)/2$

Here, J and J_0 are solute fluxes for facilitated transport with carriers and Fickian transport without carriers, respectively; $K(= k_f C_{A0}/k_r)$ is the equilibrium constant,

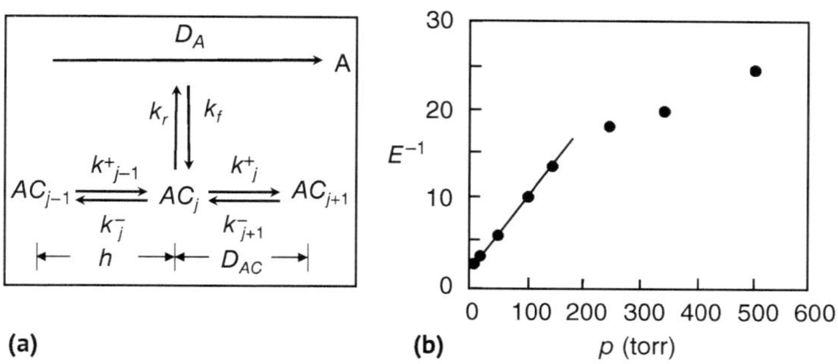

Figure 4.6 (a) Schematic of transport modes and interactions between solute A and three adjacent fixed carriers C_j, where AC represents the complex of A and C; and (b) plot of E^{-1} for PBMA/CoPIm membrane as a function of oxygen pressure p at 25 °C, [CoPIm] = 2.5 wt%, where $E = J/J_0 - 1$. *Source*: Adapted from [9].

and $\varepsilon(= D_{AC}/(k_r L^2))$ is ratio of characteristic reverse reaction to diffusion times, also frequently called inverse Damkoehler number, C_C and C_{A0} are the carrier concentration and the concentration of solute A at the interface, respectively, and D_{AC} is the effective diffusion coefficient of solute–carrier complex between two carriers with k_j^+ being the mobility at jth location, h being the spacing between carrier sites, + and – indicating direction and D_A is Fickian diffusion coefficient. Equation (4.4) is identical to that of facilitated transport in liquid membrane developed by Smith and Quinn [10].

If the reaction rate is assumed to be much faster than diffusion rate. i.e. $\tanh\varphi/\varphi \sim 0$, and then the model is reduced to the dual-mode transport model.

$$\frac{J}{J_0} = \left[1 + \frac{\alpha K}{1 + \alpha K}\right] \tag{4.5}$$

The validity of the model was examined against the experimental data for PBMA/CoPIm as shown in Figure 4.6b, where $E = J/J_0 - 1$. At low pressure, Equation (4.5) fits them well, but starts to deviate gradually as the oxygen pressure increases. This is because the excess carrier assumption is violated at high oxygen pressures [9].

4.2.3 Limited Chain Mobility Model

Cussler et al. analyzed fixed-site facilitated transport with a concept of "limited mobility of chained carriers" (Figure 4.7a) [11]. It was assumed that no uncomplexed solute can exist in the membrane, and the reaction between a carrier and solute occurs only at the surface of the membrane and is fast.

The mobility of chained carriers, pending on side chain, can allow the carrier–solute complex to encounter a second, uncomplexed carrier, resulting in a facilitated transport, which can be mathematically expressed as follows:

$$J = D_A \frac{C_C}{L}\left(\frac{Kp}{1 + Kp}\right)\chi \tag{4.6}$$

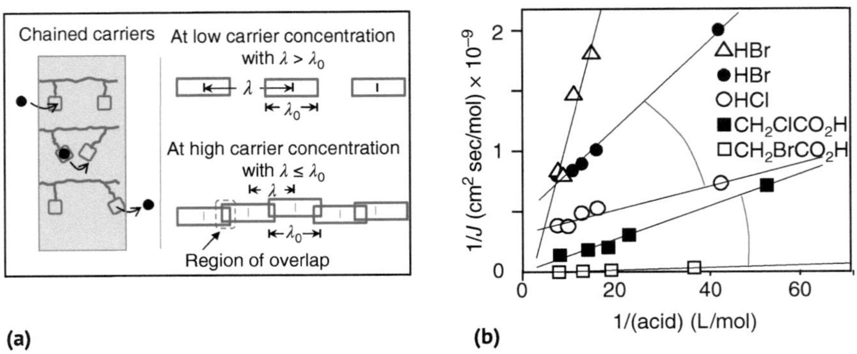

Figure 4.7 (a) Solute diffusion due to chained carriers. The membrane is assumed to consist of carriers with λ apart and mobile over a distance λ_0. Solute transfer can occur between these chained carriers only when $\lambda \leq \lambda_0$; (b) inverse ion flux [1/*J*] vs inverse solute concentration 1/[acid]. Data (△) from poly(vinyl pyridine-co-styrene), (●) and (○) from poly(*N*-vinyl imidazole-co-styrene), and (■) and (□) from poly(vinyl pyridine-co-acrylonitrile). *Source*: Adapted from [11].

where D_A is the effective diffusion coefficient through a membrane thickness L and the chained carrier correction factor χ,

$$\chi = \frac{\lambda}{\lambda_0}\left[\left(2 - \frac{\lambda_0}{\lambda}\right) + \left(\frac{\lambda_0}{\lambda} - 1\right)\left(\frac{1}{2} + \frac{1 + \cosh\psi}{\psi\sinh\psi}\right)\right]^{-1} \text{ and } \psi = \sqrt{\frac{2kC_C\lambda(\lambda_0 - \lambda)^2}{\lambda_0 D_A}}$$

called the Thiele modulus ψ with k being the exchange rate constant and C_C being the average carrier concentration. Here, λ is the distance between carriers and λ_0 represents the distance for limited chain mobility.

The key result in Equation (4.6) has two important special cases which clarify its physical meaning. The first case, that of fast reaction with $\psi >> 1$, is like the common limit of fast reaction in mobile carrier membranes, where mobile carriers diffuse freely within a liquid membrane. The second special case, that of fast diffusion with $\psi \sim 0$, will occur much more frequently in chained carrier membranes, where chained carriers have limited diffusion because they are covalently bound in the polymer chains [11].

If $\psi >> 1$, the reaction is fast, and the flux J is mostly diffusion-controlled. Then, Equation (4.6) becomes:

$$J = D_A\frac{C_C}{L}\left(\frac{Kp}{1 + Kp}\right)\left(\frac{2\lambda/\lambda_0}{3 - \lambda_0/\lambda}\right) \tag{4.7}$$

When the diffusion rate is much faster than that of chemical reaction ($\psi \sim 0$), Equation (4.6) becomes:

$$J = [kC_C\lambda^3(\lambda_0 - \lambda)/\lambda_0^2]\frac{C_C}{L}\left(\frac{Kp}{1 + Kp}\right) \tag{4.8}$$

The quantity in square bracket in Equation (4.8) is a new apparent diffusion coefficient. When $\lambda \geq \lambda_0$, this apparent diffusion coefficient is zero, indicating percolation threshold.

This model suggests that the carrier-mediated transport is only allowed over a limited distance λ due to the limited mobility of the chained carriers λ_0, which, in turn, leads to a percolation threshold at $\lambda = \lambda_0$. The experimental data for anion flux of Cl^- and $ClCH_2COO^-$ through polymeric membranes containing cations show the nearly linear relationship between the ion flux J and the carrier concentration C_c without showing the percolation threshold (actually $1/J$ vs $1/C_C$ in Figure 4.7b) [11]. The facilitated oxygen transport in fixed-site carrier membranes has been also experimentally observed at the low concentration of CoPIm in PBMA, as low as 0.6 wt% [2], suggesting the existence of the percolation threshold highly unlikely. Details for the existence of the percolation threshold are described afterwards in Section 4.3.

4.2.4 Concentration Fluctuation Model

In facilitated transport membranes, reversible complexation and decomplexation reactions between solute A and carrier C must continuously occur:

$$A + C \underset{k_2}{\overset{k_1}{\rightleftarrows}} [A \cdot\cdot C] \quad \text{and} \quad K = k_1/k_2 \tag{4.9}$$

At any given moment, the local solute concentration, C_A, may temporarily fluctuate from the average concentration due to the reversible reactions between solute A and carrier C, as illustrated in Figure 4.8a and b (an extended view of Figure 4.8a). When solute A reacts with carrier C to form the complex $[A \cdot\cdot C]$ at a rate constant k_1, C_A at that site decreases instantaneously. Conversely, C_A increases when the complex $[A \cdot\cdot C]$ dissociates, releasing A back into the matrix at a rate constant k_2. The average fluctuation in concentration is represented as C_f. Despite these fluctuations, the overall time-averaged concentration profile remains linear, as shown in Figure 4.8a.

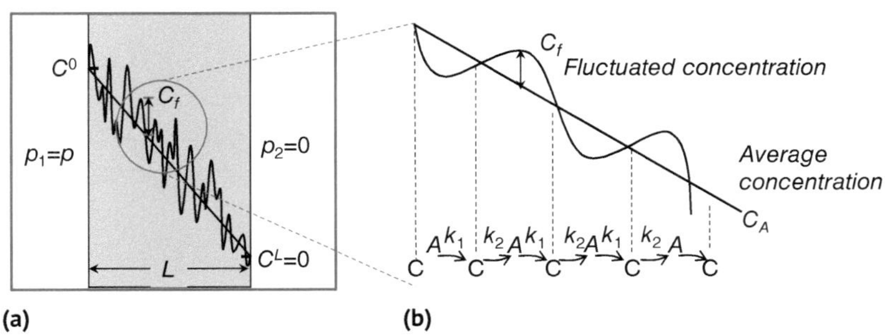

Figure 4.8 (a) Proposed concentration fluctuation of solute A for facilitated transport at the steady state; and (b) extended fluctuating concentration profile where k_1 and k_2 are forward and backward reaction rate constants of the solute A with carrier C, respectively, C_f is the extent of the fluctuating concentration. *Source*: Redrawn from [4, 5].

The concentration fluctuation arises the increase in the chemical potential μ_A of solute A due to the increased Gibbs free energy G of heterogeneous system according to the Cahn's theory [12]:

$$G = \int \left[g(C_A) + \kappa(\nabla C_A)^2\right] dV \tag{4.10}$$

where $g(C_A)$ is the free energy density of homogeneous system, $\kappa(\nabla C_A)^2$ is additional free energy due to the concentration gradient with $\kappa > 0$ and is always positive [12]. The increased Gibbs free energy G and consequently increased chemical potential μ_A will result in a higher driving force for mass transport and lead to the facilitated transport as schematically represented in Figure 4.9.

Based on above concepts and analogy between electron transport in a series of parallel resistor–capacitor circuits and mass transport in a fixed-site carrier membrane (Figure 4.10a and b), Kang et al. developed a concentration fluctuation model for facilitated transport behavior in solid state with one adjustable parameter p_f [4, 5].

$$\frac{P}{P_0} = 1 + \left(\frac{p_f}{p}\right)\sqrt{n^2 + \left\{\frac{2\pi k_2 C_C L^2}{P_0}\frac{\ln(1+Kp)}{p}\right\}^2} \quad \text{with } n = NC_C(\pi r^2 L) \tag{4.11}$$

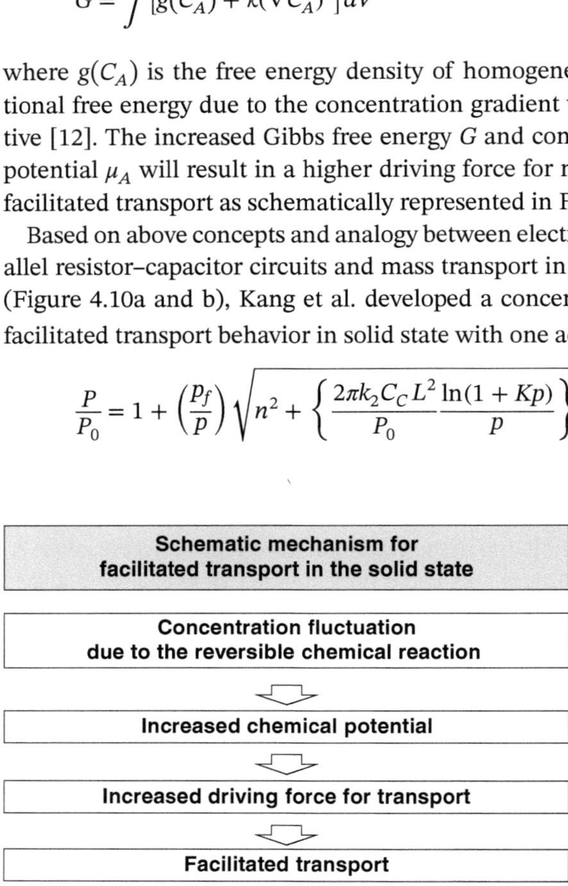

Figure 4.9 Schematic mechanism for facilitated transport in the solid. *Source*: Y.S. Kang.

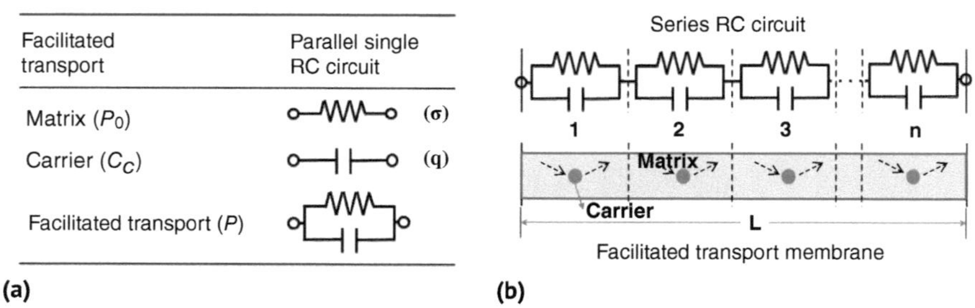

Figure 4.10 (a) Analogy of components between facilitated transport membrane and parallel single circuit for electric conduction; and (b) analogy between electron transport in the series RC circuit and mass transport in fixed-site carrier membrane. *Source*: Redrawn from [4, 5].

where P and P_0 are permeabilities of a facilitated transport membrane with carriers and a membrane matrix without carriers, respectively. p_f is the extent of pressure fluctuation, representing the extent of the concentration fluctuation C_f, as illustrated in Figure 4.8a, due to the reversible reaction between A and C (Equation (4.9), $p_1(= p)$ and $p_2(= 0)$ are the applied pressure of solute A at upstream and downstream sides, respectively, and thus the pressure difference across the membrane $\Delta p = p$. n is the number of carrier for a solute to react or interact during one dimensional diffusion pathway across a film of volume ($= \pi r^2 L$) with r being the permeant radius, L being the film thickness and N being the Avogadro number. k_2 and $K(= k_1/k_2)$ are the backward reaction rate constant and the equilibrium constant of the carrier–solute reaction, respectively, in Equation (4.9).

The concentration fluctuation model was examined against the experimental data of facilitated oxygen transport through poly(methyl methacrylate) (PMMA), poly(butyl methacrylate) (PBMA), and poly(dimethyl siloxane) (PDMS) membranes containing cobalt(II) porphyrins (CoPs, $C_C = 2.29 \times 10^{-5}$ mol/cm^3) [13], as shown in Figure 4.11a [4, 5], where the facilitation factor, F, is defined as $F = P/P_0$. All parameter values were independently measured by Nishide et al [13] except the pressure fluctuation p_f. The linear relationships between $F-1$ and $(1/p)\left\{n^2 + \left[(2\pi k_2 L^2 C_C/P_0)(\ln(1+Kp)/p)\right]^2\right\}^{1/2}$ demonstrate the validity of Equation (4.11) to all three membrane cases.

From the slope of the linear curves, p_f corresponding to C_f was estimated to be in the range of $10^{-2} < p_f < 10^{-4}$ cmHg, which is extremely small as expected, compared to the applied pressure. The agreement between the mathematical model

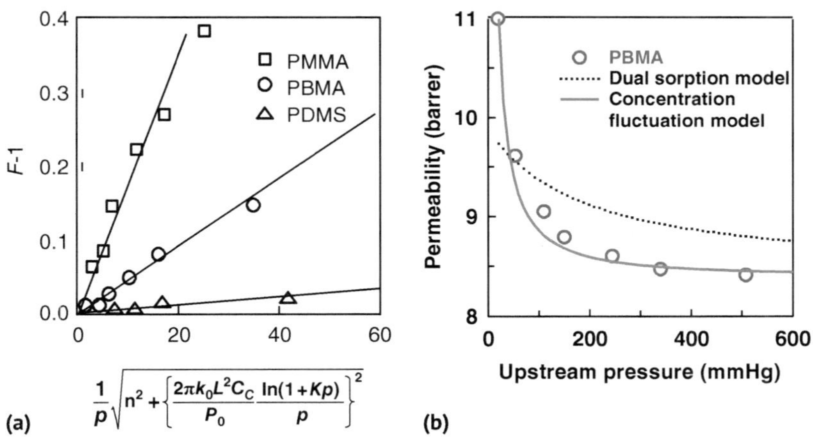

Figure 4.11 (a) Plots of $F-1$ (where $F = P/P_0$) vs $(1/p)\left\{n^2 + \left[(2\pi k_2 L^2 C_C/P_0)(\ln(1+Kp)/p)\right]^2\right\}^{1/2}$ by varying the matrix of PMMA, PBMA, and PDMS; and (b) the comparison between the concentration fluctuation model predictions and experimental data for PBMA membrane containing cobalt porphyrin: $P_0 = 8.4$ Barrer, $L = 60$ μm, $k_2 = 2.3 \times 10^{-2}$ /sec, $K = 5.3 \times 10^{-2}$ /cmHg, $C_C = 2.29 \times 10^{-5}$ mol/ cm^3, and $p_f = 5.22 \times 10^{-4}$ cmHg. *Source*: Replotted from [4, 5].

predictions and the experimental results is exceptional in the whole range of pressure, even at high pressure range (Figure 4.11b). This model also shows the facilitated transport even at low concentration range ($C_C = 2.29 \times 10^{-5}$ mol/cm^3) of the carrier, whereas the hopping models including "the dual-mode transport sorption," "the effective diffusion coefficient," and "the limited mobility of chained carrier models" might be valid only when the carrier concentration is high.

Equation (4.11) looks complicated and can be simplified as:

$$P \propto k_2 C_C L^2 \ln(1 + Kp)/p^2 \tag{4.12}$$

when $n << (2\pi k_2 L^2 C_C/P_0)(\ln(1 + Kp)/p)$, which is applicable under typical experimental conditions. Equation (4.12) clearly expresses the linear correlation between P and both C_C and k_2, emphasizing the particular importance of k_2. The significance of the solubility, represented by the equilibrium constant K, is also noteworthy, albeit following a logarithmic like $\ln(1 + Kp)$, rather than a linear pattern. The improved permeability behavior associated with decreasing the applied pressure p is likewise evident as shown in Figure 4.11b and also frequently observed experimentally [14].

Figure 4.12a illustrates the linear relationship between the pyridine concentration and the flux of CH_2ClCO^- anion through poly(4-vinyl pyridine-co-acrylonitrile) membranes, where the pyridine can play as the CH_2ClCO^- carrier [15]. Figure 4.12b shows that P_{CO_2} linearly correlates with L^2 the square of the selective layer thickness of the composite membranes, which contain a selective layer of mixed matrix of poly(vinyl alcohol)/poly(vinyl amine)/multi-walled carbon nanotubes (PVA/PAm/MWNT) coated on a microporous polysulfone support [16]. Both results are again consistent with the concentration fluctuation model (Equations (4.11) and (4.12)).

Finally, the concentration fluctuation model accounts for four different pathways for the facilitated transport (only with one diffusion coefficient D, one kinetic and equilibrium constant k_2 and K, and one adjustable parameter p_f) including the dissolved solute-to-dissolved solute ($A \rightarrow A$), dissolved solute-to-complexed solute ($A \rightarrow AC$), complexed solute-to-dissolved solute ($AC \rightarrow A$), and complexed

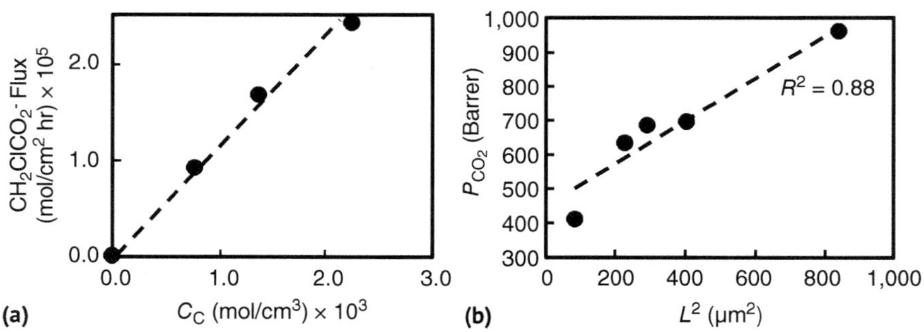

Figure 4.12 (a) Effect of pyridine carrier concentration C_C on flux of $CH_2ClCO_2^-$ through poly(4-vinyl pyridine-co-acrylonitrile) membranes; (b) correlation between CO_2 permeability and the square of the selective layer thickness L containing mixed matrix of PVA/PAm/MWNT coated on a microporous polysulfone support. *Source*: Replotted from [15, 16].

Figure 4.13 Schematic illustration of four different diffusional pathways. *Source*: Y.S. Kang.

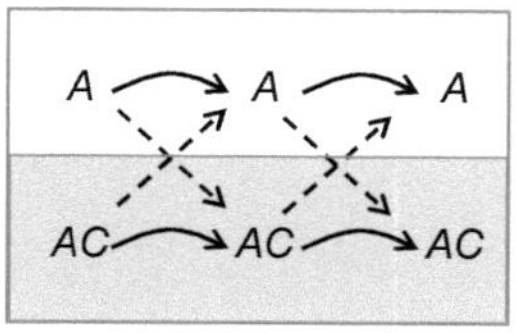

solute-to-complexed solute ($AC \rightarrow AC$), as shown in Figure 4.13, in contrast to the dual-mode transport model including only two diffusional pathways (Figure 4.5a).

4.3 Concentration Fluctuation Model vs Direct Hopping Models

The three mathematical models, including Sections 4.2.1–4.2.3, can be categorized as a hopping model with percolation threshold because facilitated transport cannot occur without direct diffusion between carriers. In contrast, Section 4.2.4 does not belong to the hopping model. In this context, the validity of the hopping models will be examined against the concentration fluctuation model.

In order for solute to directly diffuse between two carriers by hopping, the average distance, λ, between two carriers should be shorter than the average diffusional jump distance, δ. According to the hopping model, carrier-mediated transport could occur only when $\lambda < \delta$ (Figure 4.14a), whereas no carrier-mediated transport can take place when $\lambda > \delta$ (Figure 4.14b). Therefore, in this case, the threshold concentration will be present at $\lambda = \delta$.

The distance λ between two carriers is simply estimated from:

$$\lambda = \left(\frac{1}{NC_C}\right)^{1/3} \tag{4.13}$$

where C_C is the molar concentration of the carrier in a matrix with N being the Avogadro number. For instance, the λ values for 0.6, 4.5, and 30 wt% of CoPIm in PBMA are approximately 68, 35, and 18 Å, respectively, and experimental facilitated oxygen transport was observed even when the CoPIm concentration was as low as 0.6 wt% with $\lambda = 68$ Å [17].

The average diffusional jump distance δ can be obtained from diffusion activation energy E_D from the following relationship [18].

$$E_D = \delta\left(\frac{\pi d^2}{4}\right)CED * N \tag{4.14}$$

where CED is the cohesive energy density of polymer with N being the Avogadro number and d is the kinetic diameter of a solute. The δ values estimated for common polymers are plotted as a function of kinetic diameter d of gases as shown in Figure 4.15. Here, E_D and CED are from Van Krevelen [19].

The δ value for oxygen ranges from 3.0 to 7.0 Å for common polymers and is 6.3 Å for poly(dimethyl siloxane), the most gas-permeable polymer. In contrast, the λ value for the polymer membranes containing 0.6–30 wt% of CoPIm in PBMA

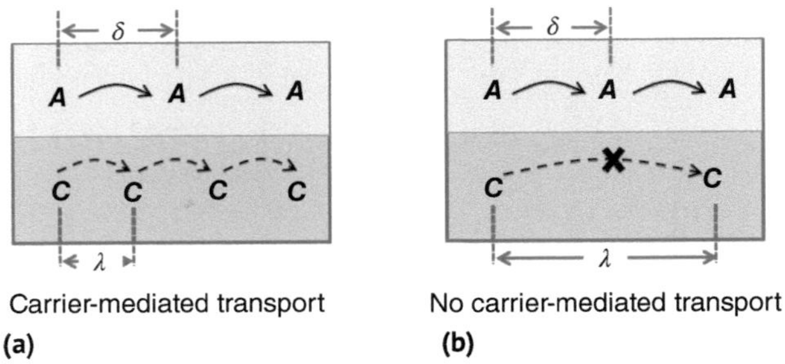

Figure 4.14 The validity of the hopping model for solute *A* with carrier *C*: (a) the carrier-mediated transport occurs when $\delta \geq \lambda$; and (b) no carrier-mediated transport when $\delta < \lambda$, where the average diffusional jump distance δ of various gases for a given polymer and the distance λ between two carriers. *Source*: Y.S. Kang.

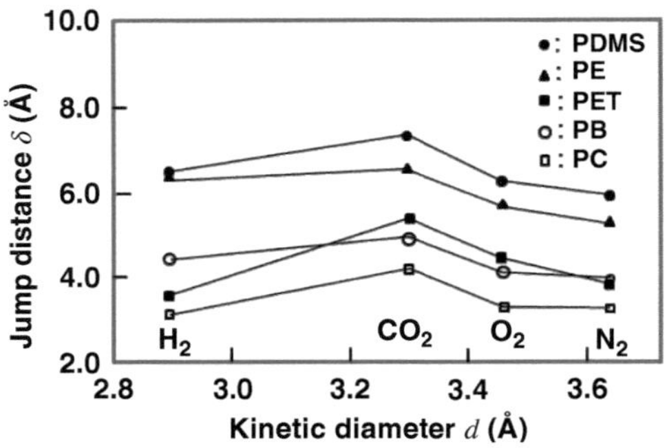

Figure 4.15 The average diffusional jump distance δ of various gases for polymers. *Source*: Y.S. Kang.

ranges from 68 to 18 Å. These results indicate the facilitated transport in those membranes were observed even though $\lambda > \delta$. If the hopping mechanism is the dominant transport mechanism, the CoPIm membranes should not exhibit the facilitated transport within the experimental range, yet the facilitated transport did occur [17]. Therefore, it can be argued that the hopping mechanism may only be valid at high carrier concentrations where $\lambda < \delta$.

Among the four mathematical models described above, only the concentration fluctuation model is not categorized under the hopping model. This model predicts facilitated transport without requiring direct hopping between two carriers. Furthermore, it emphasizes the importance of the reversible reaction rates, particularly the backward rate constant k_2 and aligns very well with experimental results.

4.4 Facilitated Oxygen Transport

Both Co(II) and Fe(II) ions have a coordination number of six in their complexes. When Co(II) or Fe(II) ion is coordinated by four nitrogens and/or oxygens from a planar porphyrin or Schiff's base, two of its coordination sites are unfilled. When the fifth coordination site is further coordinated by either an imidazole or pyridine

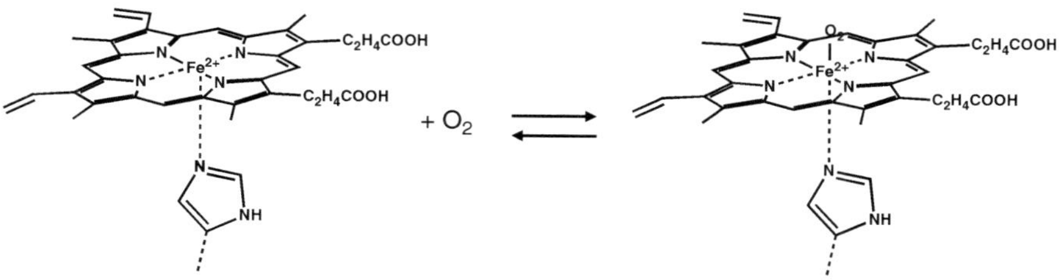

Figure 4.16 Reversible oxygen coordination to iron (II) porphyrin. *Source:* Ref. [20]/ Springer Nature.

residue covalently bound to polymeric chains, the sixth site is still left open. The open sixth site can be thus utilized to bind or interact with oxygen reversibly, to become an oxygen carrier as presented in Figure 4.16 [20].

4.4.1 Reversible Oxygen Solubility

When oxygen is in contact with a membrane containing cobalt(II) porphyrins, it forms oxygen–cobalt adducts and thus its solubility is enhanced markedly, compared to the neat polymeric matrix. For example, the oxygen solubility for the poly[(octyl methacrylate)-co-(4-vinyl pyridine)] (POMPy) membrane containing 37 wt% [α, α, α, α-meso-tetrakis(*o*-pivalamidophenyl) porphyrinato] cobalt(II) (CoP) complex was ~6.2 (cm^3(STP)/cm^3 polymer) at 30 atm, which is more than six times greater than that of the neat polymer (~1 cm^3(STP)/cm^3 polymer) as shown in Figure 4.17a [21].

The total oxygen solubility or concentration C_A in POMPy/CoP was analyzed by the dual sorption model of Equation (4.15).

$$C_A = k_D p + \frac{C'_A K}{1 + Kp} p \tag{4.15}$$

where k_D is the Henry's law solubility coefficient, p is the applied pressure, K is the equilibrium binding constant, and C'_A is the saturated amount of the gas A reversibly bound to the carrier. The first term of Equation (4.15) is for ordinary dissolution by Henry's law and the second for the complexation by the Langmuir type sorption isotherm by assuming the first-order interaction kinetics between carrier and oxygen [21].

Figure 4.17a also illustrates that the amount of oxygen sorbed is nearly identical to that desorbed, suggesting that the interactions are reversible regardless of the carrier concentration used. The reversible oxygen binding to Co(II) porphyrin/1-methylimidazole (CoPIm) complex attached to poly(*n*-butyl methacrylate) (PBMA) was again confirmed by the visible absorption spectroscopy as shown in Figure 4.17b [2]. λ_{max} at 528 nm of deoxy-CoPIm changed to 545 nm of oxy-CoPIm ($[O_2]$:[Co] = 1:1 adduct), immediately after exposure of the membrane to oxygen. The reversible oxygen binding ability in the POMPy complex membrane containing [*N*,*N*-disalicylideneethylenediamine] cobalt (CoS) was also confirmed by ESR spectroscopy [17].

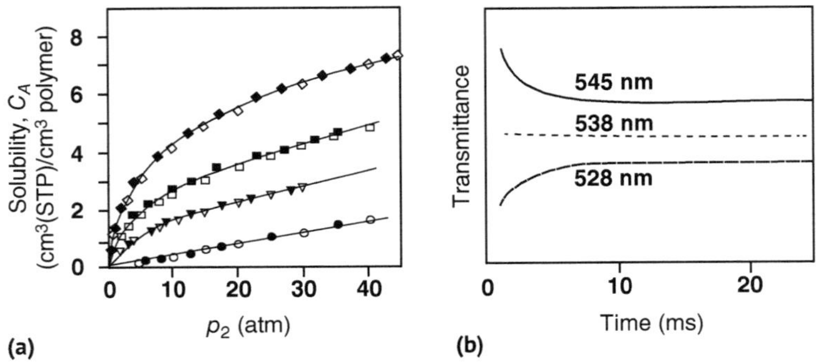

Figure 4.17 (a) Oxygen sorption and desorption isotherms for POMPy/CoP membranes at 45 °C. Open and solid symbols are sorption and desorption runs, respectively. ○, CoP = 0 wt%; ▽, 15 wt%; □, 25 wt%; ◇, 37 wt%. *Source*: Replotted from [21]. (b) Flash photolysis of the PBMA/CoPIm membrane under exposure to air and approximation to pseudo-first-order kinetics for the membrane at 25 °C [CoPIm] = 2.5 wt%. *Source*: Adapted from [2].

4.4.2 Kinetics of Reversible Interactions

The rates of the reversible oxygen binding to the PBMA/CoPIm ([CoPIm]=2.5 wt%) membrane were measured by pulse and laser flash spectroscopy as shown in Figure 4.17b [2]. The oxygen binding, k_1, and dissociation, k_2, rate constants were estimated to be 5.2×10^5 L/mol sec and 2.8×10^4 /sec by assuming pseudo-first order reaction, respectively, and $K = 18.6$ L/mol, representing that the CoPIm complex is active for reversible oxygen binding in the solid membrane and consequently acts as an effective oxygen carrier at 25 °C.

4.4.3 Facilitated Oxygen Transport

Figure 4.18 shows the permeabilities of oxygen, P_{O_2}, for the imidazolium porphyrinato-cobalt (Co(II)PIm) mixed with poly(butyl methacrylate) (PBMA) matrix. P_{O_2} increases markedly with the decrease in the oxygen pressure p, which indicates that Co(II)PIm in the solid membrane clearly plays as an oxygen carrier to facilitate oxygen transport. However, P_{O_2} with Co(III)PIm does not increase but maintains nearly unchained as varying p, demonstrating no facilitated transport,

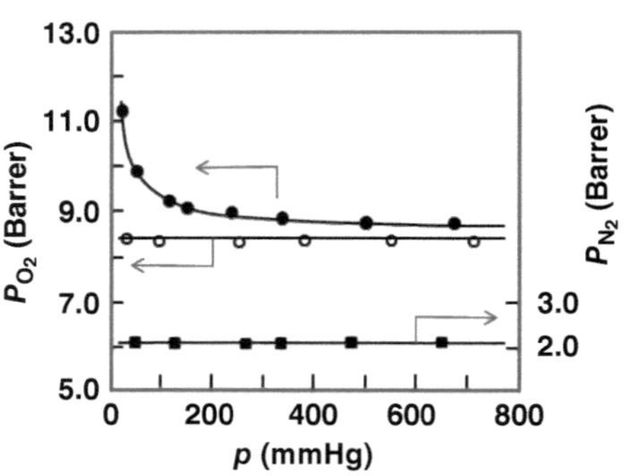

Figure 4.18 Effects of upstream gas pressure p on permeability P_{O_2} and P_{N_2} in the PBMA/Co(II)PIm membrane at 25 °C for oxygen (●); nitrogen (■); and in the inert PBMA/Co(III)PIm membrane for oxygen (○) at 25 °C. *Source*: Adapted from [2].

because Co(III)PIm does not interact with O_2. Although Co(II) and Fe(II) ions are known to reversibly interact with oxygen, they can be irreversibly oxidized to Co(III) and Fe(III) ions, respectively, and consequently lose their oxygen carrier capability, resulting in no facilitated transport as shown in Figure 4.18. Thus, the oxidation should be prevented to ensure the oxygen carrier activity. The oxidation commonly occurs through formation of dimers of two planar Schiff's bases or porphyrins [3].

4.5 Facilitated Olefin Transport

It has been known that olefins like ethylene and propylene can reversibly interact with transition metal ions, specifically Ag^+ and Cu^+ ions [22, 23] as well as metallic nanoparticles [24, 25]. These reversible interactions serve to play as an effective olefin carrier, leading to facilitated olefin transport. Here, we will begin by introducing the roles of metallic ions such as silver and copper ions, as olefin carriers, followed by an exploration of the roles of metallic nanoparticles where their surfaces are activated to generate a partial positive change with help of polarizers.

4.5.1 Metallic Ion Carriers

Silver and copper ions form complexes with olefin molecules having double bond, but not with paraffins without double bond. The interaction or bond between the metallic ion and ethylene molecule is a *specifically* and *reversibly* coordinative or dative σ bond as shown in Figure 4.19.

When these cations are integrated into a polymeric matrix, they generate membranes capable of separating olefins from their mixtures with paraffins. A simple way to incorporate silver or copper ions into a solid polymer matrix is by forming polymer electrolytes or polymer salts complexes. This involves dissolving metallic salts in a polymer solvent to produce free ions in a solid solution. Therefore, understanding the formation, structure, and reversible olefin interactions with metallic ions in polymer electrolytes is crucial for insights into facilitated transport phenomena.

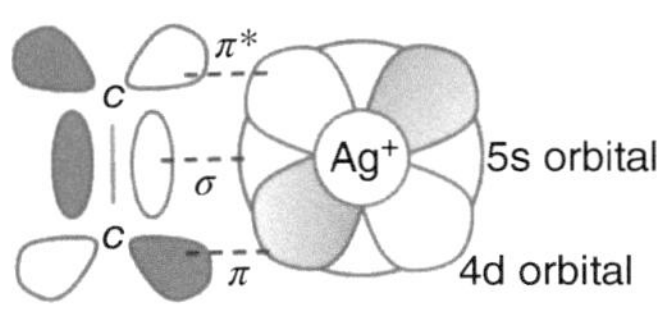

Figure 4.19 Interactions between silver ions and ethylene. *Source*: Y.S. Kang.

4.5.1.1 Formation of Polymer Electrolytes

Dissolving metallic salts, MX, such as silver and copper salts in a polymer solvent requires appropriate coordinative interactions between metallic ions, M^+, and ligands, *L*, in polymers as shown in Figure 4.20.

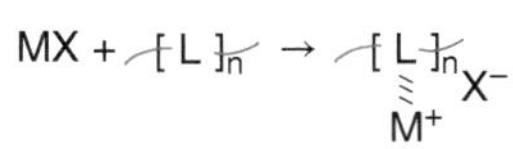

Figure 4.20 Schematic formation of polymer electrolyte by coordinative interactions between metal cation M^+ and ligand *L* in polymer solvent, where X^- is a counter anion of MX salt. *Source*: Y.S. Kang.

It is found that silver salts are readily dissolved to generate silver ions by forming polymer electrolytes containing silver ions, termed as silver-polymer electrolytes hereafter, when their lattice energies are small with large anions, such as BF_4^-, $CF_3SO_4^-$, ClO_4^-, and PF_6^- [26, 27]. On the other hand, when the anion is small like

F^-, Cl^- and NO_3^-, its interaction with Ag^+ is strong according to the Coulomb interactions between Ag^+ and the anion so that silver salts are much less soluble in a polymeric solvent and consequently the formation of silver-polymer electrolytes is limited in many cases. For instance, $AgBF_4$ is readily dissolved in a polar polymeric solvent such as poly(ethylene oxide) (PEO) to form silver-polymer electrolytes, whereas $AgNO_3$ is sparingly dissolved.

Ligands (*Ls*) in polymeric solvents for coordinative interactions with metallic cations are electron-rich polar atoms such as oxygen, nitrogen, and sulfur atoms. Typical polymeric solvents are thus poly(ethylene oxide) (PEO) [28–30], poly(2-ethyl-2-oxazoline) (POZ) [31–33], poly(*N*-vinyl pyrrolidone) (PVP) [34, 35], poly(vinyl methyl ketone) (PVMK) [36], poly(methacrylates) (PMA) [37], and cellulose acetate (CA) [38, 39]. The coordinative interaction of Ag^+ with carbonyl oxygen of POZ was confirmed by both the theoretical structure calculation based on the density functional theory [23] and the carbonyl band shift of IR spectra from 1,641 to 1,595 /cm upon incorporation of silver salts [27].

When metallic salt is dissolved in a polymeric matrix, three different ionic species will be formed depending on their environments: free ion, contact ion pair, and ion aggregate as shown in Figure 4.21a–c. Among three different ionic species, the free ion is the most active in interacting with olefins.

4.5.1.2 Structure of Silver-polymer Electrolytes

Dissolution of AgX into a polymeric solvent to form a silver-polymer electrolyte is possible by the coordinative interactions between Ag^+ and electron-rich ligands (*Ls*) in the polymer solvent, as shown in Figure 4.20. The observed coordination number changes in the range of 1–5 depending on the mole ratio of AgX and the environment.

When the coordinative bond number of Ag^+ is two or more, polymeric chains will be effectively cross-linked by Ag^+ as shown in Figure 4.22a.

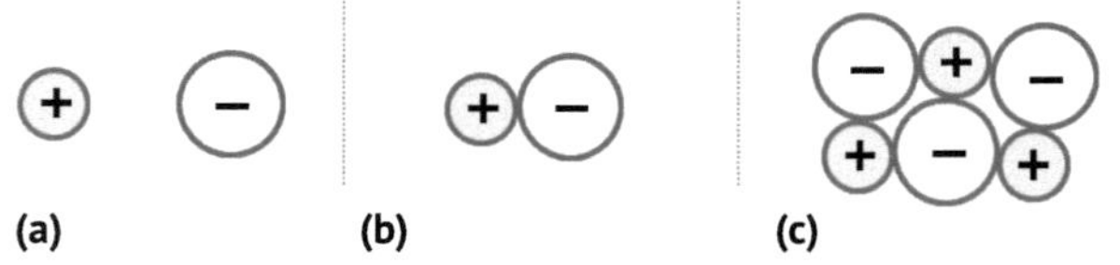

Figure 4.21 Schematic structure of (a) free ion, (b) contact ion pair, and (c) ionic aggregate in polymer electrolyte. *Source*: Y.S. Kang.

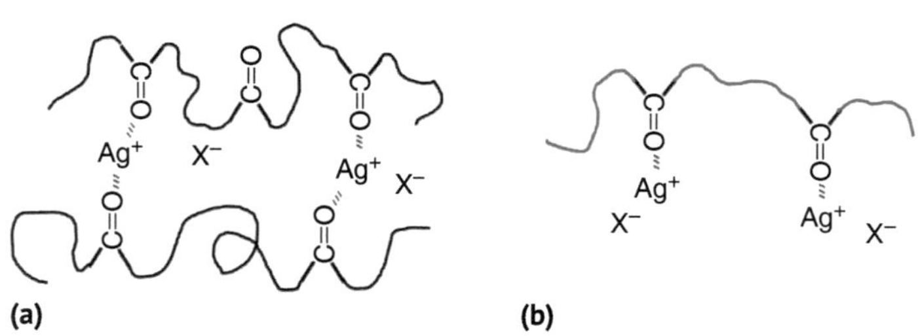

Figure 4.22 Schematic drawings of (a) cross-links among polymeric chains by Ag^+ with the coordination number of 2; and (b) dangling of heavy metallic ions with the coordination number of 1. *Source*: Y.S. Kang.

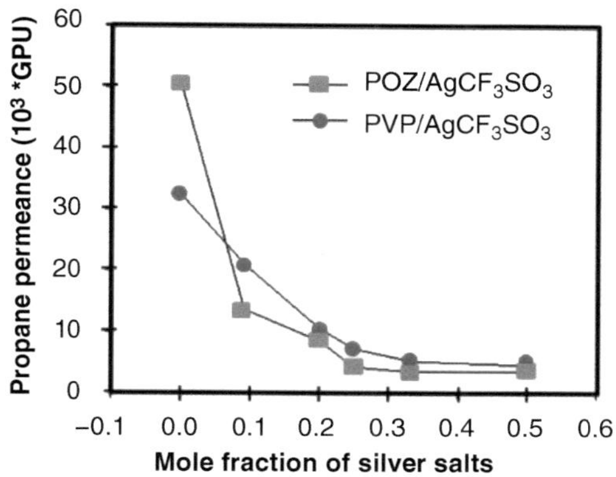

Figure 4.23 Permeances of propane through polymer electrolyte membranes containing $AgCF_3SO_3$ dissolved in poly(vinyl pyrrolidone) (PVP) and poly(2-ethyl oxazoline) (POZ) as a function of silver concentration. *Source*: Replotted from [31].

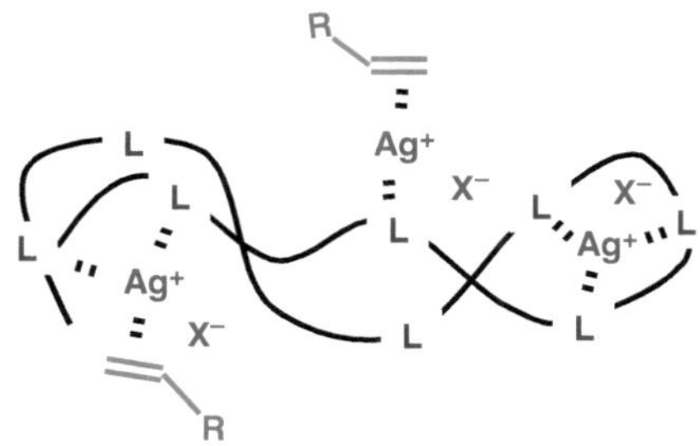

Figure 4.24 Schematic illustration of silver-polymer electrolyte consisting of *L*s and potential interaction between silver ion, Ag^+, and olefin, $CH_2{=}CH{-}R$. *Source*: Y.S. Kang.

In addition, heavy metallic ions will be dangled on polymeric chains with the coordination number of 1 (Figure 4.22b), making polymeric chains less mobile to some extent. Therefore, it can be anticipated that the chain mobility will be decreased upon incorporation of silver salt because of the cross-linking and dangling effects of Ag^+.

Therefore, the decrease in the chain mobility associated with the cross-linking and dangling effects of Ag^+ ions may reduce the permeation rates of both interactable and noninteractable solutes like propylene and propane, respectively. Figure 4.23 demonstrates that the permeance of noninteractable propane decreases with the concentration of $AgCF_3SO_3$ added, which aligns with the cross-linking and dangling effects of Ag^+ ions [31]. Consistently, the glass transition temperature, T_g, of silver-polymer electrolytes, as a measure of the chain mobility, also increases with the amount of silver salts added at their low concentrations, as shown in Figure 9.9 in Section 9.3 of Chapter 9 [40].

4.5.1.3 Olefin Sorption in Silver-polymer Electrolytes

The schematic structure of silver-polymer electrolyte and its interaction with olefin is illustrated in Figure 4.24. Therefore, extremely large amounts of olefins could be dissolved in silver-polymer electrolytes, compared to neat polymer without silver salts.

However, if the coordination sites of Ag^+ are completely filled by the ligand oxygen atoms of polymeric chains, Ag^+ may no longer be active in coordinating with

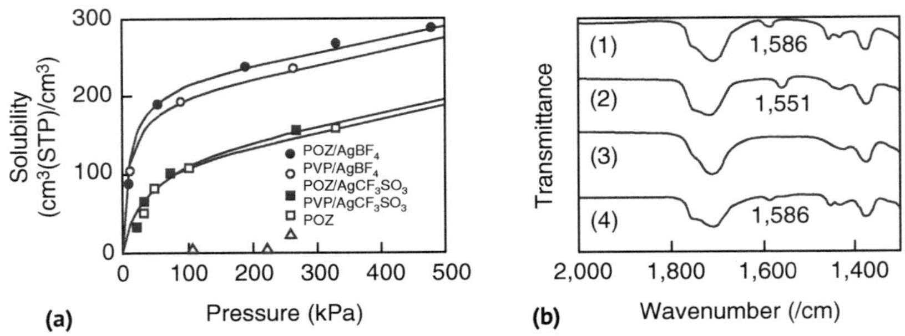

Figure 4.25 (a) Solubility of propylene in polymer electrolyte films at 25 °C. The solid lines were calculated from the dual sorption model (Equation (4.15)). *Source*: Replotted from [23, 32]. (b) IR spectra for the reversible olefin interactions of 1/3 cellulose acetate (CA)/$AgBF_4$ membrane with (1) propylene, (2) 1,4-butadiene, (3) ethylene, and (4) propylene environments. *Source*: Adapted from [39].

olefin molecule anymore. Experimentally, it was found that only when the coordination number of Ag^+ is less than 3, the silver ions are active in interacting with olefin to play as an olefin carrier, whereas they are not active when the coordination number of Ag^+ is equal or larger than 3 [31].

For example, the propylene solubility is 245 cm^3(STP)/cm^3 in 1/1 POZ/$AgBF_4$ whereas it is as low as 3 cm^3(STP)/cm^3 in neat POZ at 224 kPa (Figure 4.25a). Here, 1/1 POZ/$AgBF_4$ represents the 1/1 molar ratio of the repeating unit of POZ to $AgBF_4$. The olefin solubility is almost twofold higher for $AgBF_4$ dissolved in POZ or PVP electrolytes than that for $AgCF_3SO_3$. The results indicate that the olefin solubility also strongly depends on the anion type [32]. In the case of PEO/silver salt membranes, 1/1 PEO/$AgBF_4$ and 1/1 PEO/$AgCF_3SO_3$ sorb 8.46 and 1.76 g of propylene per 100 g of polymer electrolyte, respectively, at 30 °C and 70 cmHg. In contrast, 1/1 PEO/$AgNO_3$ sorbs only 0.52 g of propylene per 100 g of polymer electrolyte [30].

The solid lines were for the theoretical calculations from the dual sorption model (Equation (4.15)) [23, 32], indicating that the propylene solubility is mostly determined by complexing with Ag^+.

4.5.1.4 Reversible Interaction of Silver Ion with Olefin

The reversible olefin coordination to silver ions in silver-polymer electrolytes was investigated spectroscopically [36, 39]. When 1/3 cellulose acetate (CA)/$AgBF_4$ complex was exposed to 207 kPa of propylene and then purged with N_2 for a short time (~1 second), a new peak appeared at 1,586 /cm corresponding to the C=C stretching vibration of the coordinated propylene (Figure 4.25b). The peak at 1,586 /cm remained even after degassing at 10^{-5} torr for 4 hours at room temperature. However, exposure of the propylene-coordinated complex to 207 kPa of 1,4-butadiene and subsequent quick treatment with N_2 gave a new peak at 1,551 /cm along with the disappearance of the peak at 1,586 /cm. The peak at 1,551 /cm can be assigned to the C=C stretching frequency of the coordinated 1,4-butadiene. This peak at 1,551 /cm disappeared again when ethylene was introduced. The peak at 1,586 (/cm) reappeared when ethylene was replaced by propylene. These results indicate that the coordinated propylene is labile to be easily exchanged by other olefins, but strong enough not to be removed completely by vacuum. The same reversible olefin coordination behavior was also observed by UV spectroscopy [36, 39].

4.5.1.5 Reversible Interaction Mechanism of Silver Ion with Olefin

The reversible interaction or reaction mechanism between silver ion and olefin is elucidated by calculating the theoretical structure of a silver-polymer electrolytes in the gas phase and its complex with ethylene. The theoretical electrolyte structure (I) of $AgBF_4$ dissolved in ethylene glycol dimethyl ether used as a model compound of PEO was obtained by the density functional theory (DFT) calculation and is presented in Figure 4.26 [41].

When one ethylene molecule (denoted by $CH_2{=}CH_2\{1\}$) is introduced to the electrolyte (I), a new complex (II) is formed by substituting one of the two interaction sites of the anion ligand (BF_4^- here) with $CH_2{=}CH_2\{1\}$, which is referred to as the *complexation step*. This step is a diffusion-controlled type without activation as expressed by the 1st step of Figure 4.27.

The ethylene molecule, $CH_2{=}CH_2\{1\}$, already complexed to the silver ion, undergoes exchange with a newly approaching ethylene molecule, $CH_2{=}CH_2\{2\}$, by an activated process through a transition state (III) following a normal push-pull SN_2 type mechanism. This step is called the *exchange step*, as depicted in Figure 4.27. The transition state exhibits a characteristic trigonal bipyramid 5-coordinate structure, as commonly observed in SN_2 reactions. Consequently, the reversible reaction mechanism of olefins with silver cations in polymer electrolyte membranes is delineated by the two steps depicted in Figure 4.27: the complexation and exchange reaction steps [41].

The first complexation step is apparently not completely reversible and slow to form silver-olefin complexes as an intermediate complex (II); the second exchange step is rapid and reversible, which appears to be the key step in determining facilitated olefin transport. Note that the second step is only observed where the concentration gradient is present.

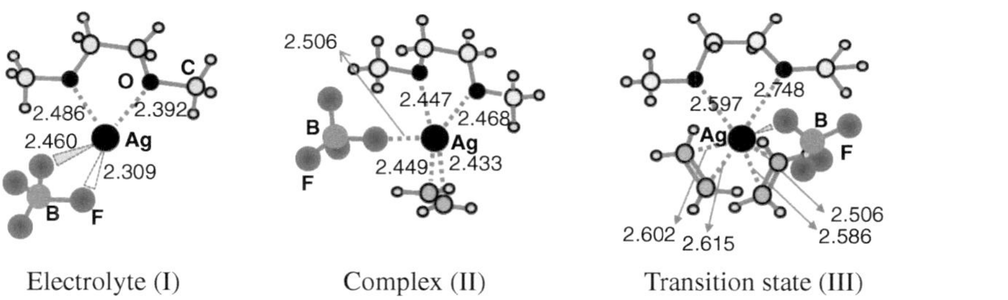

Figure 4.26 Optimized structures of (a) electrolyte (I) formed between ethylene glycol dimethyl ether used as a model compound of PEO and $AgBF_4$, and (b) complex (II) between electrolyte (I) and ethylene molecule ($CH_2{=}CH_2\{1\}$). (c) Transition state structure (III) for the exchange reaction of the bound $CH_2{=}CH_2\{1\}$ in complex (II) with another $CH_2{=}CH_2\{2\}$. Bond lengths are in Å. *Source*: Adapted from [41].

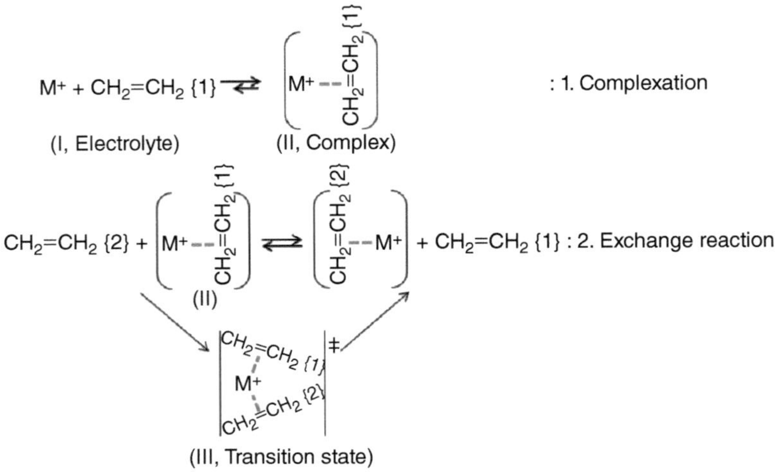

Figure 4.27 Two-step reaction mechanism of silver ion with olefin in polymer electrolyte membranes: 1. nonactivated *complexation* and 2. activated *exchange* steps. *Source*: Adapted from [41].

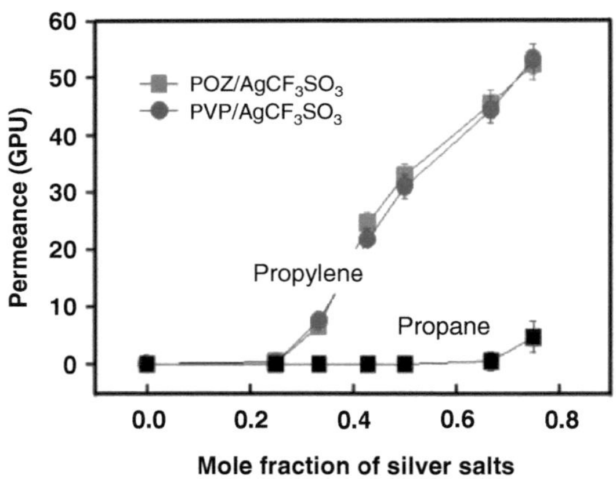

Figure 4.28 Permeances Q of propylene and propane as a function of the mole fraction of $AgCF_3SO_3$. *Source*: Adapted from [31].

4.5.1.6 Facilitated Olefin Transport

In silver-polymer electrolytes, for instance, silver ions interact with olefins such as propylene and ethylene specifically (but not with corresponding paraffins such as propane and ethane) and reversibly to act as an olefin carrier for facilitated olefin transport.

Figure 4.28 shows that the propylene permeance Q increases nearly linearly with the concentration of $AgCF_3SO_3$ dissolved in poly(vinyl *N*-pyrrolidone) (PVP) up to 50 GPU when the mole fraction of silver salt is greater than 0.25, whereas the propane permeance looks nearly unchanged up to the silver salt mole fraction of 0.5. However, it decreases slightly with the silver salt concentration up to 5×10^{-3} GPU as already shown in Figure 4.23. Interestingly, only when the coordination number of the silver ion is less than 3, corresponding to the mole fraction of the silver salt of 0.25, the silver ions are active as an olefin carrier, whereas they are not active when the coordination number of Ag^+ is equal or bigger than 3 [31]. These results suggest that the equilibrium coordination number of Ag^+ in these experimental ranges could be 3, only below which Ag^+ may have empty sites for olefin coordination.

It is worth noting that we employ the term "permeance" Q instead of "permeability" P here to characterize the permeation properties of the composite membrane. This choice is based on the readily available definition of Q as $Q = J/\Delta p = P/l$ where J represents the flux and Δp is the pressure difference across the membrane with thickness l. The Q is commonly expressed as a gas permeation unit (GPU), where $1\ \text{GPU} = 1.0 \times 10^{-6}\ \text{cm}^3(\text{STP})/\text{cm}^2\ \text{sec cmHg}$. In this context, the precise determination of the coated selective layer thickness l required for calculating permeability P is not necessary when evaluating permeance Q. Further details on this topic will be provided in Chapters 5 and 6.

The nearly linear increase in Q with the carrier concentration at mole ratios of silver salt higher than 0.25, as shown in Figure 4.28 is consistent with the concentration fluctuation model of Equations (4.11) and (4.12), demonstrating a typical behavior of facilitated transport phenomena. Here, silver ions *specifically* and *reversibly* interact with olefins to act as an olefin carrier, but not with paraffins having no double bonds. The decreased propane permeance is primarily due to the increases in the polymer chain rigidity by both cross-links of PVP or POZ chains and the dangling effects of Ag^+ as described previously and illustrated in Figures 4.22 and 4.23 [31].

4.5.2 Surface-activated Metallic Nanoparticular Carriers

Silver ion Ag^+ has positive charge (+1) so that it is coordinated by the double bond of olefin molecule to act as an olefin carrier for facilitated olefin transport as described previously. However, since metallic silver Ag^0 does not have any positive charge to interact with olefin, anticipating that Ag^0 is not capable of playing as an effective olefin carrier.

Nonetheless, the interaction enthalpies, ΔH, between small silver nanoparticles (Ag NPs) comprising 10 silver atoms and ethylene were estimated by theoretical *ab initio* calculations to be up to −4.48 kcal/mol, depending on the position of ethylene (Figure 4.29) [26, 31]. This result suggested that the Ag NPs could act as another olefin carrier when they have a finite positive charge (δ^+) on their surface by changing the chemical environment (Figure 4.30). One simple way to induce the surface positive charge δ^+ is to utilize an electron-accepting polarizer.

In this context, this section focuses primarily on silver nanoparticular (Ag NPs), as they exhibit the highest carrier activity toward olefins among the metallic nanoparticles such as gold, copper, and silver.

4.5.2.1 Metallic Nanoparticle Carriers

The first experimental evidence of the facilitated olefin transport was reported in poly(ethylene-propylene-diene) rubber (EPR) membranes containing Ag NPs activated by *p*-benzoquinone (*p*-BQ), called polarizer hereafter [24]. Since then, other

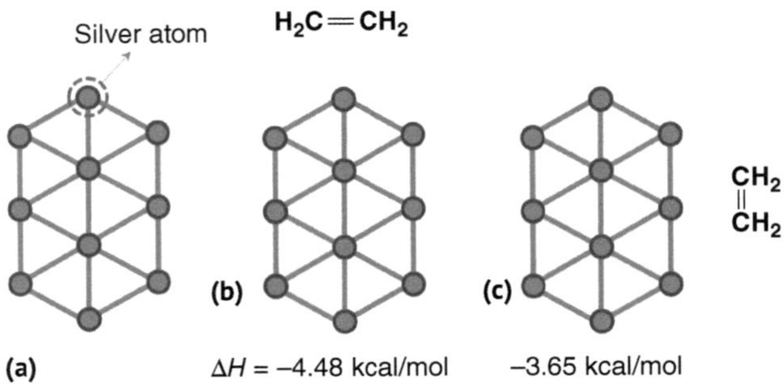

Figure 4.29 (a) The theoretical structure of a silver nanoparticle containing 10 atoms; (b–c) the two different complexation structures and corresponding interaction enthalpies with ethylene (ΔH). Complex structures with the olefin in the (b) upper edge, and (c) side of the silver nanoparticle. *Source*: Ref. [24]/John Wiley & Sons.

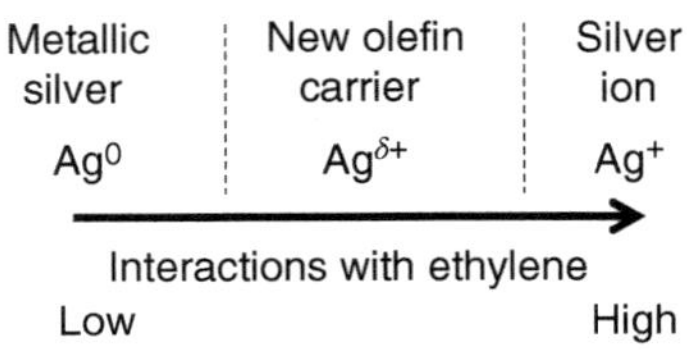

Figure 4.30 The effect of the positive charge density of metallic silver Ag^0 on its interactions with ethylene. The partial positive charge (δ^+) on silver nanoparticles is induced by utilizing a polarizer. *Source*: Y.S. Kang.

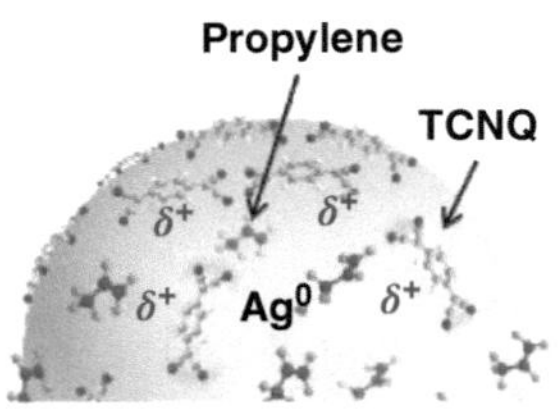

Figure 4.31 Schematic structure of surface-activated silver nanoparticle (Ag NPs) by 7,7′,8,8′-tetracyanoquinodimethane (TCNQ) and propylene adsorption on the positively charged δ^+ surface. *Source*: Y.S. Kang.

metallic nanoparticles such as copper [42, 43], and gold [44, 45] with the help of various polarizers also demonstrated the olefin carrier activity for facilitated olefin transport.

Metallic nanoparticles have additional merit as an olefin carrier, compared to metallic ions, since oxidation of metals would hardly occur whereas metallic ions, in particular silver and copper ions, are chemically labile to be reduced to become inactive [46]. In other words, they are much more stable chemically and electrochemically, which is critically important for long-term practical applications.

4.5.2.2 Polarizers for Silver Nanoparticles

Even though the interaction enthalpy between a small Ag NPs containing 10 silver atoms and ethylene calculated by the *ab initio* calculation was substantial (−4.48 to −1.42 kcal/mol), it could be further increased by incorporating polarizers on their surface to induce a partial positive charge δ^+ [24] as shown in Figure 4.31. Therefore, polarizers play critically important roles in inducing a partial positive charge δ^+ on the nanoparticular surface and thereby an olefin carrier function. Electron-accepting polarizers successfully used are *p*-BQ [24], 7,7′,8,8′-tetracyanoquinodimethane (TCNQ) [47], nitro-aromatic compounds such as dinitrotoluene and dinitrobenzene [48] and ionic liquids such as 1-butyl-4-methylimidazolium tetrafluoroborate [Bmim][BF_4] [49, 50]. Interestingly, tetrathiafulvalene (TTF),

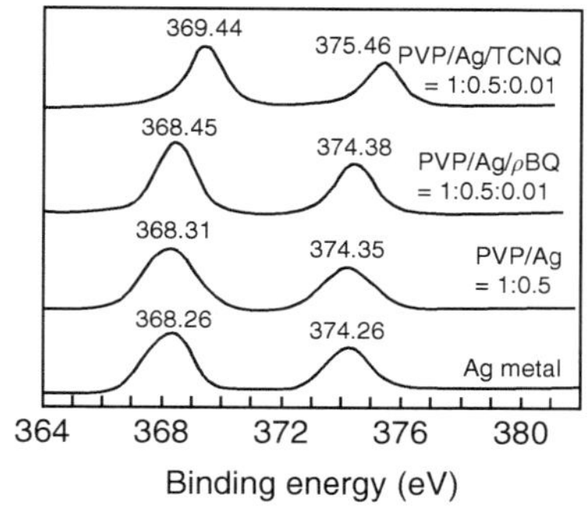

Figure 4.32 XPS spectra to measure the binding energy of silver atoms depending on their environment. *Source*: Ref. [47]/John Wiley & Sons.

which is known to be an electron donor, also acts successfully as a polarizer [51]. Therefore, the surface charge density δ^+ of Ag NPs and their chemical activity toward olefins can be readily tuned by choosing appropriate polarizer. In particular, the increased chemical activity due to the increased surface positive charge density results in favorable interactions of Ag NPs toward olefin, allowing Ag NPs to be a novel olefin carrier for facilitated transport.

The induction of positive charge on Ag NPs surface is characterized by measuring the binding energy of Ag atoms by X-ray photoelectron spectroscopy (XPS). Figure 4.32 exhibits XPS spectra informing that the binding energy increases from 368.26 eV for neat Ag NPs to 368.45 eV with *p*-BQ and to 369.44 eV with TCNQ, indicating that both *p*-BQ and TCNQ are effective in inducing partial positive charge [47].

4.5.2.3 Solubility of Propylene

Figure 4.33a shows that the amount of propylene adsorbed on the surface-activated Ag NPs increases with the concentration of the polarizer of TCNQ on the surface, and reaches a maximum value of ~250 mg/100 mg of Ag NPs, followed by the decrease at high TCNQ concentrations, seemingly due to the multiple TCNQ layer formation on their surface [52]. On the other hand, the solubility of propane, which hardly interact with the surface-activated silver nanoparticles, monotonically decreases slightly with the TCNQ concentration as expected. The binding energy of Ag $3d_{5/2}$ orbital, as a measure of the positive charge density on the Ag NP surfaces, was measured by XPS. Figure 4.33b shows that the binding energy increases sharply with the concentration of TCNQ added, reaches a maximum at around 0.3 molecules/nm^2, which is consistent with the propylene solubility data shown in Figure 4.33a [52].

4.5.2.4 Correlation Between Charge Density and Solubility

The amount of propylene adsorbed on the surface-activated Ag NPs by TCNQ has a strong linear correlation with the binding energy of Ag $3d_{5/2}$ (Figure 4.34), which represents a measure of the positive charge density δ^+ on their surface. In other words, the positive charge density characterized by the binding energy of Ag NPs mostly determines the amount of propylene adsorbed on the surface-activated Ag NPs by TCNQ.

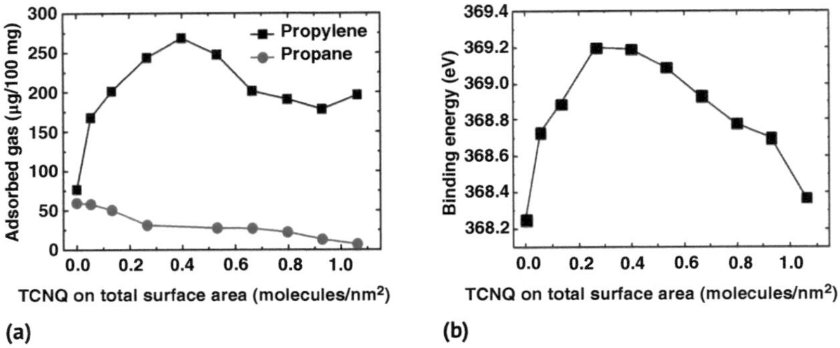

Figure 4.33 (a) The amount of the adsorbed propylene and (b) the binding energy change of Ag $3d_{5/2}$ orbital as a function of the 7,7′,8,8′-tetracyanoquinodimethane (TCNQ) concentration on surface-activated silver nanoparticles. *Source*: Adapted from [52].

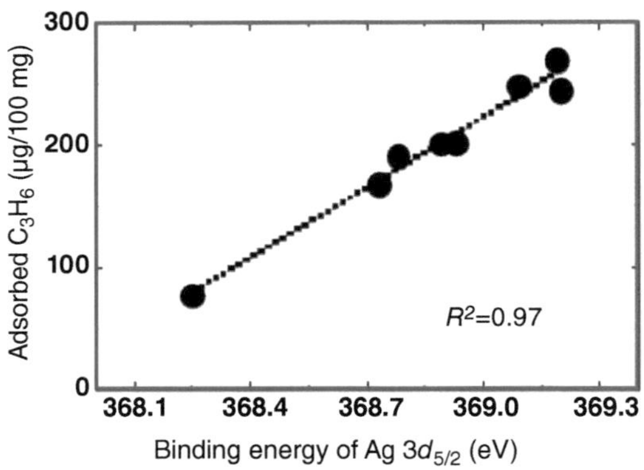

Figure 4.34 Correlation between the amount of the propylene adsorbed on the surface-activated silver nanoparticles by 7,7′,8,8′-tetracyanoquinodimethane (TCNQ) and the binding energy of Ag $3d_{5/2}$ orbital. *Source*: Adapted from [52].

4.5.2.5 Reversible Interactions

Figure 4.35 shows the Fourier transform infrared (FT-IR) spectra for 1/1/0.85 EPR/Ag NPs/*p*-BQ composite as a function of propylene sorption or desorption time along with two reference curves of pure propylene and neat EPR/Ag NP/*p*-BQ composite without propylene exposure [24]. Here, EPR stands for poly(ethylene-propylene-diene) rubber and the diameter of Ag NPs used was ~70 nm. When the composite membrane is exposed to propylene (at $t = 0$), two new peaks for the C═C stretching vibration of propylene at 1,664 and 1,640 /cm were apparent, which were in the same positions as those for pure propylene. However, their intensities decreased shortly and a new peak at 1,649 /cm became dominant with increasing propylene exposure time. In other words, the free propylene peaks at 1,664 and 1,640 /cm shifted to 1,649 /cm, presumably owing to partial electron transfer from the C═C bond of propylene to the partially positively charged surface of the silver nanoparticles. Following propylene desorption for 5 minutes, the peak at 1,649 /cm significantly decreased in its intensity but not completely disappeared in the spectra, indicating that the interaction is strong enough for propylene not to be desorbed even in a vacuum environment. Note that the propylene peaks in the EPR/Ag NP composite

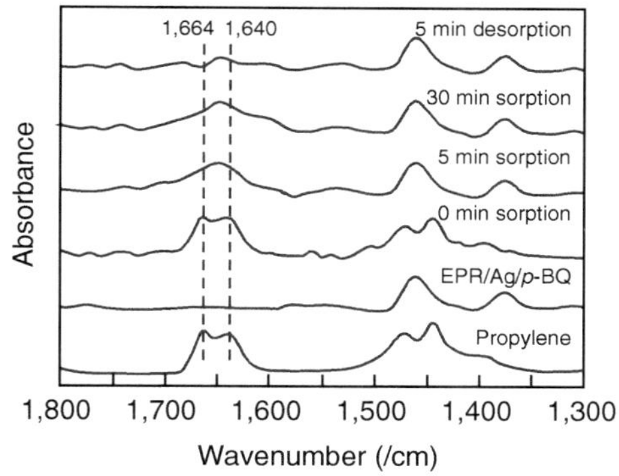

Figure 4.35 Fourier transform infrared spectra for 1/1/0.85 poly(ethylene-propylene-diene) rubber as a function of propylene sorption or desorption time. *Source*: Ref. [24]/John Wiley & Sons.

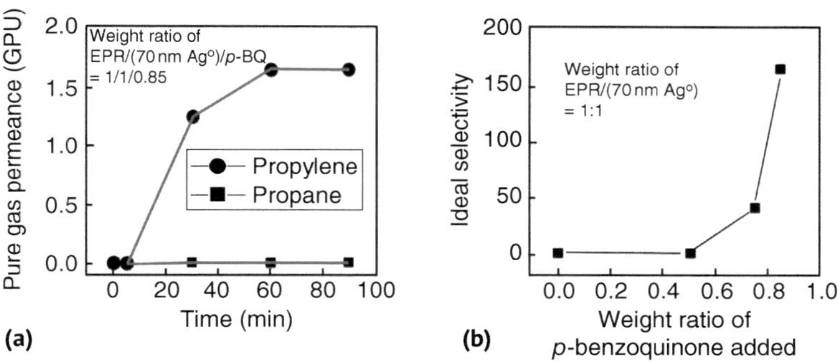

Figure 4.36 (a) Pure gas permeance through composite membranes containing metallic silver nanoparticles (Ag NPs) activated by *p*-benzoquinone (*p*-BQ) as a polarizer, subsequently dispersed in poly(ethylene-propylene-diene) rubber (EPR), and coated on an asymmetrically microporous polysulfone support; and (b) the change in the ideal selectivity as a function of the concentration of *p*-BQ. *Source*: Ref. [24]/John Wiley & Sons.

without *p*-BQ instantaneously disappeared in a vacuum environment (the data not shown). The FT-IR spectra suggested that the interactions between the surface of Ag NPs activated by *p*-BQ and propylene are reversible, expecting an olefin carrier activity for facilitated propylene transport.

4.5.2.6 Facilitated Olefin Transport

When Ag NPs activated by *p*-BQ are dispersed in EPR and subsequently coated on an asymmetrically microporous polysulfone support to make composite membranes, the pure propylene permeance Q_1 increased with time as shown in Figure 4.36a and reached the high permeance whereas the propane permeance Q_2 does not changed, but maintained very low. These results demonstrated the facilitated olefin transport due to the carrier action of the surface-activated Ag NPs by *p*-BQ toward only propylene having double bond, but not toward propane. Consequently, the ideal selectivity of propylene over propane (Q_1/Q_2) increased sharply up to higher than 150 at the *p*-BQ weight ratio of 0.8 (Figure 4.36b).

When utilizing the ionic liquid of 1-butyl-3-methyl-imidazolium tetrafluoroborate $[Bmim][BF_4]$ as both a polarizer to activate the surface of Ag NPs and a dispersion

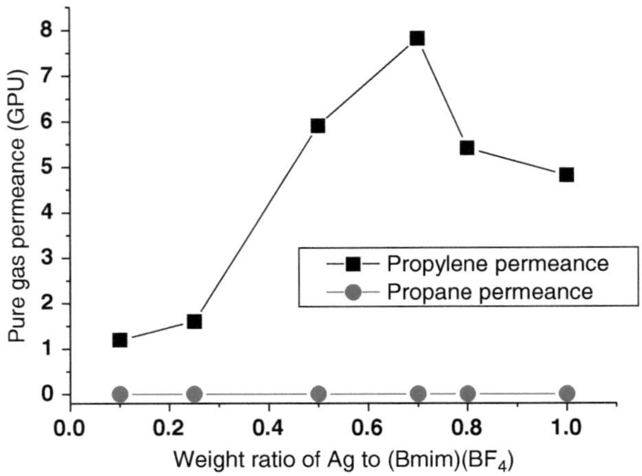

Figure 4.37 Pure gas permeances through [Bmim][BF_4] membranes with varying weight ratio of Ag nanoparticle to [Bmim][BF_4]. *Source*: Ref. [50]/American Chemical Society.

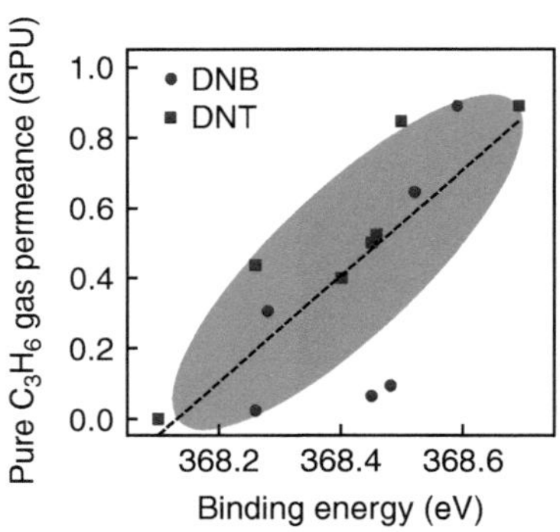

Figure 4.38 Influence of the binding energy of silver nanoparticles (Ag NPs) on the pure gas permeance by varying the surface polarizer: dinitrobenzene (DNB) and dinitrotoluene (DNT). *Source*: Ref. [48]/Springer Nature.

medium for Ag NPs, the pure propylene permeance exhibited a notable increase, reaching 7.8 GPU with a weight ratio of Ag NPs to [Bmim][BF_4] of 0.7, while the propane permeance exhibited minimal change (Figure 4.37) [49, 50]. Moreover, the ideal selectivity peaked at 780. However, at weight ratios exceeding 0.7, propylene permeance decreased as the Ag NPs content increased, likely due to Ag NPs aggregation, presumably leading to a loss in carrier activity. These findings indicate that the ionic liquid [Bmim][BF_4] can also serve as a polarizer for Ag NPs for facilitated olefin transport.

4.5.2.7 Correlation Between Charge Density and Facilitated Transport

To further investigate the influence of the positive charge density δ^+ of Ag NPs for facilitated transport, the pure propylene permeance was measured and plotted as a function of the Ag binding energy (Figure 4.38), which represents the positive charge density δ^+ of the Ag NPs surface. The pure propylene permeance exhibits a nearly linearly increase with the binding energy of Ag atoms, which suggests a rather strong correlation between them. While olefin solubility increases by a factor of 3–4 (Figure 4.33), the olefin permeability experiences a more than 10-fold increase (Figure 4.38). This discrepancy implies that the primary factor contributing to the enhancement in the propylene permeability is the carrier-mediated transport by the improved olefin carrier action of highly positively charged Ag atoms. This indicates that the facilitated transport phenomenon is a highly powerful tool for enhancing the permeation of a given gas only.

4.6 Facilitated CO_2 Transport in Solid and Quasi-solid States

CO_2 has a quadrupole moment [53], but no dipole moment, allowing it to interact with both acids and bases due to its amphoteric nature. This property makes both acids and bases promising candidates as CO_2 carriers for facilitated transport.

$$CO_2 + C \underset{k_2}{\overset{k_1}{\rightleftarrows}} [CO_2 \cdot\cdot C] \qquad K = k_1/k_2$$

Key concepts in designing advanced CO_2 carriers involve achieving both a fast backward reaction rate constant (k_2) and a favorable equilibrium constant (K), as well as maintaining a high carrier concentration (C_C), in line with the concentration fluctuation model (Equations (4.11) and (4.12) in Section 4.2.4 of this chapter) [4, 5].

For instance, Lewis bases like alkyl amines have long been utilized as CO_2 carriers. Similarly, cations (e.g. metallic ions, M^+) show the Lewis acidic characters and can also serve as CO_2 carriers, remarkably enhancing CO_2 separation performance. Interestingly, free anions (X^-) formed alongside these cations can also play as Lewis basic CO_2 carriers, though their roles and mechanisms are not yet well studied and understood.

These CO_2 carriers are incorporated into either a liquid water medium, forming supported or immobilized liquid membranes (SLMs or ILMs) or into a polymeric medium, fabricating facilitated CO_2 transport membranes in solid state. Water often plays crucial roles in enhancing separation performance, so a small amount is frequently added to polymer membranes, resulting in a quasi-solid-state membrane. One common approach to introducing water into these membranes is through feeds containing partially dehydrated flue gases.

Facilitated CO_2 transport membranes in solid and quasi-solid states offer significant advantages over SLMs for practical applications. Liquid solvents, such as water in SLMs, may evaporate during operation, compromising membrane integrity and long-term performance. In contrast, solid polymer membranes are more robust, allowing the selective layer to be made extremely thin (as thin as 1 μm or less in composite membranes), compared to the millimeter-scale thickness of SLMs. Additionally, solid-state polymer membranes enable operation under high pressure or vacuum, which is highly advantageous for practical use.

This section first examines the basic chemistry of reversible CO_2 interactions with both Lewis bases and acids in aqueous solutions. It then explores the reversible interactions and applications of these carriers in solid and quasi-solid-state membranes containing small amounts of water.

4.6.1 Interactions of CO_2 with Lewis Bases and Acids in Aqueous Solution

In a liquid membrane using a water medium, water serves dual functions: first, as a transport medium, and second, as a reactant to form bicarbonate HCO_3^-, a key CO_2 carrier for facilitated CO_2 transport. Consequently, the formation kinetics and concentration of HCO_3^-, are critical in improving the CO_2 separation performance. We will delve deeper into the effects of these factors on basic principles in conjunction with fundamental CO_2 chemistry and facilitated CO_2 transport.

4.6.1.1 Liquid Membranes of Neat Water

Figure 4.39a illustrates that the CO_2 permeability increases with the decrease in applied pressure, a characteristic feature for facilitated transport, across a neat water SLM. Facilitated CO_2 transport occurs because water reversibly reacts with CO_2 to form bicarbonate ions, HCO_3^-, and therefore, its amount is vital in improving the CO_2 permeability, which is generated by following chemical reactions:

$$CO_2 + H_2O \leftrightarrow H_2CO_3 \tag{4.16}$$

$$H_2CO_3 \leftrightarrow HCO_3^- + H^+ \tag{4.17}$$

$$HCO_3^- \leftrightarrow CO_2 + OH^- \tag{4.18}$$

$$HCO_3^- \leftrightarrow CO_3^{2-} + H^+ \tag{4.19}$$

The HCO_3^- ions are formed by two sequential reactions of Equations (4.16) and (4.17) at upstream side, which then diffuse to the opposite side of the aqueous membrane, where CO_2 is released (Equation (4.18)) [55]. Figure 4.39b shows that the CO_2 transports through two different pathways: Fickian diffusion of freely dissolved CO_2 and carrier-mediated transport through HCO_3^-, consequently resulting in facilitated CO_2 transport. Therefore, the formation and transport of HCO_3^- limit the overall kinetics of CO_2 transport inevitably [56].

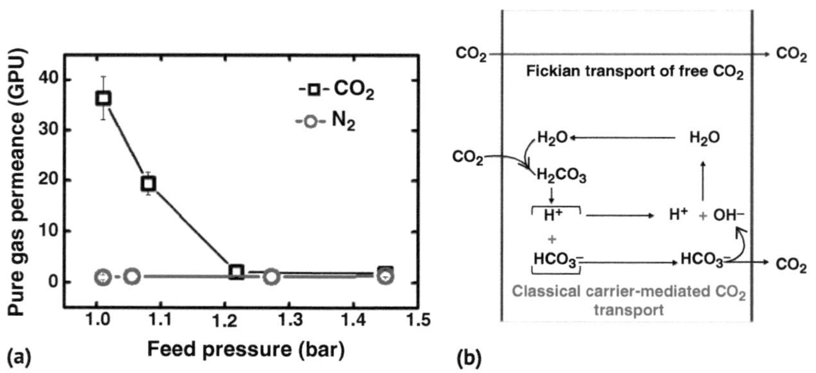

Figure 4.39 (a) Facilitated transport of CO_2 and Fickian transport of N_2 through neat water; and (b) schematic mechanism of facilitated CO_2 transport in neat water. *Source:* Adapted from [54].

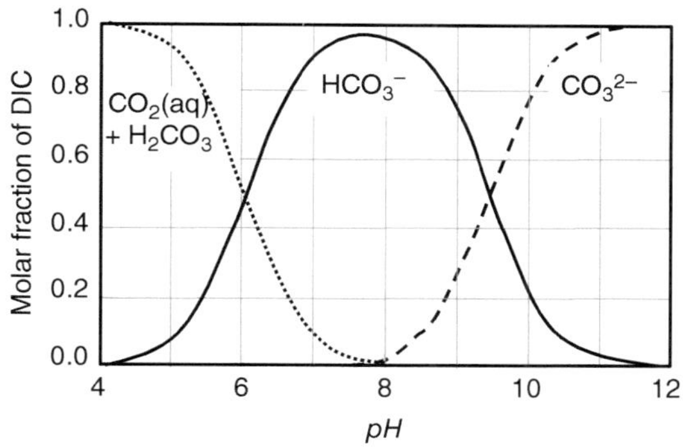

Figure 4.40 The mole fraction of carbonate species vs pH at 25 °C with 5,000 ppm salinity (Bjerrum plot) [Bjerrum plot, Wikipedia]. Note that the concentration of H_2CO_3 is assumed to be negligibly small at every pH.

Since CO_2 in aqueous solution is converted into four different species: dissolved bare CO_2, bicarbonate HCO_3^-, carbonate CO_3^{2-}, and carbonic acid H_2CO_3 and their concentrations depend on pH of the solution. Note that the concentration of H_2CO_3 is commonly negligibly small at every pH, compared to the concentration of dissolved free CO_2. The concentration change depending on pH is illustrated in the Bjerrum plot (Figure 4.40), indicating the high HCO_3^- concentration at a pH range between 6 and 9. Here, dissolved inorganic carbon (DIC) refers to the total concentration of carbon species in water, including aqueous bare CO_2(aq), bicarbonate (HCO_3^-), carbonate $\left(CO_3^{2-}\right)$, and carbonic acid (H_2CO_3).

Consequently, CO_2 transport can be mostly determined by both dissolved bare CO_2 and HCO_3^- with the Fickian and carrier-mediated transport mechanisms, respectively, as shown in Figure 4.39b, where the latter becomes dominant in neutral pH solution (~7–8) by the formation of a large amount of HCO_3^- [57].

Interestingly, when the pH level becomes much higher (pH > 8), the production rate of HCO_3^- will be markedly improved (0.03–8,500 L/sec mol) by the direct reaction of OH^- and CO_2 ($OH^- + CO_2 \rightarrow HCO_3^-$) [55]. It has been thus demonstrated that the enhancement in the overall kinetics improves the CO_2 permeation properties, even though the HCO_3^- concentration is nearly unchanged [58].

4.6.1.2 Liquid Water Membranes with Lewis Bases

Among diverse Lewis bases, the use of amine compounds has emerged as the most prevalent and effective approach [58]. The reactivity of amine compounds against CO_2 in an aqueous solution is inherently associated with their chemical structure. For instance, primary and secondary amines exhibit distinct reaction mechanisms (Figure 4.41a), compared to tertiary amines (Figure 4.41b) where water is unavoidably required.

Primary and secondary amines may involve the following chemical reactions, as shown in Reactions (4.20)–(4.22), to facilitate the formation of bicarbonate HCO_3^-.

$$R_1R_2NH + CO_2 \leftrightarrow R_1R_2NH^+CO_2^-(\text{or } R_1R_2NCO_2H)\ (\text{carbamic acid}) \quad (4.20)$$

$$R_1R_2NCO_2H + R_1R_2NH \leftrightarrow R_1R_2NCO_2^-\ (\text{carbamate}) + R_1R_2NH_2^+ \quad (4.21)$$

$$R_1R_2NCO_2^- + H_2O \leftrightarrow R_1R_2NH + HCO_3^- \quad (4.22)$$

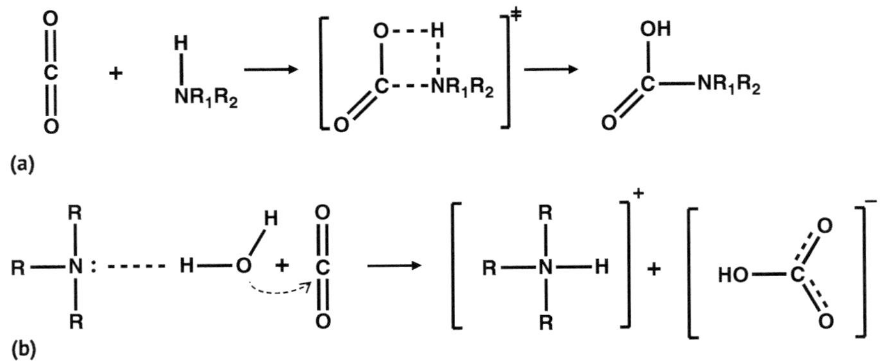

Figure 4.41 (a) Direct formation of carbamic acid via a four-membered mechanism. *Source*: Adapted from [59]. (b) Formation mechanism of bicarbonate using a tertiary amine with water. *Source*: Adapted from [60].

Reaction (4.20) serves as a rate-determining step regardless of whether water is present. In contrast, Reactions (4.21) and (4.22) represent swift proton transfer reactions [58, 61]. Here, HCO_3^- is derived from the carbamate anion $R_1R_2NCOO^-$ (Reaction (4.22)). It is worth noting that secondary amines exhibit enhanced basicity and greater reactivity toward CO_2 compared to primary amines, owing to the electron-donating effect of the alkyl group [62].

Conversely, tertiary amine groups cannot directly react with CO_2 to form a carbamate. Instead, an alternative reaction takes place to generate bicarbonate ions with the assistance of water molecules, as described in Reactions (4.23) and (4.24).

$$R_1R_2R_3N + H_2O \leftrightarrow R_1R_2R_3NH^+ + OH^- \tag{4.23}$$

$$CO_2 + OH^- \leftrightarrow HCO_3^- \tag{4.24}$$

$$R_1R_2R_3N + CO_2 + H_2O \leftrightarrow R_1R_2R_3NH^+ + HCO_3^- \tag{4.25}$$

Furthermore, under the influence of water, a tertiary amine can also react with dissolved CO_2, leading to its conversion into HCO_3^- according to Reaction (4.25), as visually represented in Figure 4.41b [63]. Figure 4.42a and b provides a schematic overview of the creation of HCO_3^-, which serves as a carrier for CO_2 in carrier-mediated transport.

Furthermore, amine compounds act as proton ion sinks as seen in Reaction (4.27) that originate from the dissociation of water through Reaction (4.26). This leads to a favorable shift in equilibrium, resulting in a greater concentration of OH^- ions, which, in turn, promotes the formation of HCO_3^- through Reaction (4.28) [57]:

$$H_2O \leftrightarrow H^+ + OH^- \tag{4.26}$$

$$R_1R_2R_3N + H^+ \leftrightarrow R_1R_2R_3HN^+ \tag{4.27}$$

$$CO_2 + OH^- \leftrightarrow HCO_3^- \tag{4.28}$$

As a short summary, the inclusion of amine compounds in an aqueous membrane is expected to enhance the production of HCO_3^-, a crucial factor in enhancing CO_2 permeability. Despite the typically strong CO_2 separation performance displayed by

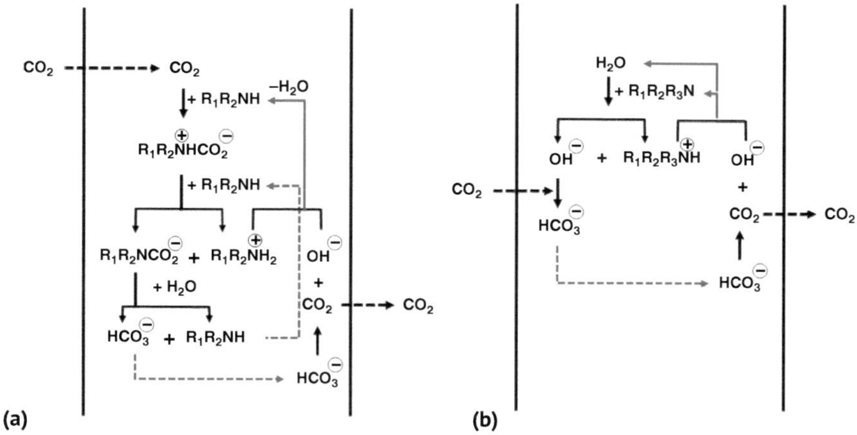

Figure 4.42 Schematic illustrations of the formation of HCO_3^- and the carrier-mediated CO_2 transport with the help of (a) primary and secondary amines; and (b) tertiary amine. *Source*: Y.S. Kang.

traditional aqueous amine-based SLMs, their usability is severely constrained due to the loss and deterioration of amine compounds and water solvent within these membranes [64, 65].

4.6.1.3 Liquid Water Membranes with Lewis Acids

Metallic cations in an aqueous solution form hydrated structures with amphoteric water, which exhibit increased acidity due to polarization of the water molecules. This acidity enables interactions with another amphoteric CO_2. For example, K^+ [A] in aqueous solutions acts as a Lewis acid, binding to the Lewis basic oxygens [B] of water to form a Lewis acid–base [AB] complex. Subsequently, the hydrogens of the bound water in the [AB] complex exhibit increased acidity [A], characterized by an enhanced partial positive charge (δ^+) as depicted in Figure 4.43a. This partial positive charge enables the bound water to interact with the Lewis basic oxygens [B with δ^-] of CO_2, leading to the formation of an [ABAB] complex [54].

This unique interaction for the formation of the [ABAB] complex allows the bound water to serve as a novel carrier for the facilitated CO_2 transport. The [ABAB] complex formation arises from the amphoteric functions of both water and CO_2, which helps play water as an effective novel CO_2 carrier for the additional carrier-mediated transport, thereby significantly improving the CO_2 permeance as shown in Figure 4.43b.

A significant CO_2 permeance increase from 40 GPU for neat water (Figure 4.39a) to 150 GPU (Figure 4.43b) is observed upon introducing metallic salts such as CF_3CO_2M (M^+ is Li^+, Na^+, or K^+) into an aqueous solution as a selective layer for SLMs at a CO_2 pressure of about 1 bar [54]. Such markedly enhanced CO_2 permeance up to fourfold is mostly attributed to the additional carrier-mediated transport by the formation of reversible [ABAB] complexes in addition to the classical carrier-mediated transport with the help of the bicarbonate HCO_3^- as described previously.

The formation of the [ABAB] complexes was confirmed through various nuclear magnetic resonance (NMR) spectroscopic techniques suggested by refs. [66, 67].

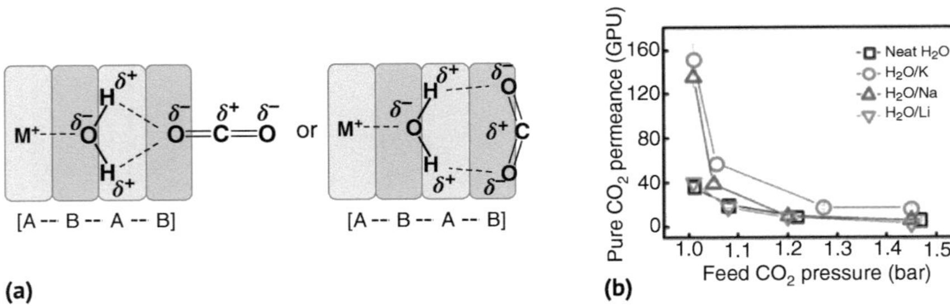

Figure 4.43 (a) Schematic illustration of the molecular structure of the [ABAB] complex involving a Lewis acidic metallic cation [A], Lewis basic oxygen [B] and Lewis acidic hydrogen [A] of water, and the Lewis basic CO_2 [B]. This diagram highlights the water coordinatively bound to metallic cations, serving as a novel CO_2 carrier for facilitated transport. *Source*: Y.S. Kang. (b) Pure CO_2 permeance through aqueous solutions containing various metallic salts tested as a function of applied pressure. *Source*: Adapted from [54].

The ^{39}K NMR spectra (Figure 4.44a) of the aqueous CF_3CO_2K electrolyte indicated shielding (lower ppm/higher field) of the K peak upon CO_2 addition, suggesting increased electron density of the K^+ ion [A] due to its coordination by oxygen atoms [B] of water molecule [54]. Similarly, the ^{1}H NMR spectra (Figure 4.44b) revealed CO_2 [B] complexation with hydrogen atoms of water [A] through shielding of the hydrogen signal. Conversely, the ^{13}C peak deshielding in CO_2-bubbled CF_3CO_2M (M = Li, Na, K) electrolytes highlighted the reduced electron density of carbon in CO_2 (Figure 4.44c), attributed to interactions between water's hydrogen [A] and CO_2's oxygen [B], functioning Lewis acidic bound water [A] as a CO_2 [B] carrier. Notably, K^+ ions showed the greatest deshielding, reflecting stronger hydrogen–oxygen interactions, which may be attributed to the different Lewis acidities toward the water molecules and their different spatial accessibilities to CO_2 [68]. The absence of CO_2 signal shielding implies negligible $CF_3CO_2^-$ and CO_2 interactions (Figure 4.44c). Additionally, CO_2 peak intensity in the ^{13}C NMR spectra of CO_2-saturated CF_3CO_2K (Figure 4.44d) decreased over time in ambient air, indicating the formation of specific and reversible [ABAB] complexes.

The formation of the [ABAB] complex is further confirmed by estimating self-diffusion coefficient of CO_2 using ^{13}C DOSY NMR spectroscopy (^{13}C diffusion ordered NMR spectroscopy). The diffusion coefficient for the dissolved bare CO_2 ($D = 1.91 \times 10^{-5}$ cm^2/sec) was higher than that ($D = 0.38 \times 10^{-5}$ cm^2/sec) for complexed CO_2 of the [ABAB] complex in the aqueous CF_3CO_2K electrolyte, suggesting the larger effective size due to the formation of the [ABAB] complex. These observations are consistent with the formation of a new class of CO_2 complexes, the [ABAB] complexes [54].

Based on these findings, a new facilitated CO_2 transport mechanism has been proposed for an aqueous solution containing CF_3CO_2M. In this case, two different CO_2 carriers are formed for two different carrier-mediated transports. Figure 4.45 illustrates that CO_2 transports across a membrane with three different mechanisms: (1) Fickian diffusion, (2) classical carrier-mediated transport aided by bicarbonate,

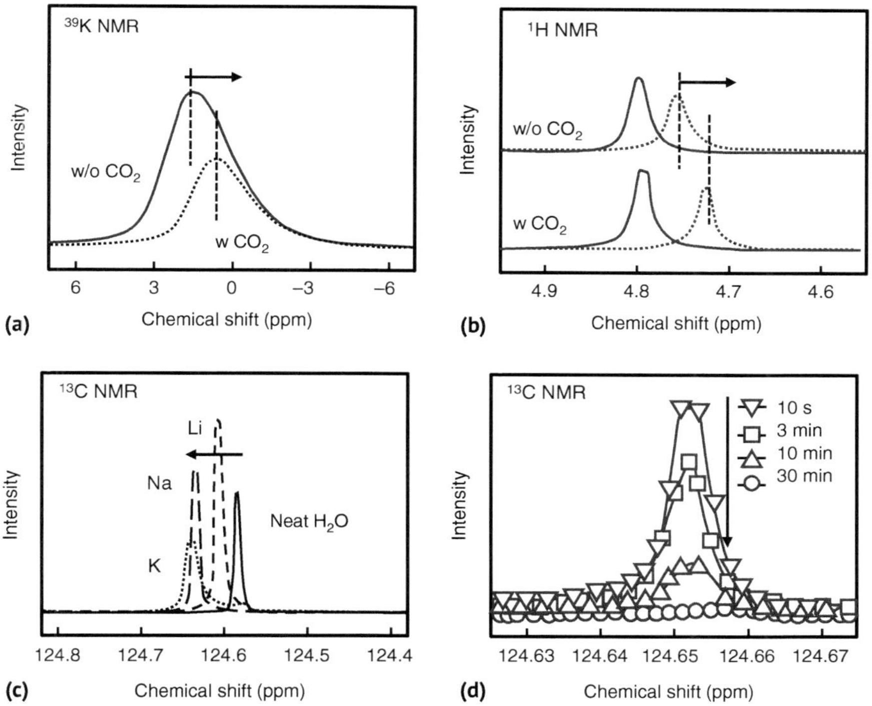

Figure 4.44 Nuclear magnetic resonance (NMR) spectra for the formation of a reversible [ABAB] complex between CO_2 and bound water. (a) ^{39}K NMR spectra of CF_3CO_2K electrolytes with and without CO_2; (b) ^1H NMR spectra of CF_3CO_2K electrolytes with and without CO_2 with a reference of neat water peak; (c) ^{13}C NMR spectra of CO_2-bubbled CF_3CO_2M (M = Li, Na, and K) aqueous electrolytes from 124.4 to 124.8 ppm; and (d) ^{13}C NMR spectra of CO_2-saturated aqueous CF_3CO_2K electrolyte (the solution thickness of 5 mm) as a function of the exposure time in ambient atmosphere (298 K, 40% RH). The spectrum of CO_2-bubbled aqueous CF_3CO_2K electrolyte was magnified five times in that range for clarity. *Source*: Adapted from [54].

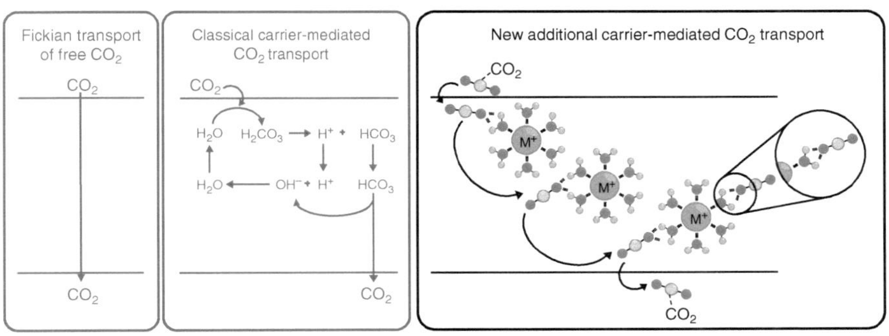

Figure 4.45 New facilitated CO_2 transport mechanism proposed for aqueous CF_3CO_2M membrane with two different CO_2 carriers, including a new additional carrier-mediated transport by formation of reversible [ABAB] complex. *Source*: Adapted from [54].

and (3) new additional carrier-mediated transport involving a water molecule coordinatively bound to the metallic cation M^+ by the formation of the reversible [ABAB] complex as a novel CO_2 carrier. Here, it is worth noting that the concept of water bound to metallic cations acting as a novel carrier for facilitated CO_2 transport was originally proposed by Park et al. [54].

Roles of water to improve CO_2 permeation properties in aqueous solution SLMs are summarized here. The most importantly, water involves in the formation of bicarbonate HCO_3^- directly and indirectly to facilitate the CO_2 transport. For instance, water and CO_2 directly react to generating HCO_3^-, which diffuses across SLM and is subsequently dissociated into CO_2 at the downstream side for facilitating CO_2 transport. Formation of HCO_3^- can be further indirectly accelerated through forming carbamates as intermediates obtained by reacting water with alkyl amines. In the presence of strong Lewis acid such as K^+, water bound to K^+ reversibly interacts with CO_2 to form the reversible [ABAB] complexes, rendering a new CO_2 carrier function to enhance CO_2 transport. As a result, water bound to K^+ plays as a new CO_2 carrier for facilitated CO_2 transport.

4.6.2 Polymer Membranes with Lewis Bases

Alkyl amines have been most commonly used as bases for CO_2. Their chemical structures dictate both basicity and steric hindrance, and consequently the CO_2 carrier properties, represented by K and k_2. Since the steric hindrance correlates with the basicity, they should be optimized to give high CO_2 permeability along with high selectivity of CO_2 over other gases such as nitrogen and methane. Moreover, steric hindrance, rather than amine basicity, significantly affects their reaction with CO_2 [61]. These intertwined factors make predicting amine basicity or even the order of basicity challenging.

4.6.2.1 Basicity

The basicity can be frequently determined by *pKa* in aqueous solution, where higher *pKa* indicates weaker acid and strong conjugate base, and also by the donor number (DN) (also referred to as "donicity") proposed by Gutmann, a quantitative measure for the tendency to donate electron pairs to counterpart acceptors [69]. The *pKa* and DN values of common alkyl amines are listed in Table 4.1, suggesting the alkyl group substitution effect.

For instance, in the ethylamine family, the basicity represented by *pKa* is influenced by both the ethyl substitution due to the electron-donating effect as well as the steric hindrance of the alkyl groups, leading to the order $NH_3 < CH_3CH_2NH_2 < (CH_3CH_2)_3N < (CH_3CH_2)_2NH$ (Table 4.1). It is worth noting that the steric hindrance substantially impacts basicity, with the secondary amine being more basic than the tertiary amine due to moderate steric hindrance in the former and high steric hindrance in the latter.

4.6.2.2 Steric Hindrance

The steric hindrance of amine compounds can be quantitatively expressed as the Taft's steric hindrance parameter, denoted as $\sum E_s$, which can be calculated by the

Table 4.1 *pKa* and donor number (DN) of amine compounds.

Amine	EA	DEA	TEA	DPEA	MEtA	DEtA	TEtA	PD	PZ	NH_3
pKa[a]	10.75	10.98	10.76	11.4	9.45	8.88	7.73	11.1	9.83	9.26
DN[b]	55.3	50.0	61.0	—	—	—	—	—	—	59.0
T_b(°C)	16.6	55.5	89.3	127	170	217	190	106	145	—

DEA, diethyl amine; DEtA, diethanol amine; DPEA, diisopropyl amine; EA, ethyl amine; MEtA, monoethanol amine; PD, piperidine; PZ, piperazine; TEA, triethyl amine; TEtA, triethanol amine [70].
[a]https://chem.libretexts.org/@go/page/36441.
[b]Gutmann defined the donor number, DN, as a quantitative measure for the tendency to donate electron pairs to acceptors, and its counterpart [69].

Table 4.2 The Taft's steric hindrance parameter $\sum E_s$ and amino group concentration of poly(vinyl amine) and its derivatives.

Polyamine	$\sum E_s$	Amino group (wt%)
Poly(vinyl amine)	0.93	37
Poly(*N*-methyl-*N*-vinyl amine)	1.00	26
Poly(*N*-isopropyl-*N*-vinyl amine)	1.86	18
Poly(*N*-*t*-butyl-*N*-vinyl amine)	2.67	15

Source: Adapted from [72].

group contribution method [71]. Table 4.2 lists the $\sum E_s$ value of poly(vinyl amine) and *N*-substituted poly(vinyl amine)s. As the attached alkyl group becomes bulkier, $\sum E_s$ is in an ascending trend, indicating an increased steric hindrance degree. Meanwhile, the amino group concentration in the polyamine is in a descending trend, caused by the added alkyl groups as listed [72].

4.6.2.3 Chemical Structure Effects on CO_2 Absorption and Desorption

The CO_2 absorption and desorption by varying amine structure are measured to quantify the CO_2 solubility and its reversibility, and their kinetics, including the forward k_1 and backward k_2 interaction constants. The CO_2 sorption isotherms by varying amine structure reveal that the CO_2 absorption increases PA-Si < TA-Si << SA-Si as illustrated in Figure 4.46a, where PA-Si stands for primary amine ((triethoxysilyl) propan-1-amine), SA-Si for secondary amine ((triethoxysilyl)-*N*-methyl-propan-1-amine), and TA-Si for tertiary amine ((triethoxysilyl)-*N*,*N*-dimethyl-propan-1-amine) [62]. The CO_2 absorption trend of alkyl amines is consistent with their basicity. Notably, sterically hindered TA-Si exhibits nearly complete reversibility in sorption and desorption behavior, crucial for facilitated transport, while PA-Si and SA-Si show partially irreversible behavior [62].

Figure 4.46b illustrates the sorption/desorption kinetics of amine-functionalized organosilica xerogel powders [61, 62, 73]. SA-Si exhibits higher solubility and faster

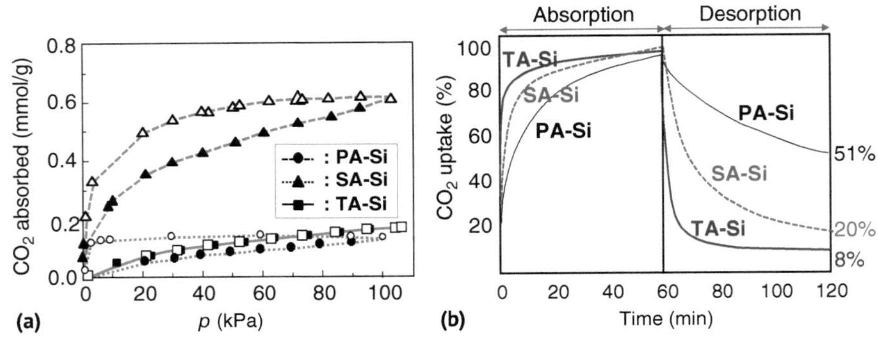

Figure 4.46 (a) Sorption (filled symbols) and desorption (open symbols) of CO_2 in amine-silica (Si) membranes on porous alumina (50% porosity) as a function of applied pressure at 35 °C for primary (PA-Si), secondary (SA-Si) amines, and tertiary amine (TA-Si). *Source*: Adapted from [62]. (b) Time course for the first cycle of normalized CO_2 absorption/desorption for amine-functionalized organosilica xerogel powders performed on thermogravimetric analysis/differential thermal analysis (TG/DTA) equipment. *Source*: Adapted from [61].

sorption and desorption rates compared to PA-Si. Interestingly, the sterically hindered TA-Si demonstrated much faster sorption and desorption rates, despite lower sorption capability than the secondary amine. As previously mentioned, the high desorption rate, represented by the high k_2 value, plays a vital role in enhancing separation performance.

4.6.2.4 Chemical Structure Effects on Separation Performance

Highly soluble secondary amines are often selected for CO_2 separation, with their steric hindrance tailored by modifying the structure of one alkyl group to enhance separation performance. A common strategy involves converting primary amines into secondary amines by introducing sterically controlled alkyl groups on the nitrogen atom. For example, poly(allyl amine) (PAA), which contains primary amines, was modified into secondary amines by adding alkyl groups ranging from moderately hindered isopropyl to highly hindered isobutyl and tert-butyl.

The CO_2 permeability and CO_2/N_2 selectivity results revealed the impact of steric hindrance on the separation performance, particularly when a small amount of water was introduced into the feed stream as shown in Figure 4.47. All secondary amines outperformed the primary amine PAA in both CO_2 permeability and CO_2/N_2 selectivity, emphasizing the role of steric hindrance. Among the tested alkyl amines, moderately hindered isobutyl amine showed the best overall performance [74].

Both results are consistent with the fact that the basicity and steric hindrance of the alkyl compounds should be balanced to optimize the separation performance. Practically, membranes containing moderately hindered amines appear to be a promising choice for achieving efficient separation of CO_2 molecules, primarily due to the optimized values for both thermodynamic K and kinetic k_2 parameters.

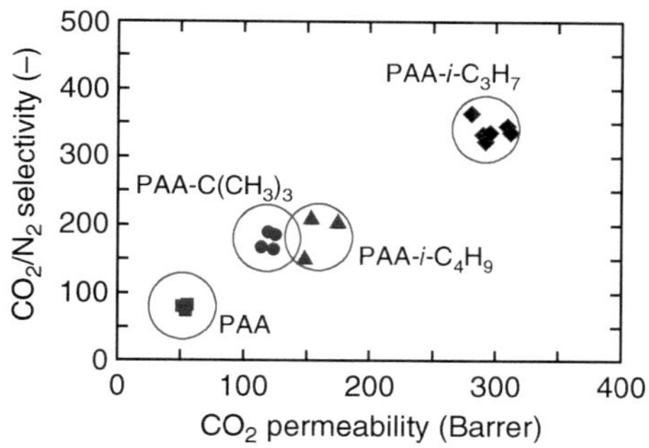

Figure 4.47 CO_2/N_2 selectivity versus CO_2 permeability for membranes containing 70 wt% polyamines/30 wt% cross-linked poly(vinyl alcohol) (PVA) at 110 °C and a feed pressure of 2 atm with water injection rates (0.03/0.03 cm^3/min: feed/sweep) using a feed gas composition of 20% CO_2, 40% H_2, and 40% N_2 (on dry basis). *Source*: Adapted from [74].

Table 4.3 Relative acidity of representative metallic ions.

	Li^+	Na^+	K^+	Mg^{2+}	Ca^{2+}	Al^{3+}	Ga^{3+}
z	1	1	1	2	2	3	3
r (pm)[a]	90	116	152	86	114	68	76
z^2/r (/nm)	11	9	7	47	35	132	118
Hydration number	4 and 6[b]	5 and 6[b]	6 and 7[b]	—	12	6	6

[a]Wikipedia, ionic radius.
[b]Ref. [68].

4.6.3 Polymer Electrolyte Membranes with Lewis Acids

Strong Lewis acidic species such as metallic cations or surface-polarized metallic nanoparticles can also directly interact with CO_2 even under dry environment without water. In this case, CO_2 can act as a Lewis base owing to the overlap between HOMO of CO_2 and LUMO of the strong Lewis acid, which facilitates the donation of an electron pair from CO_2 to the strong Lewis acidic species [58]. As a result, the acidity and concentration of the Lewis acid species can significantly influence the CO_2 transport properties, where the acidity of metallic ions is proportional to z^2/r with z and r being the charge and radius, respectively, as listed in Table 4.3. Among alkaline metals, the radius increases from Li^+, Na^+ to K^+ and consequently the z^2/r value decreases. Therefore, the z^2/r value of cations will be optimized to maximize the CO_2 carrier activity.

One simple and effective way to incorporate metallic cations in a polymeric matrix for solid-state membranes is to make polymer electrolyte where metallic salts are homogeneously dissolved in a polymeric matrix. For instance, polymer electrolyte membranes consisting potassium bis(trifluoromethanesulfonyl)imide KTf_2N dissolved in Pebax 2533 showed the typical facilitated CO_2 transport behavior as illustrated in Figure 4.48. Here, Pebax 2533 is a thermoplastic elastomer made of flexible polyether (poly(tetramethylene ether)) blocks and rigid polyamide 12 blocks. In

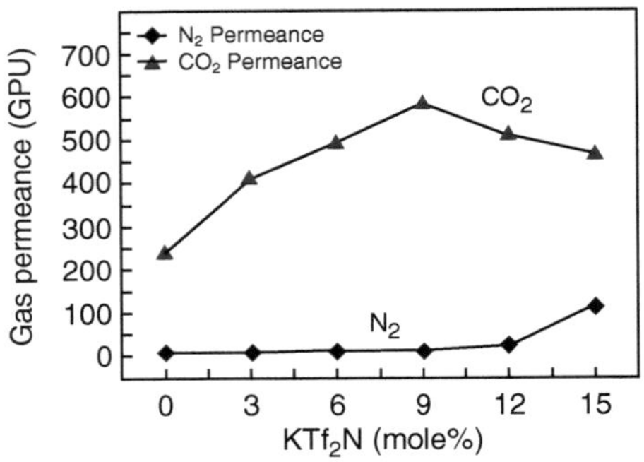

Figure 4.48 Separation performance of composite membranes consisting of KTf$_2$N/Pebax 2533 selective layer by varying concentration of KTf$_2$N. *Source*: Adapted from [75].

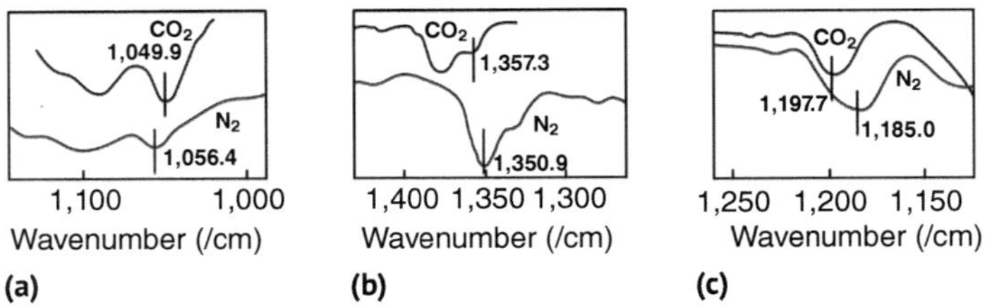

Figure 4.49 Fourier transform infrared (FT-IR) spectra of (a) S—N—S; (b) S=O; and (c) C—F bonds of KTf$_2$N dissolved in Pebax 2533 in the CO_2 (top) and N_2 environments (bottom). *Source*: Adapted from [75].

order to characterize the carrier activity for the facilitated CO_2 transport, the interactions of CO_2 with both K^+ and Tf_2N^- ions and their reversibility were directly characterized by FT-IR spectroscopy since CO_2 is known to be amphoteric to interact with both Lewis acids, such as metallic cations and Lewis bases, including the corresponding anion [75].

The interactions of CO_2 with the Tf_2N^- anion were investigated by FT-IR spectroscopy (Figure 4.49) [75]. The peaks in N_2 environment assigned for asymmetric stretching of S—N—S, S=O, and C—F bonds of the Tf_2N^- anion were observed at 1,055, 1,300–1,400, and 1,200 /cm, respectively [76, 77]. The FT-IR band at 1,056.4 /cm was redshifted to 1,049.9 /cm in the CO_2 environment (Figure 4.49a), indicating weakness of the S—N—S bond stretching. This suggests a weakened ionic interaction between the N^- atom and K^+ due to the interactions between CO_2 and Tf_2N^-. Blue shifts were observed for the S=O bond (Figure 4.49b) and the C—F bond (Figure 4.49c) of Tf_2N^- in the CO_2 environment, which suggests complexing multiple interactions of CO_2 with the various functional groups of [Tf_2N], most probably with the N atom.

The interaction of CO_2 with a negatively charged fluorinated anion of ILs based on the 1-alkyl-3-methylimidazolium [Bmim] cation in conjunction with either [PF_6] or [BF_4] was also verified by ATR-IR (attenuated total reflectance infrared spectroscopy) [78]. Furthermore, the band shift of the S—N—S bond in FT-IR was found to be reversible (the data not shown here). It was thus proposed that the interactions between [Tf_2N] and CO_2 could be possible and reversible, as illustrated in Figure 4.50. However, more basic studies on the direct interactions between CO_2 and Tf_2N^- (X^-) should be further carried out.

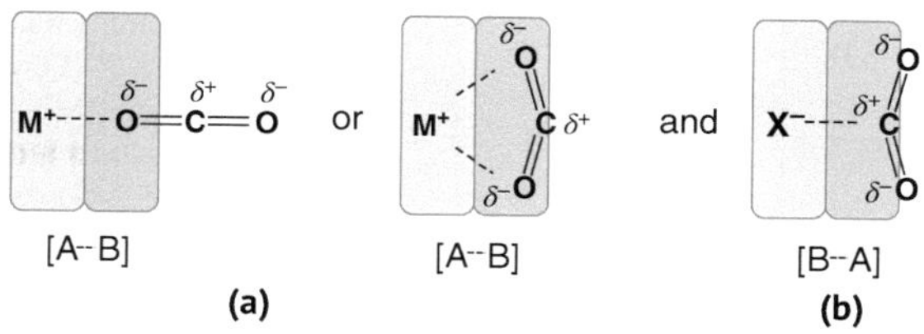

Figure 4.50 Schematic structures for interactions of CO_2 with both (a) cation M^+ and (b) anion X^- by forming [AB] complex. *Source*: Y.S. Kang.

In addition to metallic cations, quaternary ammonium cations, known to be highly interactive toward CO_2, have been used for CO_2 absorbents. For example, polyelectrolyte membranes containing quaternary ammonium groups show the extremely high CO_2/gas selectivity along with a very low CO_2 permeance for CO_2 separation [79]. In addition, these polyelectrolyte membranes have been used for removing other acid gases such as H_2S from gas mixtures, and the transport of H_2S by reversible reaction was analogous to that for CO_2 [80].

4.6.3.1 Effects of Water in Quasi-solid Membranes

When the small amount of water is present inside of the membranes containing metallic salts such as polymer electrolyte membranes (PEMs), the CO_2 permeability markedly increases [54, 63, 81]. Water plays several crucial roles within solid polymer electrolyte membranes (see Section 4.6.1): (1) it serves as a novel innovative CO_2 carrier when it binds to metallic cations by forming reversible [ABAB] complexes, (2) it forms an efficient CO_2 carrier, HCO_3^-, through its reaction with CO_2, (3) it enhances salt dissociation, leading to a higher concentration of free ions, and (4) it acts as a swelling agent for PEMs to increase their free volumes for improving CO_2 permeability. Therefore, the presence of water inside the membrane always helps increase the CO_2 separation performance. This is particularly advantageous in treating flue gas (around 10–25 % v/v water) and natural gas sweetening (15–60 mg/m^3 water) containing water.

Polymer electrolyte membrane consisting of a metallic salt such as CF_3CO_2K (0.7 molar ratio of K^+ to the monomeric unit of PEO) dissolved in poly(ethylene oxide) (PEO) matrix showed the extremely high CO_2 permeance up to 4,600 GPU with feed gas of 70% relative humidity at a feed pressure of 1.01 bar, whereas it maintained the low N_2 permeance of about 3 GPU, resulting in the ideal separation factor of about 1,500 (Figure 4.51) [54]. Note that the CO_2 permeance through the reference neat PEO membrane exhibited a typical CO_2 permeance of 350 GPU and the ideal separation factor of 3, regardless of the applied pressure in this experimental range. These results suggest that water inside the membranes, induced by the feed gas with 70% relative humidity, plays very important roles in improving the permeance of CO_2, but not N_2.

In this case, metal ions having positive charge (strong Lewis acid [A]) will interact with oxygen of water molecule (Lewis base [B]) to induce partial positive charge δ^+ on hydrogen atom (weak Lewis acid [A]) of water for the CO_2 (Lewis base [B]) absorption, forming the [ABAB] complexes as described previously in Section 4.6.1 (Figure 4.43a). Consequently, the bound water can also act as a novel CO_2 carrier for the facilitated transport in the quasi-solid-state membrane containing small amounts of water.

Research has further explored the roles of bound water in membranes, particularly its impact on CO_2 permeance. Specifically, poly(ether-*b*-amide 6, weight ratio

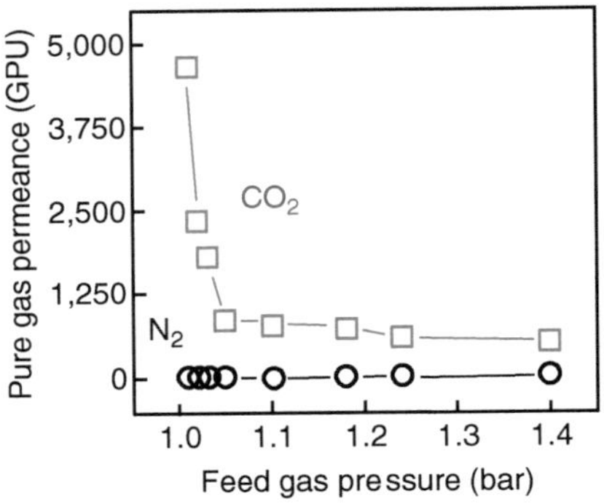

Figure 4.51 Pure gas permeance through PEO/CF_3CO_2K) (0.7 mole ratio of K^+ to the monomeric unit of PEO) as functions of the absolute feed pressure with 70% RH gas at 298 K. *Source*: Adapted from [54].

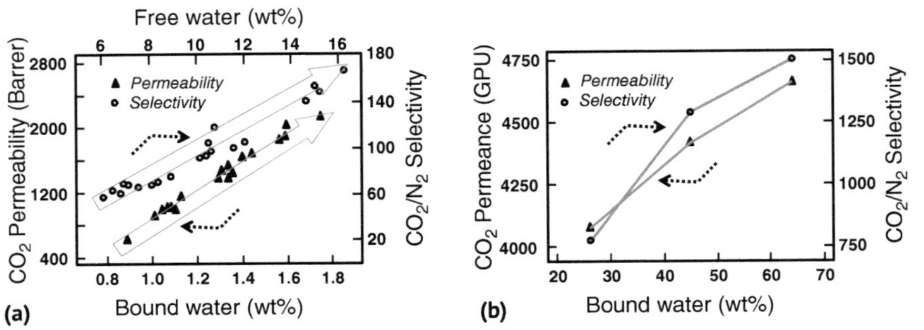

Figure 4.52 (a) CO_2 permeability as a function of the bound water content through Pebax 1657 – salt electrolyte membranes along with the CO_2/N_2 selectivity. *Source*: Replotted from [63]. (b) Correlation between the CO_2 permeance and the CO_2/N_2 selectivity as a function of the concentration of bound water (carrier concentration). *Source*: Adapted from [54].

6/4) (Pebax 1657) membranes mixed with alkali or alkaline-earth metal salts significantly boost both the CO_2 permeance and the CO_2/N_2 selectivity nearly linearly with the amount of the bound water (Figure 4.52a) [63]. Such improvement becomes more pronounced with higher bound water content, also showing a nearly linear increase in the CO_2 permeance as well as the CO_2/N_2 selectivity in relation to the bound water concentration to K^+ (Figure 4.52b) [54]. This phenomenon supports the concentration fluctuation model for facilitated transport, suggesting bound water to K^+ acts as an effective CO_2 carrier. These findings are consistent across studies, reinforcing the roles of the bound water in facilitating CO_2 transport in polymer electrolyte membranes [5]. This mechanism, involving the selective and reversible complexation of CO_2 by forming [ABAB] complexes with the bound water, is effective even in quasi-solid polymer electrolyte membranes.

4.6.4 Polymer Membranes with Ionic Liquids

Room temperature ionic liquids (ILs) consist of cation and anion with melting points less than 100 °C. Such cation and anion are large-sized soft Lewis acid and Lewis base, respectively, and consequently they can reversibly interact with amphoteric

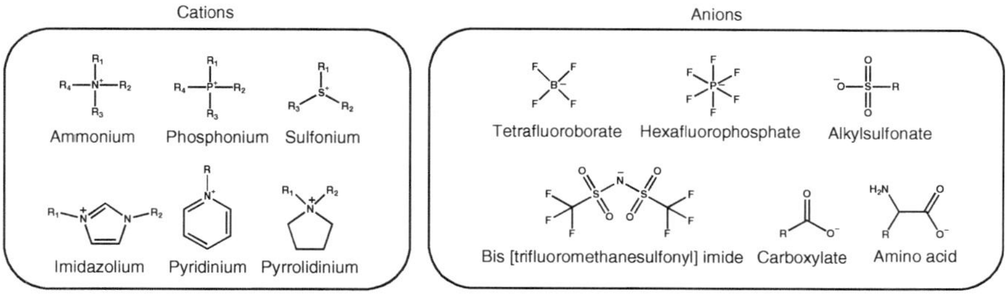

Figure 4.53 Chemical structures of representative cations and anions for ionic liquids. *Source*: Y.S. Kang.

CO_2 to act as CO_2 carriers for facilitated CO_2 transport, i.e. an IL molecule is likely to offer two distinct types of CO_2 carriers: one via the cation as a Lewis acid and the other through the anion as a Lewis base. This dual functionality makes ILs very attractive for CO_2 separation and capture [82]. Chemical structures of typical cations and anions of ILs are drawn in Figure 4.53.

The thermodynamics and kinetics of interactions of ILs with CO_2 are crucial for determining their CO_2 separation performance. Thermodynamically, this interaction is described by the equilibrium constant K, which reflects the solubility or absorption capacity of CO_2 in ILs. Kinetically, it involves the reaction rate constants for both the forward and reverse reactions, denoted as k_1 and k_2, respectively.

Here, the interaction properties of CO_2 with various ILs depending on their chemical structures will be dealt first. Subsequently, transport behavior will be briefly explored by comparing the facilitated transport mechanism and the conventional solution–diffusion mechanism based on acid–base theory.

4.6.4.1 Interactions of CO_2 with ILs

In ILs, the cation plays a minor role and the basicity of the anion is observed to primarily affect the CO_2 solubility [83]. Interestingly, a quite good linear relationship between CO_2 solubility, absorption enthalpy ΔH, and the pKa value of the various cyclic anions including pyrazole, imidazole, triazole, tetrazole, and oxazole with a fixed cation of P_{66614} in DMSO was observed as shown in Figure 4.54 [84]. A similar linear relationship between CO_2 absorption and pKa of anion of ILs was also observed [85]. These results indicate that highly basic anion of ILs is essential to have high CO_2 absorption capacity.

Therefore, controlling the basicity of anions by incorporating functional groups such as amine, carboxylate, phenolate, halogen, hydroxyl, or their combinations can significantly enhance CO_2 permeability. Notably, amino acids naturally combine with both amine and carboxylate groups within a single molecule, offering a distinct advantage in improving CO_2 solubility.

Among many functional groups mentioned above, amine groups are particularly attractive because they interact with CO_2 both directly and indirectly, facilitating the CO_2 transport as detailed in Sections 4.6.1 and 4.6.2. ILs with highly basic anions, sometimes also called superbase, have been tested to maximize the CO_2 solubility [86]. Chemical structures of representative superbase anions containing

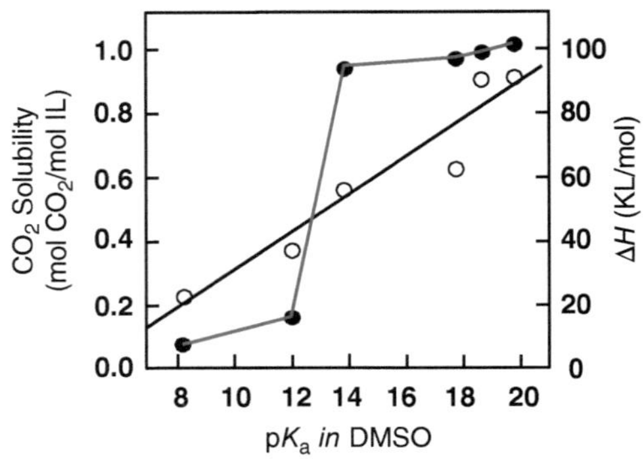

Figure 4.54 The relationship between CO_2 solubility (mol CO_2/mol IL) (●), absorption enthalpy (ΔH, ○), and the *pKa* value of the anion, including pyrazole, imidazole, triazole, tetrazole and oxazole in DMSO (dimethyl sulfoxide) with P_{66614}. The linear fit for the relationship between ΔH and *pKa* of the anion in DMSO is shown; $R^2 = 0.930$. *Source*: Adapted from [84].

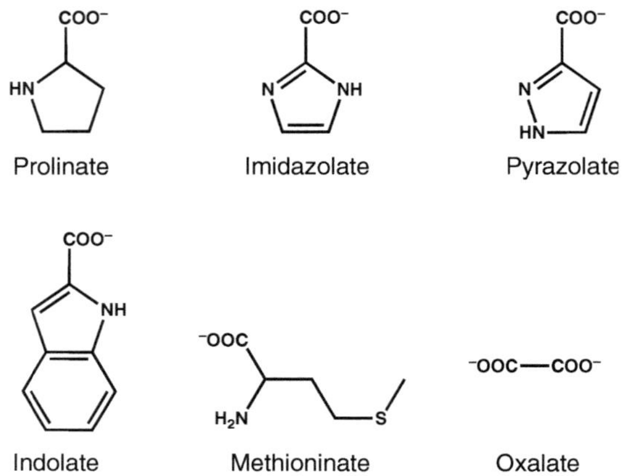

Figure 4.55 Chemical structures of representative superbase anions. *Source*: Y.S. Kang.

both amino and carboxylate groups [87] are illustrated in Figure 4.55. In addition, phenolate [85, 88] and oxalate [84] have also demonstrated high CO_2 absorption capability.

ILs with superbase anions commonly exhibit their enhanced CO_2 interactions by combining amino and carboxylate groups. The carboxylate group in the vicinity of NH-moiety stabilizes $-CO_2^-$ at the carbamic acid intermediate, thereby enabling the multimolar absorption. For instance, the amine group in iminodiacetate, IDA, acts as a nucleophile, attacking the electrophilic carbon in CO_2 to form carbamic acid intermediate. Meanwhile, the negatively charged carboxylate groups ($-COO^-$) interact additionally with the slightly positive carbon atom in CO_2 due to the polarization of the C=O bonds as shown in Figure 4.56 [89]. This dual interaction mechanism allows CO_2 solubility to reach up to 1.69 mol CO_2 per mol IDA [89].

In addition, the acetate group alone of [Bmim][TFA] with [Bmim] being 1-butyl-3-methyl-imidazolium and [TFA] being trifluoroacetate is a strong base and therefore strengthen their interactions with CO_2. DFT calculations demonstrate such enhanced interactions due to a weak charge transfer between the carbon atom of CO_2 (acting as a Lewis acid) and an oxygen atom of the COO^- group of [TFA] (as a Lewis base) [90]. Therefore, [Bmim][TFA] is able to accommodate a large amount of carbon dioxide.

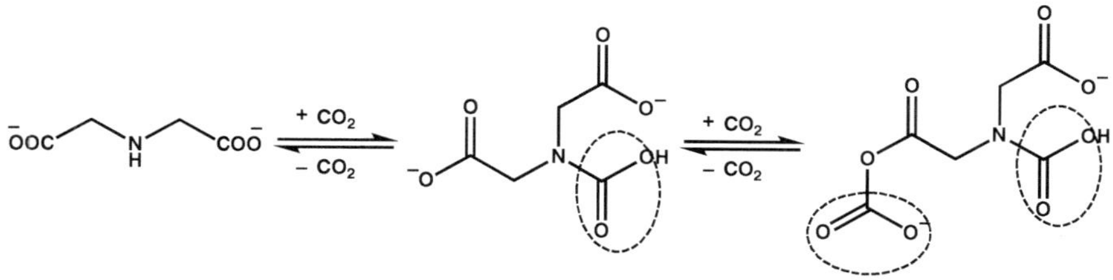

Figure 4.56 Multiple-site interaction of iminodiacetate, IDA, anion with CO_2. *Source*: Adapted from [89].

2

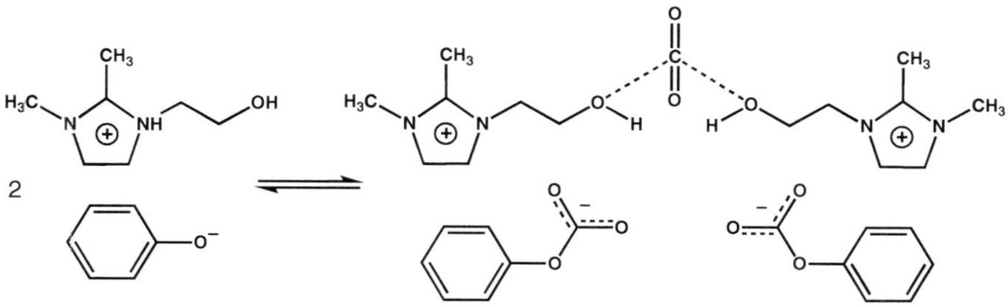

Figure 4.57 The proposed mechanism for CO_2 absorption of 1-(2-hydroxyethyl)-2,3-dimethylimidazolium phenoxide IL. *Source*: Adapted from [88].

Phenolate anions also contribute to enhance the CO_2 absorption capacity by forming phenol carbonate. For instance, ILs with P_{66614} cation and substituted phenolate anion showed a good linear correlation with the CO_2 uptake with *pKa* of the anion by forming phenol carbonates [85]. Here, the cation–anion interaction strength affects the CO_2 uptake in phenolate ILs [91]. When hydroxyl groups are incorporated in cation, ILs with phenolate anion yielded 1.5 equivalent CO_2 uptake as both hydroxyl and phenoxide groups participated in the CO_2 absorption [83, 88] as shown in Figure 4.57.

The interaction of CO_2 with a negatively charged fluorinated anion of $[PF_6]$ and $[BF_4]$ was evidenced by an acid–base interaction with the 1-butyl-3-methylimidazolium [Bmim] cation [78]. $[BF_4]$ is a stronger Lewis base in its interaction with CO_2 than $[PF_6]$. The CO_2 capture mechanism in 1-*n*-hexyl-3-methylimidazolium PF_6 [Hmim]$[PF_6]$ and tris(pentafluoroethyl)trifluorophosphate [Hmim][FEP] showed the importance of electrostatic interaction in the CO_2 uptake with smaller and symmetrical $[PF_6]$ anion whereas the van der Waals interaction has a key role in large and asymmetrical [FEP] anion [92].

Apart from these, weak interactions such as halogen bonding (X—O bonding), a net attractive interaction between an electrophilic halogen atom X and a nucleophilic atom O, and hydrogen bonding are shown to affect the CO_2 uptake [93, 94]. In the case of [Bmim]-CO_2 complex, CO_2 interacts with [Bmim] cation by the formation of hydrogen bond on the C_2 carbon [95].

4.6.4.2 Transport Mechanism

The enhanced CO_2 transport through both SLMs with ILs and poly(IL)s has been successfully interpreted by the facilitated transport due to the reversible interactions

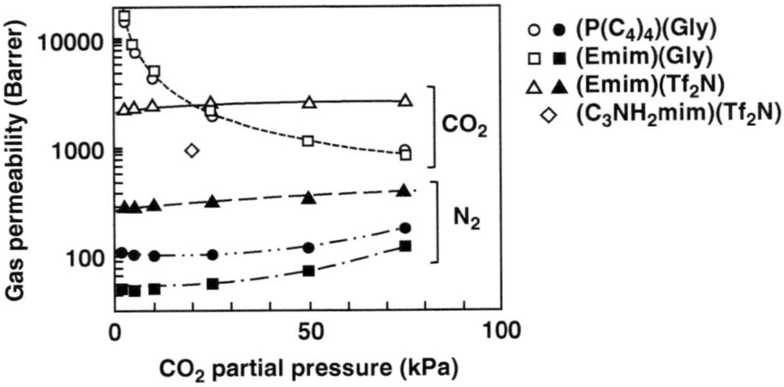

Figure 4.58 CO_2 (open symbols) and N_2 (closed symbols) permeabilities of [$P(C_4)_4$][Gly], [Emim][Gly] and [Emim][Tf_2N] as a function of CO_2 partial pressure at $T = 373.15$ (K) and humidity = 0%. *Source*: Adapted from [98].

of CO_2 with anions mostly and cations, to some extent, of ILs. The facilitated transport mechanism in the solid state was described in Section 4.2 in this Chapter.

In the meantime, the improved CO_2 transport has been also interpreted by the conventional solution–diffusion mechanism. In the context of the solution–diffusion mechanism, the improvement of the CO_2 permeability is mostly due to the increase in the solubility, while the diffusion coefficient of CO_2 and other relevant gases in ILs often differs only by less than a factor of 2 [96, 97].

Figure 4.58 illustrates the CO_2 permeability as a function of the applied CO_2 pressure for completely dry [P_{4444}][Gly], [Emim][Gly], [Emim][Tf_2N], and [C_3NH_2Mim][Tf_2N] membranes without water at $T = 373.15$ K where [P_{4444}] stands for tetrabutyl phosphonium, [Gly] for glycine amino acid, [Emim] for 1-ethyl-3-methylimidazolium, [C_3NH_2Mim] for *N*-aminopropyl-3-methylimidazolium and [Tf_2N] for bis(trifluoromethylsulfonyl)-imide [98]. The CO_2 permeability sharply increases with decreasing the applied pressure through the two membranes ([P_{4444}][Gly] and [Emim][Gly]) having [Gly] anions without water whereas it slightly increases with increasing pressure through [Emim][Tf_2N]. The former two membranes demonstrate typical facilitated transport behavior, resulting from the CO_2 carrier action of glycine amino acid as described previously, and the latter would apparently suggest the conventional solution–diffusion mechanism. Of course, detailed mechanistic studies should be further conducted to verify the valid transport mechanism. Nonetheless, these results suggest the strong chemical structure-dependent transport mechanism.

4.6.5 Surface-activated Metallic Nanoparticle

Ag NPs with surfaces activated by electron polarizers interact reversibly and specifically with olefins, not paraffins, allowing them to function as olefin carriers for facilitated olefin transport. This results in exceptionally high separation performance, as described in Section 4.5.2 of Chapter 4 [24, 47].

Similarly, copper nanoparticles (Cu NPs) can serve as CO_2 carriers by interacting with amphoteric CO_2. This is achieved when their surfaces are partially positively

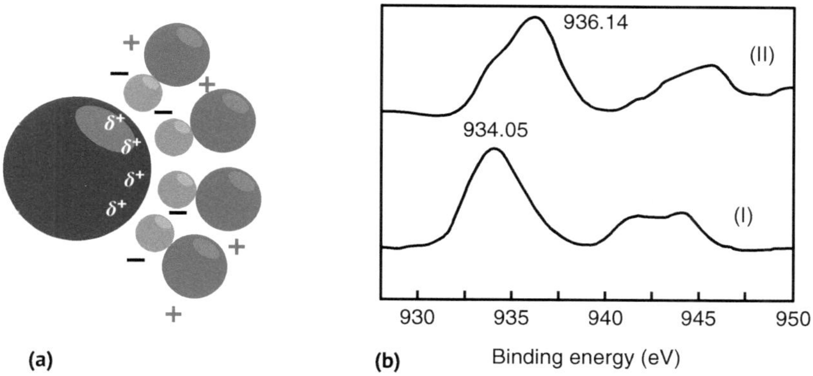

Figure 4.59 (a) Adsorption of ionic liquid on the surface of copper nanoparticles, inducing a partial positive charge (δ^+); and (b) XPS spectra showing copper binding energy in (I) neat copper NP flakes and (II) 1/0.004 [$BMIM^+ BF_4^-$]/Cu NPs. *Source*: Adapted from [99].

charged using ionic liquids (ILs) as electron polarizers (Figure 4.59a). The synergistic effect of surface-activated Cu NPs and ILs enhances CO_2 transport by utilizing three CO_2 carriers: the surface-activated Cu NPs, and the cations and anions of the IL [99].

Incorporating Cu NPs into an IL layer of [$BMIM^+ BF_4^-$] induces a partial positive charge on their surface, increasing their activity in metallic copper-CO_2 complexation. XPS analysis confirms the partial positive charge δ^+, showing an increase in the $d_{5/2}$ orbital binding energy from 934.05 to 936.14 eV after IL adsorption (Figure 4.59b). As a result, supported liquid membranes (SLMs) with neat [$BMIM^+ BF_4^-$] achieve a CO_2 permeance of 17 GPU, which increases to 25 GPU when [$BMIM^+ BF_4^-$] is combined with Cu NPs (1,000/4 w/w). The ideal selectivities improve from 5.0 to 11.0 for CO_2/N_2 and from 4.8 to 11.0 for CO_2/CH_4. These improvements are primarily attributed to the reversible interactions of CO_2 with three distinct CO_2 carriers. However, further research is needed to elucidate the interaction mechanisms and to design more advanced CO_2 carriers.

Nanocomposites with [$Hmim^+ NO_3^-$]/Cu NPs exhibit CO_2/CH_4 and CO_2/N_2 selectivities of 6.2 and 7.4, respectively [100, 101]. Further enhancements are achieved by incorporating graphene oxide (GO) into Cu NP/IL systems, such as [$Hmim^+ NO_3^-$] or 1-methyl-3-octylimidazolium tetrafluoroborate, significantly improving separation performance [102].

4.7 Challenges and Prospects

Facilitated transport, which combines both Fickian and carrier-mediated transports, uses a specific chemical as a carrier to selectively and reversibly interact with a specific solute. This method has proven to be effective in improving the performance of separation processes in solid or quasi-solid states, especially by increasing both permeability and selectivity with low driving force or low pressure gradient across the membrane. Several mathematical models, such as the concentration fluctuation model, show how properties like the kinetic parameter, k_2, thermodynamic parameter, K, and carrier concentration, C_C, affect transport. Thus, it is crucial to optimize k_2 and K because they are often in conflicts, to maximize efficiency.

While there has been notable success, especially in olefins including ethylene and propylene and CO_2 separations, research in other applications has been somewhat restrained, primarily due to concerns about the chemical stability of common carriers like Co(II) for oxygen, Ag(I) for olefins, alkyl amines for CO_2. Despite these challenges, progress in developing stable and effective carriers like surface-activated metallic Ag NPs for olefin separation continues, demonstrating their potential in enhancing transport properties and sustainable operations. For CO_2 separation, amphoteric property of CO_2 can be effectively and successfully utilized to interact with both cations and anions of, for instance, amino acids and ILs, which is very interesting and challenging in designing future polymeric materials, giving high permeability and high selectivity simultaneously.

References

1 H. Nishide, M. Ohyanagi, O. Okada, E. Tsuchida, Macromolecules 19(2) (1986) 494–496.
2 H. Nishide, M. Ohyanagi, O. Okada, E. Tsuchida, Macromolecules 20(2) (1987) 417–422.
3 H. Nishide, T. Suzuki, H. Kawakami, E. Tsuchida, Journal of Physical Chemistry 98(19) (1994) 5084–5088.
4 Y.S. Kang, J.M. Hong, J. Jang, U.Y. Kim, Journal of Membrane Science 109(2) (1996) 149–157.
5 J.M. Hong, Y.S. Kang, J. Jang, U.Y. Kim, Journal of Membrane Science 109(2) (1996) 159–163.
6 M. Ohyanagi, H. Nishide, K. Suenaga, E. Tsuchida, Macromolecules 21(6) (1988) 1590–1594.
7 R.M. Barrer, Journal of Membrane Science 18(Mar) (1984) 25–35.
8 G.H. Fredrickson, E. Helfand, Macromolecules 18(11) (1985) 2201–2207.
9 R.D. Noble, Journal of Membrane Science 50(2) (1990) 207–214.
10 D.R. Smith, J.A. Quinn, AIChE Journal 25(1) (1979) 197–200.
11 E.L. Cussler, R. Aris, A. Bhown, Journal of Membrane Science 43 (2–3) (1989) 149–164.
12 J.W. Cahn, Journal of Chemical Physics 42(1) (1965) 93–99.
13 H. Nishide, M. Ohyanagi, O. Okada, E. Tsuchida, Macromolecules 21(10) (1988) 2910–2913.
14 S.U. Hong, J.H. Jin, J. Won, Y.S. Kang, Advanced Materials 12(13) (2000) 968–971.
15 M. Yoshikawa, S. Shudo, K. Sanui, N. Ogata, Journal of Membrane Science 26(1) (1986) 51–61.
16 L. Ansaloni, Y.N. Zhao, B.T. Jung, K. Ramasubramanian, M.G. Baschetti, W.S.W. Ho, Journal of Membrane Science 490 (2015) 18–28.

17 E. Tsuchida, H. Nishide, M. Ohyanagi, H. Kawakami, Macromolecules 20(8) (1987) 1907–1912.
18 J. Crank, G.S. Park, Diffusion in Polymers, Academic Press, 1968.
19 D.W. van Krevelen, P.J. Hoftyzer, 1990, 198–203.
20 E. Tsuchida, H. Nishide, Topics in Current Chemistry 132 (1986) 63–99.
21 H. Nishide, H. Kawakami, S. Toda, E. Tsuchida, Y. Kamiya, Macromolecules 24(21) (1991) 5851–5855.
22 J.H. Kim, B.R. Min, H.S. Kim, J. Won, Y.S. Kang, Journal of Membrane Science 212 (1–2) (2003) 283–288.
23 J.H. Kim, Y.S. Kang, J. Won, Macromolecular Research 12 (2004) 145–155.
24 Y.S. Kang, S.W. Kang, H. Kim, J.H. Kim, J. Won, C.K. Kim, K. Char, Advanced Materials 19(3) (2007) 475–479.
25 S.W. Kang, Membrane Journal 26(3) (2016) 173–178.
26 C.K. Kim, J.G. Won, H.S. Kim, Y.S. Kang, H.G. Li, C.K. Kim, Journal of Computational Chemistry 22(8) (2001) 827–834.
27 J.H. Jin, S.U. Hong, J. Won, Y.S. Kang, Macromolecules 33(13) (2000) 4932–4935.
28 I. Pinnau, L.G. Toy, C. Casillas, 1997. No. US 5670051
29 I. Pinnau, L.G. Toy, Journal of Membrane Science 184(1) (2001) 39–48.
30 S. Sunderrajan, B.D. Freeman, C.K. Hall, I. Pinnau, Journal of Membrane Science 182 (1–2) (2001) 1–12.
31 J.H. Kim, B.R. Min, C.K. Kim, J. Won, Y.S. Kang, Macromolecules 35(13) (2002) 5250–5255.
32 S.U. Hong, C.K. Kim, Y.S. Kang, Macromolecules 33(21) (2000) 7918–7921.
33 Y. Yoon, J. Won, Y.S. Kang, Macromolecules 33(9) (2000) 3185–3186.
34 J.H. Kim, B.R. Min, C.K. Kim, J. Won, Y.S. Kang, The Journal of Physical Chemistry B 106(10) (2002) 2786–2790.
35 J.H. Kim, B.R. Min, J. Won, Y.S. Kang, Chemistry: A European Journal 8(3) (2002) 650–654.
36 H.S. Kim, J.H. Ryu, H. Kim, B.S. Ahn, Y.S. Kang, Chemical Communications (14) (2000) 1261–1262.
37 K.J. Lee, J.Y. Jho, J. Won, Y.S. Kang, Macromolecular Rapid Communications 23(14) (2002) 839–843.
38 H.S. Kim, Y.J. Kim, J.J. Kim, S.D. Lee, Y.S. Kang, C.S. Chin, Chemistry of Materials 13(5) (2001) 1720–1725.
39 J.H. Ryu, H. Lee, Y.J. Kim, Y.S. Kang, H.S. Kim, Chemistry: A European Journal 7(7) (2001) 1525–1529.
40 J.Y. Kim, S.U. Hong, J. Won, Y.S. Kang, Macromolecules 33(8) (2000) 3161–3165.
41 C.K. Kim, C.K. Kim, B.S. Lee, J. Won, H.S. Kim, Y.S. Kang, The Journal of Physical Chemistry A 105(39) (2001) 9024–9028.
42 J. Kim, S.W. Kang, S.H. Mun, Y.S. Kang, Industrial and Engineering Chemistry Research 48(15) (2009) 7437–7441.
43 K.I. Han, S.W. Kang, J. Kim, Y.S. Kang, Journal of Membrane Science 374 (1–2) (2011) 43–48.
44 S.W. Kang, J. Hong, J.H. Park, S.H. Mun, J.H. Kim, J. Cho, K. Char, Y.S. Kang, Journal of Membrane Science 321(1) (2008) 90–93.

45 D.I. Gittins, F. Caruso, Angewandte Chemie-International Edition 40(16) (2001) 3001–3004.

46 J.H. Kim, J. Won, Y.S. Kang, Journal of Membrane Science 241(2) (2004) 403–407.

47 I.S. Chae, S.W. Kang, J.Y. Park, Y.G. Lee, J.H. Lee, J. Won, Y.S. Kang, Angewandte Chemie-International Edition 50(13) (2011) 2982–2985.

48 Y. Eum, B.S. Kim, I.S. Chae, G.H. Moon, S.C. Park, J. Jang, Y.S. Kang, Macromolecular Research 28(11) (2020) 1026–1031.

49 S.W. Kang, D.H. Lee, J.H. Park, K. Char, J.H. Kim, J. Won, Y.S. Kang, Journal of Membrane Science 322(2) (2008) 281–285.

50 S.W. Kang, K. Char, Y.S. Kang, Chemistry of Materials 20(4) (2008) 1308–1311.

51 H. Choi, J.H. Lee, Y.R. Kim, D. Song, S.W. Kang, S.S. Lee, Y.S. Kang, Chemical Communications 50(24) (2014) 3194–3196.

52 J.H. Lee, S.W. Kang, M.S. Yeom, Y.R. Kim, H. Choi, D. Song, J. Won, Y.S. Kang, Journal of Materials Chemistry A 2(19) (2014) 6987–6993.

53 A. Buckingham, R. Disch, Proceedings of the Royal Society A 273(1353) (1963) 275–289.

54 S.C. Park, I.S. Chae, G.H. Moon, B.S. Kim, J. Jang, M. Wessling, Y.S. Kang, Journal of Materials Chemistry A 7(10) (2019) 5190–5194.

55 D.M. Kem, Journal of Chemical Education 37 (1960) 14.

56 T. Enns, Science 155(3758) (1967) 44–47.

57 J.H. Meldon, Y.S. Kang, AIChE Symposium Series 79(227) (1983).

58 Y.F. Li, S.F. Wang, G.W. He, H. Wu, F.S. Pan, Z.Y. Jiang, Chemical Society Reviews 44(1) (2015) 103–118.

59 R. Ben Said, J.M. Kolle, K. Essalah, B. Tangour, A. Sayari, ACS Omega 5(40) (2020) 26125–26133.

60 T.L. Donaldson, Y.N. Nguyen, Industrial and Engineering Chemistry Research 19(3) (1980) 260–266.

61 L. Yu, M. Kanezashi, H. Nagasawa, T. Tsuru, Applied Sciences-Basel 8(7) (2018).

62 L. Yu, M. Kanezashi, H. Nagasawa, T. Tsuru, Journal of Membrane Science 541 (2017) 447–456.

63 J.T. Wang, S.F. Wang, Q.P. Xin, Y.F. Li, Journal of Materials Chemistry A 5(15) (2017) 6794–6816.

64 W. Zhao, G.H. He, F. Nie, L.L. Zhang, H.C. Feng, H.J. Liu, Journal of Membrane Science 411 (2012) 73–80.

65 W. Zhao, G.H. He, L.L. Zhang, J. Ju, H. Dou, F. Nie, C.N. Li, H.J. Liu, Journal of Membrane Science 350 (1–2) (2010) 279–285.

66 T.R. Eichhorn, Y. Takado, N. Salameh, A. Capozzi, T. Cheng, J.N. Hyacinthe, M. Mishkovsky, C. Roussel, A. Comment, Proceedings of the National Academy of Sciences of the United States of America 110(45) (2013) 18064–18069.

67 M.H. Oleary, R.J. Jaworski, F.C. Hartman, Proceedings of the National Academy of Sciences of the United States of America 76(2) (1979) 673–675.

68 J. Mähler, I. Persson, Inorganic Chemistry 51(1) (2012) 425–438.

69 V. Gutmann, Electrochimica Acta 21(9) (1976) 661–670.

70 A.V. Rayer, K.Z. Sumon, L. Jaffari, A. Henni, Journal of Chemical and Engineering Data 59(11) (2014) 3805–3813.

71 D.F. Detar, Journal of Organic Chemistry 45(25) (1980) 5174–5176.

72 Z. Tong, W.S.W. Ho, Journal of Membrane Science 543 (2017) 202–211.
73 L. Yu, M. Kanezashi, H. Nagasawa, J. Oshita, A. Naka, T. Tsuru, Separation and Purification Technology 178 (2017) 232–241.
74 Y.N. Zhao, W.S.W. Ho, Journal of Membrane Science 415 (2012) 132–138.
75 S. Lee, S.C. Park, T.Y. Kim, S.W. Kang, Y.S. Kang, Journal of Membrane Science 548 (2018) 358–362.
76 O. Höfft, S. Báhr, V. Kempter, Langmuir 24(20) (2008) 11562–11566.
77 I. Rey, P. Johansson, J. Lindgren, J.C. Lassègues, J. Grondin, L. Servant, Journal of Physical Chemistry A 102(19) (1998) 3249–3258.
78 S.G. Kazarian, B.J. Briscoe, T. Welton, Chemical Communications (20) (2000) 2047–2048.
79 R. Quinn, D.V. Laciak, Journal of Membrane Science 131 (1–2) (1997) 49–60.
80 R. Quinn, Journal of Membrane Science 139(1) (1998) 97–102.
81 L. Liu, A. Chakma, X.S. Feng, Journal of Membrane Science 310 (1–2) (2008) 66–75.
82 S.F. Wang, X.Q. Li, H. Wu, Z.Z. Tian, Q.P. Xin, G.W. He, D.D. Peng, S.L. Chen, Y. Yin, Z.Y. Jiang, M.D. Guiver, Energy & Environmental Science 9(6) (2016) 1863–1890.
83 S.K. Shukia, S.G. Khokarale, T.Q. Bui, J.P.T. Mikkola, Front Mater 6 (2019).
84 C.M. Wang, X.Y. Luo, H.M. Luo, D.E. Jiang, H.R. Li, S. Dai, Angewandte Chemie-International Edition 50(21) (2011) 4918–4922.
85 C.M. Wang, H.M. Luo, H.R. Li, X. Zhu, B. Yu, S. Dai, Chemistry - A European Journal 18(7) (2012) 2153–2160.
86 S.F.R. Taylor, C. McCrellis, C. McStay, J. Jacquemin, C. Hardacre, M. Mercy, R.G. Bell, N.H. de Leeuw, Journal of Solution Chemistry 44 (3–4) (2015) 511–527.
87 B.E. Gurkan, J.C. de la Fuente, E.M. Mindrup, L.E. Ficke, B.F. Goodrich, E.A. Price, W.F. Schneider, J.F. Brennecke, Journal of the American Chemical Society 132(7) (2010) 2116–2117.
88 M. Vafaeezadeh, J. Aboudi, M.M. Hashemi, RSC Advances 5(71) (2015) 58005–58009.
89 F.F. Chen, K. Huang, Y. Zhou, Z.Q. Tian, X. Zhu, D.J. Tao, D.E. Jiang, S. Dai, Angewandte Chemie-International Edition 55(25) (2016) 7166–7170.
90 M.I. Cabaço, M. Besnard, Y. Danten, J.A.P. Coutinho, Journal of Physical Chemistry B 115(13) (2011) 3538–3550.
91 T.B. Lee, S. Oh, T.R. Gohndrone, O. Morales-Collazo, S. Seo, J.F. Brennecke, W.F. Schneider, Journal of Physical Chemistry B 120(8) (2016) 1509–1517.
92 X.C. Zhang, F. Huo, Z.P. Liu, W.C. Wang, W. Shi, E.J. Maginn, Journal of Physical Chemistry B 113(21) (2009) 7591–7598.
93 X. Zhu, Y.X. Lu, C.J. Peng, J. Hu, H.L. Liu, Y. Hu, Journal of Physical Chemistry B 115(14) (2011) 3949–3958.
94 J. Palomar, M. Gonzalez-Miquel, A. Polo, F. Rodriguez, Industrial & Engineering Chemistry Research 50(6) (2011) 3452–3463.
95 S. Zhao, X.Z. Tian, J.N. Liu, Y.L. Ren, J.J. Wang, Computational and Theoretical Chemistry 1052 (2015) 12–16.
96 Z.D. Dai, R.D. Noble, D.L. Gin, X.P. Zhang, L.Y. Deng, Journal of Membrane Science 497 (2016) 1–20.

97 R. Condemarin, P. Scovazzo, Journal of chemical engineering 147(1) (2009) 51–57.
98 S. Kasahara, E. Kamio, T. Ishigami, H. Matsuyama, Chemical Communications 48(55) (2012) 6903–6905.
99 J.H. Lee, J. Hong, J.H. Kim, Y.S. Kang, S.W. Kang, Chemical Communications 48(43) (2012) 5298–5300.
100 J.H. Lee, I.S. Chae, D. Song, Y.S. Kang, S.W. Kang, Separation and Purification Technology 112 (2013) 49–53.
101 G.H. Hong, J.H. Oh, D. Ji, S.W. Kang, Chemical Engineering Journal 252 (2014) 263–266.
102 I.S. Chae, J.H. Lee, J. Hong, Y.S. Kang, S.W. Kang, Chemical Engineering Journal 251 (2014) 343–347.

5

Selective Transport Membranes

The selective transport of gas or liquid mixtures through membranes is critical in many separation processes, such as chemical and petrochemical processing, natural gas sweetening, and environmental applications. The performance of these membranes is typically characterized by two key parameters: permeability and selectivity.

Permeability, P_i, of component i is a measure of how efficiently component i can pass through a membrane as shown in Figure 5.1. It is usually quantified as the gas volume or liquid weight that can permeate per a specific area of the membrane, under a unit pressure gradient or concentration gradient, respectively, over a unit of time as described in Section 3.1 of Chapter 3. This parameter is crucial because it determines the efficiency with which a membrane can process a gas stream.

Selectivity, on the other hand, refers to the ability of the membrane to discriminate between different gases and liquids in a mixture of components i and j. It is usually expressed as the ratio of the permeabilities of two components, P_i/P_j, indicating how much faster one component permeates through the membrane compared to another. High selectivity along with high permeability is essential for applications such as the production of high-purity oxygen from air, the removal of carbon dioxide from natural gas or olefin separation from its mixtures with corresponding paraffins.

The polymer materials used to make these membranes can vary widely, each with different permeability and selectivity characteristics. The chemical structure of the polymer membrane mostly determines these properties. Researchers continuously seek polymer materials that can offer both high permeability and high selectivity, aiming to develop membranes that are both highly efficient and effective in separation processes.

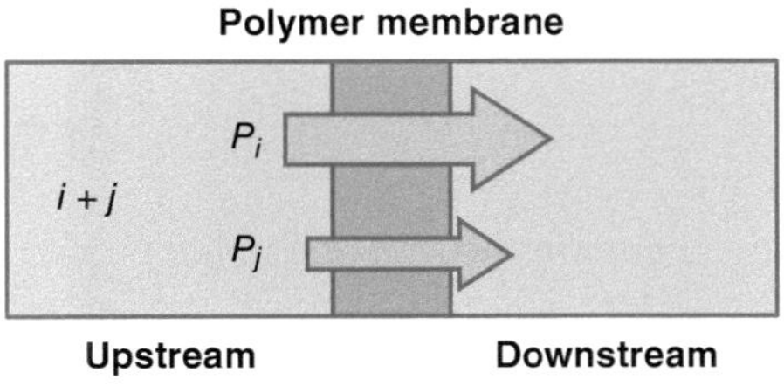

Figure 5.1 Schematic permeation characteristic of components i and j in terms of permeabilities P_i and P_j. *Source*: Y.S. Kang.

This chapter will first explore the theoretical basis for the relationship between permeability and selectivity, often referred to as the "upper-bound curve," after defining permeability and selectivity. It will then discuss how the structure of polymers affects both permeability and selectivity, followed by time-dependent separation performance by physical aging in glassy polymers. Finally, special attention will be given to facilitated transport membranes in the solid state because they are demonstrated to be a powerful tool to improve both permeability and selectivity simultaneously.

5.1 Definitions of Permeability and Selectivity

Membrane materials may be characterized by both how fast and how selectively a specific molecule from its mixtures permeates across a membrane. The former is commonly represented by the permeability and the latter by the selectivity. The permeability P_i of component i, defined earlier in Chapter 3, is expressed as:

$$P_i = D_i \times S_i \tag{5.1}$$

where D_i and S_i are the diffusion and the solubility coefficients, respectively, of component i. The Barrer is the commonly used unit of the gas permeability where 1 Barrer $= 1.0 \times 10^{-10}$ cm^3 (STP) cm/cm^2 sec cmHg.

Another key parameter to characterize the separation performance of a gas or liquid mixture is the selectivity. The ideal selectivity $\alpha_{i/j}$ is defined as:

$$\alpha_{i/j} = \frac{P_i}{P_j} = \left(\frac{D_i}{D_j}\right)\left(\frac{S_i}{S_j}\right) \tag{5.2}$$

Here, component i is more permeable than component j, indicating $\alpha_{i/j} \geq 1$. $\alpha_{i/j}$ is also called the "ideal separation factor" [1]. The ideal selectivity $\alpha_{i/j}$ is readily obtained from P_i and P_j for pure gas or liquid. Since $P_i = D_i \times S_i$, the ideal selectivity $\alpha_{i/j}$ can be partitioned into diffusivity selectivity (D_i/D_j) and solubility selectivity (S_i/S_j). Thus, the ideal selectivity $\alpha_{i/j}$ is solely dependent on the material property at a given temperature and pressure.

When gas, vapor, or liquid mixtures permeate across a membrane, the presence of a component i can, in some circumstances, influence the transport of other molecules, such as component j in the membrane. In this case, the ideal selectivity $\alpha_{i/j}$ may not correctly represent the real selectivity of mixtures through the membrane. Therefore, a new definition of selectivity is commonly used: the mixed-gas selectivity or gas selectivity $\alpha^*_{i/j}$, representing the ability of a membrane to separate a binary feed gas or liquid mixture,

$$\alpha^*_{i/j} = \frac{(y_i/y_j)}{(x_i/x_j)} \tag{5.3}$$

where x_i and y_i are the mole fraction of component i in the feed and in the permeate, respectively. The mixed-gas selectivity $\alpha^*_{i/j}$ and the ideal selectivity $\alpha_{i/j}$ are mathematically interrelated as [2]:

$$\alpha^*_{i/j} = \alpha_{i/j}\frac{p_2 - p_1(y_i/x_i)}{p_2 - p_1(y_j/x_j)} \tag{5.4}$$

where p_2 and p_1 are the feed side and the permeate side pressures, respectively. Thus, the mixed-gas selectivity depends not only on the properties of gas–polymer system but also on process parameters such as upstream and downstream pressures and compositions. When the upstream pressure is much greater than the downstream pressure so that $p_2 \gg p_1$, Equation (5.4) reduces to [2]:

$$\alpha_{i/j} = \alpha^*_{i/j} \tag{5.5}$$

Therefore, the separation performance of gas, vapor, or liquid mixtures through polymeric membrane is commonly characterized in terms of the two parameters of the permeability P_i and the ideal selectivity $\alpha_{i/j}$ or the mixed-gas selectivity $\alpha^*_{i/j}$.

5.2 Overview for Separation Performance of Gas Mixtures

Separation performance of gas, vapor, or liquid mixtures through polymeric membranes is commonly characterized by the permeability and the ideal selectivity. Figure 5.2 shows a typical permeability and ideal selectivity curve for propylene/propane mixtures across polymeric membranes. It can be seen that the permeability and the selectivity are commonly in conflict, resulting in trade-off relationship between them: highly permeable polymer would be less selective and *vice versa* [3].

Materials with the best performance would be in the upper right-hand corner of Figure 5.2, represented by "Goal." However, materials with permeability/selectivity combinations above and to the right of the line drawn in this figure are rare. This line defines the so-called "upper-bound" combinations of permeability and selectivity of many polymer materials for a particular gas pair. Therefore, there has a large body of researches to develop membrane materials overcoming the *upper-bound curve*. Such goal can be achieved by three main research avenues: (1) polymeric structure-property relationship, (2) facilitated transport, and (3) mixed matrix

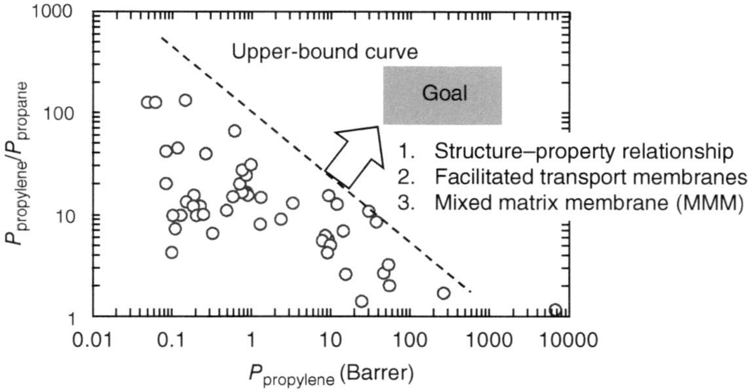

Figure 5.2 Relationship between propylene permeability $P_{propylene}$ and the ideal selectivity of propylene over propane $P_{propylene}/P_{propane}$ through polymeric materials at ambient temperature, and the empirical "upper-bound curve." *Source*: Y.S. Kang.

Table 5.1 Kinetic diameter of common gases [Wikipedia] [4–6].

Gas	H_2	O_2	N_2	CO_2	CH_4	C_3H_8	C_3H_6
Kinetic diameter (nm)	0.289	0.346	0.364	0.330	0.380	0.430	0.450

membranes. In the following sections, the first two avenues except the mixed matrix membrane studies will be focused.

Since the kinetic diameter plays very important roles in determining both the diffusion and solubility coefficients, they are listed in Table 5.1.

5.3 Theoretical Basis for Relationship Between Permeability and Selectivity (Upper-bound Curve)

On the basis of an exhaustive literature survey, Robeson quantified the upper-bound curve by graphing the available data as shown in Figure 5.3, which presents hydrogen permeabilities P_{H_2} and its selectivities over nitrogen α_{H_2/N_2} for many polymers along with the upper-bound curve [7, 8]. The theoretical background of the upper-bound curve was subsequently provided by Freeman [9].

The upper-bound performance characteristics were best described by Robeson [7] and Freeman [9] as:

$$\alpha_{A/B} = \beta_{A/B} / P_A^{\lambda_{A/B}} \tag{5.6}$$

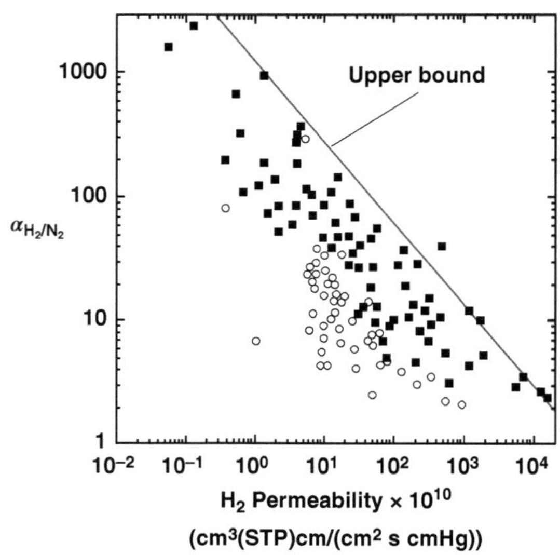

Figure 5.3 Relationship between hydrogen permeability and H_2/N_2 selectivity for rubbery (○) and glassy (■) polymers and the empirical upper-bound relation. *Source*: Adapted from [7].

which indicates that as the permeability, P_A, of an upper-bound polymer to gas A increases, selectivity, $\alpha_{A/B}$, of the polymer for gas A over gas B decreases. Robeson reported values of $\lambda_{A/B}$ and $\beta_{A/B}$ for many common gas pairs. $\lambda_{A/B}$ for representing the slope of the upper-bound curve is linearly correlated with the molecular (kinetic) diameter difference between higher and lower permeability gases [7].

The diffusion of small molecules is an activated process, and at temperatures away from thermal transitions in the polymer (e.g. glass transition, melting, etc.), the Arrhenius equation is obeyed:

$$D_A = D_{0A} \exp\left(-\frac{E_{DA}}{\mathrm{RT}}\right) \tag{5.7}$$

where D_{0A} is a front factor, E_{DA} is the activation energy for diffusion of component A, R is the gas constant, and T is the absolute temperature.

The front factor D_{0A} is expressed as a function of E_{DA} according to the linear free energy relation:

$$\ln(D_{0A}) = a\frac{E_{DA}}{RT} - b \tag{5.8}$$

where a and b are independent of gas type. Additionally, a is independent of polymer type and has a universal value of 0.64 [10]. b has a value of 9.2 for rubbery polymers and 11.5 for glassy polymers [11].

The relation between activation energy E_{DA} and the penetrant diameter d_A is given as:

$$E_{DA} = cd_A^2 - f \tag{5.9}$$

where c and f are constants which depend on the polymer. For light gases, d_A is taken to be the kinetic diameter of the penetrant. Van Krevelen reports c values from 250 for extremely flexible poly(dimethyl siloxane) (PDMS) to approximately 1,100 cal/mol Å^2 for stiff chain, glassy poly(vinyl chloride) [11].

Substitution of Equations (5.8) and (5.9) into Equation (5.7) gives the following expression for the diffusion coefficient [9]:

$$\ln(D_A) = -c\left(\frac{1-a}{RT}\right)d_A^2 + f\left(\frac{1-a}{RT}\right) - b \tag{5.10}$$

Then the permeability P_A becomes:

$$\ln(P_A) = -c\left(\frac{1-a}{RT}\right)d_A^2 + f\left(\frac{1-a}{RT}\right) - b + \ln(S_A) \tag{5.11}$$

and the selectivity $\alpha_{A/B}$ is

$$\ln(\alpha_{A/B}) = \ln(S_A/S_B) + \ln(D_A/D_B) = \ln(S_A/S_B) + c\left(\frac{1-a}{RT}\right)(d_B^2 - d_A^2) \tag{5.12}$$

Combining Equations (5.11) and (5.12) yields:

$$\begin{aligned} \ln(\alpha_{A/B}) &= -\left[\left(\frac{d_B}{d_A}\right)^2 - 1\right]\ln P_A \\ &+ \left\{\ln\left(\frac{S_A}{S_B}\right) - \left[\left(\frac{d_B}{d_A}\right)^2 - 1\right]\cdot\left[b - f\left(\frac{1-a}{RT}\right) - \ln S_A\right]\right\} \end{aligned} \tag{5.13}$$

Therefore, $\lambda_{A/B}$ and $\beta_{A/B}$ in Equation (5.13) are expressed as:

$$\lambda_{A/B} = \left(\frac{d_B}{d_A}\right)^2 - 1 \tag{5.14}$$

and

$$\beta_{A/B} = \frac{S_A}{S_B} S_A^{\lambda_{A/B}} \exp\left\{-\lambda_{A/B}\left[b - f\left(\frac{1-a}{RT}\right)\right]\right\} \tag{5.15}$$

Polymers that have high diffusivity selectivity (D_A/D_B) should have high values of c (Equation (5.12)). In this regard, for a high-performance glassy polyimide (synthesized from 3,3′,4,4′-biphenyltetra-carboxylic dianhydride and 4,4′-diaminodiphenyl ether), the value of c may be as high as 2,400 cal/mol Å^2. The value of f ranges from 0 for rubbery polymers and low-performance glassy polymers to approximately 14,000 cal/mol for the polyimide [12]. The ratio $(f/c)^{1/2}$ is a crude measure of the average distance between polymer chains. For the polyimide just described, $(f/c)^{1/2}$ is 2.5 Å [9].

Within the scope of this simple theory, there is no influence of polymer structure on the slope $\lambda_{A/B}$ of the upper bound or trade-off curves, because this parameter depends only on penetrant size ratio (Equation (5.14)). If this is true, then the slopes of the upper-bound lines are unlikely to change with further polymer development efforts. In contrast, $\beta_{A/B}$ contains variables that can be systematically tuned by rational polymer structure manipulation to simultaneously improve permeability and selectivity characteristics. In this regard, the most fruitful pathway for development of higher performance polymeric membranes for the separations is to systematically increase $\beta_{A/B}$, either through solubility selectivity enhancement and/or increases in chain stiffness (i.e. increasing c), while simultaneously increasing interchain spacing (i.e. fractional free volume) to increase permeability. Increasing interchain separation to increase permeability without sacrificing selectivity should only be effective as long as the interchain separation is not so large that penetrant diffusion coefficients are no longer governed by thermally stimulated polymer segmental motions [9].

Materials that are beyond this limit already exist. Poly(1-trimethylsilyl-1-propyne) (PTMSP), the most permeable polymer known and the polymer that has the highest free volume among all hydrocarbon-based polymers, also has the lowest selectivities (e.g. α_{O_2/N_2} ~0.92) for permanent gases among all polymers and exhibits many permeation characteristics similar to those of microporous materials. Simply increasing polymer backbone stiffness to the extent that this increases c in Equations (5.12) and (5.13) results in higher selectivity but lower diffusivity and, therefore, permeability. So, backbone stiffness increases should be coupled with increases in interchain separation to achieve both higher permeability and higher selectivity. These simple considerations suggest that *simultaneous chain stiffness and interchain separation increases can be used to systematically improve separation performance until the interchain separation becomes large enough that the polymer segmental motion no longer governs penetrant diffusion* [9].

In conclusion, Freeman developed a theory to explain the observed trade-off behavior and establish an empirical principle for achieving optimal permeability and selectivity properties. According to this principle, to maximize these properties up to a certain limit, one must design a polymer structure with a rigid backbone (i.e. increase c, which increases $\alpha_{A/B}$), which enhances diffusivity selectivity at the cost of diffusivity, while simultaneously disrupting interchain packing to increase f, thereby increasing diffusivity and, in turn, permeability. This principle is exemplified to the extreme in the case of polymers with intrinsic microporosity (PIMs) and thermally rearranged polymers (TRs).

5.4 Polymeric Structure–Properties Relationship

Understanding the relationship between polymeric structure and transport properties would be critically important in designing membrane materials having both the high permeability and the high selectivity simultaneously or to overcome the upper-bound curve as shown in Figure 5.2. It has been reported that the major physicochemical factors influencing the gas permeability and selectivity of polymers are (1) the mobility of polymer chains, as frequently reflected by glass and/or subglass transition temperatures, (2) the fractional free volume (FFV) or intersegmental d-spacing, and (3) the penetrant-polymer interactions, as reflected by the solubility of the penetrant gases in polymers [9, 13, 14], whose effects were described in more detail in Chapter 3.

Separation performances in terms of the selectivity of CO_2/CH_4 as a function of CO_2 permeability of representative polymers are shown in Figure 5.4. As can be seen, the separation performances are widely spread, in particular, depending on the polymeric chain mobility. In this regard, the polymer materials are categorized here: (1) polymers with flexible chains and high free volumes like rubbery polymers; (2) ductile and tough glassy polymers, including engineering plastics; (3) polymers with extremely rigid chains and high fractional free volumes, exemplified by PIMs,

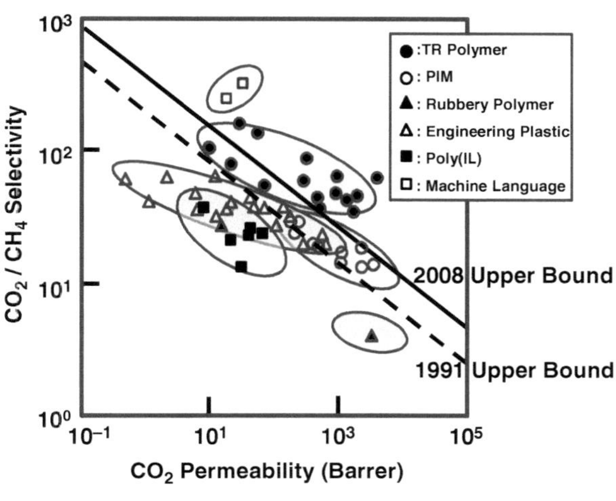

Figure 5.4 CO_2/CH_4 separation properties of various polymer materials reported in the literature. *Source*: Replotted from [15–20].

and (4) emerging polymer materials including electronically conductive polymers, perfluorinated polymers and novel polymers designed by machine learning technologies will also be explored in some detail, focusing their physicochemical structure effects. It is also noteworthy that these efforts result in the enhancement of the upper-bound curve from 1991 to 2008 as shown in Figure 5.4.

5.4.1 Polymers with Flexible Chains and High Free Volume

Polymers with flexible chains and high free volume such as rubbery polymers are generally very highly permeable, but their selectivities are quite low. The high gas permeability is strongly associated with high chain mobility associated with high fractional free volume. The low selectivity is mostly because of the small diffusivity selectivity (D_i/D_j) with the nearly fixed solubility selectivity (S_i/S_j), where the ideal selectivity $\alpha_{ij} = (D_i/D_j)(S_i/S_j)$. Low ($D_i/D_j$) is mostly because of the highly flexible polymeric chains, associated with the large fractional free volume. It is worthy to note here again that the solubility is mostly determined by the condensability of gas itself, frequently characterized by the critical temperature or the Lennard-Jones temperature, and consequently (S_i/S_j) is mostly fixed for a given gas pair regardless of the polymer material chosen. In this regard, it would be rather difficult to improve both the permeability and the selectivity simultaneously.

Typical rubbery polymers used for gas separation applications are PDMS, poly(ethylene oxide) (PEO), and poly(phosphazene) and their derivatives. PDMS show the oxygen permeability of up to 400–500 Barrer, the highest permeability among common polymers, mostly owing to the flexible Si—O bond and the large fractional free volume, and the oxygen selectivity over nitrogen is only around 1.8 [16]. Therefore, PDMS has been frequently employed as a plugging material for surface pores of asymmetrically microporous supports for fabricating thin film composite membranes.

Polymers containing polar moieties, such as ether groups, have an affinity for CO_2 due to dipole–quadrupole interactions [19, 21, 22]. Polyethers have oxygen atoms having partially negative charge due to their electron withdrawing capability and carbon atoms with partial positive charges, which are unevenly distributed in the ether linkage. The symmetric CO_2 molecule has a permanent electrical quadrupole moment and tends to align its oxygen atoms near the positively charged regions (carbon) of the ether-containing polymer while positioning its carbon atom near the negatively charged regions (oxygen) [23]. This net attraction promotes the solvation of CO_2 and the CO_2 solubility could increase with a greater content of ether oxygen [5]. Therefore, polymers and copolymers containing ethylene oxide segments have been also intensively investigated for CO_2 permeation with a very good balance of high selectivity and permeability [24]. Interestingly, two rubbery polymers of PDMS and PEO have been combined together to get their advantages into efficient CO_2 gas separation membranes [18].

However, there have been challenges to improve separation performance and film properties: Reduction in crystallinity and improvement in mechanical strength. Therefore, there have been several approaches to suppress the crystallinity of ether oxygen-rich polymers: (1) copolymerization; (2) blending with low molecular weight (MW) moiety to disrupt chain packing; and (3) formation of secondary polymer chains by crosslinking or branching [5].

Poly(1,3-dioxolane), a polyether but noncrystallizable, with a high O:C ratio of 0.67 exhibited a high CO_2 permeability of 1,400 Barrer and a CO_2/N_2 selectivity of 64 at 70 °C. This might be the first polyether that demonstrated a high CO_2/N_2 selectivity at a temperature considerably higher than room temperature [23].

Poly(phosphazene)s having two substituent sites are also highly permeable and therefore chemically designed functional groups can be readily introduced into the main chains to improve the selectivity [25, 26].

5.4.2 Ductile and Tough Glassy Polymers

Glassy polymers often exhibit low permeability and somewhat high selectivity, which is mostly associated with high diffusive selectivity (D_i/D_j). This high (D_i/D_j) is mainly due to the sieving effect of the rather rigid polymer chains. The solubility selectivity (S_i/S_j) does not vary significantly depending on the polymer used, as previously described. Therefore, from a materials perspective, it has been suggested that the mobility or stiffness of the polymer chains and the FFV or intersegment distance are two major factors that determine the transport properties in glassy polymers. In general, chain stiffness increases selectivity but lowers permeability, while higher free volume fraction increases permeability but lowers selectivity. Therefore, to achieve higher permeability and selectivity simultaneously, chain stiffness must be combined with higher fractional free volume or increased interchain separation [13]. In addition, the distribution of the free volume would also play important roles in both the permeability and the selectivity.

In this regard, many studies have attempted to design new chemical structures to simultaneously improve permeability and selectivity by controlling chain mobility as well as interchain separation or fractional free volume. These typical strategies are (1) to design a new main-chain chemical structure of the polymer, and (2) to introduce pendants, particularly large substituents, into the main chain to increase the free volume or spacing between polymer segments and control the chain stiffness.

5.4.2.1 Effects of Main-chain Chemical Structure

Common glassy polymers used in gas separation applications are aromatic polymers including polysulfone (PSF), polycarbonate (PC), poly(ether ether ketone), and polyimide, which have adequate mechanical toughness and are also frequently called engineering plastics. These polymers exhibit free volume fractions in the range of ~0.15 and high glass transition temperatures, T_g, mostly above 150 °C. The pure-gas permeability, ideal selectivity, and fractional free volume f, according to changes in main-chain structure of representative glassy polymers are summarized in Tables 5.2 and 5.3, respectively, along with their chemical structures (Figure 5.5) [17].

PSFs containing bisphenol A in the main backbone (Figure 5.5a) have been commercialized first for syngas ratio adjustment by Monsanto Co. in the 1970s. Since then, many different polysulfones having different chemical structures including poly(ether sulfone)s have been also developed.

Table 5.2 Pure-gas permeability (Barrer), free volume fraction *f*, and glass transition temperature T_g of representative polymers.

Polymer	CO_2	H_2	O_2	N_2	CH_4	*f*	T_g (°C)
PSF[a]	5.6	14	1.4	0.25	0.25	0.144	185
PPO[b]	61	61	16.8	4.1	4.3	0.183	215
TB-PC[c]	4.2	—	1.4	0.18	0.18	0.092	—
Polyimide[d]	10	18	2.1	0.32	0.28	0.170	320

[a]polysulfone; [b]poly(3,5-dimethyl-1,4-phenylene oxide); [c]tetrabromo-polycarbonate; [d]Matrimid
Source: Adapted from [17].

Table 5.3 Ideal selectivity through representative polymers.

Polymer	H_2/CH_4	H_2/N_2	CO_2/CH_4	CO_2/N_2	CO_2/H_2	O_2/N_2	N_2/CH_4
PSF[a]	56	56	32	32	0.4	5.5	1
PPO[b]	14	15	14	15	1	4.1	0.95
TB-PC[c]	—	—	32	23	—	7.8	1.4
Polyimide[d]	64	56	36	31	0.56	6.6	1.1

Source: Adapted from [17].

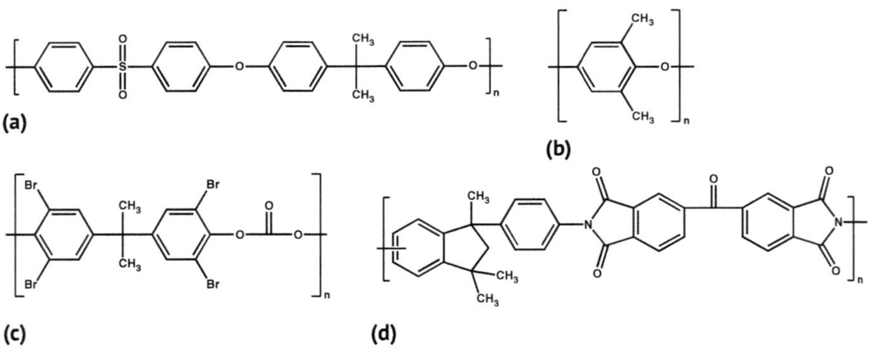

Figure 5.5 Chemical structures of (a) polysulfones (PSF), (b) poly(phenylene oxide) (PPO), (c) tetrabromo-polycarbonate (TB-PC), and (d) polyimide (Matrimid). *Source*: Y.S. Kang.

Both poly(3,5-dimethyl-1,4-phenylene oxide) (PPO) (Figure 5.5b) and tetrabromo-polycarbonate (TB-PC) (Figure 5.5c) also show good separation performance along with mechanical strength necessary for high-pressure operations [17]. In particular, TB-PC shows extremely high O_2/N_2 selectivity up to 7.8 and becomes attractive for air separation.

Aromatic polyimides have been investigated to improve the separation performance over a wide range of gas mixtures, probably due to the relatively easy availability of diverse monomers. Matrimid is a commercial aromatic polyimide, consisting of 3,3′-4,4′-benzophenone tetracarboxylic dianhydride and 5(6)-amino-1-(4′-aminophenyl-1,3-trimethylindane) (Figure 5.5d). Matrimid exhibits a good combination of CO_2 permeability and CO_2/CH_4 selectivity even at high temperatures. It is also interesting to note that the replacement of the isopropylidene ($-C(CH_3)_2-$) bridging moiety between two phenylenes in the main chain with the bulkier one, like hexafluoro isopropylidene ($-C(CF_3)_2-$), makes the polymers much more permeable to CO_2 and O_2, mainly due to the enhanced free volume [27].

5.4.2.2 Effects of Substituents

Introduction of a bulky substituent, in particular, into the polymer main chains is expected to result in significant changes in both the mobility and intersegment distance of the polymer chains. Pixton and Paul reviewed studies concerned with the effect of bulky side groups in glassy polymers on their gas transport properties, including the silylation of poly(2,6-dimethyl-1,4-phenylene oxide) (PPO) and PSF [28]. Such changes occur more specifically by the size and mobility of the substituent, degree of substitution (DS), the substitution site and spacer group.

For example, the presence of bulky trimethylsilyl groups in polymer chains is of particular interested in gas separation properties, mainly due to the increase in fractional free volume, associated with their size and mobile character. The high mobility of the trimethylsilyl group results from longer and thus more flexible Si—C bond, where the Si—C bond length (~0.185 nm) is longer than the C—C bond length (~0.145 nm). The incorporation of bulky and mobile silyl groups is expected to improve both the permeability and selectivity of the polymer membrane by suppressing main-chain mobility and concomitantly providing more open-chain packing [13].

When the methyl groups, rather small compared to trimethylsilyl group, were introduced at the phenyl rings of PSFs and PCs, their gas permeability was in general increased with a small decrease in the selectivity. Conversely, bromine and chlorine atoms resulted in increased selectivities [29].

Symmetry of substitution in bisphenol A polysulfones was also found to have a strong effect on the gas transport properties of the modified polymers [30, 31]. Analogous modification studies have also been conducted for other glassy polymers like polyarylates and poly(2,6-dimethyl-1,4-phenylene oxide) [32, 33].

5.4.3 Polymers with Extremely Stiff Chains and High Free Volume

Our recent focus has been on whether polymers with extremely stiff chains can provide a significantly large fractional free volume f (or FFV) by inhibiting effective chain packing, thereby further enhancing separation performance. One simple way to make the polymeric main chains extremely stiff is to introduce a backbone of alternating single and double bonds, like polyacetylenes, but the double bonds are not significantly conjugated [3]. For example, poly(1-trimethylsilyl-1-propyne) (PTMSP) has a rigid backbone of alternating single and double bonds and a large f due to the bulky and flexible trimethyl-silyl groups (Figure 5.6a) [3], resulting in extremely high gas permeability, compared to the reference polyimide (Table 5.4).

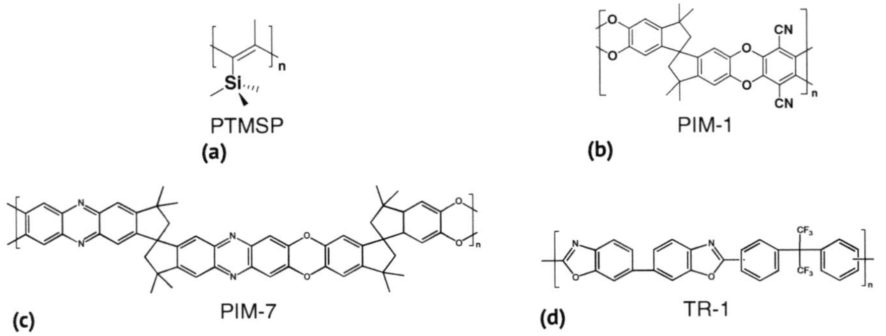

Figure 5.6 Chemical structures of (a) poly(1-trimethylsilyl-1-propyne) (PTMSP), (b) polymers with intrinsic microporosity (PIM-1), (c) PIM-7, and (d) thermally rearranged-1 (TR-1) polymer. *Source*: Y.S. Kang.

Table 5.4 Fractional free volume *f* of PTMSP, PIM-1 and TR-1 polymers.

Polymer	T_g	P_{O_2}	P_{CO_2}	R_3	R_4	f	Ref.
Polyimide[a]	304	260	—	0.319	—	0.164	[3]
PTMSP	>250	900	7,700	0.341	0.681	~0.25	[3]
PIM-1	>250	>1000	—	0.473	—	~0.2	[34]
TR-1	>250	500–1200	1,610	0.377	—	0.263	[35, 36]

Units: T_g [°C]; P [Barrer]; and R_3 and R_4 [nm].
[a]6FDA-ODA.

Furthermore, the barrier for rotation about the single bonds is as high as 170 kJ/mol at normal temperatures, it cannot easily change its conformation. These twisted, inflexible polymer chains trap a considerable amount of excess free volume. Sorption measurements suggest that PTMSP is essentially porous, with a void fraction of about 0.25, leading to the suggestion that free volume elements are connected to give a micropore structure [37]. Thus, a freshly prepared PTMSP membrane exhibits extremely high P_{O_2}, higher than 900 Barrer, with the O_2/N_2 selectivity of 1.4. However, P_{O_2} was reduced by 17.5 times after 60 days at room temperature. This decrease in P_{O_2} may be explained in part to changes in the amount and distribution of free volume by an inter- and intra-molecular rearrangement of the polymer chains [38], frequently called relaxation, accompanied by the reduction in specific volume [39, 40], which is critically important and should be overcome for practical applications. PTMSP possesses a high glass transition temperature (T_g) in excess of 250 °C and degrades thermally before it reaches T_g [41].

Since then, many different polymers having fused aromatic rings at main chains have been developed to exhibit extremely stiff chains and high *f* for the improvement of the gas permeability as well as its long-term stability by preventing the rapid relaxation of polymeric chains. Typical examples include PIMs by Budd et al. [38] and TR polymers such as TR poly(benzoxazole)s and TR polyimides by Park et al.

[35]. Their representative chemical structures are shown in Figure 5.6b–d, and their properties are summarized in Table 5.4.

In general, strategies for improving microporosity in polymeric membrane materials could be classified into two categories: (1) introducing intrinsically microporous units; and (2) maximizing polymer chain rigidity to hinder efficient chain configurations. Considering practical productions, it is recommended that manipulating commercially available monomers with bulky fluorinated groups or noncoplanar structures is the most viable and traditional approach to induce intrinsically microporous characteristics. Note that commercially available noncoplanar monomers mainly have kink moieties, which result in twisted and crank-like structures. Thus, many researchers have extensively reported on the effects of changes in kink structures of commercial monomers on microporosity and gas transport behaviors [42].

Separation properties of PIMs and TR polymers will be revisited in a little more detail in the following subsection. In general, of course, such highly porous polymers are highly permeable but less selective in separation of common gas mixtures.

5.4.3.1 Polymers with Intrinsic Microporosity

Polymers with intrinsic microporosity (PIMs) have been introduced by Budd et al. [43–46]. The "microporous" structure in PIMs is due to the highly rigid and contorted molecular structures, which prevent efficient packing of the macromolecular chains in the solid state. A key feature of PIMs is their kinked backbone (spiro-type structure) and consequently highly restricted backbone rotational movements to make high fractional free volume. The "microporosity" of PIMs is termed "intrinsic" as it arises solely from their molecular structures but not derived from the thermal processing at hot temperatures. The chemical structures of the most studied PIM-1 and PIM-7 (Figure 5.6b and c) show the spirocyclic feature giving a kinked (contorted) conformation, and the formation of ladder-like dibenzodioxane structure for rigid backbones, contributing to the "intrinsic microporosity." PIM-1 and PIM-7 were the only two PIMs originally reported to form films sufficiently strong enough for membrane tests. However, low molecular weights and low yields have been issues in the synthesis of PIMs, which can make film formation difficult and contribute to poor mechanical properties [17].

PIMs have a density in the range 1.06–1.09 g/cm^3, large fractional free volume $f \sim 0.2$, and high specific surface areas (600–900 m^2/g, by nitrogen BET). Both PIM-1 and PIM-7 showed transport properties near the 2008 upper bound for CO_2/CH_4 separations as shown in Figure 5.4.

PIMs with varying backbones and pendant groups [47, 48] and PIMs-based systems, i.e. PIM-polyimides, recently, have been developed to get further improved separation performance [17].

5.4.3.2 Thermally Rearranged (TR) Polymers

Recently, Park et al. reported a new family of polymeric materials, termed "thermally rearranged, TR, polymers," for gas separations with an intermediate cavity size, a narrow cavity size distribution such as those found elegantly in nature in the form of ion channels with the chemical structure (Figure 5.6d) [35]. These polymers exhibit extremely high CO_2 permeability, high CO_2/CH_4 ideal selectivity, and excellent resistance to CO_2-induced plasticization. For example, polybenzoxazole (PBO, TR-1 polymer), the most widely studied TR polymer, based on a fluorinated diamine (6FDA) and *o*-hydroxy dianhydride, exhibits a high CO_2 permeability of roughly 2,000 Barrer and a high CO_2/CH_4 selectivity of roughly 40, with no evidence of plasticization up to 15.2 bar. PBO is formed via molecular thermal rearrangement of aromatic polyimides containing hydroxyl groups *ortho* to the imide ring as shown in Figure 5.7. A noticeable advantage of the TR approach is that it circumvents the typical insolubility of PBOs by starting from a soluble precursor hydroxy polyimide (Figure 5.7), which makes industrial processing such as hollow fiber spinning possible [17, 35].

Following initial publications, additional studies have continued to explore the structure-property relationship of TR polymers [36, 49]. However, most research efforts and some earlier fundamental studies on imide-to-benzoxazole conversion [50, 51] require high-temperature treatments (generally > 400 °C) to produce PBOs with good separation properties, and thermal degradation may overlap with the TR process after long treatment times, resulting in the poor mechanical properties of the TR membranes [52]. Therefore, reducing the TR temperature can be a key issue for industrial applications, which can be achieved by using a flexible bisphenol A type dianhydride. Reductions in conversion temperature of up to 100 °C (from ~450 to ~350 °C) have been reported using this approach [17, 53].

Polyimides with *ortho*-functional groups other than hydroxyl groups showed a much higher permeability after the TR process than TR polymers derived from its poly(hydroxyimide) analogs [52–54]. An increase in gas permeability as the size of the *ortho*-functional group increased was obtained with some loss in selectivity [53]. This *ortho*-position group is lost during thermal rearrangement, resulting in a large increase in permeability as thermal rearrangement temperature increases. Thus, the nature of the *ortho*-position group of the TR polymer precursor can be used to tailor permeability and selectivity in TR polymers [17].

As next-generation gas separation membranes, TR polymers have three major benefits: high combinations of permeability and selectivity, resistance to plasticization, and a high chemical resistance. TR polymers are promising candidates for high-performance membrane materials for gas separations. Their practical applications can be significantly advanced if the conversion temperature could be reduced to a more energy-efficient level and commercially available monomers could be used. Table 5.4 summarizes separation performance of PIM-1 and TR-1 polymers for CO_2/CH_4 separation. The TR polymers show generally higher selectivities at similar permeabilities when compared to PIMs [17].

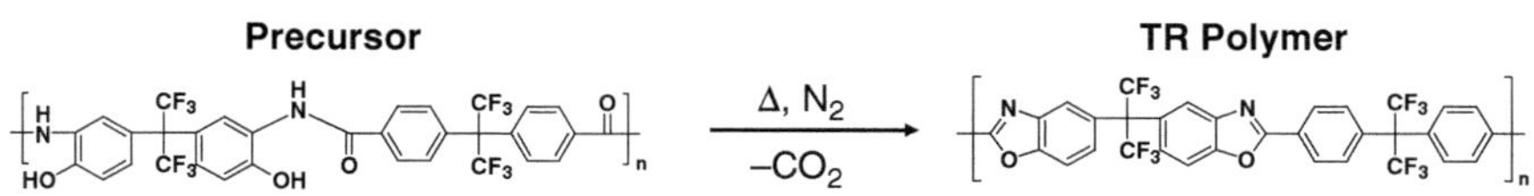

Figure 5.7 Thermal rearrangement of precursor of hydroxyl polyimide to thermally rearranged (TR) polymer of polybenzoxazole (PBO) at high temperature (350–450 °C). *Source*: Adapted from [35].

5.4.3.3 Transport Mechanism Through Polymers with Extremely Stiff Chains and High Free Volume

Gas transport through common polymers can be mostly explained by the solution–diffusion mechanism based on the free volume theory as described previously. The fundamental assumption for the free volume theory was that the redistribution of the free volume readily occurs even without extra-energy input [55] and consequently the chain flexibility and diffusion coefficient are directly correlated with the amount of the free volume [56], as evidenced in rubbery polymers. In the case of glassy state polymers, however, it is highly unlikely to see the redistribution of the free volume without extra-energy input simply because of the practically nearly frozen macroscopic chain mobility at glassy state or subglass temperatures. Nonetheless, the free volume theory (Equation (5.16)) has been successfully applied for describing both diffusion coefficient and permeability even in glassy polymers including extremely rigid backbone chains such as PIMs and TR polymers (Figure 5.8) as:

$$D = D_0 \exp\left(-\frac{E_d}{f}\right) \text{ and } P = P_0 \exp\left(-\frac{E_d}{f}\right) \tag{5.16}$$

In other words, small molecule transport through polymers having extremely rigid backbone chains and high fractional free volume f also follows the conventional solution–diffusion model, commonly represented by the simple free volume theory.

The exceptional increases in both permeability and selectivity in TR polymers have been attributed to an increase in f and a narrowing of the free volume distribution [17, 35]. Positron annihilation lifetime spectroscopy (PALS) and molecular modeling studies have shown that thermal rearrangement of polymeric chains at high temperatures up to 450 °C increases the average size of free volume elements and also makes the size distribution of these elements more uniform [35, 57].

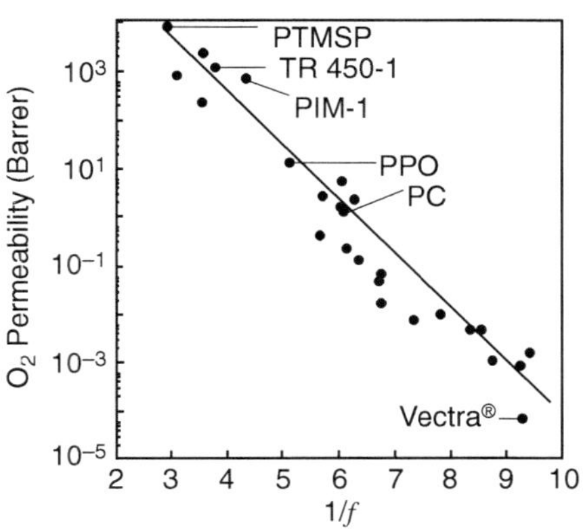

Figure 5.8 Oxygen permeability for polymers including poly(1-trimethylsilyl-1-propyne) (PTMSP), 6F-containing thermally rearranged polymer (TR450-1), polymer of intrinsic microporosity (PIM-1), polycarbonate (PC), poly(2,6-dimethyl-1,4-phenylene oxide) (PPO), Vectra (a commercial polyester liquid crystalline polymer), and various polymers. *Source*: Adapted from [17].

The conversion of the initial polyimide to PBO and the resulting increase in free volume lead to both an increase in diffusion and solubility coefficients [52, 54, 58]. However, diffusion coefficient provides a much larger contribution than solubility one to gas permeability increases. Thus, the rigid nature of PBO and large increase in *f* after thermal rearrangement are most likely critical to the large increases in the permeability.

However, it has been recently argued that such extremely high permeability observed may be attributed to a flow-like or sieving mechanism through rather large sized free volumes than the traditional solution–diffusion process. Of course, this idea is contentious and warrants additional research to clarify the transport mechanism. It is based on the notion that the free volume theory may not theoretically align with the observations that (1) polymers with highly rigid chains have a significant fractional free volume, and (2) high diffusion coefficient is observed in polymers with extremely rigid chains. In other words, the redistribution of the free volume could be highly unlikely in glassy state polymers, expecting low diffusion coefficient. However, high diffusion coefficient is experimentally observed in PIMs and TR polymers. Therefore, the gas transport could occur mostly through the fixed-sized free volume without the redistribution of the free volume, and consequently their transport behavior could be similar to the sieve mechanism to some extent, usually observed through carbon molecular sieve mechanism [17].

Pressure-driven convective pore-like flow or sieving transport mechanism is found for mass transport systems with free volume elements that exceed ~1 nm in characteristic dimension, and are fixed in size and position relative to the time scale of permeant motion [59]. Owing to the presence of relatively large free volume elements with the radius $r = 0.6 - 0.7$ nm in addition to the common-sized free volume element $r = 0.3$ nm [60], PTMSP emerges as a transitional polymer, with molecular transport occurring via a combination of convective or sieving permeation and solution–diffusion mechanisms [61]. Due to its high free volume, PTMSP behaves as an ultramicroporous membrane, with interconnected microvoids, resulting in pore flow-dominated gas transport [62]. Therefore, it can be proposed that, with a larger free volume size, the transport might resemble pore flow, where permeation through the membrane resembles similarly to flow through capillary tubes. This pore-flow-like transport might also contribute to the overall enhancement of permeability, albeit to a lesser extent, in addition to the solution–diffusion model through smaller-sized free volumes.

5.4.4 Emerging Polymer Materials

In addition to the polymer membrane materials described previously, several emerging polymers are being developed, each tailored for specific applications. For example, electronically conductive polymers, which exhibit specific interactions with oxygen, are well-suited for oxygen separation. Perfluoropolymers, known for their low solubility for hydrocarbons like methane, offer advantages in separating challenging gas pairs such as N_2/CH_4. In addition, machine learning technology has been recently used to design exceptional polymer membranes, and the resulting materials were highly successful. Thus, this subsection will explore these three types of emerging polymer membranes in some more detail.

5.4.4.1 Conductive Polymers

Electronically conductive polymers containing π-conjugated double bonds have shown interesting and high separation performance for gas mixtures [63–65]. For instance, remarkably high selectivity of O_2/N_2 up to 30 was able to obtain due to the reversible interactions between triplet state oxygen and polarons of polyaniline, as evidenced by electron spin resonance (ESR) measurements [66]. In addition, the high selectivity of 3,590 for H_2/N_2 and 336 for CO_2/CH_4 were reported through doped polyanilines [63–65]. The process for tailoring gas selectivity of the polyaniline membrane involves first enhancing the permeabilities of gases by doping with small diameter compounds (<0.35 nm) and undoping the polymer film with counter-ions of an appropriate size. High selectivities are then achieved by decreasing the permeabilities of larger gases (>0.35 nm in diameter) through controlled redoping of the polymer.

5.4.4.2 Perfluoropolymers

The strong C—F bond (485 kJ/mol) of perfluoropolymers and the protective sheath of fluorine atoms around the carbon backbone result in the extremely high chemical resistance and thermo-oxidative stability for these polymers. Therefore, they are highly attractive for applications in harsh environments. A major breakthrough in the use of perfluoropolymers for gas and vapor separation membranes is the development of a new family of amorphous glassy perfluoropolymers. Poly(tetrafluoroethylene) is one of benchmark perfluoropolymers but it is highly crystalline, so that it may not be suitable for gas separation applications. These new perfluoropolymers are mostly copolymers based on tetra-fluoroethylene and 2,2-bis(trifluoromethyl)-4,5-difluoro-1,3-dioxole, and the chemical structures of these completely fluorinated (i.e. perfluoro-) copolymers are shown in Figure 5.9a–c.

The bulky dioxole group disrupts chain packing in these materials, so the commercial grades of perfluoro-copolymers are wholly amorphous, unlike poly(tetrafluoroethylene). Therefore, high gas permeability and much improved processability, related to their high solubility in perfluorinated solvents, are obtained for these polymers. Similarly, other amorphous perfluoropolymers have been developed, such as Hyflon AD (produced by Solvay), and Cytop (by the Asahi Glass Company) as shown in Figure 5.9.

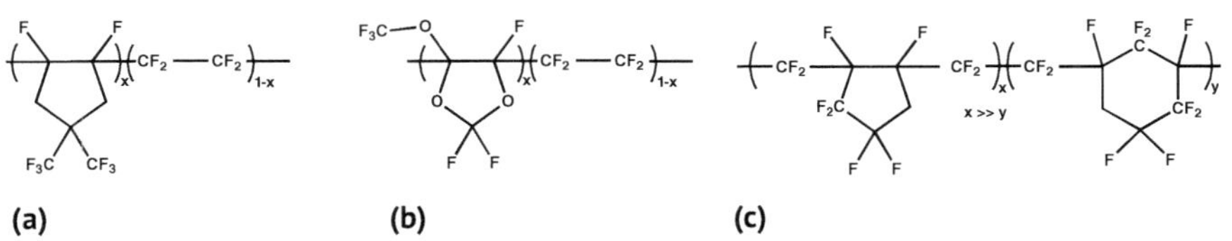

Figure 5.9 Chemical structure of commercial amorphous perfluoropolymers: (a) Teflon AF 1600 with $x = 0.65$ and Teflon AF 2400 with $x = 0.87$; (b) Hyflon AD60X with $x = 0.6$ and Hyflon AD80X with $x = 0.8$; and (c) Cytop. *Source*: Adapted from [17].

The gas transport properties of these materials have been extensively studied [67, 68]. Perfluoropolymers show gas separation properties well below the 2008 upper bound for CO_2/CH_4 and O_2/N_2. However, one potential use for this class of polymers is the separation of nitrogen from natural gas. This separation, simplified to N_2 and CH_4, would be helpful to improve the heating quality of natural gas. Perfluoropolymers, along with TR polymers, exhibit performance near the 2008 upper bound for this difficult separation. This improved performance for N_2/CH_4 separations is attributed to fluorinated polymers having high light gas such as N_2 solubility relative to hydrocarbon like CH_4 solubility. The low solubility of hydrocarbons in perfluoropolymers has been implicated in improved plasticization resistance for hydrocarbon-based separation (e.g. olefin/paraffin separations). The reason for the low solubilities of hydrocarbons in perfluoropolymers is not well-understood [17].

5.4.4.3 Machine Learning Technology

Interestingly, the machine learning technology has been highly successful in designing novel membrane materials to exceed the so-called upper bound, opening a future research direction [15]. The two polyimides containing sulfone groups designed by the machine learning technology are poly[(1,3-dioxoisoindoline-2,5-diyl)sulfonyl(1,3-dioxoisoindoline-5,2-diyl)-1,4-phenyleneoxy-1,4-phenylene] (poly(imide-sulfone-ether) and poly[(1,3-dioxoisoindoline-2,5-diyl)sulfonyl (1,3-dioxoisoindoline-5,2-diyl)-1,4-phenylenemethylene-1,4-phenylene] (poly (imide-sulfone-methylene) (Figure 5.10a,b). Both poly(imide-sulfone-ether) and poly(imide-sulfone-methylene) exhibit selectivities ~7 and 5.5 times higher, respectively, than the upper bound at the same permeability value. Furthermore, they found than the experimental and predicted data points are in relatively good agreement with each other, indicating that the machine learning model may be used as a predictive tool in designing previously unexplored polymers for gas separations.

Machine learning technology has been recently utilized to design membrane materials by controlling the fractional free volume f or FFV [69]. The strong correlation has been demonstrated between machine learning predicted f and molecular dynamic (MD) simulated f with the correlation factor of $R^2 = 0.92$ for polyamides and $R^2 = 0.94$ for homopolymers.

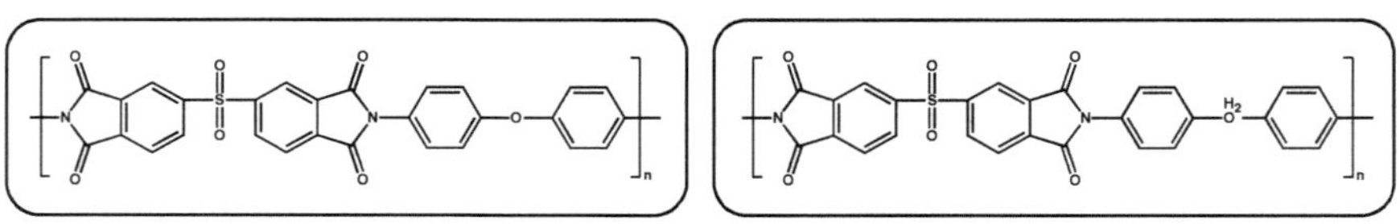

(a) (b)

Figure 5.10 Chemical structures of two polymers designed by the machine learning technology for advanced CO_2/CH_4 gas transport performance: (a) poly[(1,3-dioxoisoindoline-2,5-diyl)sulfonyl(1,3-dioxoisoindoline-5,2-diyl)-1,4-phenyleneoxy-1,4-phenylene] (poly(imide-sulfone-ether); and (b) poly[(1,3-dioxoisoindoline-2,5-diyl) sulfonyl(1,3-dioxoisoindoline-5,2-diyl)-1,4-phenylenemethylene-1,4-phenylene] (poly(imide-sulfone-methylene). *Source*: Adapted from [15].

5.5 Time-dependent Separation Performance: Physical Aging and Plasticization Effects in Glassy Polymers

Since polymers in glassy state are not in equilibrium, polymer chains relax extremely slowly over time, gradually approaching an equilibrium state even at room temperature. This slow relaxation process is known as physical aging. As a result, glassy polymers continuously undergo subtle structural or conformational changes, which in turn cause their physical properties to evolve gradually over time [70]. Consequently, the separation performance of these materials will be time-dependent.

When polymeric membranes are exposed to highly soluble diluents such as CO_2 and *n*-hexane vapor, their specific volume decreases with time by physical aging at low chemical activities (low pressures) of the diluent as shown in Figure 5.11. However, at high chemical activities (high pressures) beyond a critical point, the specific volume increases. In this case, the polymer with diluent becomes rubbery with the critical point marking the glass transition, where the diluent acts as a plasticizer – this process is known as plasticization.

Therefore, both physical aging and plasticization accompany the changes in structure and corresponding physical properties, including the specific volume, which significantly affects the transport properties of gas and diluent. In this context, the impact of physical aging process in neat polymers, as well as the combined effects of physical aging and plasticization in polymers with highly soluble diluents on permeation properties will be explored.

5.5.1 Physical Aging and Permeability Changes

PTMSP is amorphous and highly permeable, which shows serious time-dependent permeability. Typical structural changes in PTMSP are exemplified from the O_2 permeability decrease about 17.5 times after 64 days at room temperature because of the specific volume decrease from 1.1811 to 1.1670 cm^3/g (decrease about 1.2%). From the experimental data of PALS along with the specific volume and the CO_2 permeability, a model of a channel network is proposed for PTMSP with large free

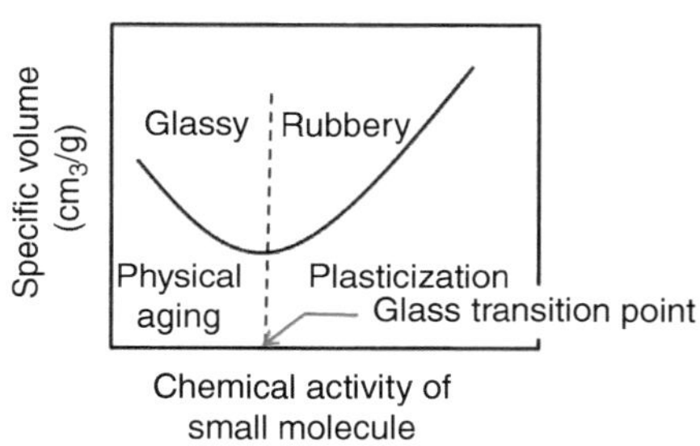

Figure 5.11 Schematic drawing of specific volume change as a function of chemical activity of small molecules, including physical aging at the glassy state and plasticization at rubbery state. *Source*: Y.S. Kang.

volumes (mean radius $r = 0.75$ nm) connected by channel-like free volumes (mean radius $r = 0.45$ nm). The number of large free volumes or holes decreases by aging, but not their size, whereas the number of small holes does not change but their size decreases [39, 40].

Additionally, the physical aging rate, A, has been observed to be roughly correlated with the fractional free volume f (or FFV) for a number of polymers as shown in Figure 5.12. Here, the physical aging rate A is defined as $A = -(\mathrm{d}\ \ln V/\mathrm{d}\ln\ t)_{p,T}$ at a given gas pressure p and temperature T with V and t being the specific volume and time, respectively. All films were approximately 400 nm thick. These results indicate that higher free volume glassy polymers tend to age more rapidly as expected than lower free volume polymers, all other factors (e.g. thickness) being equal [17, 71]. Interestingly, PSF has the lowest A value due to the lowest f value, representing that gas transport properties do not change appreciably with time, which is critical for practical applications.

5.5.2 Physical Aging and Plasticization with Highly Sobule Diluents

When a glassy polymer at nonequilibrium state is in contact with highly soluble diluent such as CO_2 or organic vapor including n-hexane, the physical aging will be accelerated to decrease the specific volume as described above and consequently the permeability of the diluent as shown in Figure 5.13a. The decrease in the specific volume and CO_2 permeability is strongly associated with the increased micro-segmental mobility of polymeric chains by the highly mobile diluent. As increasing CO_2 pressure, the permeability decreases and reaches a minimum at a certain activity, which is called as a glass transition point.

When the CO_2 pressure is further increased beyond the glass transition point, the polymer is plasticized by the diluent CO_2 to become a rubbery state and

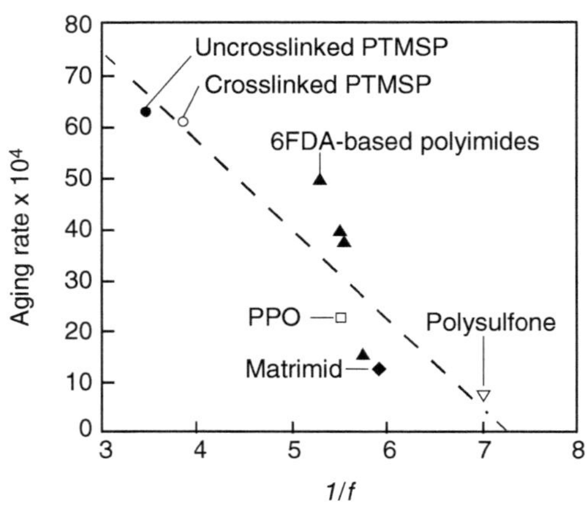

Figure 5.12 Effect of fractional free volume f on the physical aging rate A of selected polymers including cross-linked and uncross-linked poly(1-trimethylsilyl-1-propyne) (PTMSP), 6FDA-base polyimide, poly(2,6-dimethyl-1,4-phenylene oxide) (PPO), Matrimid, and polysulfone. *Source*: Replotted from [17, 71].

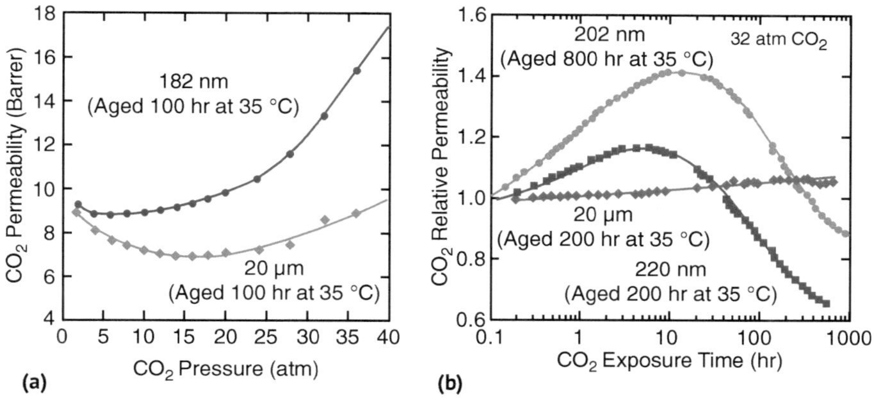

Figure 5.13 (a) CO_2 plasticization pressure curves for thin and thick films with the identical prior thermal history of Matrimid. The thin and thick films spent 3 and 10 min, respectively, at each intermediate pressure. (b) Effect of aging time and thickness on long-time CO_2 exposure behavior. *Source*: Ref. [76]/with permission of Elsevier.

consequently the CO_2 permeability increases with the applied pressure, referred to as a plasticization effect as shown in Figure 5.13a. Thus, it can be noted that the diluent permeability change is strongly associated with structure and mobility changes depending on chemical activity of the gas applied by either physical aging or plasticization effects: The CO_2 permeability decreases at low pressures, representing the physical aging effects, reaches a minimum at the CO_2 pressure of about 14 atm for the 20 μm sample, corresponding to the glass transition point, followed by the increase in the CO_2 permeability by the plasticization effects [17, 72–74]. In addition, the diluent permeability also abnormally changes with storage time, sample thickness, and history.

Figure 5.13b shows experimental data for the thickness-dependent permeation behavior. Interestingly, the thinner samples (202 and 220 nm) show much higher the CO_2 permeability than the thicker sample (20 μm), and the maximum point is also different, which is called the thickness-dependent permeation behavior, a typical non-Fickian behavior. During the first 5 hours, the transport properties of the thin films (202 and 220 nm) are dominated by plasticization and permeability increases. However, at longer times, the transport properties are dictated by physical aging, and a decrease in permeability is observed. The difference in shape of the curves suggests that the thin film experiences substantially more plasticization in short times, even at lower pressures [75, 76]. The CO_2 permeability in polymers, including polyimide (6FDA-TeMDP) [77, 78], cellulose acetate, polyimide (Matrimid), tetramethyl bisphenol A polycarbonate (TM-PC), and poly(2,6-dimethyl-1,4-phenylene oxide) (PPO) [79] also shows the similar physical aging and plasticization effects. These data suggest that thin films respond relatively more quickly and intensely to plasticizing gases, such as CO_2, than thick films.

Glassy polymers undergoing plasticization may also show hysteresis in permeation properties, with transport properties at a given pressure being different depending on whether permeability was measured during pressurization or depressurization. This effect has been observed in many different polymers, but the degree of hysteresis often varies from polymer to polymer [17, 80].

As a short summary, when highly soluble gas such as CO_2 is used for separation in glassy polymers, both physical aging and plasticization effects depending on the applied pressure used should be carefully concerned on the CO_2 permeability and the selectivity. In any circumstances, to maintain stable long-term membrane performance for industrial applications, we have to take a care of the physical aging and the plasticization effects too as much as possible in addition to the intrinsic material properties.

5.6 Facilitated Transport Membranes in the Solid and Quasi-solid States

Applying facilitated transport phenomena to gas separation membranes is a very attractive and effective approach to improve permeability and selectivity simultaneously, which are usually in conflict among common polymer materials. Solid-state facilitated transport membranes consist of a carrier C dissolved or dispersed in a solid-state polymer matrix, where the carrier C interacts reversibly only with a specific solute or gas A in the gas mixture ($A + B$). The *reversible* and *specific* interaction of carrier C with solute A results in carrier-mediated transport in addition to normal Fickian transport, thereby facilitating transport. Therefore, carrier properties would be critically essential in improving the separation performance.

$$A + C \underset{k_2}{\overset{k_1}{\rightleftarrows}} [A \cdot\cdot C] \qquad K = k_1/k_2 \tag{5.17}$$

where k_1 and k_2 are the forward and the backward reaction rate constants, respectively, and K is the equilibrium constant ($K = k_1/k_2$) and $[A \cdot\cdot C]$ is the solute–carrier complex.

The mathematical equation for the concentration fluctuation model is adapted from Chapter 4 as:

$$\frac{P}{P_0} = 1 + \left(\frac{p_f}{p}\right)\sqrt{n^2 + \left\{\frac{2\pi k_2 C_C L^2}{P_0}\frac{\ln(1 + K_p)}{p}\right\}^2} \tag{5.18}$$

where P and P_0 are permeabilities of the facilitated transport membrane in the solid state and of the membrane matrix without carrier, respectively. p_f and p are the pressure fluctuation due to the reversible reaction and the applied pressure with vacuum at downstream side, respectively. n is the number of carrier for the solute to meet during one dimensional diffusion pathway across a membrane, calculated from $n = NC_c(\pi r^2 L)$ with N being the Avogadro number, L the film thickness, and r the permeant radius. k_2 and K $(= k_1/k_2)$ are the backward reaction rate constant and the equilibrium constant of the reversible carrier-solute reaction, respectively. For detailed information of Equation (5.18), go back to Equations (4.11) and (4.12) in Chapter 4 or refs. [81, 82].

Equation (5.18) can be simplified by assuming that when $n \ll (2\pi k_2 L^2 C_c / P_0)$ $(\ln(1 + Kp)/p)$, which is applicable under typical experimental conditions as:

$$P \propto k_2 C_C L^2 \ln(1 + Kp)/p^2 \qquad (5.19)$$

This equation states that the carrier concentration C_C and the carrier properties, in particular k_2 and K, should be high to improve the permeability P of component i, at given P_0, L, and T, and thereby concomitantly to improve the selectivity. Another important thing is that k_2 is more efficient in influencing the permeability than K for a given carrier concentration C_C. This is because k_2 is directly proportional to the permeability P while K is inside of logarithm $(\ln(1 + K/p))$. Therefore, we have to keep in mind that the carrier should have high values of both k_2 and K, but k_2 is more efficient in improving the permeability.

In this context, much research has been conducted to develop efficient carriers for specific molecules such as O_2, CO_2, olefin, etc. For example, new carriers have recently been developed for the separation of olefins from their mixtures with the corresponding paraffins, providing excellent separation performance and durability [83–86]. In this respect, the field of facilitated transport membranes in solid-state will attract great interest in the near future. In the following three sections, the separation performance of O_2/N_2, olefins/paraffins such as ethylene/ethane, propylene/propane, and finally CO_2/N_2 and CO_2/CH_4 mixtures will be briefly introduced.

5.7 Oxygen Separation with Facilitated Transport Membranes

Air separation has been attractive industrially because it can be separated into either nitrogen- or oxygen-enriched air by membrane process. In fact, a membrane process to generate the nitrogen-enriched air (the nitrogen concentration up to around 95%) has been commercialized already a long time ago. In the case of the oxygen-enriched air, the membrane process has been also commercialized, but mostly limited to medical applications yet, primarily because of the poor separation performance of membrane materials. Therefore, a new membrane material delivering the high oxygen permeability as well as the high oxygen selectivity over nitrogen has been strongly required to expand the air separation market.

It has been known that cobalt(II) (Co^{++}) and ion(II) (Fe^{++}) coordinated by 4 nitrogen atoms of porphyrin interact *specifically* and *reversibly* with oxygen to form complexes and therefore they can be utilized as an oxygen carrier for facilitated oxygen transport.

$$\mathbf{Co^{++} + O_2} \underset{k_2}{\overset{k_1}{\rightleftarrows}} [\mathbf{Co^{++} \cdot\cdot O_2}] \qquad K = k_1/k_2$$

A representative facilitated oxygen transport membrane containing cobalt(II) porphyrin/imidazole (CoPIm) in poly(butyl methacrylate) (PBMA) (PBMA/CoPIm) showed that the oxygen permeability and its ideal selectivity over nitrogen were improved simultaneously to ~8.7 Barrer and 4.3 [87, 88], respectively, at 70 mmHg

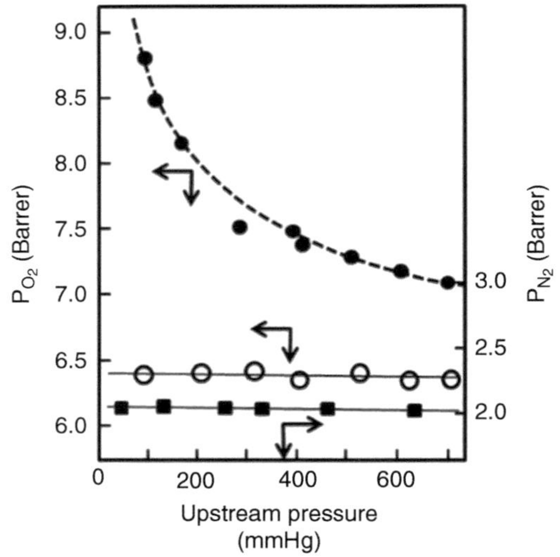

Figure 5.14 Effect of upstream pressure on oxygen (●) and nitrogen (■) permeabilities in the PBMA membrane containing 2.5 wt% CoPlm and oxygen permeability (○) in the membrane without CoPlm. *Source*: Ref. [88]/American Chemical Society.

(Figure 5.14). The neat PBMA membrane without CoPIm exhibits the oxygen permeability of 6.4 Barrer and the selectivity of $O_2/N_2 \sim 3.0$ independent of the applied pressure.

It is interesting to note that when Co(II) or Fe(II) is oxidized to Co(III) or Fe(III), respectively, Co(III) or Fe(III) is no longer active as an oxygen carrier and consequently the facilitation effect is nullified. Therefore, such oxidation reaction of Co(II) or Fe(II) should be prevented for sustainable high oxygen separation performance.

5.8 Olefin Separation with Facilitated Transport Membranes

Separation of olefins from its mixture with paraffins has been mostly conducted by cryogenic distillation process, which is known to be very energy-intensive. Therefore, there have been a lot of attempts to develop a membrane separation process to compete with and ultimately to replace current cryogenic processes. The most essential properties of membrane process are to have a membrane material exhibiting both high permeability and high selectivity simultaneously. However, conventional polymeric materials show the so-called trade-off relationship between the permeability and the selectivity.

In this respective, facilitated transport membranes in the solid state become very attractive, which critically require efficient olefin carriers. Quite recently, two different olefin carriers have been developed to demonstrate high separation performance as well as long-term sustainability. The first olefin carrier is metallic ions such as silver or copper ions [86, 89], and the second one is the metallic nanoparticle such as silver, copper, and gold nanoparticles with an induced partial positive charge on their surface by appropriate electron polarizers [83–85]. In this particular case, the polarizer becomes very important in determining the olefin carrier activity of metallic nanoparticles and consequently the separation performance of olefin/paraffin mixtures. In Chapter 4, fundamentals on facilitated transport in the solid state with metallic ion and metallic nanoparticle carriers are thoroughly reviewed. Here, their selective properties and potential applications will be briefly reviewed.

5.8.1 Polymer Electrolyte Membranes Containing Metallic Ion Carriers

Metallic ions (M^+) such as silver (Ag^+) and copper (Cu^+) ions have been known to interact *specifically* and *reversibly* with olefin ($CH_2{=}CH{-}R$), but not with paraffin, and consequently are an active olefin carrier.

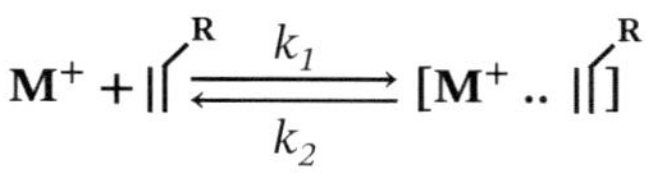

$K=k_1/k_2$ (5.20)

Such metallic ions can be readily generated by dissolving metallic salts into a polar polymeric matrix to form a polymer electrolyte. For instance, common silver salts such as $AgBF_4$ are readily dissolved into a polymeric matrix such as PEO, poly(2-ethyl oxazoline) (POZ) and poly(*N*-vinyl pyrrolidone) (PVP) to produce silver polymer electrolytes [90]. Therefore, there has been a large body of researches on polymer electrolyte membranes (PEMs) containing metallic salts such as silver and copper salts.

5.8.1.1 Separation Performance

The silver polymer electrolyte membrane (Ag-PEM) containing either $AgBF_4$ or $AgCF_3SO_3$ dissolved in either POZ or PVP was coated onto an asymmetrically microporous PSF support for preparing a thin film composite membrane. The Ag-PEM layer becomes a selective layer for the composite membrane and its thickness was estimated to be ~1.5 μm by TEM images. The composite membrane showed that the propylene permeance $Q_{\text{propylene}}$ nearly linearly increased with the salt concentration added in the Ag-PEM layer. The propylene permeance $Q_{\text{propylene}}$ through the composite membrane of POZ/$AgBF_4$ and PVP/$AgBF_4$ was achieved up to 45 GPU (1 GPU = 1.0×10^{-6} (cm^3 (STP)/(cm^2 s cmHg))), whereas the propane permeance Q_{propane} was maintained nearly unchanged only about 0.05 GPU (Figure 5.15a), resulting the ideal selectivity up to 900, which is far above the performance of conventional polymers as shown in Figure 5.15b. Here, the permeability was estimated from $P = Q \times L$ with L is the estimated thickness of the selective layer of Ag-PEM in the composite membrane. As can be seen, the propylene permeability and the ideal selectivity of propylene over propane were improved simultaneously approximately 50- and 200-fold, respectively, compared to common polymeric materials without silver salts. $AgBF_4$ is more effective as an olefin carrier than $AgCF_3SO_3$ in this experimental range. With these facilitated transport membranes, the propylene concentration higher than 99% by volume was able to obtain from the 50% feed mixture with propane by a single pass of the membrane (Figure 5.16).

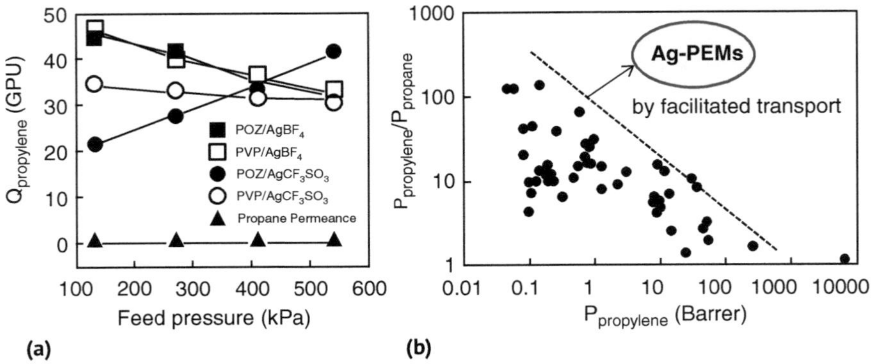

Figure 5.15 (a) Propylene permeance change with the feed pressure through silver polymer electrolyte membranes (Ag-PEMs). *Source*: Replotted from [86].
(b) Separation performance of facilitated transport membranes of Ag-PEMs.
Source: Y.S. Kang.

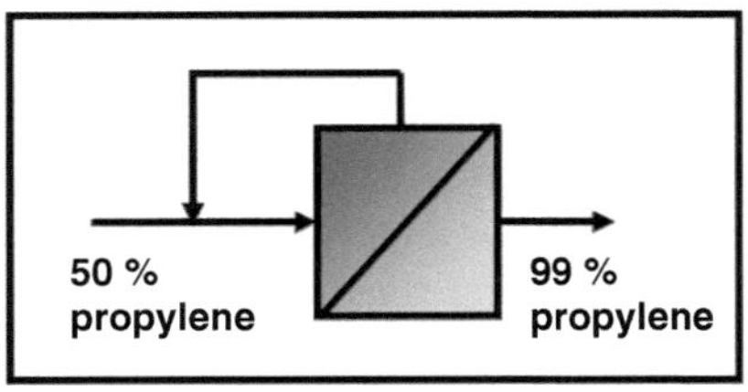

Figure 5.16 Schematic separation performance of propylene/propane mixture (50/50 v/v) and the permeate concentration at the downstream side of up to higher than 99%. *Source*: Y.S. Kang.

In the case of PEMs of PEO/$AgBF_4$ ([EO]:[Ag] = 1:1, ~80 wt% of silver salt) the pure ethylene and propylene permeances were 55 and 48 GPU, respectively, whereas the ethylene permeance through a neat PEO without $AgBF_4$ was only about 0.4 GPU. The ideal selectivities of ethylene/ethane through PEO/$AgBF_4$ and neat PEO were nearly 150 and only 1.2, respectively [89, 91, 92]. The results indicate that the olefin permeance increases 50–100 times higher than that of the pure PEO membrane, and the ideal selectivity increases up to 100-fold. In the case of mixed ethylene/ethane (70/30), the ethylene and ethane permeances were ~10 and 0.1 GPU. However, the separation performance decreased slightly with time up to 40 days, respectively [89, 91, 92].

5.8.1.2 Long-term Operational Stability and Safety

Ag-PEMs exhibited an extremely high separation performance for olefin/paraffin mixtures in the solid state, but a question remains about the sustainability of the carriers for practical applications. The facilitated transport membranes frequently suffer from a lack of long-term durability because of the deactivation of the carriers. For example, Ag-PEMs consisting of $AgBF_4$ dissolved in POZ, PVP, or PEO showed a continuous decrease in the separation performance during use with time. Figure 5.17 illustrates that the propylene selectivity over propane in POZ/$AgBF_4$ membrane continuously decreases with operation time t [93]. This is primarily due to the reduction of silver ions (Ag^+) to metallic silver (Ag^0) or nanoparticles.

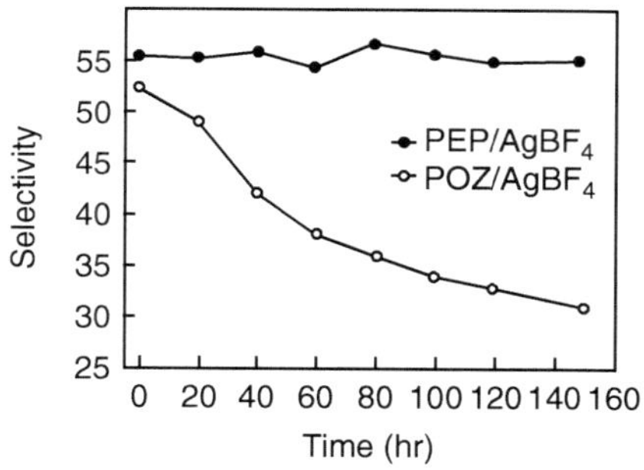

Figure 5.17 Selectivity of propylene/propane in 1/1 POZ/$AgBF_4$ (○) and 1/1 PEP/$AgBF_4$ (●) membranes as a function of time *t* at 40 psig and 20 °C, where POZ and PEP are poly(2-ethyl oxazoline) and poly(ethylene phthalate), respectively. *Source*: Adapted from [93].

Therefore, there have been several different approaches to prevent the reduction of silver ions as much as possible to improve sustainability: the use of (1) noninteractive polymer matrix with silver salts, (2) strongly interactive polymer matrix, and (3) additives has been reviewed briefly as follows.

The first approach is to utilize noninteractive polymer matrix with silver salts. Reduction of silver ions to metallic silver always involves one electron transfer to silver ion. If we eliminate all the potential sources of electrons, the reduction reaction of silver ion to metallic silver may hardly occur. Therefore, it is anticipated that polymers, noninteractive with silver salts, would not provide the electrons necessary for the reduction reaction. On this basis, a noninteractive, inert PDMS matrix was employed for silver salts to be dispersed, resulting in a significantly improved long-term durability along with high separation performance [94].

When silver salts are dispersed in a PDMS matrix to make a selective layer, silver ions will not be initially generated due to the immiscibility between the silver salt and PDMS, and therefore no olefin carrier is present for facilitated transport. However, once such salts are exposed to propylene, they would be dissolved to be an olefin carrier due to the interactions between the silver salt and the double bond of propylene.

Time evolution wide-angle X-ray scattering (WAXS) spectra of PDMS/$AgCF_3SO_3$ (1/0.73 w/w) membranes at a propylene pressure of 138 kPa show that the crystalline peaks of $AgCF_3SO_3$ aggregates were clearly seen initially but completely disappeared after 2 hours of the propylene exposure time as shown in Figure 5.18a. Correspondingly the propylene/propane selectivity increases with the expose time and reaches a high steady state value after approximately two hours (Figure 5.18b). These results suggest that silver salts dispersed in the hydrophobic PDMS matrix were not initially active to generate silver ions as an olefin carrier without propylene exposure, but became active after 2 hours operation time at 138 kPa to facilitate the propylene transport, consequently improving the permeability and the selectivity, simultaneously [94, 95].

This is the first observation of this type of separation performance in solid state facilitated transport membranes. Note also the unusually high separation performance of the PDMS/$AgBF_4$ membrane with a silver salt weight fraction of 0.73.

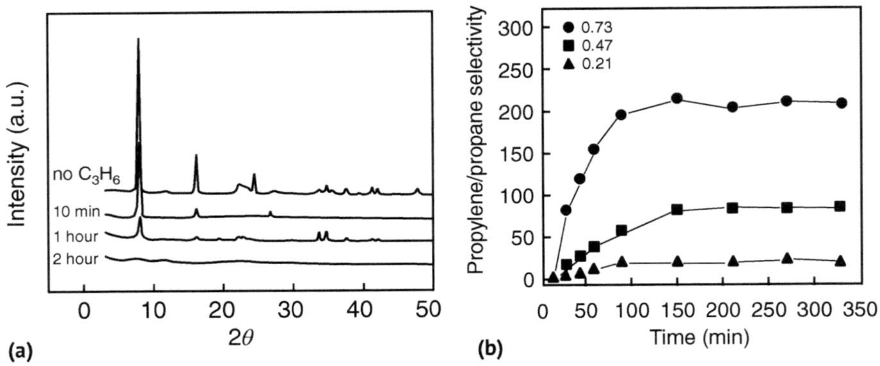

Figure 5.18 (a) Time evolution wide-angle X-ray scattering (WAXS) spectra for poly(dimethyl siloxane) (PDMS)/$AgCF_3SO_3$ membranes containing 0.73 of silver weight fraction by increasing the propylene exposure time at 138 kPa. *Source*: Reference [94]/ with permission of Elsevier; and (b) the changes in propylene/propane selectivity through PDMS/$AgBF_4$ membranes with different weight fractions of $AgBF_4$ as a function of the experiment time. *Source*: Adapted from [94].

The mixed-gas selectivity reached nearly 200 with a gas permeance of ~15 GPU. As a result, the reduction of silver ions to metallic silver atoms can be effectively prevented when a noninteractive matrix such as PDMS was used to fabricate a composite membrane.

The second approach is to utilize a polymer matrix having strong interactions with silver ions, which is conceptually opposite to the use of noninteractive, hydrophobic polymer PDMS as described previously. The complexation energies (ΔE_C), a representative of the interaction energy for polymer/silver complexes, were calculated by the density functional theory at the B3LYP level at 298 K. ΔE_C of polymer/Ag^+ is −50, −51, −43, −42, and −94 kcal/mol for POZ, PVP, PVMK (poly(vinyl methyl ketone), PMMA, and PEP (poly(ethylene phthalate)), respectively. The PEP/Ag^+ complex has the ΔE_C value almost two-fold higher than the others. Thus, the PEP/Ag^+ complex membrane exhibited much better long-term durability in the separation of propylene/propane mixtures than the POZ/Ag^+ membrane. The selectivity and permeance of the PEP/$AgBF_4$ membrane were nearly unchanged for the duration of 150 hours (Figure 5.17). This improvement is due to the suppression of the reduction of silver ions to silver metal by the formation of stable chelates between the phthalic groups of PEP and the silver ions [93].

Using additives for Ag-PEMs is another interesting approach. $Al(NO_3)_3$ is also an effective additive to prevent the reduction of silver ions in PEMs to metallic silver particles [96]. For instance, when $Al(NO_3)_3$ was added into the POZ/$AgBF_4$ polymer electrolyte, the reduction of silver ion was suppressed for longer than 14 days. When an ionic liquid (IL) of 1-butyl-3-methylimidazolium [Bmim][BF_4] was additionally incorporated as an additive, $AgNO_3$ can also be effectively utilized as an olefin carrier. Note that $AgNO_3$ itself does not act as an olefin carrier even dissolved in polar polymers.

It has been also noticed that silver ions may readily react with acetylene and sulfur compounds to reduce the carrier activity, which are commonly present as impurities in olefin/paraffin mixture. Therefore, such chemicals should be carefully treated to be removed as much as possible before use. Finally, operational safety could be the most important issue we have to consider for practical applications. Silver acetylide, possibly formed by reaction of silver ions with gaseous acetylene, is known to be an explosive and therefore the acetylene concentration should be controlled to be as small as possible to avoid the formation of silver acetylide [97].

5.8.2 Surface-activated Metallic Nanoparticle Carriers

When surface of silver nanoparticles (Ag NPs) is positively charged using a proper electron accepting polarizer, they can act as an excellent olefin carrier to facilitate the olefin transport markedly, while maintaining the paraffin transport nearly unchanged, as described in Section 4.5 of Chapter 4. Common polarizers used include *p*-benzoquinone (*p*-BQ), tetra-cyanoquinodimethane (TCNQ), nitro-aromatics, and tetrathiafulvalene (TTF) [98]. Then the surface charge density on Ag NPs, characterized by the binding energy of silver atom, is found to be linearly correlated with the olefin permeability as shown in Figure 4.38. Therefore, controlling the surface charge would be a major concern in improving olefin transport.

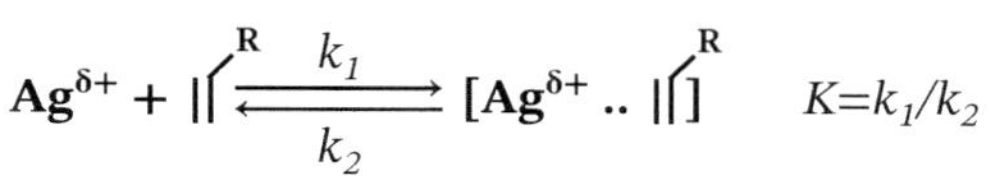

(5.21)

For instance, *p*-BQ induces the surface positive charge of Ag NPs (diameter ~70 nm) and results in an ideal selectivity of propylene to propane reached almost 170 when they are dispersed in poly(ethylene-co-propylene) rubber (EPR) for EPR/Ag NPs/*p*-BQ (1/1/0.85 by weight) (Figure 5.19a) [85], which was deposited on an asymmetrically microporous PSF support to form a composite membrane. These results first demonstrate that *p*-BQ can make the surface of silver nanoparticles positively charged δ^+ and consequently act as an olefin carrier for facilitating olefin transport.

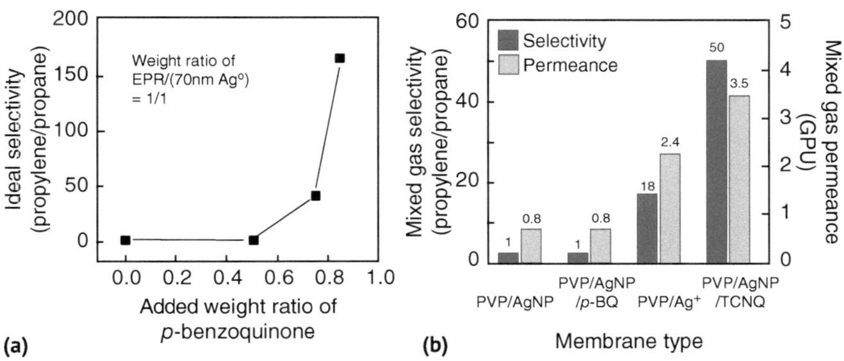

Figure 5.19 (a) Ideal selectivity through 1/1 EPR/Ag°/*p*-BQ membranes with varying weight ratios of *p*-BQ to EPR. *Source*: Ref. [85]/John Wiley & Sons. (b) The separation performances of various facilitated transport membranes for 50:50 (v/v) propylene/propane mixture. *Source*: Ref. [84]/John Wiley & Sons.

TCNQ is a stronger electron acceptor than *p*-BQ. Consequently, silver nanoparticles modified with TCNQ are expected to exhibit a higher surface charge density compared to those modified with *p*-BQ. This enhanced charge density suggests improved separation performance, as demonstrated in Figure 5.19b [84]. The membrane containing silver nanoparticles activated by TCNQ dispersed in poly(*N*-vinyl pyrrolidone) (PVP) (PVP/AgNPs/TCNQ [1/0.5/0.01]) increases the mixed-gas selectivity to almost 50, the highest value ever reported [84]. Most notably, the mixed-gas selectivity increased with the surface charge density δ^+ of Ag NPs as expected, characterized by the binding energy of Ag $3d_{5/2}$ orbital, regardless of the polarizers used including TTF [98], dinitro-toluene (DNT) and dinitro-benzene (DNB) (Figure 5.20) [83]. Interestingly, the insulating polarizer such as DNT and DNB appears to be more sensitive than the semi-conductor polarizer such as TCNQ and TTF. These results indicate that the separation performance can be readily improved by increasing the charge density of Ag NPs, which is a huge advantage for designing the molecular structure of olefin separation materials. Gas permeance is even better, compared to the membranes with silver ion carriers.

The PVP/Ag NPs/TCNQ (1:0.5:0.01) membrane delivers outstanding separation performance for a commercial propylene/propane (95/5 v/v) mixture (Figure 5.21). It achieves a propylene concentration exceeding 99.8% in a single pass and sustains this level for over 450 hours (Figure 5.22). This demonstrates the high efficiency and long-term stability of the membrane, making it a sustainable solution for olefin/paraffin separation. Notably, its performance surpasses the 99.7% propylene purity

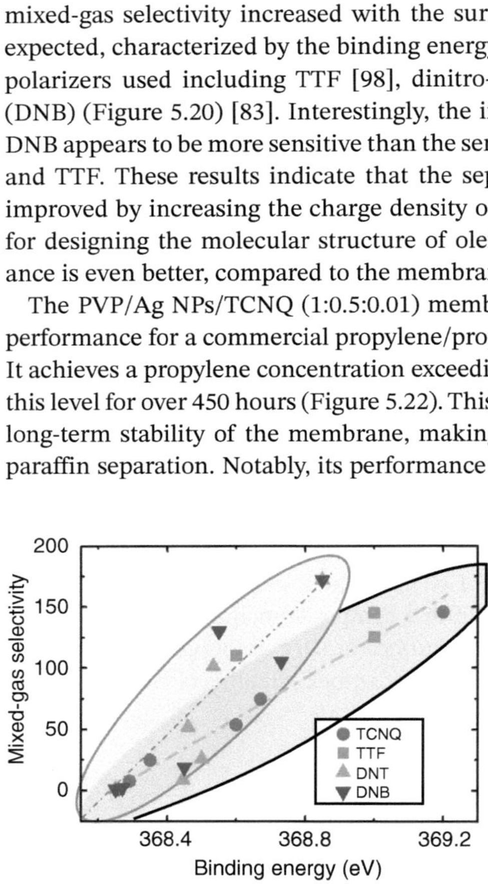

Figure 5.20 The relationship between the mixed-gas selectivity and the binding energy of $3d_{5/2}$ orbital of a silver atom. *Source*: Adapted from [99].

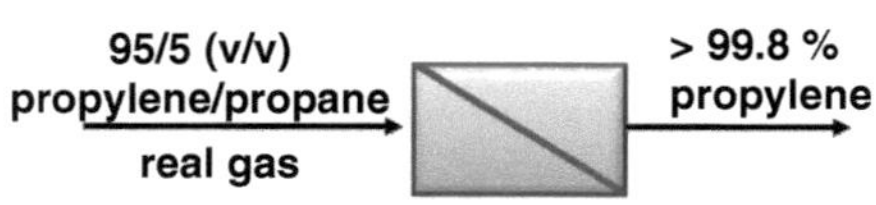

Figure 5.21 Separation performance of poly(*N*-vinyl pyrrolidone/silver nanoparticles/tetra-cyanoquinodimethane (1:0.5:0.01) membrane for real gas containing 95% propylene as a feed to increase the propylene concentration up to 99.8%. *Source*: Y.S. Kang.

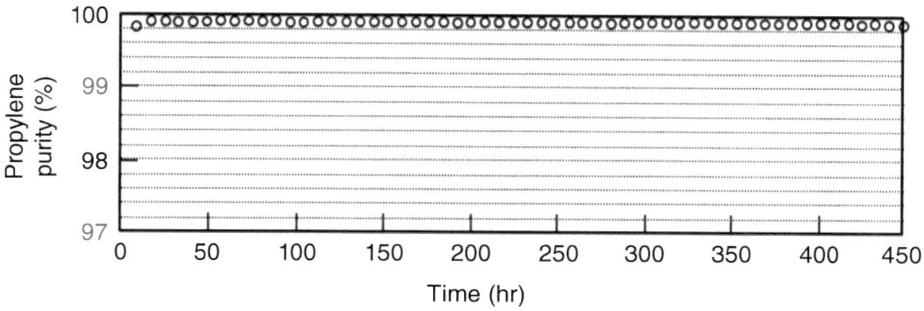

Figure 5.22 The propylene concentration at the permeate side through the poly(*N*-vinyl pyrrolidone/silver nanoparticles/tetra-cyanoquinodimethane (1:0.5:0.01) membrane as a function of time. Feed (real gas of propylene/propane (94.88/4.87 v/v)), flow rate of 5 NL/h at the temperature of 23 °C and pressure of 5.0 kg/cm^{2}g. *Source*: Y.S. Kang.

required for polymerization-grade applications, eliminating the need for further treatment.

In summary, solid-state facilitated transport membranes containing Ag NP carriers activated by *p*-BQ, TCNQ, TTF, DNT, or DNB, and then dispersed in a polymer matrix of EPR or PVP are used to improve the separation performance of olefin/paraffin mixtures such as ethylene/ethane and propylene/propane mixtures at room temperature. Additionally, the sustainability of separation performance has been exceptional. For example, the propylene feed concentration of 95% increased to higher than 99.7% after one pass through the membrane, and this high separation performance remained unchanged up to 450 hours, suggesting very high sustainable separation performance. Silver acetylide formation was also not evidenced during an exposure time of 72 hours at acetylene pressure of 1 atm and 25 °C [97]. It can be concluded that the high charge density of the silver nanoparticles plays the most important role in enhancing the separation performance of olefin/paraffin mixtures [100].

5.9 Carbon Dioxide Separation with Facilitated Transport Membranes

Membrane separation processes for CO_2 have gained significant attention due to their potential in applications, such as post-combustion carbon capture of power plant flue gases (CO_2/N_2), natural gas sweetening or processing (CO_2/CH_4), and hydrogen purification for fuel production (CO_2/H_2). The amphoteric nature of CO_2 allows it to interact with both basic and acidic compounds, enabling their roles as a CO_2 carrier. This section focuses on three types of polymer membranes that incorporate Lewis bases, Lewis acids, and ILs as CO_2 carriers. In addition, CO_2 carrier activities of counter-ions in ion exchange membranes will also be introduced. The discussion is limited to solid or quasi-solid polymer membranes, excluding liquid membranes. A quasi-solid membrane refers to a slightly swollen membrane containing a small amount of liquid, such as water.

5.9.1 Polymer Membranes with Lewis Bases

Lewis base compounds can interact directly or indirectly with CO_2 to facilitate its transport, as discussed in Section 4.6.2 of Chapter 4. These compounds are incorporated into polymer membranes through chemical bonding or simple mixing to create solid or quasi-solid membranes. This subsection provides an overview of polymers containing Lewis bases such as amine or carboxylate groups, and amino acid salts containing both amine and carboxylate groups together.

5.9.1.1 Polymers Consisting of Amines

Amine compounds can interact or react reversibly with CO_2 in a dry state to act as CO_2 carriers. In the presence of water, they can promote the formation of bicarbonate, enhancing facilitated CO_2 transport. For simplicity, these compounds are referred to as "amine carriers" regardless of whether they interact directly or indirectly with CO_2.

The efficiency of amine compounds as CO_2 carriers is primarily determined by their basicity and steric hindrance. These effects are discussed in detail in Section 4.6.2 of Chapter 4 and reviewed in references [5, 77]. The basicity of alkylamines is commonly measured by their *pKa* values and donor numbers (DN), as detailed in Table 4.1, along with their boiling temperatures T_b [101]. Among alkylamines, secondary amines achieve an ideal balance between CO_2 affinity and steric hindrance, providing better permeability and selectivity compared to primary and tertiary amines, as discussed in Sections 4.6.1 and 4.6.2 of Chapter 4.

Alkyl amins are used to prepare a selective layer for composite membranes by either being attached covalently to polymer chains or mixed with a matrix polymer. Representative examples of the former include poly(vinyl amine) (PVAm), poly(allyl amine) (PAA), poly(ethylene imine) (PEI), and poly(amidoamine) dendrimer (PAMAM) [102]. Polymers with primary amines, such as PVAm, PAA and PAMAM are frequently modified to convert into secondary amines by incorporating alkyl groups like methyl, ethyl, iso-propyl or tert-butyl. This modification enhances basicity and introduces steric hindrance, improving overall performance.

PVAm is a primary amine and highly crystalline, and its use in CO_2 capture from flue gas is undergoing scale-up [103]. However, it has been chemically modified to further improve the separation performance by converting it into a secondary amine through alkyl group substitution. The separation performances of PVAm and its derivatives containing secondary amines including poly(*N*-*t*-butyl-*N*-vinyl amine), poly(*N*-iso-propyl-*N*-vinyl amine), and poly(*N*-methyl-*N*-vinyl amine) are compared in Figure 5.23 [104, 105]. As expected, moderately hindered secondary (*N*-methyl) amines exhibit higher permeability and mixed-gas selectivity compared to highly hindered secondary (*N*-*t*-butyl, *N*-iso-propyl) and primary amines, aligning with results for PAA derivatives [102, 106]. Similarly, moderately hindered PAA with iso-propyl groups outperforms heavily hindered *t*-butyl and iso-butyl derivatives [102].

Another efficient way to improve separation efficiency is to mix with low molecular weight alkyl amines into a matrix polymer. In this case, new functional groups

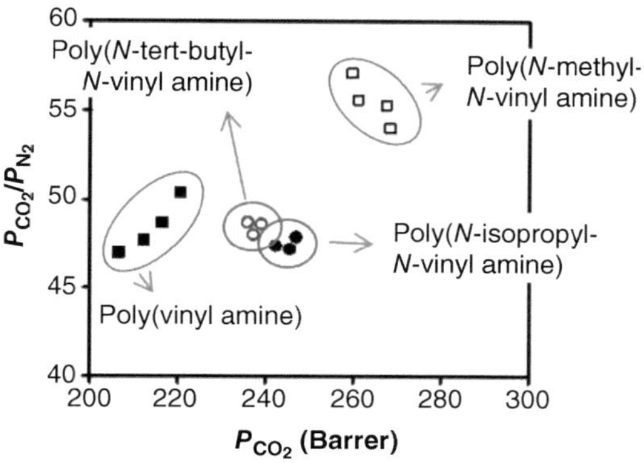

Figure 5.23 CO_2 permeabilities and mixed-gas selectivities of CO_2/N_2 through facilitated transport membranes of poly(*N*-vinyl amine) (PVAm) and its derivatives containing secondary amines at 57 °C with humidified feed and sweep gas. *Source*: Adapted from [104].

like hydroxyl, carboxylate or carbonate [105] groups have been often incorporated to further improve the separation performance and sustainability.

Typical low molecular weight alkylamine compounds include branched amines like iso-propyl amine and tert-butyl amine, cyclo-amines such as piperidine PD and piperazine PZ, and alkanol amines such as monoethanolamine MEA, diethanolamine DEA, and triethanolamine TEA and their derivatives. Ethylene diamine EDA has been also utilized occasionally.

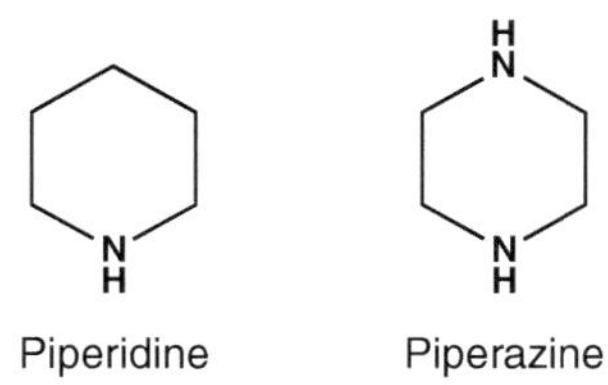

Cyclic amines including piperidine, PD, and piperazine, PZ, having secondary amine in a six-membered ring, show very high rate constants for its carboxylation reaction with CO_2 and decarboxylation reaction of carbamates $R_2NCO_2^-$, which play important roles in bicarbonate HCO_3^- formation [107]. In addition, PZ with two secondary amines in a six-membered ring seems to be highly effective in forming HCO_3^- as a CO_2 carrier.

Alcohol OH groups in alkylamines also improve CO_2 separation performance mostly because of the improvements in converting CO_2 into HCO_3^-, an effective CO_2 carrier, compatibilizing with a hydrophilic polymer matrix such as PVA, plasticizing effects, and reducing the volatility. For instance, PVA membranes with monoethanolamine MEA having OH groups show P_{CO_2} of 600 Barrer while PVA membranes with ethyl amine and PAMAM without OH groups showed ~1 and 100 Barrer, respectively, at 40 °C, 80% relative humidity, applied pressure difference ($\Delta p(CO_2) = 63$ kPa) with 80 wt% of amine [106, 108]. These results are consistent with the PVA/PZ derivative membranes of, for instance, PVA/(3-(1-piperazinyl)-1,2-propanediol) membranes having OH groups [109].

Effects of the steric hindrance of alkanolamines were examined for their suitability with the PVA matrix as a CO_2 carrier, which include representative primary, secondary, and tertiary alkanolamines (i.e. monoethanolamine (MEA), 2-amino-2-methyl-1-propanol (AMP), diethanolamine (DEA) and *N*-methyldiethanolamine (MDEA). Among the amines tested, DEA was found to be the most effective in improving the CO_2/N_2 selectivity, up to higher than 100 [110].

5.9.1.2 Polymers Consisting of Carboxylates or Carbonates

Carboxylates, as Lewis bases, also interact reversibly with CO_2, serving as CO_2 carriers to a lesser extent than alkylamines. However, when carboxylates are located near amine groups, such as in amino acids, their reversible interactions with CO_2 become more efficient and faster. This enhanced carrier activity, combined with their superior chemical stability compared to alkylamines, makes carboxylates highly attractive for practical applications, especially in humid environments.

Numerous studies have explored combining carboxylates with other functional groups to improve CO_2 separation performance and durability. For example, polymeric membranes containing both carboxylate and amide groups demonstrated exceptionally high separation performance (Q_{CO_2} = 5,831 GPU and α_{CO_2/N_2} = 86) along with excellent operational stability [111]. Additionally, copolymer membranes of poly(diallyldimethylammonium carbonate-co-vinylamine), which contain carbonate groups, primary amino groups, and quaternary ammonium groups, exhibited synergistic effects among these functionalities, significantly enhancing CO_2 separation performance [105].

5.9.1.3 Amino Acid Salts

Amino acid salt has both amino (or amine) and carboxylate groups together. Therefore, it reacts with CO_2 to form carbamate, which will in turn produce HCO_3^- by reacting with water as described in Section 4.6 of Chapter 4. The carboxylate group in amino acid salts leads to fast reaction kinetics for carboxylation and decarboxylation, resulting in the improvement of the carrier activity for the facilitated CO_2 transport, compared to simple alkylamines without carboxylate groups [112]. Chemical structures of representative amino acid salts frequently used for CO_2 carriers are shown in Figure 5.24.

Four different potassium amino acid salts with prolinate (ProK), argininate (ArgK), glycinate (GlyK), and cysteinate (CysK), embedded in a PVA matrix, significantly

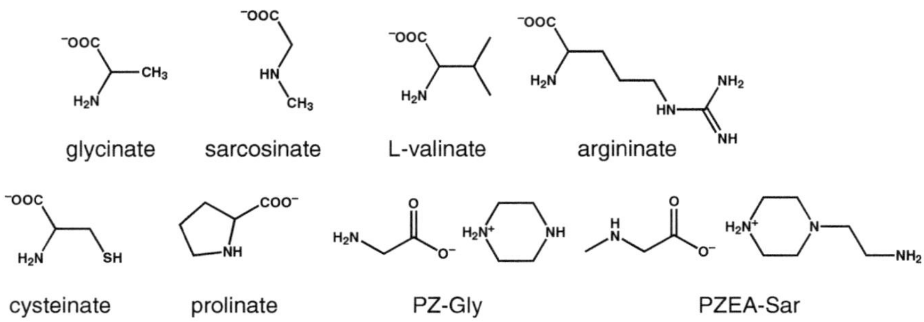

Figure 5.24 Chemical structures of representative amino acid anions. *Source:* Y.S. Kang.

increases the CO_2 permeance with little influence on the CO_2/N_2 selectivity. The addition of 40% ProK, the most effective among 4 salts used, into the PVA matrix nearly doubled the CO_2 permeance (from 399 to 791 GPU) [5, 112]. In addition, potassium 2-aminoisobutyrate (AIBA-K) containing a sterically hindered primary amino group has been often blended with various polymers such as sulfonated polybenzimidazole to improve separation performance. [5, 113].

5.9.1.4 Polymer Matrix Effects

When low molecular weight Lewis bases are used as CO_2 carriers, a polymeric matrix is needed not only to dissolve them homogeneously but also to provide good film-forming properties and mechanical strength. Consequently, some degree of hydrophilicity is required, as common Lewis base compounds are often hydrophilic. Poly(vinyl alcohol) (PVA) and poly(ethylene oxide) (PEO) are two commonly used matrix polymers. However, both PVA and PEO are highly crystalline, which inhibits gas permeation. When mixed with low molecular weight amines, their crystallinities are reduced, leading to improved gas permeability—an important advantage. PVA contains many hydroxyl groups that interact with amine groups, increasing the local concentration of alkyl amines. PEO dissolves common Lewis base compounds well but has relatively poor film-forming properties, which are often improved by crosslinking the polymer chains to form a network structure.

The effects of the polymer matrix between PVA and PEO have been directly investigated to show better separation performance with PEO. For instance, the PEG/MEA membrane shows much more improved separation performance in terms of both the CO_2 permeability and the CO_2/H_2 selectivity than the PVA/MEA membrane (Table 5.5) [108].

PEI is another attractive polymer containing secondary amine, which is readily blended with PVA matrix. The PEI/PVA blend selective layer for thin film composite membranes gave the CO_2 permeance of 1.4×10^{-7} mol/m^2/s/Pa with a CO_2/N_2 selectivity of up to 300 at a low pressure (1 bar) [114].

In addition, poly(amidoamine) dendrimers (PAMAMs) having a large amount of primary and secondary amines have been frequently incorporated into a polymeric matrix including PVA and PEO [115, 116].

Table 5.5 CO_2 separation properties of MEA-containing PVA and PEO membranes over H_2 at 40 °C and 90% relative humidity.

Membrane[a]	CO_2 permeability (Barrer)[b]	Selectivity (CO_2/H_2)
PVA/MEA	92.3	11.1
PEO/MEA	416.3	99.3

MEA, monoethanolamine; PEO, poly(ethylene oxide); PVA, poly(vinyl alcohol).
[a]Matrix/MEA = 50/50 wt.
[b]Estimated from the product of the gas permeance and thickness of the selective layer [108].
Source: Adapted from [108].

5.9.1.5 Issues and Challenges

Issues for utilizing alkylamine compounds as CO_2 carrier would be the oxidation of amine groups by oxidant gases (mainly O_2) contained in the flue gas, resulting in the deterioration of the separation performance with operation time. The degradation chemistry of amine compounds has been studied and reviewed in literatures [117].

Therefore, there have been several approaches to develop new materials, delivering long-term separation performance for practical applications while maintaining the separation performance change minimized. One such example is to utilize the carboxylate group as a fixed CO_2 carrier for separation, rather than amine groups. For instance, the poly(sodium acrylate-co-acrylamide) membrane shows the prolonged anti-oxidizability up to about 350 hours along with high separation performance [118].

Relatively high volatility, represented by low T_b, of alkylamines is another drawback (Table 4.1 in Chapter 4), which may result in carrier loss and hence a reduction in separation performance. Therefore, it is recommended to utilize anti-oxidative alkylamines with less volatility or high boiling temperature.

5.9.2 Polymer Membranes with Lewis Acids

Lewis acid compounds including metallic cations and organic cations like quaternary ammonium cations can reversibly interact with amphoteric CO_2, forming complexes depending on their environments. This interaction enables them to function as CO_2 carriers that facilitate its transport. This process is detailed at Figure 4.43 (Section 4.6.1 of Chapter 4) with water and Figure 4.50 (Section 4.6.3 of Chapter 4) without water. A straightforward yet effective method to integrate metal ions into a polymer solvent involves generating a solid polymer electrolyte. This is achieved by dissolving metallic or organic salts in a polar polymer solvent. Quaternary ammonium cations are structurally diverse to be also used as a CO_2 carrier along with metallic cations. Additionally, incorporating a small amount of water into polymeric electrolyte membranes to form quasi-solid state has been recognized as highly effective in enhancing separation performance.

5.9.2.1 PEMs Without Water

PEMs containing alkali or alkaline-earth metal salts dissolved in a polymer solvent showed higher CO_2 permeability than the control membrane without salts while maintaining the permeabilities of other gases such as the CH_4 and N_2 nearly unchanged, resulting in the simultaneous increase in both the CO_2 permeability and the selectivity of CO_2 over other gases.

Pebax + MTf_2N: Figure 5.25a shows the CO_2 permeance through PEMs containing 3 mol% of metal bis(trifluoromethanesulfonyl)imide MTf_2N dissolved in Pebax 2533 matrix by varying the metal ion M^+ from Li^+, Na^+, and K^+. Here, Pebax 2533 is a thermoplastic elastomer of flexible poly(tetramethylene ether) and rigid polyamide 12 blocks. Both the CO_2 permeance and the CO_2/N_2 selectivity are simultaneously improved to a large extent, indicating facilitated CO_2 transport due to the

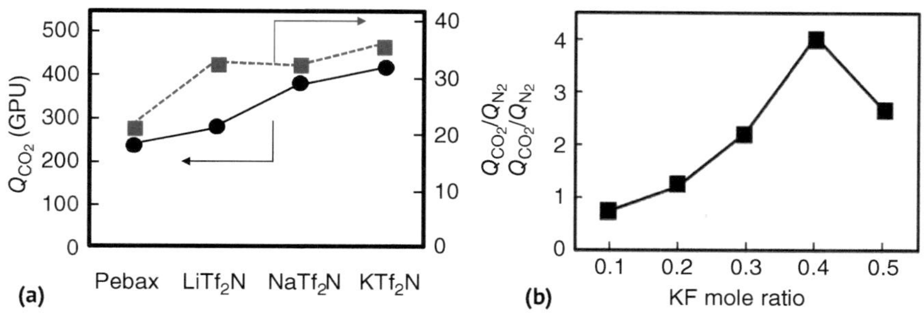

Figure 5.25 (a) The CO_2 permeance Q_{CO_2} and the CO_2/N_2 selectivity through Pebax 2533/ MTf_2 N membranes with 3 mol% of MTf_2N by varying the metal ion M^+ from Li^+, Na^+, and K^+. *Source*: Ref. [119]/with permission of Elsevier. (b) Selectivity of poly(vinyl pyrrolidone/KF (1/0.4) electrolyte membranes as a function of the KF concentration used. Selectivity of PVP/KF (1/0.4) electrolyte membranes as a function of the KF concentration used. *Source*: Ref. [120]/Royal Society of Chemistry.

carrier action of the metallic ions, as described in Section 4.6.3 of Chapter 4. In addition, K^+ shows the highest results among Li^+, Na^+, and K^+; that is, K^+ shows the best CO_2 carrier activity [119]. The linear relationship between the CO_2 permeance and the concentration of KTf_2N up to 9 mol% is illustrated in Figure 4.48 (Section 4.6.3 of Chapter 4).

PVP + KF: PEMs of KF dissolved in the poly(vinyl pyrrolidone) (PVP) matrix also exhibit much higher CO_2 permeance from nonmeasurably small to 28.0 GPU and the improved CO_2/N_2 selectivity from nearly 1 to 4.1, compared to neat PVP without salt in the dry state (Figure 5.25b), due to the carrier action of K^+ [120].

This result is surprising because the metal ions usually lead to transient crosslinks due to the coordinative bond formation (Figure 4.22 [Section 4.5.1 of Chapter 4]), resulting in higher polymer chain rigidity and lower free volume for inhibiting molecular transport. In fact, the glass transition temperature T_g of 183 °C for neat PVP is markedly increased to 237 °C for the *PVP*/KF polymer electrolyte. It is therefore claimed that the increase in the CO_2 permeance is attributed mostly to the carrier action of K^+ for CO_2 even without water. The polymer electrolyte formation is evidenced by the slight decrease in the binding energy of K atom from 292.6 to 292.28 eV along with the red shift of the C=O bond of PVP [120].

PEO + organic acid: Interestingly, when 5-hydroxyisophthalic acid (HIA), an organic acid, is dissolved in a PEO matrix to generate PEMs, the CO_2 permeability nearly linearly increases with increasing the concentration of HIA at lower levels. This culminates in a maximum permeability of about 573 Barrer and a maximum selectivity of CO_2/N_2 higher than 30 at the 0.005 mole ratio of HIA to PEO, as illustrated in Figure 5.26a and b. This exemplifies the characteristic facilitated transport process [121]. It is also noteworthy that the permeability of N_2 remains nearly constant despite the variations in HIA concentration within the scope of this experiment.

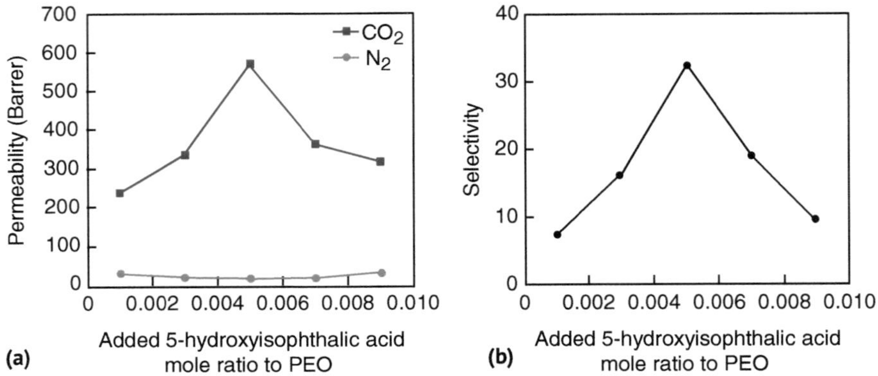

Figure 5.26 Separation performance of poly(ethylene oxide) (PEO)/5-hydroxyisophthalic acid (HIA) polymer electrolyte membranes: (a) pure-gas permeability and (b) ideal selectivity of CO_2/N_2. *Source*: Ref. [121] / with permission of Elsevier.

5.9.2.2 PEMs with Water for Quasi-solid State

When small amount of water is present in a PEM containing metallic salts dissolved in a polar polymeric solvent, water molecules could play several different functions in facilitating CO_2 transport: the increase in both CO_2 solubility and free ion concentration, the molecular water channel formation, plasticizer or swelling agent and the formation and transport of HCO_3^-, in addition to the CO_2 carrier activity by forming [ABAB] complexes as described in Section 4.6.1 in Chapter 4. Here, only the CO_2 carrier effects will be explained [122].

The CO_2 permeance through PEMs of PEO/CF_3CO_2K (0.7 mole ratio of K^+ to the monomeric unit of PEO) with 70% RH feed exhibit 4,650 GPU and the ideal CO_2/N_2 selectivity >1,500 at $p_{CO_2} = 1.01$ atm and 298 K (Figure 4.51 in Section 4.6.3 of Chapter 4) [123], which lies far beyond the state-of-the-art upper bound (Figure 5.27a). More importantly, such high CO_2 permeance and N_2 permeance lasted over

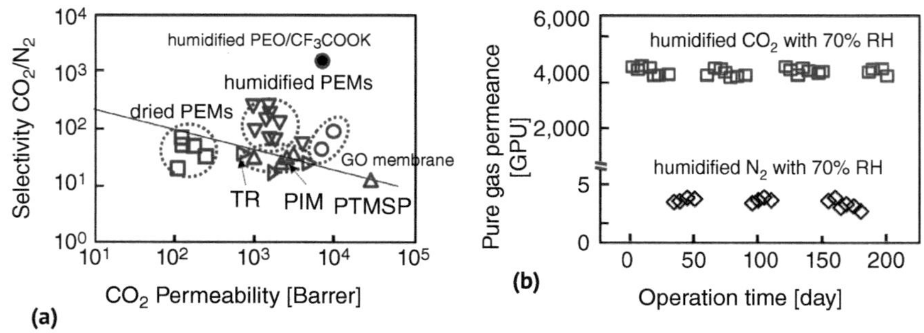

Figure 5.27 (a) Trade-off relation between CO_2 permeability and the CO_2/N_2 ideal selectivity including PEO/CF_3CO_2K PEM (●, 0.7 mole ratio of K^+ to the monomeric unit of PEO) with 70% relative humidity as a function of operation time at $p_{CO_2} = 1.01$ atm and 298 K with the conventional upper bound [polymer electrolyte membranes (PEMs], thermally rearranged polymers (TR), polymers of intrinsic microporosity (PIM), and triazole-based PIM (TZPIM), graphene oxide (GO)]; and (b) pure-gas permeances of CO_2 and N_2 through PEO/CF_3CO_2K PEMs as a function of operation time. *Source*: Adapted from [123].

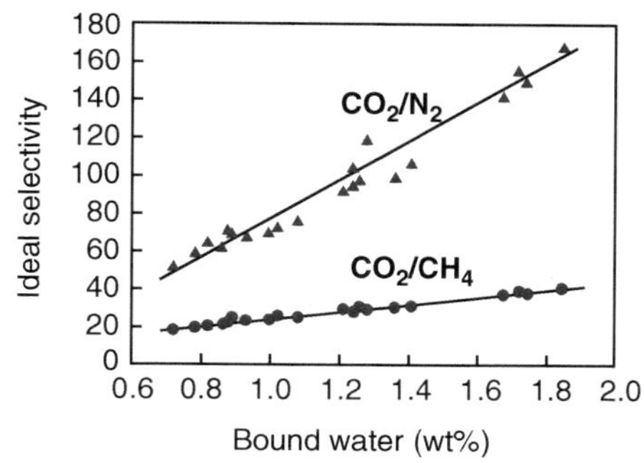

Figure 5.28 Correlations between pure-gas CO_2/CH_4, CO_2/N_2 selectivity and bound water content of the Pebax 1657 membrane. *Source*: Adapted from [124].

a test period of 6 months as shown in Figure 5.27b. Considering the high content of bound-water to cation in the membrane (64 wt%), the outstanding performance of the membrane can be primarily attributed to the carrier action of the water molecules bound to K^+, where the carrier action of the bound water was described by the formation of reversible [ABAB] complexes in detail at Section 4.6.1 of Chapter 4.

Pebax 1657 (a block copolymer of polyamide 6 and PEO about 40:60) membranes containing metallic salts MCl such as LiCl, NaCl, KCl, $CaCl_2$, and $MgCl_2$ along with small amount of water showed both CO_2 permeability and CO_2/gas selectivity significantly elevated [124]. These results can be obtained primarily due to the well-controlled total water and bound water content within the membrane (Figure 5.28).

Transport Mechanism: Such huge increase in both the permeance (or permeability) and selectivity could primarily be due to the carrier action of the bound water coordinating alkaline metal cations by reversible [ABAB] complex formation as described in Section 4.6.1 and Figure 4.41a in Chapter 4. Water molecules in a PEM may coordinate to Lewis acidic metallic cations [A] by electron-shifting from the electron-rich oxygen [B] atom of the water to the electron-lean metallic atom. This process may induce the hydrogen atom of the water molecule to be slightly positively charged, becoming stronger acidic [A], and consequently interacts with electron-rich oxygen [B] atoms of CO_2 as a Lewis base to form reversible [ABAB] complexes according to the results of the NMR and FT-IR spectroscopies [123]. In other words, the bound water coordinating to alkaline metal cations and CO_2 would form reversible [ABAB] Lewis acid–base complexes and consequently the bound water can act as a new CO_2 carrier for the facilitated CO_2 transport.

Challenges: The presence of water inside membranes offers significant benefits, such as reducing separation costs by enhancing performance and enabling the use of partially dehydrated feeds for flue gas treatment and natural gas conditioning. However, it also poses challenges. Water in the permeate requires additional removal steps, increasing costs depending on the application. Furthermore, operational stability can be a concern due to water evaporation on the permeate side.

5.9.3 Polymer Membranes with Ionic Liquids (ILs)

Room-temperature ILs are salts composed of soft cations and anions that remain in a liquid state under ambient conditions. They have melting points below 100 °C and are typically nonvolatile and nonflammable. ILs are particularly attractive as

CO_2 carriers because they generate both free cations and anions, enabling dual carrier functionality. Their liquid state enhances gas permeability, and their chemical stability supports long-term operational reliability.

The interactions between the cations and anions of ILs and CO_2, which are critical for their effectiveness as CO_2 carriers, were explored in Section 4.6.4 of Chapter 4, along with the relationship between their chemical structures and transport properties. Chemical structures of frequently used ILs are illustrated in Figure 5.29.

For membrane applications, ILs can be used either in their liquid state (as in supported liquid membranes, SLMs) or incorporated into polymers to form solid or quasi-solid-state membranes. Here, quasi-solid state means polymer membrane containing small amounts of water as described in Section 4.6 of Chapter 4. In the solid or quasi-solid state, ILs are either:

1. Incorporated into the main or side chains of polymers by covalent bondings, referred to as "poly(IL)s," or
2. Dissolved in a polymeric solvent to form PEMs, called "PEM-ILs."

Developing efficient ILs and incorporating them into polymer matrices to form solid- or quasi-solid state membranes is a promising strategy. Solid- or quasi-solid state polymer membranes offer several advantages over liquid-state SLMs, including: the ability to create thinner selective layers (as thin as 1 μm or less) with higher IL content, compatibility with high-pressure or vacuum operations, and improved long-term operational stability.

This section will first discuss efficient ILs for CO_2 separation, followed by the development of solid or quasi-solid polymer membranes aimed at enhancing CO_2 flux and selectivity for practical applications.

5.9.3.1 ILs for High Separation Performance

ILs with diverse chemical structures have been attempted to improve the CO_2 permeability while maintaining the permeability of other gases such as H_2, N_2, and CH_4 unchanged, leading to high CO_2 selectivity. For example, incorporating functional groups such as amines and/or carboxylates into the cations and/or anions has proven effective (see Section 4.6.4 of Chapter 4). Cyanide groups in anions also result in high CO_2 separation performance.

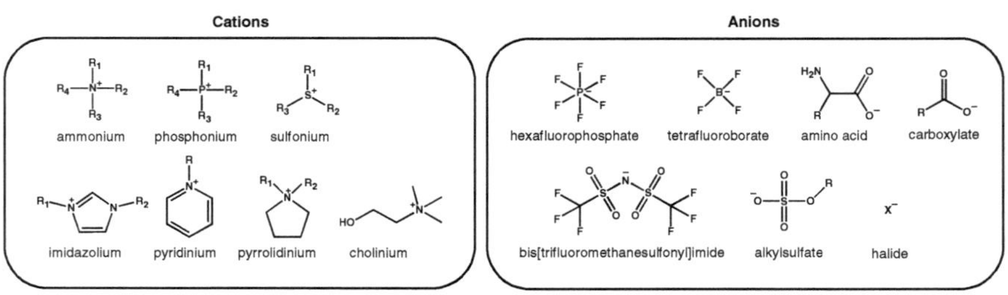

Figure 5.29 Chemical structures of representative cations and anions for ionic liquids (ILs). *Source*: Y.S. Kang.

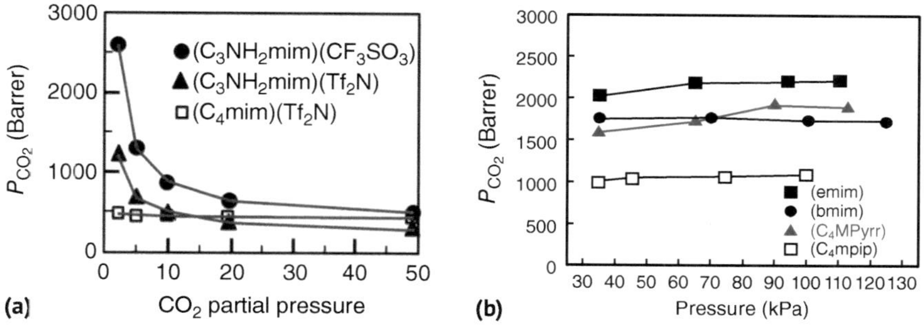

Figure 5.30 Effects of CO_2 partial pressure on (a) P_{CO_2} and (b) P_{CO_2}/P_{CH_4} under humidified conditions at 25 °C and the total pressure of 1.01×10^5 Pa (1 atm). *Source:* Adapted from [125]. (b) The effect of pressure difference across SLMs with $B(CN)_4^-$ anion on the CO_2 permeability P_{CO_2}. *Source*: Ref. [127] / Royal Society of Chemistry.

Amino groups in cations: Figure 5.30a shows the separation properties of CO_2/CH_4 mixtures as a function of a CO_2 partial pressure using three ILs under humidified condition [125]. Two ILs containing amino groups of [C_3NH_2mim][CF_3SO_3] and [C_3NH_2mim][Tf_2N] exhibit much higher P_{CO_2} and P_{CO_2}/P_{CH_4} than [C_4mim][Tf_2N] without amino group, where [C_3NH_2mim] and [C_4mim] are *N*-aminopropyl-3-methylimidazolium and 1-butyl-3-methylimidazolium, respectively [125]. Notably, [C_3NH_2mim][Tf_2N] showed the CO_2/H_2 selectivity greater than 15 even under dry conditions at elevated temperature (85 °C), demonstrating its facilitated CO_2 transport capability [126].

Amino acid anions: Amino acid anions containing both amino and carboxylate groups enhance the CO_2 interactions further as discussed previously in Section 4.6.4 of Chapter 4. For example, glycinate [Gly], the simplest amino acid containing carboxylate vicinity to NH bond, demonstrates excellent CO_2-facilitating properties [128]. ILs like [P_{4444}][Gly] and [Emim][Gly] in SLMs achieved CO_2 permeabilities of up to 5,000 and 8,300 Barrers, respectively, at high temperature 373.15 K and dry condition without water (Figure 4.58 in Section 4.6.4 of Chapter 4), where [P_{4444}] and [Emim] are tetrabutylphosphonium and 1-ethyl-3-methylimidazolium, respectively. The selectivity of CO_2/N_2 reached nearly 400 at low CO_2 partial pressures. This remarkable performance is attributed to the dual carrier action of the amine and carboxylate groups in glycine, which facilitates CO_2 transport even in dry environments.

Cyanide groups in anions: ILs containing many cyanide -CN groups without amino or carboxylate groups show extremely high separation performance. SLM of 1-ethyl-3-methyl imidazolium [Emim][$CF_3SO_2C(CN)_2$] shows a CO_2 permeability of 1095 Barrer and selectivities varying from 16.9 to 22.2 for CO_2/CH_4 and 37.0 to 44.4 for CO_2/N_2 gas pairs at 293.15 K and 1 bar of feed pressure [129]. In addition, ILs of [$B(CN)_4$] anion with diverse cations including imidazolium, pyrrolidinium, and piperidinium cations demonstrate the remarkable potential of these ILs for CO_2 separation [127].

Exceptionally high CO_2 solubility of ILs containing the [B(CN) $_4$] anion with several cations including [Emim], [Bmim], [Emmim], [C4mpyr], and [C4mpip] is obtained, which is larger than the prototypical IL, [Emim][Tf_2N], having a CO_2 solubility of 0.103 (mol/L atm). Here, [C4mpyr] and [C4mpip] are 1-butyl-1-methylpyrrolidin-1-ium and 1-butyl-1-methylpiperidin-1-ium, respectively. The [Emmim][B$(CN)_4$] IL has the highest solubility of 0.148 (mol/L atm). Interestingly, [Emim][B(CN) $_4$] with the lowest CO_2 solubility give the highest CO_2 permeability of 2,040 Barrer with the selectivity for CO_2/N_2 of 53 (Figure 5.30b). Perhaps, this is primarily due to the highest carrier activity of [Emim][B$(CN)_4$] demonstrating the importance of the reaction rate, particularly k_2, than the thermodynamic solubility term K, for facilitated transport [127].

Effects of water and temperature: When the feed gas is dry, one mole of [P_{4444}][Gly] or [Emim][Gly] reacts with half a mole of CO_2 to form a complex. In the presence of even a small amount of water, equimolar amounts of [P_{4444}][Gly], CO_2, and H_2O react with one another to form bicarbonate, known as an effective CO_2 carrier. Therefore, the performance of CO_2 separation is further increased. Another possibility for the increase in CO_2 permeability is a change in viscosity of ILs with water absorption [128].

The permeability of gases normally increases with increasing temperature due to an increase in the fractional free volume f, leading to lower viscosity and consequently higher CO_2 permeability. In the meantime, the CO_2 solubility generally decreases dramatically with increasing temperature, compared to other gases like CH_4 and N_2, leading to lower selectivity values at higher temperature. Therefore, a low operating temperature is suggested to be favored for applications [130].

Transport Mechanism: Based on the solution–diffusion mechanism, Scovazzo developed a mathematical model for predicting CO_2 permeability and selectivity accounting for the viscosity and molar volume of ILs [131]. This model indicates that the ILs with smaller molecular size and low viscosity tend to have both high permeability and high selectivity.

5.9.3.2 Solid-state Poly(IL)s and PEM-ILs

When highly efficient ILs are transformed into solid-state polymers via poly(IL) s or PEM-ILs, their CO_2 permeabilities will be significantly reduced, compared to liquid-state ILs. However, much enhanced CO_2 flux or permeance will be obtained by reducing the selective layer thickness up to 1 μm or less for composite membranes. Poly(IL)s or PEM-ILs have another attractive point containing high concentrations of IL groups, resulting in high carrier concentration to markedly facilitate CO_2 transport. The solid-state selective layer also helps make high pressure or vacuum process possible and operation stability much longer, which are huge advantages for practical applications. Comprehensive reviews of the synthesis of various poly(IL)s and PEM-ILs, and their membrane applications can be found [17, 132–136].

5.9.3.3 Poly(IL) Membranes

Poly(IL)s are polyelectrolyte polymers whose repeating unit bear an IL electrolyte. The most studied poly(IL)s for CO_2 absorption/separation are on vinyl polymers

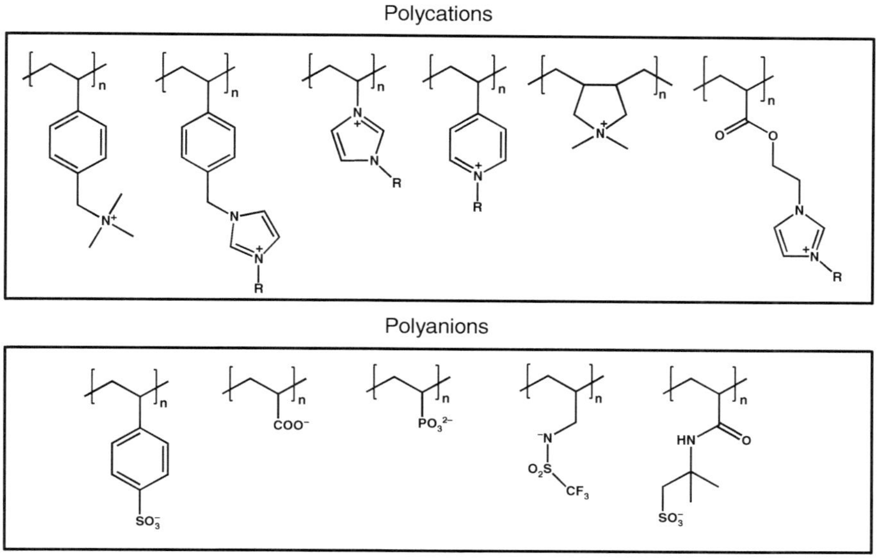

Figure 5.31 Chemical structures of representative polycations and polyanions in vinyl-based poly(IL)s. *Source*: Adapted from [133].

including polystyrene or polyacrylate backbones as shown in Figure 5.31. They have better CO_2 separation performance due to the high CO_2 sorption capacity and therefore, it has been usually selected as the base material. In addition, poly(IL)s with a PEO backbone have also been paid much attention because of the interactions of the ether oxygen with CO_2 [17].

Imidazolium-based Poly(IL)s: Poly(IL) membranes derived from styrene and acrylate monomers and incorporating alkyl imidazolium bis(trifluoromethylsulfonyl) imide [Tf_2N] were used to measure the permeabilities, solubilities, and diffusivities of CO_2, N_2, and CH_4 by varying the alkyl group from methyl to *n*-butyl and to *n*-hexyl. The chemical structure of the polymer backbone chains has little effect on the gas permeabilities of the membranes, while the length of the alkyl chain on the imidazole cation plays a significant role in both gas permeability and selectivity [133, 137, 138]. The CO_2 permeability increases with the increase in the alkyl group chain length whereas the CO_2 selectivity decreases to much less extent, indicating a distinct "permeability–selectivity trade-off." For instance, while increasing the length of the alkyl substituent, the CO_2 permeability increased from 9.2 to 32 Barrers along with the decreased selectivity from 32 to 28 for CO_2/N_2 and from 39 to 17 for CO_2/CH_4. These results support the hypothesis that the short alkyl side chain leads to a tighter ion domain whereas the long alkyl side chain in poly(IL)s directly reduces the density of the ionic groups and thus reduces the ionic interactions inside the membrane [133, 137, 138].

Another notable work is to modify imidazolium cations by introducing linear-, branched-, and cyclo-alkyl groups on styrene-based poly(IL) membranes. Relative to their *n*-alkyl counterparts, poly(ILs) with branched and cyclic functionalities exhibit ~20% higher CO_2/N_2 and CO_2/CH_4 selectivities, the gas permeabilities drop by more than 50%. The improvements in selectivity is suggested to be associated with smaller fractional free volumes, f (or FFV), of branched and cyclic functional polymers, compared to n-alkyl chains, which is supported by computational studies [139]. Additionally, incorporation of oligo(ethylene glycol) and nitrile-terminated alkyl groups on the imidazole cations of the IL monomers further improves CO_2/N_2 and CO_2/CH_4 selectivities (39–44 and 29–37, respectively) [136, 140]. Two novel polyurethane-based poly(IL)s containing imidazolium cations show the CO_2 sorption of 75.7 mol% at 20 bar. The presence of nitrogenated polyether structures in their backbone polyurethanes allied to the imidazolium ring has proved to be interesting for CO_2 capture [141].

Polybenzimidazole (PBI): PBIs possessing two imidazole groups per repeat unit show good film-forming abilities and thermomechanical properties, while most of the known poly(IL) membranes are brittle and difficult to make film and withstand high pressure. For instance, [DPyDBzPBI-BuI] and [DAnDBzPBI-BuI] cations exhibit relatively high CO_2 permeability and selectivity, compared with other PBI-based poly(IL) membranes. However, most membranes show the CO_2/N_2 and CO_2/CH_4 selectivity of 20–30, which is still too low yet. Additionally, it has been suggested that a mere variation of the anion in the poly(IL) resulted in a significant effect on the physical properties [133, 142].

Ammonium-based poly(IL)s: Ammonium-based poly(IL)s have been also investigated for CO_2 separation membranes. Among ammonium-based poly(IL)s, the CO_2 selectivity increases with the higher positive charge on the cation species in the following order: hydroxyethyl-dimethylammonium > trimethylammonium > pyrrolidinium [133, 143]. Poly(vinyl benzyltrimethylammonium) P[VBTMA][PF_6] has high CO_2 sorption capacity of 10.66 mol%, suggesting a potentially good CO_2 separation membrane material [133, 144, 145].

Phosphonium-based poly(IL)s: Among the phosphonium-based poly(IL)s, poly([tri-n-alkyl)vinylbenzylphosphonium][Tf_2N] shows the gas permeability linearly increase with increasing alkyl chain length appended to the phosphonium group. For instance, the poly([P_{888VB}][Tf_2N])-based membranes present higher CO_2 permeability of 186 Barrer, compared to other pure poly(IL) materials. On the contrary, the gas selectivity decreases with the phosphonium alkyl chain length increasing from C_4 to C_6, then remain approximately constant between C_6 and C_8 [133].

Cellulose-based poly(IL)s: A series of novel cellulose-based poly(IL)s with imidazolium and ammonium cations are prepared by tuning cation–anion interaction to improve CO_2 sorption. It is concluded that the CO_2 sorption capacity of the poly(IL) s is directly influenced by anion. The CO_2 sorption capacity at 3 MPa of CO_2 pressure follows the order: $[Cl]^- < [BF]^- < [Tf_2N]^- < [PF_6]^-$ [146].

5.9.3.4 PEM-ILs Membranes

PEM-ILs are a PEM containing ILs dissolved in a polar polymeric solvent such as PEO and even poly(IL)s. Incorporating ILs into a polymeric solvent simply adds CO_2 carrier for facilitated CO_2 transport and additionally helps soften the solid polymer solvent for improvement of free volume and consequently CO_2 permeability.

Here, selection of the polymeric solvent is critically important because the amount of dissolved ILs is mostly determined by their interactions. In this regard, poly(IL)s as a polymeric solvent are attractive because of their intrinsic polarity to accommodate ILs and also their CO_2 carrier function.

PEM-ILs: Cross-linked poly(ethylene glycol) (PEG) membranes by a tris(2-aminoethyl)amine crosslinker are mixed with four ILs having different anions (i.e. [Bmim][BF_4], [Bmim][PF_6], [Bmim][CCl_4], and [Bmim][Tf_2N]). The base membranes are an ether-rich rubber with a glass transition temperature below −40 °C. With a 80% loading of ILs, [Bmim][Tf_2N] demonstrates the best CO_2 permeability of 135 Barrer and a CO_2/N_2 selectivity of 45 at 35 °C [147].

Poly(IL)-ILs: Separation performance of poly(IL)s is further improved by adding task-specific ILs to form poly(IL)-ILs composites [148]. Therefore, the degree of ionic interactions between the poly(IL)s and ILs is much stronger than any interactions between conventional polymers and ILs, through which high-pressure operation could be possible. For instance, poly(IL)-ILs composite containing 20 mol% [Emim][Tf_2N] exhibited CO_2 permeability of 44 Barrer, a gain of ~400% relative to the analogous neat poly(IL). Selectivities in the composite were 39 for CO_2/N_2 and 27 for CO_2/CH_4, respective differences of +33% and −25% compared with the neat poly(IL).

Poly(IL)-IL composites of pyrrolidinium-poly(IL)s having [$C(CN)_3$] as counter-anion and different amounts (20, 40, and 60 wt%) of free [Emim][$C(CN)_3$] display that the incorporation of high free IL contents increases both CO_2 permeability and CO_2/N_2 selectivity. Consequently, the finest CO_2/N_2 separation performances are obtained with the CO_2 permeabilities of 439 and 542 Barrer and CO_2/N_2 selectivities of 64.4 and 54.0 [149]. The effects of the cyano-functionalized anions ([$N(CN)_2$], [$C(CN)_3$] or [$B(CN)_4$]) on the CO_2 separation performance of pyrrolidinium-poly(IL)-ILs composites were further investigated. The membranes containing the [$C(CN)_3$] and [$B(CN)_4$] anions show the best performance, enlightening the promise these anions entail for future high-performance membranes [150].

5.9.4 Bicontinuous Structures Including Ion Exchange Membranes

Ion-exchange membranes (IEMs) are nanometer-scale bicontinuous structures consisting of two interconnected domains: hydrophilic ion channels and a hydrophobic matrix. The continuously interconnected hydrophilic ion channels, containing fixed charged groups and counter-ions, are formed. These ion channels are swollen by water, creating pathways for ion transport. Meanwhile, the hydrophobic matrix provides mechanical strength, ensuring structural integrity and durability. Cation exchange membranes have fixed anionic groups (e.g. $-SO_3^-$) at the channel interfaces, balanced by positively charged counter-ions. Conversely, anion exchange membranes feature fixed cationic groups (e.g. $-NR_3^+$) at the interfaces, balanced by negatively charged counter-ions. These arrangements allow efficient ion transport while maintaining the membrane's mechanical stability.

Ethylene diamine monopositive (EDA^+) cations, incorporated into a cation exchange membrane made of sulfonated polystyrene grafted onto a poly(tetrafluoroethylene) backbone, exhibit typical CO_2 carrier activity for facilitated transport [151]. The presence of EDA^+ significantly increases CO_2 permeability, reaching approximately 6,000 Barrer, about 20-fold improvement compared to the reference membrane without EDA^+ at $p_{CO_2} = 22$ mmHg and 22 °C. The N_2 permeability was about 10 Barrer, resulting in an ideal CO_2/N_2 selectivity of approximately 600. Perfluorosulfonic acid cation exchange membranes with EDA^+ demonstrated a CO_2/CH_4 selectivity of 551 [152]. This facilitated CO_2 transport is attributed to the reversible reaction between the unprotonated amine group of EDA^+ and CO_2, forming a carbamate:

$$CO_2 + 2H_3^+N - C_2H_4 - NH_2 = H_3^+N - C_2H_4 - NHCOO^- + H_3^+N - C_2H_4 - NH_3^+ \tag{5.22}$$

Extensive research has been conducted to enhance CO_2 separation performance using ion exchange membranes [153]. Studies also investigated the effects of membrane thickness and thermal annealing temperature on perfluorosulfonic acid membranes with EDA^+ carriers [154].

Carbonate and hydroxide ions in aqueous solutions are known to facilitate the transport of carbon dioxide by virtue of the chemical reaction:

$$CO_3^{2-} + CO_2 + H_2O \leftrightarrow 2\ HCO_3^- \tag{5.23}$$

$$OH^- + CO_2 \leftrightarrow HCO_3^- \tag{5.24}$$

An anion-exchange membrane composed of quaternized poly(vinyl pyridine), for instance, has counter anion of either CO_3^{2-} or OH^- by soaking it in aqueous K_2CO_3 or KOH solution, respectively. The facilitated CO_2 transport was evidenced in both cases. However, the effect was small for the CO_3^{2-} membrane as expected considering the slow kinetics of CO_2 hydrolysis of Equation (5.23) [151]. In the case of the OH^- form of an anion exchange membranes with a polysulfone-methylene-quaternary phosphonium cations showed a high CO_2 permeability of 1,090 Barrer and a CO_2/N_2 selectivity of 275 when it was fully hydrated [155]. Increasing the basicity of the functional group in hydroxide-exchange membranes shows an improvement in CO_2 separation performance. The newly introduced phosphonium functional group exhibits better performance than imidazolium and ammonium functional groups tested.

At high IL content in PEM-IL composites, phase separation often occurs, forming an IL-rich phase and a polymer-rich phase. By carefully controlling the phase-separated morphology, such as achieving nanoscopically bicontinuous structures, these membranes can exhibit improved separation performance while maintaining mechanical strength, similar to ion exchange membranes (IEMs).

For example, incorporating [Bmim][Tf_2N] into sulfonated polyimide at an IL loading of 75 wt% creates a bicontinuous nanostructure with IL-rich and IL-lean phases as shown in Figure 5.32. This structure delivers a high CO_2 permeability of 431 Barrer, a CO_2/N_2 selectivity of 30 at 30 °C, and a high film modulus. In general, a softer, free IL layer enhances gas permeability, while the solid polymer matrix ensures mechanical strength [156].

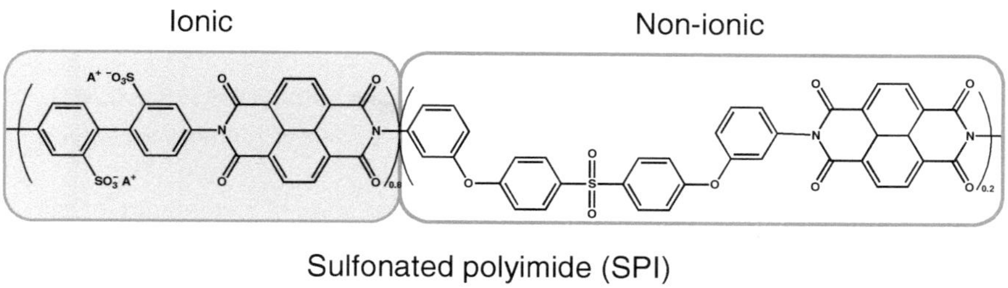

Figure 5.32 Chemical structure of sulfonated polyimide for forming bicontinuous nanostructure. *Source*: Ref. [156]/American Chemical Society.

5.9.5 Effects of Water in Quasi-solid Membranes

Water within membranes significantly enhances both CO_2 permeability and selectivity due to its diverse functionalities. Most importantly, water facilitates CO_2 transport through several mechanisms. First, water reacts directly with CO_2 to form bicarbonate ions (HCO_3^-), enabling facilitated CO_2 transport regardless of the presence of other chemical species. Bicarbonate ions are well-known as essential components in improving CO_2 permeability.

Additionally, Lewis bases, such as alkylamines, enhance the formation of HCO_3^- in aqueous environment. This occurs either through a direct reaction, where tertiary amines interact with CO_2 and water to produce HCO_3^- ($R_3N + CO_2 + H_2O \rightarrow R_3N^+ + HCO_3^-$), or indirectly via the formation of carbamate intermediates ($R_2NCO_2^-$) that are later converted to bicarbonate.

Lewis acids also play a significant role in aqueous environment. For instance, metallic cations bind water molecules, and these water–cation complexes interact with CO_2 to form reversible [ABAB] complexes. These interactions strongly enhance the facilitated transport of CO_2.

Beyond these mechanisms, water contributes to the overall performance of the membrane by dissociating metallic salts, which generates more free ions, and by swelling and plasticizing the membrane. These effects increase the diffusion and permeation coefficients, further improving membrane efficiency.

From a practical perspective, water-rich feed gas mixtures, such as flue gas containing 15–20% water by volume or natural gas with 15–60 mg/m^3 water, can be separated more economically and effectively. This makes the process particularly advantageous for real-world applications.

However, there are challenges associated with water in membranes. For one, water in the permeate may require additional steps for removal, adding to operational costs. Additionally, water evaporation on the permeate side can impact the membrane's stability and performance over time.

In conclusion, water plays a crucial role in enhancing CO_2 separation by facilitating bicarbonate formation and acting as a carrier through interactions with Lewis acids and bases. While its presence introduces some operational challenges, the benefits it offers in terms of performance and efficiency often outweigh these issues.

References

1 R.W. Baker, Membrane Technology and Applications, John Wiley & Sons, 2023.
2 B.D. Freeman, Y. Yampolski, I. Pinnau, Materials Science of Membranes for Gas and Vapor Separation, Wiley Online Library, 2006.
3 Y. Yampolskii, I. Pinnau, B.D. Freeman, Materials science of membranes for gas and vapor separation, Wiley, 2006.
4 R.L. Burns, W.J. Koros, Journal of Membrane Science 211(2) (2003) 299–309.
5 Y. Han, W.S.W. Ho, Journal of Membrane Science 628 (2021).
6 W.J. Koros, C. Zhang, Nature Materials 16(3) (2017) 289–297.
7 L.M. Robeson, Journal of Membrane Science 62(2) (1991) 165–185.
8 L.M. Robeson, Journal of Membrane Science 320(1-2) (2008) 390–400.
9 B.D. Freeman, Macromolecules 32(2) (1999) 375–380.
10 R.M. Barrer, G. Skirrow, Journal of Polymer Science 3(4) (1948) 549–563.
11 D. W. Van Krevelen, K. Te Nijenhuis, Properties of polymers: their correlation with chemical structure; their numerical estimation and prediction from additive group contributions. Elsevier, 2009.
12 K. Haraya, K. Obata, T. Hakuta, H. Yoshitome, Membrane 11(1) (1986) 48–52.
13 I.W. Kim, K.J. Lee, J.Y. Jho, H.C. Park, J. Won, Y.S. Kang, M.D. Guiver, G.P. Robertson, Y. Dai, Macromolecules 34(9) (2001) 2908–2913.
14 H. Yamamoto, Y. Mi, S.A. Stern, A.K. Stclair, Journal of Polymer Science Part B: Polymer Physics 28(12) (1990) 2291–2304.
15 J.W. Barnett, C.R. Bilchak, Y.W. Wang, B.C. Benicewicz, L.A. Murdock, T. Bereau, S.K. Kumar, Science Advances 6(20) (2020), eaaz4301.
16 K. Berean, J.Z. Ou, M. Nour, K. Latham, C. McSweeney, D. Paull, A. Halim, S. Kentish, C.M. Doherty, A.J. Hill, K. Kalantar-zadeh, Separation and Purification Technology 122 (2014) 96–104.
17 D.E. Sanders, Z.P. Smith, R.L. Guo, L.M. Robeson, J.E. McGrath, D.R. Paul, B.D. Freeman, Polymer 54(18) (2013) 4729–4761.
18 S.R. Reijerkerk, M.H. Knoef, K. Nijmeijer, M. Wessling, Journal of Membrane Science 352 (1–2) (2010) 126–135.
19 H. Lin, B.D. Freeman, Journal of Membrane Science 239(1) (2004) 105–117.
20 T.C. Merkel, R.P. Gupta, B.S. Turk, B.D. Freeman, Journal of Membrane Science 191 (1–2) (2001) 85–94.
21 F. Rindfleisch, T.P. DiNoia, M.A. McHugh, Journal of Physical Chemistry 100(38) (1996) 15581–15587.
22 K. Ghosal, R.T. Chern, B.D. Freeman, W.H. Daly, I.I. Negulescu, Macromolecules 29(12) (1996) 4360–4369.
23 J.Y. Liu, S.Z. Zhang, D.E. Jiang, C.M. Doherty, A.J. Hill, C. Cheng, H.B. Park, H.Q. Lin, Joule 3(8) (2019) 1881–1894.
24 S. Shishatskiy, J.R. Pauls, S.P. Nunes, K.V. Peinemann, Journal of Membrane Science 359(1–2) (2010) 44–53.
25 E.S. Peterson, M.L. Stone, R.R. Mccaffrey, D.G. Cummings, Separation Science and Technology 28(1–3) (1993) 423–440.
26 R.R. Mccaffrey, D.G. Cummings, Separation Science and Technology 23(12–13) (1988) 1627–1643.

27 C.L. Aitken, W.J. Koros, D.R. Paul, Macromolecules 25(14) (1992) 3651–3658.
28 M.R. Pixton, D.R. Paul, Relationships between structure and transport properties for polymers with aromatic backbones, Chapter 3 in "Polymeric Gas Separation Membranes", D. R. Paul, Yuri. P. Yampolskii (Ed.), CRC Press (1994), 83–153.
29 M.S. McCaig, E.D. Seo, D.R. Paul, Polymer 40(12) (1999) 3367–3382.
30 J.S. Mchattie, W.J. Koros, D.R. Paul, Polymer 32(5) (1991) 840–850.
31 C.L. Aitken, W.J. Koros, D.R. Paul, Macromolecules 25(13) (1992) 3424–3434.
32 J. Zhang, X.H. Hou, Journal of Membrane Science 97 (1994) 275–282.
33 G. Perego, A. Roggero, R. Sisto, C. Valentini, Journal of Membrane Science 55(3) (1991) 325–331.
34 T. Emmler, K. Heinrich, D. Fritsch, P.M. Budd, N. Chaukura, D. Ehlers, K. Rätzke, F. Faupel, Macromolecules 43(14) (2010) 6075–6084.
35 H.B. Park, C.H. Jung, Y.M. Lee, A.J. Hill, S.J. Pas, S.T. Mudie, E. Van Wagner, B.D. Freeman, D.J. Cookson, Science 318(5848) (2007) 254–258.
36 H.B. Park, S.H. Han, C.H. Jung, Y.M. Lee, A.J. Hill, Journal of Membrane Science 359(1–2) (2010) 11–24.
37 R. Srinivasan, S.R. Auvil, P.M. Burban, Journal of Membrane Science 86(1–2) (1994) 67–86.
38 P.M. Budd, N.B. McKeown, D. Fritsch, Journal of Materials Chemistry 15(20) (2005) 1977–1986.
39 G. Consolati, I. Genco, M. Pegoraro, L. Zanderighi, Journal of Polymer Science Part B: Polymer Physics 34(2) (1996) 357–367.
40 T. Nakagawa, T. Saito, S. Asakawa, Y. Saito, Gas Separation & Purification 2(1) (1988) 3–8.
41 L.S. Kocherlakota, D.B. Knorr, L. Foster, R.M. Overney, Polymer 53(12) (2012) 2394–2401.
42 W.H. Lee, J.G. Seong, X.F. Hu, Y.M. Lee, Journal of Polymer Science 58(18) (2020) 2450–2466.
43 N.B. McKeown, P.M. Budd, K.J. Msayib, B.S. Ghanem, H.J. Kingston, C.E. Tattershall, S. Makhseed, K.J. Reynolds, D. Fritsch, Chemistry: A –European Journal 11(9) (2005) 2610–2620.
44 P.M. Budd, K.J. Msayib, C.E. Tattershall, B.S. Ghanem, K.J. Reynolds, N.B. McKeown, D. Fritsch, Journal of Membrane Science 251 (1–2) (2005) 263–269.
45 P.M. Budd, E.S. Elabas, B.S. Ghanem, S. Makhseed, N.B. McKeown, K.J. Msayib, C.E. Tattershall, D. Wang, Advanced Materials 16(5) (2004) 456–459.
46 P.M. Budd, B.S. Ghanem, S. Makhseed, N.B. McKeown, K.J. Msayib, C.E. Tattershall, Chemical Communications (2) (2004) 230–231.
47 M. Carta, R. Malpass-Evans, M. Croad, Y. Rogan, J.C. Jansen, P. Bernardo, F. Bazzarelli, N.B. McKeown, Science 339(6117) (2013) 303–307.
48 N.Y. Du, H.B. Park, G.P. Robertson, M.M. Dal-Cin, T. Visser, L. Scoles, M.D. Guiver, Nature Materials 10(5) (2011) 372–375.
49 S.H. Han, N. Misdan, S. Kim, C.M. Doherty, A.J. Hill, Y.M. Lee, Macromolecules 43(18) (2010) 7657–7667.
50 G.L. Tullos, J.M. Powers, S.J. Jeskey, L.J. Mathias, Macromolecules 32(11) (1999) 3598–3612.
51 G.L. Tullos, L.J. Mathias, Polymer 40(12) (1999) 3463–3468.

52 D.F. Sanders, Z.P. Smith, C.P. Ribeiro, R.L. Guo, J.E. McGrath, D.R. Paul, B.D. Freeman, Journal of Membrane Science 409 (2012) 232–241.

53 R.L. Guo, D.F. Sanders, Z.P. Smith, B.D. Freeman, D.R. Paul, J.E. McGrath, Journal of Materials Chemistry A 1(2) (2013) 262–272.

54 Z.P. Smith, D.F. Sanders, C.P. Ribeiro, R.L. Guo, B.D. Freeman, D.R. Paul, J.E. McGrath, S. Swinnea, Journal of Membrane Science 415 (2012) 558–567.

55 M.H. Cohen, D. Turnbull, The Journal of Chemical Physics 31(5) (1959) 1164–1169.

56 J. Crank, G. S. Park (Ed.), Diffusion in polymers, Academic Press (1968) 75–105.

57 Y.Y. Jiang, F.T. Willmore, D. Sanders, Z.P. Smith, C.P. Ribeiro, C.M. Doherty, A. Thornton, A.J. Hill, B.D. Freeman, I.C. Sanchez, Polymer 52(10) (2011) 2244–2254.

58 Z.P. Smith, R.R. Tiwari, T.M. Murphy, D.F. Sanders, K.L. Gleason, D.R. Paul, B.D. Freeman, Polymer 54(12) (2013) 3026–3037.

59 J.G. Wijmans, R.W. Baker, Journal of Membrane Science 107 (1–2) (1995) 1–21.

60 K. Nagai, T. Masuda, T. Nakagawa, B.D. Freeman, I. Pinnau, Progress in Polymer Science 26(5) (2001) 721–798.

61 T.C. Merkel, Z.J. He, I. Pinnau, B.D. Freeman, P. Meakin, A.J. Hill, Macromolecules 36(18) (2003) 6844–6855.

62 D. Hofmann, L. Fritz, J. Ulbrich, D. Paul, Polymer 38(25) (1997) 6145–6155.

63 J. Pellegrino, Annals of the New York Academy of Sciences 984 (2003) 289–305.

64 B.R. Mattes, M.R. Anderson, J.A. Conklin, H. Reiss, R.B. Kaner, Synthetic Metals 57(1) (1993) 3655–3660.

65 M.R. Anderson, B.R. Mattes, H. Reiss, R.B. Kaner, Science 252(5011) (1991) 1412–1415.

66 L. Rebattet, E. Genies, J. Allegraud, M. Pineri, M. Escoubes, Polymers for Advanced Technologies 4(4) (1993) 32–37.

67 J.C. Jansen, M. Macchione, E. Drioli, Journal of Membrane Science 287(1) (2007) 132–137.

68 I. Pinnau, L.G. Toy, Journal of Membrane Science 109(1) (1996) 125–133.

69 L. Tao, J.L. He, T. Arbaugh, J.R. McCutcheon, Y. Li, Journal of Membrane Science 665 (2023).

70 X.-Y. Wang, F.T. Willmore, R.D. Raharjo, X. Wang, B.D. Freeman, A.J. Hill, I.C. Sanchez, The Journal of Physical Chemistry B 110(33) (2006) 16685–16693.

71 S.D. Kelman, B.W. Rowe, C.W. Bielawski, S.J. Pas, A.J. Hill, D.R. Paul, B.D. Freeman, Journal of Membrane Science 320 (1–2) (2008) 123–134.

72 H. Eguchi, D.J. Kim, W.J. Koros, Polymer 58 (2015) 121–129.

73 C.H. Ma, W.J. Koros, Journal of Membrane Science 428 (2013) 251–259.

74 J.S. Mchattie, W.J. Koros, D.R. Paul, Polymer 33(8) (1992) 1701–1711.

75 N.R. Horn, D.R. Paul, Polymer 52(24) (2011) 5587–5594.

76 N.R. Horn, D.R. Paul, Polymer 52(7) (2011) 1619–1627.

77 S.F. Wang, X.Q. Li, H. Wu, Z.Z. Tian, Q.P. Xin, G.W. He, D.D. Peng, S.L. Chen, Y. Yin, Z.Y. Jiang, M.D. Guiver, Energy & Environmental Science 9(6) (2016) 1863–1890.

78 S. Kanehashi, T. Nakagawa, K. Nagai, X. Duthie, S. Kentish, G. Stevens, Journal of Membrane Science 298 (1–2) (2007) 147–155.

79 A. Bos, I.G.M. Pünt, M. Wessling, H. Strathmann, Journal of Membrane Science 155(1) (1999) 67–78.
80 A.C. Puleo, D.R. Paul, S.S. Kelley, Journal of Membrane Science 47(3) (1989) 301–332.
81 Y.S. Kang, J.M. Hong, J. Jang, U.Y. Kim, Journal of Membrane Science 109(2) (1996) 149–157.
82 J.M. Hong, Y.S. Kang, J. Jang, U.Y. Kim, Journal of Membrane Science 109(2) (1996) 159–163.
83 J.H. Lee, S.W. Kang, M.S. Yeom, Y.R. Kim, H. Choi, D. Song, J. Won, Y.S. Kang, Journal of Materials Chemistry A 2(19) (2014) 6987–6993.
84 I.S. Chae, S.W. Kang, J.Y. Park, Y.G. Lee, J.H. Lee, J. Won, Y.S. Kang, Angewandte Chemie International Edition 50(13) (2011) 2982–2985.
85 Y.S. Kang, S.W. Kang, H. Kim, J.H. Kim, J. Won, C.K. Kim, K. Char, Advanced Materials 19(3) (2007) 475–479.
86 S.U. Hong, J.H. Jin, J. Won, Y.S. Kang, Advanced Materials 12(13) (2000) 968.
87 E. Tsuchida, H. Nishide, M. Ohyanagi, H. Kawakami, Macromolecules 20(8) (1987) 1907–1912.
88 H. Nishide, M. Ohyanagi, O. Okada, E. Tsuchida, Macromolecules 19(2) (1986) 494–496.
89 I. Pinnau, L.G. Toy, Journal of Membrane Science 184(1) (2001) 39–48.
90 J.H. Kim, Y.S. Kang, J. Won, Macromolecular Research 12(2) (2004) 145–155.
91 I. Pinnau, L.G. Toy, C. Casillas, Olefin separation membrane and process (No. US 5670051), Membrane Technology and Research, Inc. (Menlo Park, CA), 1997.
92 S. Sunderrajan, B.D. Freeman, C.K. Hall, I. Pinnau, Journal of Membrane Science 182 (1–2) (2001) 1–12.
93 S.W. Kang, J.H. Kim, D. Ko, C.K. Kim, J. Won, K. Char, Y.S. Kang, Journal of Polymer Science Part B: Polymer Physics 42(18) (2004) 3344–3350.
94 J.H. Kim, J. Won, Y.S. Kang, Journal of Membrane Science 241(2) (2004) 403–407.
95 B. Jose, J.H. Ryu, B.G. Lee, H. Lee, Y.S. Kang, H.S. Kim, Chemical Communications (20) (2001) 2046–2047.
96 Y.S. Park, Y.S. Kang, S.W. Kang, Journal of Membrane Science 495 (2015) 61–64.
97 Y.R. Kim, J.H. Lee, H. Choi, W. Cho, Y.S. Kang, Journal of Membrane Science 463 (2014) 11–16.
98 H. Choi, J.H. Lee, Y.R. Kim, D. Song, S.W. Kang, S.S. Lee, Y.S. Kang, Chemical Communications 50(24) (2014) 3194–3196.
99 Y. Eum, B.S. Kim, I.S. Chae, G.H. Moon, S.C. Park, J. Jang, Y.S. Kang, Macromolecular Research 28(11) (2020) 1026–1031.
100 I.S. Chae, S.W. Kang, Y.S. Kang, RSC Advances 4(57) (2014) 30156–30161.
101 A.V. Rayer, K.Z. Sumon, L. Jaffari, A. Henni, Journal of Chemical & Engineering Data 59(11) (2014) 3805–3813.
102 Y.N. Zhao, W.S.W. Ho, Journal of Membrane Science 415 (2012) 132–138.
103 M. Sandru, T.J. Kim, W. Capala, M. Huijbers, M.B. Hägg, Energy Procedia 37 (2013) 6473–6480.
104 Z. Tong, W.S.W. Ho, Journal of Membrane Science 543 (2017) 202–211.
105 P.Y. Li, Z. Wang, Y.N. Liu, S. Zhao, J.X. Wang, S.C. Wang, Journal of Membrane Science 476 (2015) 243–255.

106 L. Yu, M. Kanezashi, H. Nagasawa, T. Tsuru, Applied Sciences-Basel 8(7) (2018).
107 M. Caplow, Journal of the American Chemical Society 90(24) (1968) 6795–6803.
108 I. Taniguchi, K. Kinugasa, M. Toyoda, K. Minezaki, Science and Technology of advanced Materials 18(1) (2017) 950–958.
109 I. Taniguchi, K. Kinugasa, M. Toyoda, K. Minezaki, H. Tanaka, K. Mitsuhara, Polymer Journal 53(1) (2021) 129–136.
110 G.J. Francisco, A. Chakma, X.S. Feng, Journal of Membrane Science 303(1–2) (2007) 54–63.
111 M.M. Wang, Z. Wang, S.C. Li, C.X. Zhang, J.X. Wang, S.C. Wang, Energy & Environmental Science 6(2) (2013) 539–551.
112 Z.D. Dai, J. Deng, L. Ansaloni, S. Janakiram, L.Y. Deng, Journal of Membrane Science 578 (2019) 61–68.
113 H. Bai, W.S.W. Ho, Industrial & Engineering Chemistry Research 48(5) (2009) 2344–2354.
114 M.S. Abd Rahaman, L. Zhang, L.H. Cheng, X.H. Xu, H.L. Chen, RSC Advances 2(24) (2012) 9165–9172.
115 S.H. Duan, I. Taniguchi, T. Kai, S. Kazama, Journal of Membrane Science 423 (2012) 107–112.
116 I. Taniguchi, S. Duan, T. Kai, S. Kazama, H. Jinnai, Journal of Materials Chemistry A 1(46) (2013) 14514–14523.
117 A. Sayari, A. Heydari-Gorji, Y. Yang, Journal of the American Chemical Society 134(33) (2012) 13834–13842.
118 M.M. Wang, Z. Wang, J.X. Wang, Y.Q. Zhu, S.C. Wang, Energy & Environmental Science 4(10) (2011) 3955–3959.
119 S. Lee, S.C. Park, T.-Y. Kim, S.W. Kang, Y.S. Kang, Journal of Membrane Science 548 (2018) 358–362.
120 J.H. Oh, Y.S. Kang, S.W. Kang, Chemical Communications 49(86) (2013) 10181–10183.
121 K.W. Yoon, S.W. Kang, Chemical Engineering Journal 334 (2018) 1749–1753.
122 S.C. Feng, X.B. Du, J.Q. Luo, Y.B. Zhuang, J.F. Wang, Y.H. Wan, Separation & Purification Technology 309 (2023).
123 S.C. Park, I.S. Chae, G.H. Moon, B.S. Kim, J. Jang, M. Wessling, Y.S. Kang, Journal of Materials Chemistry A 7(10) (2019) 5190–5194.
124 Y.F. Li, Q.P. Xin, H. Wu, R.L. Guo, Z.Z. Tian, Y. Liu, S.F. Wang, G.W. He, F.S. Pan, Z.Y. Jiang, Energy & Environmental Science 7(4) (2014) 1489–1499.
125 S. Hanioka, T. Maruyama, T. Sotani, M. Teramoto, H. Matsuyama, K. Nakashima, M. Hanaki, F. Kubota, M. Goto, Journal of Membrane Science 314 (1–2) (2008) 1–4.
126 C. Myers, H. Pennline, D. Luebke, J. Ilconich, J.K. Dixon, E.J. Maginn, J.F. Brennecke, Journal of Membrane Science 322(1) (2008) 28–31.
127 S.M. Mahurin, P.C. Hillesheim, J.S. Yeary, D.E. Jiang, S. Dai, RSC Advances 2(31) (2012) 11813–11819.
128 S. Kasahara, E. Kamio, T. Ishigami, H. Matsuyama, Chemical Communications 48(55) (2012) 6903–6905.
129 B.F. Soares, D.R. Nosov, J.M. Pires, A.A. Tyutyunov, E.I. Lozinskaya, D.Y. Antonov, A.S. Shaplov, I.M. Marrucho, Molecules 27(2) (2022) 413.

130 Z.D. Dai, R.D. Noble, D.L. Gin, X.P. Zhang, L.Y. Deng, Journal of Membrane Science 497 (2016) 1–20.

131 P. Scovazzo, Journal of Membrane Science 343 (1–2) (2009) 199–211.

132 Y.J. Sun, J. Wu, X.S. Zhu, Journal of Molecular Structure 1304 (2024).

133 M.M. Zhang, R. Semiat, X.Z. He, Separation & Purification Technology 299 (2022) 121784.

134 S. Zulfiqar, M.I. Sarwar, D. Mecerreyes, Polymer Chemistry 6(36) (2015) 6435–6451.

135 D. Mecerreyes, Progress in Polymer Science 36(12) (2011) 1629–1648.

136 J.E. Bara, D.E. Camper, D.L. Gin, R.D. Noble, Accounts of Chemical Research 43(1) (2010) 152–159.

137 G. Bernardo, H. Gaspar, Polymers 15(3) (2023).

138 J.E. Bara, S. Lessmann, C.J. Gabriel, E.S. Hatakeyama, R.D. Noble, D.L. Gin, Industrial & Engineering Chemistry Research 46(16) (2007) 5397–5404.

139 W.J. Horne, M.A. Andrews, M.S. Shannon, K.L. Terrill, J.D. Moon, S.S. Hayward, J.E. Bara, Separation & Purification Technology 155 (2015) 89–95.

140 J.E. Bara, C.J. Gabriel, E.S. Hatakeyama, T.K. Carlisle, S. Lessmann, R.D. Noble, D.L. Gin, Journal of Membrane Science 321(1) (2008) 3–7.

141 T.O. Magalhaes, A.S. Aquino, F. Dalla Vecchia, F.L. Bernard, M. Seferin, S.C. Menezes, R. Ligabue, S. Einloft, RSC Advances 4(35) (2014) 18164–18170.

142 S.V. Shaligram, A.S. Rewar, P.P. Wadgaonkar, U.K. Kharul, Polymer 93 (2016) 30–36.

143 D. Nikolaeva, I. Azcune, E. Sheridan, M. Sandru, A. Genua, M. Tanczyk, M. Jaschik, K. Warmuzinski, J.C. Jansen, I.F.J. Vankelecom, Journal of Materials Chemistry A 5(37) (2017) 19808–19818.

144 J.B. Tang, H.D. Tang, W.L. Sun, M. Radosz, Y.Q. Shen, Polymer 46(26) (2005) 12460–12467.

145 M.T. Huang, J.Y. Cao, Y.Z. Hong, Y.Z. Su, H.T. Wang, J. Li, Industrial & Engineering Chemistry Research 58(19) (2019) 8317–8322.

146 F.L. Bernard, R.B. Duczinski, M.F. Rojas, M.C.C. Fialho, L.A. Carreño, V.V. Chaban, F. Dalla Vecchia, S. Einloft, Fuel 211 (2018) 76–86.

147 J. Deng, J.B. Yu, Z.D. Dai, L.Y. Deng, Industrial & Engineering Chemistry Research 58(13) (2019) 5261–5268.

148 J.E. Bara, R.D. Noble, D.L. Gin, Industrial & Engineering Chemistry Research 48(9) (2009) 4607–4610.

149 L.C. Tomé, D.C. Guerreiro, R.M. Teodoro, V.D. Alves, I.M. Marrucho, Journal of Membrane Science 549 (2018) 267–274.

150 R.M. Teodoro, L.C. Tomé, D. Mantione, D. Mecerreyes, I.M. Marrucho, Journal of Membrane Science 552 (2018) 341–348.

151 O.H. Leblanc, W.J. Ward, S.L. Matson, S.G. Kimura, Journal of Membrane Science 6(3) (1980) 339–343.

152 R.D. Noble, Journal of Membrane Science 50(2) (1990) 207–214.

153 J.D. Way, R.D. Noble, D.L. Reed, G.M. Ginley, L.A. Jarr, AIChE Journal 33(3) (1987) 480–487.

154 H. Matsuyama, K. Matsui, Y. Kitamura, T. Maki, M. Teramoto, Separation & Purification Technology 17(3) (1999) 235–241.

155 L. Xiong, S. Gu, K.O. Jensen, Y.S.S. Yan, ChemSusChem 7(1) (2014) 114–116.

156 A. Ito, T. Yasuda, T. Yoshioka, A. Yoshida, X. Li, K. Hashimoto, K. Nagai, M. Shibayama, M. Watanabe, Macromolecules 51(18) (2018) 7112–7117.

6

Measurement of Molecule Transport Properties

Measuring transport parameters, such as diffusion coefficient, D, solubility coefficient, S, and permeability, P, of gases, vapors, and liquids in polymers, is crucial for understanding the relationship between polymer structure and transport properties, as well as for various applications.

There are three main approaches to measuring D, S, and P as summarized in Table 6.1. Each method typically provides information on only one or two of these parameters, and no single method can measure all three simultaneously. The choice of method depends on the nature and geometry of the polymer, the state of the permeant, and the required accuracy.

This chapter focuses on manometric and gravimetric methods along with a brief introduction to pulsed field gradient nuclear magnetic resonance (PFG-NMR). But FT-IR and UV-visible spectroscopic methods are not covered.

Polymer samples come in various shapes like slabs (films), spheres, and cylinders, each requiring specific mathematical approaches to estimate transport parameters. This chapter focuses on experimental methods and mathematical solutions for one-dimensional, thin slabs and briefly touches on spheres, excluding cylinders. Due to its experimental and mathematical simplicity, the one-dimensional thin slab method is widely used to characterize transport parameters in polymer samples.

Table 6.1 Experimental techniques to measure diffusion, D, solubility (or partition), S (or k), coefficients, and permeability, P, in polymers.

Method	Technique	Permeant	D	S	P
Permeation	Time-lag, transient	Gas	O	—	O
	Permeation cell	Gas and vapor	—	—	O
Sorption	Gravimetric, transient	Vapor	O	O	—
	Pressure-decay	Gas and vapor	—	O	—
Spectroscopy	PFG-NMR	Gas and vapor	O	O	—
	FT-IR, UV, transient	Gas and vapor	O	O	—

Molecule and Ion Transport through Polymer Membranes, First Edition. Yong Soo Kang.

The diffusion coefficient, D, is generally constant for gas permeation under normal pressures but often becomes concentration-dependent for vapor and liquid permeations. Thus, this chapter will briefly discuss the implications of concentration-dependent D. In glassy polymers, equilibrium surface concentration can take time to achieve, a phenomenon referred to as time-dependent surface concentration, which will also be briefly introduced.

For more detailed mathematical derivations and solutions, refer to *The Mathematics of Diffusion* [1] and *Diffusion in Polymers* [2].

6.1 Definitions of Diffusion Coefficient, Solubility Coefficient, and Permeability

Fick's first law and transport parameters of diffusion coefficient, solubility coefficient, and permeability are described in detail in Chapter 3. However, they will be briefly described here again for convenience. Unidirectional mass flux, J, will be treated in the x-direction only through a semi-finite slab with a thickness L.

Fick's first law is expressed as:

$$J = -D\frac{\mathrm{d}C}{\mathrm{d}x} \tag{6.1}$$

where the diffusion coefficient, D (cm^2/sec), is defined as the proportionality constant between the molar or volumetric flux, J (mol/cm^{2}sec or cm^3 (STP)/cm^2 sec), and its driving force ($\mathrm{d}C/\mathrm{d}x$) (mol/cm^3 cm or cm^3 (STP)/cm^3 cm). At the steady state, Equation (6.1) can be converted into

$$J = -D\frac{\Delta C}{L} \tag{6.2}$$

through a slab membrane with the thickness of L (cm), where $\Delta C = C_1 - C_2$ (Figure 6.1).

According to Equation (6.2), C_1, C_2, and D should be known to estimate J for a given thickness L of the slab membrane at steady state. However, it will be cumbersome to measure the concentrations of C_1 and C_2. One convenient way to get the C value is thus to utilize a linear relationship of C with the applied pressure p in gas or with C^B (the concentration in the bulk contacting with the slab membrane) in liquid phase. Thus, their relationships are expressed as:

$$\begin{aligned} C &= Sp \quad \text{for gas phase} \\ C &= KC^B \quad \text{for liquid phase} \end{aligned} \tag{6.3}$$

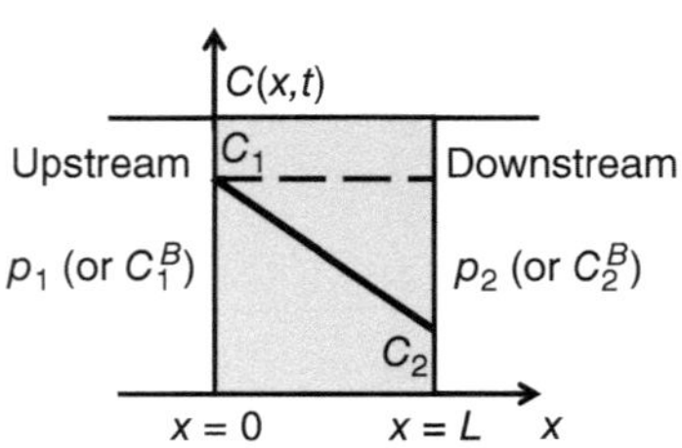

Figure 6.1 Scheme of concentration profile at steady state in gas or liquid phase. *Source*: Y.S. Kang.

where S (mol/cm^3 cmHg or cm^3 (STP)/cm^3 cmHg) and K (−) are the solubility coefficient in gas phase and the partition coefficient in liquid phase, respectively. If the S and K values do not change significantly in the range of ΔC, in other words, maintain to be nearly unchanged at the constants, substitution of Equation (6.3) into Equation (6.2) yields at steady state as:

$$J = P\frac{\Delta p}{L} \quad \text{for gas}$$
$$J = P\frac{\Delta C^B}{L} \quad \text{for liquid phase} \tag{6.4}$$

where the permeability P is defined for both gas and liquid phases as:

$$P = D \times S \quad \text{for gas phase}$$
$$P = D \times K \quad \text{for liquid phase} \tag{6.5}$$

Equation (6.5) indicates that P is only parameter representing the properties of material used in determining flux of a component, which is defined as the product of D and S (or K). Therefore, there are three transport parameters D, S (or K), and P in evaluating the transport properties. In other words, the mass flux J can be readily estimated from P and Δp (or ΔC^B) along with the slab thickness L. Table 6.2 lists the units of three transport parameters for convenience.

Fick's second law can be derived by considering a slab membrane with a thickness of dx between x and x + dx in a time interval dt as shown in Figure 6.2.

$$\frac{\partial C}{\partial t} = -\frac{\partial J}{\partial x} = \frac{\partial}{\partial x}\left(D\frac{\partial C}{\partial x}\right) \tag{6.6}$$

When the diffusion coefficient is constant, or not a function of concentration or position, Equation (6.6) reduces to:

$$\frac{\mathrm{d}C}{\mathrm{d}t} = D\frac{\mathrm{d}^2C}{\mathrm{d}x^2} \tag{6.7}$$

This equation describes the concentration change over time in a diffusion process characterized by D. It is second-order with respect to thickness dx, linear, and homogeneous. As a result, its solution must satisfy one initial condition and two boundary conditions.

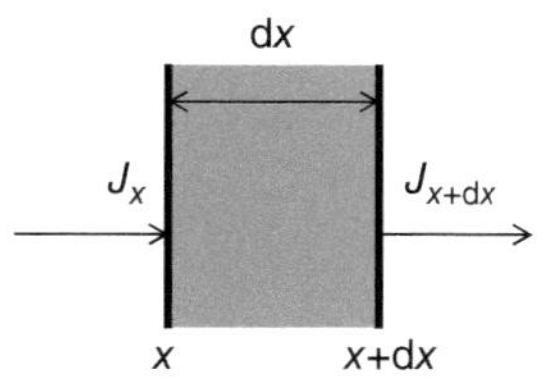

Figure 6.2 Mass transport across dx between x and x + dx. *Source*: Y.S. Kang.

Table 6.2 Units of transport parameters.

	J	D	S	K	P[a]
SI unit	mol/m^{2}s	m^2/s	mol/m^3 Pa	Dimensionless	mol m/m^{2}s Pa
Conventional unit	cm^3(STP)/cm^{2}s	cm^2/s	cm^3(STP)/ cm^3 cmHg	Dimensionless	cm^3(STP) cm/ cm^{2}s cmHg

[a] 1 Barrer = 1.0 × 10^{-10} cm^3(STP)cm/cm^2 s cmHg.
Source: Y.S. Kang.

6.2 Permeation and Sorption Features in a Slab Membrane

Permeation and sorption methods are frequently used to measure the three transport parameters, i.e. D, S (or K), and P. The transient permeation method by measuring pressure change gives direct information of both D and P, and is typically used for fast permeable molecules, such as permanent gases, including O_2 and CO_2. On the other hand, the transient sorption experiment gives D and S (or K) directly, favorable for soluble but rather slowly permeable molecules such as organic vapors or liquids to measure the weight change. It is very important to note that the permeation experiment does not give any direct information on S, whereas the sorption experiment does not give any direct information on P. Therefore, such P and S are frequently required to be independently measured because the relation of $P = D \times S$ (or K) is strictly based on the Fick's law with D and S being unchanged at the experimental conditions, which may not be always valid in glassy polymers as described in Chapter 3.

6.3 Permeation Method for a Slab Membrane: Manometric and Time-lag Methods

6.3.1 Theoretical Backgrounds

When the surface concentrations at the upstream and downstream sides are different from each other ($C_1 > C_2$) in a semi-infinite slab membrane of a thickness of L, as shown in Figure 6.3a, the mass transport occurs from high pressure side to low pressure side, resulting in permeation. The corresponding initial and boundary conditions are mathematically stated as:

$C(x, 0) = C_0$ for initial condition where $0 \leq x \leq L$ at $t < 0$
$C(0, t) = C_1$ and $C(L, t) = C_2$ with $C_1 > C_2$ for boundary conditions at $t \geq 0$

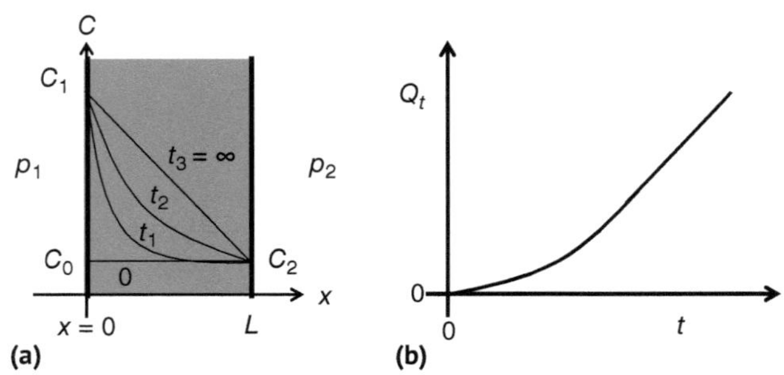

Figure 6.3 Schematic changes in (a) the concentration $C(x, t)$ profile with a constant D for permeation experiment when p_1 and p_2 are maintained constant at the upstream and downstream sides, respectively; and (b) the total amount of the species Q_t permeated per unit area A with time t. *Source:* Y.S. Kang.

The integration of Equation (6.7) subject to the above initial and boundary conditions gives [1]:

$$C = C_1 + (C_2 - C_1)\frac{x}{L} + \frac{2}{\pi}\sum_{n=1}^{\infty}\frac{C_2\cos n\pi - C_1}{n}\sin\frac{n\pi x}{L}\exp\{-Dn^2\pi^2t/L^2\}$$
$$+\frac{4C_0}{\pi}\sum_{m=0}^{\infty}\frac{1}{2m+1}\sin\frac{(2m+1)\pi x}{L}\exp\{-D(2m+1)^2\pi^2t/L^2\} \tag{6.8}$$

The concentration changes with respect to x and t based on Equation (6.8) are schematically drawn in Figure 6.3a.

The total amount of the species, Q_t (cm^3 (STP)/cm^2), permeated per unit area A (cm^2) of a slab membrane with the thickness L (cm) can be then expressed as [1]:

$$Q_t = \int_{0.}^{t} J|_{x=L}\,dt = -\int_0^t D\frac{dC}{dx}\bigg|_{x=L} dt \tag{6.9}$$

From Equations (6.8) and (6.9), Q_t can be obtained as [1]:

$$Q_t = D(C_1 - C_2)\frac{t}{L} + \frac{2L}{\pi^2}\sum_{n=1}^{\infty}\frac{C_1\cos n\pi - C_2}{n^2}\{1 - \exp\{-Dn^2\pi^2t/L^2\}$$
$$+\frac{4C_0L}{\pi^2}\sum_{m=0}^{\infty}\frac{1}{(2m+1)^2}\{1 - \exp\{-D(2m+1)^2\pi^2t/L^2\} \tag{6.10}$$

In common experiments, both C_0 and C_2 are arranged to be zero, i.e. the slab membrane is initially at zero concentration ($C_0 = 0$) and the concentration at the downstream surface should be maintained effectively at zero ($C_2 = 0$). In these conditions, we find [1]:

$$Q_t = \frac{DC_1}{L}\left[t - \frac{L^2}{6D}\right] - \frac{2LC_1}{\pi^2}\sum_{n=1}^{\infty}\frac{(-1)^n}{n^2}\exp\left(\frac{-Dn^2\pi^2t}{L^2}\right) \tag{6.11}$$

The results are schematically illustrated in Figure 6.3b.

As $t \sim \infty$, Q_t increases linearly with time, representing to reach the steady state, and then Equation (6.11) reduces to [1]:

$$Q_t = \frac{DC_1}{L}\left[t - \frac{L^2}{6D}\right] = \frac{Pp_1}{L}\left[t - \frac{L^2}{6D}\right] \tag{6.12}$$

6.3.2 Experimental Methods for Permeation

Figure 6.4 illustrates (a) the schematic structure of a permeation cell and (b) a whole permeation apparatus under a temperature-controlled condition.

A slab membrane with the thickness of L and the effective surface area of A was installed between upstream and downstream sides as shown in Figure 6.4a. Initially, the total system is evacuated to make the slab sample concentration zero ($C_0 = 0$). Then, the upstream pressure p_1 should be maintained constant during whole experiment and thus it connects to big gas tanks (5) in Figure 6.4b. The important thing is that the downstream pressure p_2 should be close to zero for $C_2 = 0$, but high enough to measure p_2 precisely with time. During experiments, p_2 is plotted as a function of time t, as shown in Figure 6.5. When you see the linearly increasing straight curve

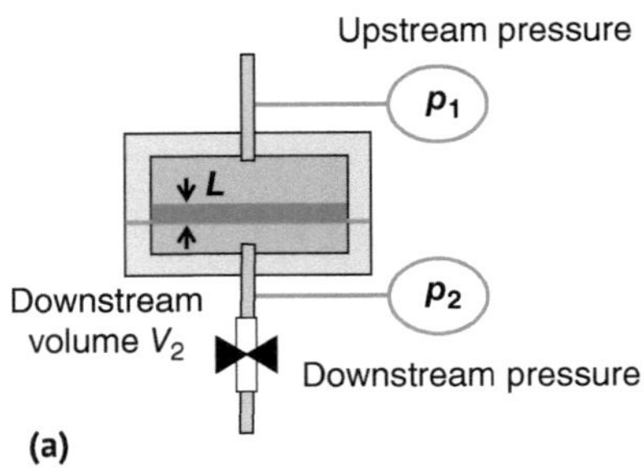

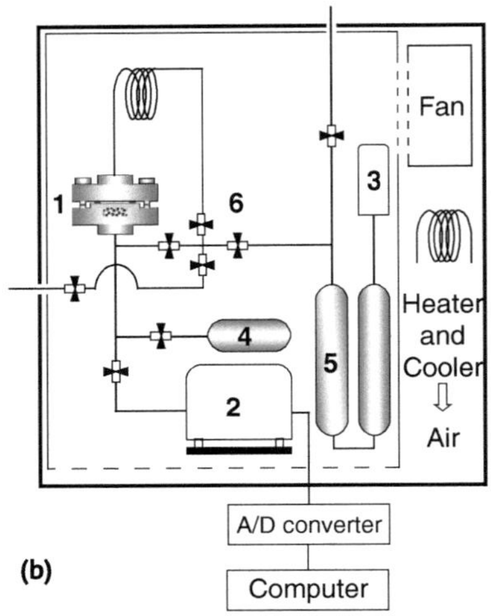

Figure 6.4 (a) Schematic gas permeation cell; and (b) constant pressure permeation apparatus with the number 1 for permeation cell, 2 for downstream pressure gauge (p_2), 3 for upstream pressure gauge (p_1), 4 for gas cylinder to increase the downstream volume, 5 for upstream gas tank, and 6 for computer-controlled valves. *Source*: Y.S. Kang.

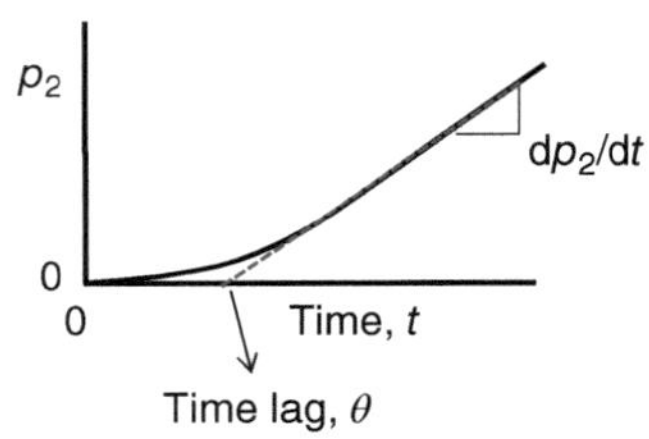

Figure 6.5 Downstream pressure p_2 as a function of t and the time lag θ. *Source*: Y.S. Kang.

at long times, reaching a steady state, you may stop the experiment. The permeability coefficient and diffusion coefficient can be obtained from the steady-state slope, $\mathrm{d}p_2/\mathrm{d}t$, and the time lag, θ, the intercept at $p_2 = 0$ extrapolated with the linearly increasing straight line, respectively, as shown in Figure 6.5.

According to Equation (6.12), the steady-state slope $(\mathrm{d}Q_t/\mathrm{d}t) = Pp_1/L$ at long times. Q_t is related with p_2 in the downside volume of V_2 at temperature T as:

$$Q_t = \frac{p_2 V_2}{AzRT} \tag{6.13}$$

where A, z, and R are the effective surface area for the permeation, the gas compressibility factor, representing the non-ideality for gas, and the gas constant, respectively. From the steady-state slope $(\mathrm{d}p_2/\mathrm{d}t)$ in Figure 6.5, P can be obtained as:

$$P = \frac{L}{AzRT}\frac{V_2}{p_1}\left(\frac{\mathrm{d}p_2}{\mathrm{d}t}\right) \tag{6.14}$$

Additionally, the extrapolated time with Q_t zero, called the time lag θ in the plot p_2 vs t (Figure 6.5), the diffusion coefficient D can be obtained according to Equation (6.15) as:

$$D = L^2/6\theta \tag{6.15}$$

Therefore, from permeation experiment, both P and D can be obtained from Equations (6.14) and (6.15), respectively.

In summary, both permeability, P, and diffusion coefficient, D, are measured by the slope and time lag, respectively, from a single transient permeation experiment. In the case of requiring the solubility coefficient, an independent experiment is suggested for both gas and liquid permeation. Here, it is also very important to note the assumptions to solve Fick's second law for permeation: (1) constant D; (2) constant S or k_D for Henry's law; and (3) instantaneous equilibrium at both interfaces.

6.4 Sorption Method for a Slab Membrane: Gravimetric Methods

6.4.1 Theoretical Backgrounds

The initial concentration C_0 and the two boundary conditions C_1 at both surfaces are kept at constant concentrations in a semi-infinite slab membrane with the thickness of $2L$ for a sorption experiment as shown in Figure 6.6a:

$C(x, 0) = C_0$ for initial condition where $-L \leq x \leq L$
$C(-L, t) = C(L, t) = C_1$ or $\mathrm{d}C(0, t)/\mathrm{d}x = 0$ for boundary conditions with $t \geq 0$

Simply because of the symmetric boundary condition, $\mathrm{d}C(0, t)/\mathrm{d}x = 0$ and the thickness of the slab membrane is conveniently used by $2L$. In this case, mass transport occurs inwardly from both interfaces.

The solution of Equation (6.7) subject to the above initial and boundary conditions for the sorption experiments becomes [1]:

$$\frac{C - C_0}{C_1 - C_0} = 1 - \frac{4}{\pi} \sum_{n=0}^{\infty} \frac{(-1)^n}{2n+1} \exp\left\{-D(2n+1)^2\pi^2 t/4L^2\right\} \cos\frac{(2n+1)\pi x}{2L} \tag{6.16}$$

Equation (6.16) gives concentration changes by varying the time (Figure 6.6a).

The total amount of penetrant M_t sorbed per unit area of the slab at time t is calculated by [1]:

$$M_t = \int_{-L}^{L} C(x, t)\,\mathrm{d}x = -2\int_0^t D\frac{\mathrm{d}C(L, t)}{\mathrm{d}x}\,\mathrm{d}t \tag{6.17}$$

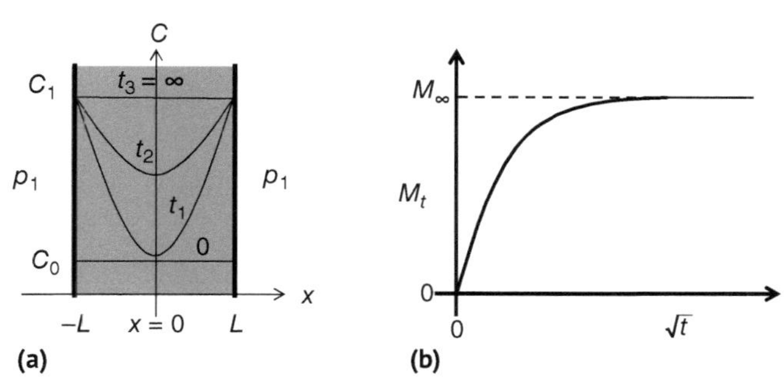

Figure 6.6 Schematic changes in transient sorption experiment: (a) the concentration $C(x, t)$; and (b) the amount M_t of species i sorbed at time t as a function of square-root of time t. *Source*: Y.S. Kang.

Then M_t becomes [1]:

$$\frac{M_t}{M_\infty} = 1 - \sum_{n=0}^{\infty} \frac{8}{(2n+1)^2\pi^2} \exp\{-D(2n+1)^2\pi^2 t/4L^2\} \tag{6.18}$$

where M_∞ is the amount of permeant sorbed after infinite time, representing equilibrium solubility. Figure 6.6b illustrates the amount M_t of species i sorbed at time t as a function of square-root of time t. Equation (6.18) states that M_t/M_∞ is a function of only (Dt/L^2).

6.4.2 Experimental Methods for Transient Sorption

There are two experimental sorption methods: transient sorption and equilibrium sorption. The former transient sorption can give us both the diffusion coefficient and solubility coefficient, whereas the latter equilibrium sorption can give us only the solubility coefficient but not the diffusion coefficient. Experimentally, the geometry of the sample for transient sorption is very important to understand the transport behavior. However, for the equilibrium sorption case, the equilibrium solubility is only measured at a given applied pressure regardless of the sample geometry, and therefore experimental procedure is very simple, which will be discussed at the end of this section.

Transient sorption experiments for a penetrant into a slab membrane require to measure the weight change M_t of the penetrant as a function of time t at given pressure p_1 and T. The weight change M_t measurement is readily obtained by quartz spring microbalance (Figure 6.7) with measuring the spring extension Δl with time t or quartz crystal microbalance (Figure 6.8) with measuring frequency change Δf of quartz crystal with t.

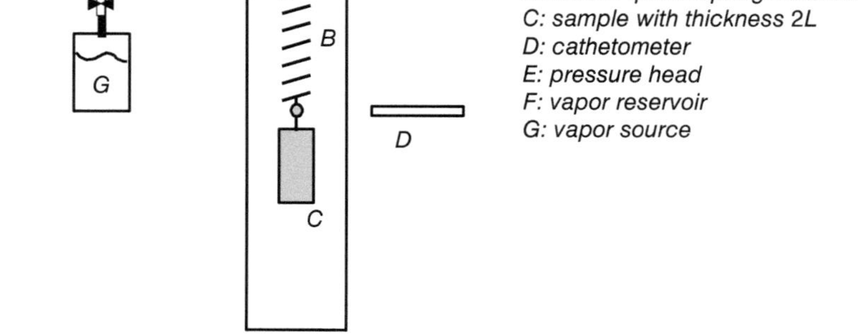

Figure 6.7 Schematic drawing of quartz spring microbalance apparatus for sorption experiments at given pressure p_1 and T. *Source*: Y.S. Kang.

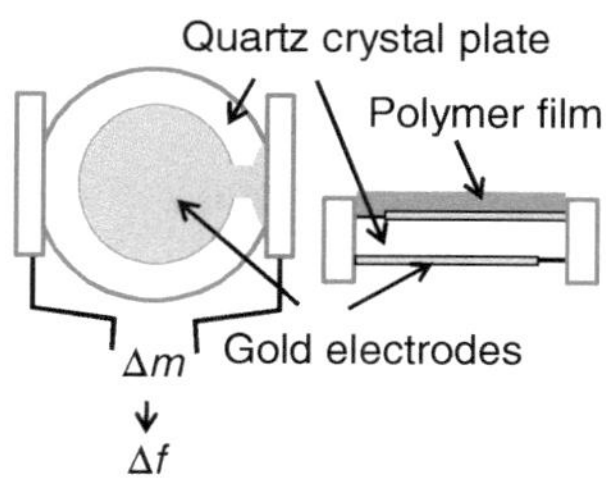

Figure 6.8 Schematic structure of quartz crystal microbalance for sorption experiments at given pressure p_1 and T. *Source:* Y.S. Kang.

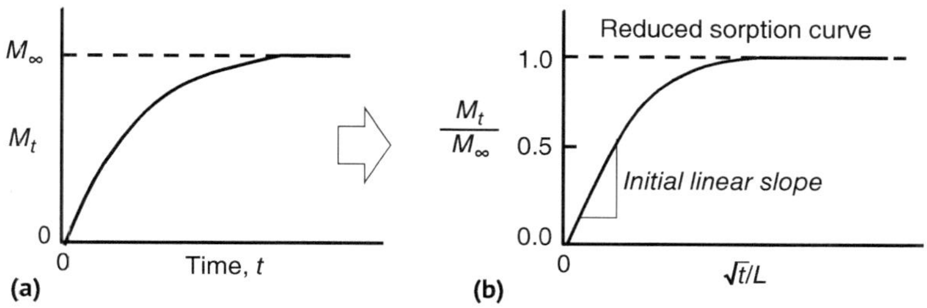

Figure 6.9 (a) Original sorption curve of M_t vs t; and (b) reduced sorption curve of M_t/M_∞ vs $t^{1/2}/L$. *Source:* Y.S. Kang.

In typical experiments, M_t is measured over increasing sorption time t, as shown in Figure 6.9a. The equilibrium uptake, M_∞, can be used to determine the solubility coefficient S from $S = M_\infty/p_1$. To verify Fickian behavior and determine the diffusion coefficient, the experimental data of M_t vs. t are often replotted as M_t/M_∞ against $(t/L^2)^{1/2}$, forming a reduced sorption curve (Figure 6.9b).

In the reduced sorption curve, M_t/M_∞ increases linearly up to a value of approximately 0.5–0.6, after which it deviates from linearity and gradually approaches the equilibrium value of 1, as shown in Figure 6.9b. This is a typical evidence for the Fickian diffusion behavior. In contrast, the M_t vs t curve shows little to no linear region, or at most a very short one, even in the case of Fickian diffusion.

For short times, Equation (6.18) is converted to [1]:

$$\frac{M_t}{M_\infty} = \frac{4}{L}\left[\frac{Dt}{\pi}\right]^{1/2} = \frac{4\sqrt{D}}{\sqrt{\pi}}\left[\frac{t}{L^2}\right]^{1/2} \tag{6.19}$$

which indicates M_t/M_∞ increases linearly with $\left[t/L^2\right]^{1/2}$. The diffusion coefficient, D, can be then readily obtained from the initial slope $\mathrm{d}(M_t/M_\infty)/\mathrm{d}(t/L^2)^{1/2}$:

$$D = \frac{\pi}{16}\left[\frac{\mathrm{d}(M_t/M_\infty)}{\mathrm{d}(t/L^2)^{1/2}}\right]^2 \tag{6.20}$$

Another method to obtain D is from the half time $[t/L^2]_{0.5}$ at $M_t/M_\infty = 0.5$ by utilizing Equation (6.18). It is directly related with the diffusion coefficient as [1]:

$$D = \frac{0.049}{(t/L^2)_{0.5}} \tag{6.21}$$

The diffusion coefficient can also be evaluated from sorption data at a late stage or long times. At the late stage, Equation (6.18) becomes:

$$\ln\left[1 - \frac{M_t}{M_\infty}\right] = \ln\frac{8}{\pi^2} - \frac{\pi^2 D}{L^2}t \tag{6.22}$$

At long times as shown in Figure 6.10, the diffusion coefficient D can also be obtained from the linear slope [1]:

$$D = -\frac{L^2}{\pi^2}\frac{\mathrm{d}\ln[1 - M_t/M_\infty]}{\mathrm{d}t} \tag{6.23}$$

In the case of liquid system, the basic concept is equivalent to the gas system described above, and the experiment is much simpler than that with gas. When a plain slab is introduced in a liquid solvent solution, the solvent penetrates into the slab. The slab is taken out from the solution and wiped out to measure its weight change as function of time to obtain M_t and M_∞ at the equilibrium. Thus, the experimental data will show similar shape with Figure 6.9, from which the partition coefficient and diffusion coefficient are estimated using the same mathematical procedures as described previously, only when swelling of the slab by the liquid penetrant is small enough to be negligible.

6.4.3 Experimental Methods for Equilibrium Sorption

Figure 6.11a shows a schematic equilibrium sorption apparatus for gas, and Figure 6.11b shows an original data set to measure the solubility coefficient S. In the case of gas sorption, initially the total system is evacuated to make the sample concentration zero ($C_0 = 0$). Then, gas can be introduced into a V_0 volume with controlled pressure p_0. Then, the central valve between V_0 and V_1 will open to introduce the gas into chamber V_1 containing the polymer sample with the volume of V_s. Then, measure the pressure change (p_1) with time as shown in Figure 6.11b to obtain the equilibrium pressure p_f.

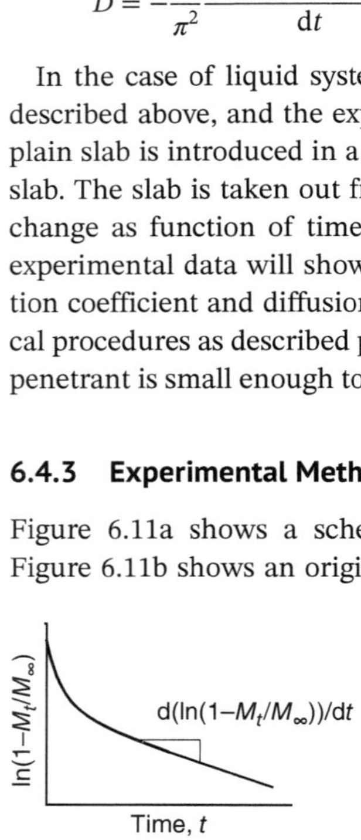

Figure 6.10 $\ln[1 - M_t/M_\infty]$ as a function of time at a late stage. *Source*: Adapted from [1].

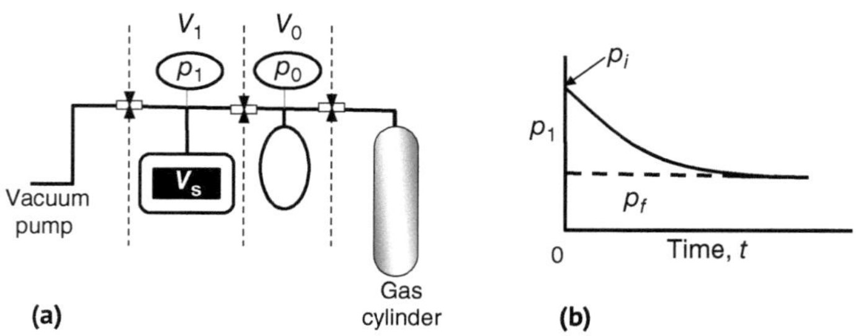

Figure 6.11 (a) Schematic equilibrium sorption apparatus for gas and (b) changes in pressure as a function of time during equilibrium sorption experiment. *Source*: Y.S. Kang.

The initial amount of the gas penetrant in V_0 is $n_0 = p_0 V_0/zRT$ and that at equilibrium state $n_1 = p_f(V_0 + V_1 - V_s)/zRT$. The total amount of the gas sobbed by the sample volume of V_s will be Δn $(= n_0 - n_1)$, and therefore the concentration C_1 of gas absorbed can be obtained as $C_1 = \Delta n/V_s$. Then, the solubility coefficient, S, is readily calculated from:

$$S = \frac{C_1}{p_f} = \frac{(\Delta n/V_s)}{p_f} \tag{6.24}$$

One important assumption to note is no volume change of the sample upon the absorption to calculate the solubility coefficient.

6.5 Concentration-dependent Diffusion Coefficient

When D is a function of concentration, the average diffusion coefficient $\overline{D}$ in the range of from C to C_0 is mathematically related as [1]:

$$\bar{D} = \frac{1}{C_0 - C}\int_{C}^{C_0} D\mathrm{d}C \tag{6.25}$$

When $C = 0$, it follows that:

$$\bar{D} = \frac{1}{C_0}\int_{0}^{C_0} D\mathrm{d}c \tag{6.26}$$

When $\overline{D}C_0$ is plotted against C_0 as shown in Figure 6.12, the average diffusion coefficient $\overline{D}$ between 0 to C_0 may be obtained from the slope [1]:

$$\frac{\mathrm{d}}{\mathrm{d}C_0}(\bar{D}C_0) = \bar{D} \tag{6.27}$$

When the diffusion coefficient can be expressed as a function of the penetrant concentration C as:

$$D(C) = D^{\circ}(1 + \omega C) \tag{6.28}$$

where ω is a constant representing the interactions between penetrant and polymeric matrix. Transient permeation method can be applied to estimate the time-lag θ, from which $D(C)$ can be calculated as [1]:

$$\theta = \frac{L^2}{6D^{\circ}}\left[\frac{1 + \omega C/4}{1 + \omega C/2}\right] \tag{6.29}$$

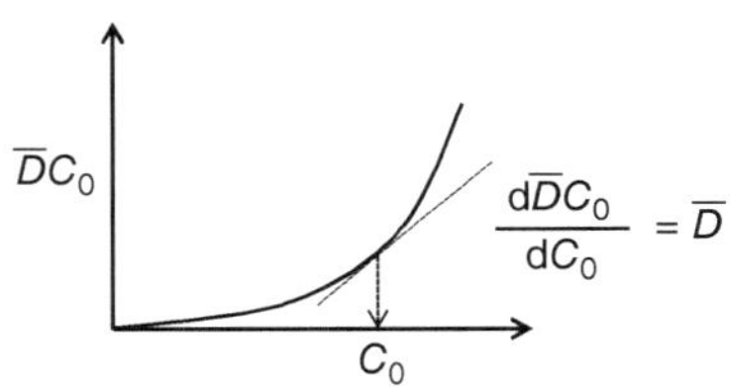

Figure 6.12 $\overline{D}C_0$ is plotted as a function of C_0 for concentration-dependent diffusion coefficient. *Source*: Adapted from [1].

6.6 Variable Surface Concentration

It is important to note that mathematical solutions of the Fick's second law, such as Equations (6.8) for permeation and (6.16) for sorption are obtained by assuming that C_1 and

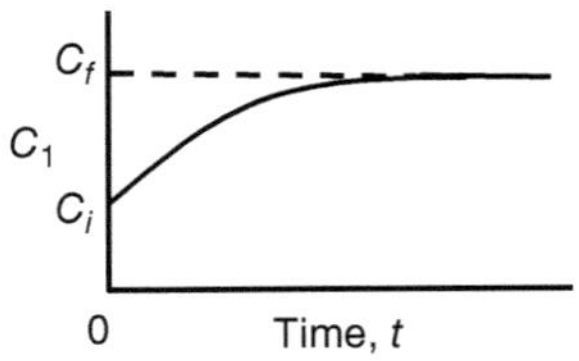

Figure 6.13 The concentration changes on the front surface with time in glassy state. *Source*: Y.S. Kang.

C_2 are constant but not function of time. In other words, the constant boundary conditions are commonly accepted in the case of a rubbery state, whereas they are not usually in the case of a glassy state material. This is mostly due to the fact that the rubbery state is in equilibrium, whereas the glassy state is not in equilibrium; in fact, slowly relaxing into the equilibrium state with time, which is commonly called relaxation or physical aging, conveniently characterized by relaxation constant τ.

For instance, the concentration at the front surface C_1 at glassy state can be schematically shown as Figure 6.13.

Here, C_1 increases with time from the initial, instantaneous concentration, C_i, and reaches the final equilibrium value, C_f, after long experimental time. This process can be mathematically expressed as:

$$C_1 = C_i + (C_f - C_i)\ \exp\left(\frac{-t}{\tau}\right) \tag{6.30}$$

Detailed mathematical solutions associated with the variable surface concentration are present in the book [1].

6.7 Sorption in a Sphere

For sorption in a sphere, the diffusion equation for a constant diffusion coefficient takes the form [1]:

$$\frac{\partial C}{\partial t} = D\left(\frac{\partial^2 C}{\partial r^2} + \frac{2}{r}\frac{\partial C}{\partial r}\right) \tag{6.31}$$

On putting $u = Cr$ where $0 \le r \le R$, Equation (6.31) becomes:

$$\frac{\partial u}{\partial t} = D\frac{\partial^2 u}{\partial r^2} \tag{6.32}$$

Equation (6.32) for a sphere with radius of R is solved subject to:
initial condition: $u(r, 0) = 0$
boundary conditions: $u(0, t) = 0$ and $u(R, t) = RC_0$, at $t \ge 0$

$$\frac{M_t}{M_\infty} = 1 - \frac{6}{\pi^2}\sum_{m=0}^{\infty}\frac{1}{m^2}\exp\left(\frac{-Dm^2\pi^2 t}{R^2}\right) \tag{6.33}$$

This equation in early stages becomes:

$$\frac{M_t}{M_\infty} = \frac{6}{\sqrt{\pi}}\left(\frac{Dt}{R^2}\right)^{1/2} \tag{6.34}$$

and therefore

$$D = \frac{\pi}{36}\ \mathrm{d}(M_t/M_\infty)/\mathrm{d}(t/R^2)^{1/2} \tag{6.35}$$

Thus, the diffusion coefficient may be obtained, provided that the initial slope is accurately measured.

6.8 Pulsed Field Gradient (PFG)-NMR for Diffusion Coefficient

NMR spectroscopy has been successfully utilized to study static properties of the matter (i.e. structure) and dynamic properties like self-diffusion, flow, and relaxation. In his pioneering work on spin echo, Hahn [3] pointed out that the echo amplitude is influenced by molecular diffusion (Brownian motion) because of the fluctuations of the local magnetic field. The pulsed field gradient (PFG)-NMR method was first theoretically and experimentally demonstrated by Stejskal and Tanner [4]. Accordingly, PFG-NMR sequences with different gradient strengths or durations is repeated to measure the attenuation of the NMR signal. The signal intensity decreases as the gradient strength increases due to molecular diffusion. The signal attenuation can be analyzed by using the Stejskal and Tanner equation and is still commonly used to obtain dynamic information, such as the self-diffusion coefficient. Since then, PFG-NMR is a widely used tool for determining the diffusion coefficient of molecules and ions both in bulk and even in confined geometry (e.g., within porous materials). Experimentally, when a sample containing small species of polymers is exposed to the stimulated echo pulse sequence with a field gradient strength g, corresponding echo peaks $A(g)$ are generated [5].

The diffusion coefficient is correlated to the echo amplitude by following the Stejskal and Tanner equation [4]:

$$A(g) = A(0)\ \exp\left(- \gamma^2 g^2 \delta^2 \left(\Delta - \frac{\delta}{3}\right) D\right) \tag{6.36}$$

where $A(g)$ and $A(0)$ are the echo amplitudes with and without application of the pulsed gradient, respectively. γ is the gyromagnetic ratio of the nucleus, g (G/cm) is the strength of the field gradient pulse, δ (ms) is its duration time, and Δ (ms) is the time between the two gradient pulses. By plotting $\ln[A(g)/A(0)]$ against g^2 (or $\gamma^2 g^2 \delta^2 (\Delta - \delta/3)$) as shown in Figure 6.14 [6], the diffusion coefficient, D, can be derived from the slope of the resulting straight line according to the Stejskal and Tanner equation [4]. Here, Δ is typically in the range of hundreds ms, and δ varies from 100 μs to 10 ms. The error in the mean diffusion coefficient is mostly estimated to be ±5%.

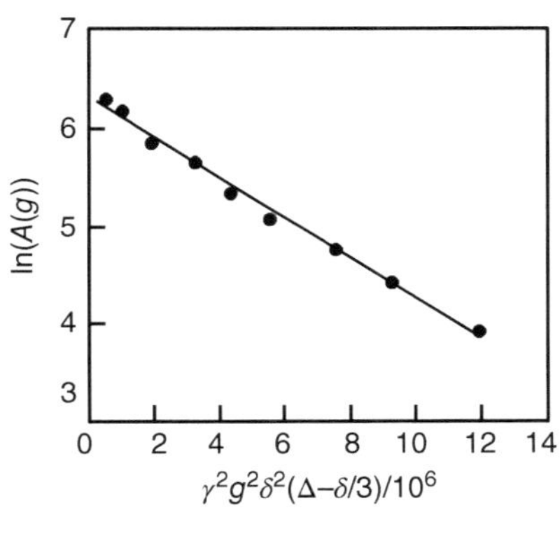

Figure 6.14 Typical plot of $\ln[A(g)]$ against $\gamma^2 g^2 \delta^2 (\Delta - \delta/3)\, 10^6\, \mathrm{Scm^{-2}}$ in the electrolyte of PEO/$LiCF_3SO_3$ with an ethylene oxide to lithium ratio of 20/1 and a molecular weight of 2,000 at 90 °C. *Source*: Adapted from [6].

PFG-NMR allows the determination of the self-diffusion coefficient of gases in porous materials [7], and also to measure the varying contributions of different diffusional pathways (such as intrapore or interpore) dependent on the diffusion analysis time [8, 9]. Multicomponent diffusion may also be measured easily with PFG-NMR when adsorbates possess different NMR active nuclei [10]. For example, CO_2 and NO_x diffusion may be decoupled by measurement of the ^{13}C and ^{15}N nuclei, respectively [11, 12].

It is important to note that the mutual diffusion coefficient, D_m, obtained from Fick's law in the presence of a concentration gradient is distinct from the self-diffusion coefficient, D_s, which is typically measured using techniques like PFG-NMR. The self-diffusion coefficient, D_s, arises from random Brownian motion and can be observed experimentally using tracer methods (e.g., isotopic labeling). It is linked to the mean square displacement of particles by the relation $\langle r^2 \rangle \sim 6D_s t$ in three-dimensional systems.

In real systems with interactions between particles, D_m depends on the thermodynamic factor, Φ, which accounts for non-ideal behavior. This relationship can be expressed by Darken equation as:

$$D_m = \Phi D_s \tag{6.37}$$

where $\Phi = 1 + \text{dln}(\gamma)/\text{dln}(c)$ with γ and c being the activity coefficient and concentration, respectively. Therefore, in ideal and dilute systems, $\Phi \sim 1$ and $D_m \sim D_s$.

References

1 J. Crank, The Mathematics of Diffusion, Oxford University Press, 1975.
2 J. Crank, G.S. Park, Diffusion in Polymers, Academic Press, 1968.
3 E.L. Hahn, Physical Review 80(4) (1950) 580–594.
4 E.O. Stejskal, J.E. Tanner, The Journal of Chemical Physics 42(1) (1965) 288–292.
5 C.S. Johnson Jr., Progress in Nuclear Magnetic Resonance Spectroscopy 34(3–4) (1999) 203–256.
6 J. Shi, C.A. Vincent, Solid State Ionics 60(1–3) (1993) 11–17.
7 K. Díaz, L. Garrido, M. López-González, L.F. Del Castillo, E. Riande, Macromolecules 43(1) (2010) 316–325.
8 R. Mueller, R. Kanungo, M. Kiyono-Shimobe, W.J. Koros, S. Vasenkov, Langmuir 28(27) (2012) 10296–10303.
9 S.M. Pugh, A.C. Forse, Journal of Magnetic Resonance 346 (2023) 107343.
10 F. Rittig, C.G. Coe, J.M. Zielinski, The Journal of Physical Chemistry B 107(19) (2003) 4560–4566.
11 A. Baniani, S.J. Berens, M.P. Rivera, R.P. Lively, S. Vasenkov, Adsorption 27 (2021) 485–501.
12 J. Kärger, M. Avramovska, D. Freude, J. Haase, S. Hwang, R. Valiullin, Adsorption 27(3) (2021) 453–484.

7

Applications of Selective Transport Membranes

Separation performance of gas or liquid mixtures (i and j) by membrane can be characterized by the production rate of component i separated per unit time, i.e. N_i (g/sec), which can be expressed as a product of the mass flux J_i (g/cm^2 sec) of component i and the membrane area A (cm^2):

$$N_i = J_j \times A \tag{7.1}$$

The mass flux J_i can be, in turn, determined by three properties: separation material (P_i and α_{ij}), membrane structure (specifically its thickness, L), and pressure difference (Δp_i) across the membrane, as shown in Figure 7.1. Membrane area A for a given separation system can be determined by membrane module configuration and system design, which also play important roles in determining final performance. According to Fick's law at a steady state:

$$J_i = P_i \frac{\Delta p_i}{L} \tag{7.2}$$

To maximize the separation performance of a membrane for a given material (with specified P_i, and α_{ij}) and operational conditions (specified Δp_i and T), the thickness (L) of the membrane must be optimized. This optimization aims to achieve both high separation efficiency and adequate mechanical strength. Composite membrane structures, which consist of a thin top coating layer on an asymmetrically microporous support, are often used for this purpose, as shown in Figure 7.2.

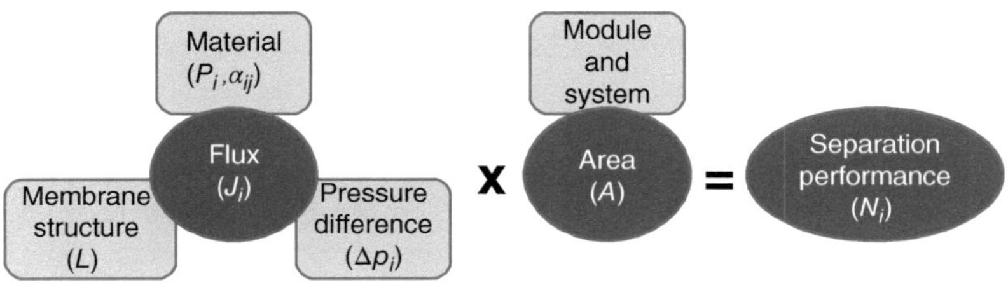

Figure 7.1 Overall membrane separation performance N_i depending on mass flux J_i and area A of membrane. *Source*: Y.S. Kang.

Figure 7.2 Schematic structure of composite membrane. *Source*: Y.S. Kang.

In summary, separation performance depends on membrane materials, membrane structure, module configuration, and system design. While the material properties (P_i and α_{ij}) are discussed in Chapters 3–5, this chapter explores membrane configurations, such as composite structures, their transport properties, and their integration into membrane modules and systems briefly for practical applications [1].

7.1 Structure and Transport Through Composite Membranes

To optimize the separation characteristics of a given material, its thickness should be as thin as possible because mass flux is inversely proportional to its thickness according to Fick's first law (Equation (7.2)). However, the decrease in the membrane thickness is limited because of the mechanical strength necessary for practically high-pressure operation, up to 30 atm, and also for handling endurance during module fabrication. Providing the high separation performance as well as mechanical strength requires a composite structure as schematically illustrated in Figure 7.2, configuring a top coating layer and an asymmetrically microporous support. Since the thickness of the top coating layer should be thin to improve the performance, the composite membrane is also frequently called a thin film composite membrane; however, hereafter, it will be called simply composite membrane just for convenience.

In composite membranes, the asymmetrically microporous support consists of a dense skin layer on the top of an asymmetrically microporous sublayer (Figure 7.2). Electron microscopic images of a hollow fiber support fabricated via phase inversion process are given in Figure 7.3 to show the skin and asymmetrically microporous sublayer structures with diameters of ~300 μm.

Asymmetrically microporous supports made by phase inversion process can exhibit three different types of skin layers: (a) defect-free, (b) slightly microporous or defective, and (c) highly porous, as shown in Figure 7.4. The defect-free membrane may give high gas separation performance because the skin layer is defect-free and extremely thin. However, achieving defect-free skin is extremely challenging. In the case of microporous skins (Figure 7.4b and c), post-plugging of the pores with other polymer is necessary for gas or liquid separation by forming composite membranes as illustrated in Figure 7.5a and b. For instance, while polymers can make a skin layer with low porosity (below 0.01%), they are still too porous for effective gas separation. To mitigate this, pores are sealed with highly permeable polymers like poly(dimethyl siloxane) (PDMS), forming a top coating layer for composite membrane as depicted in Figure 7.5a and ensuring that such small pores hardly affect

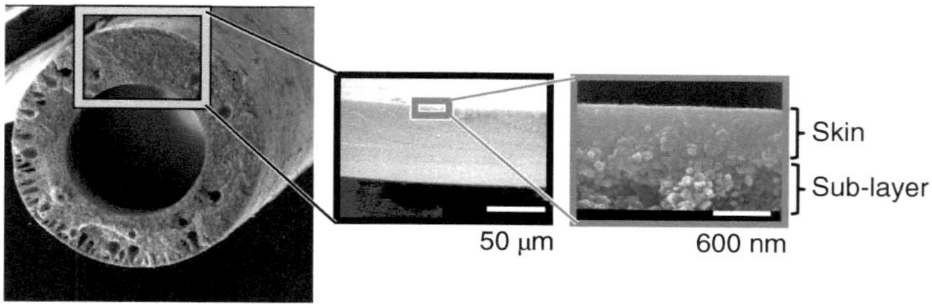

Figure 7.3 Hollow fiber support with asymmetric micropores prepared by phase inversion process. *Source*: Y.S. Kang.

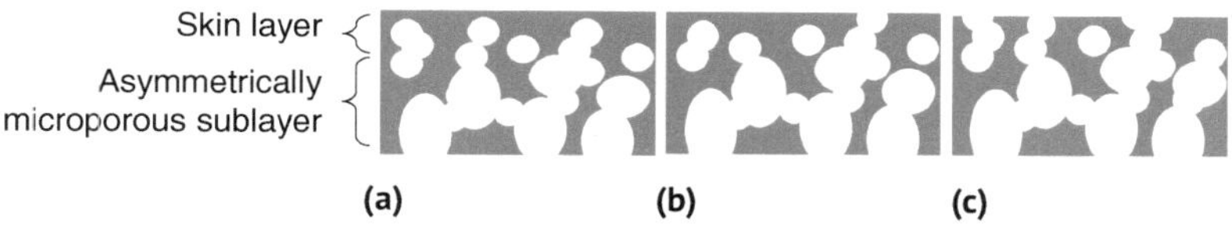

Figure 7.4 Asymmetrically microporous support with (a) defect-free skin; (b) minor microporous skin; and (c) highly porous skin. *Source*: Y.S. Kang.

Figure 7.5 Structures of composite membranes configuring a top coating layer on asymmetrically microporous support with (a) minor microporous skin; and (b) highly porous skin. *Source*: Y.S. Kang.

the separation performance of membranes. Its permeance and selectivity depend heavily on the properties and configuration of the top and support layers, covering aspects like porosity, pore size, pore connectivity, and thickness.

The selectivity $\alpha_{\text{composite}}$ of a composite membrane in Figure 7.5a is estimated to lie within the range $\alpha_{\text{coating}} \leq \alpha_{\text{composite}} \leq \alpha_{\text{sublayer}}$, where α_{coating} and α_{sublayer} are the selectivities of the coating and sublayer materials, respectively. However, predicting permeance and selectivity are particularly complex and will be discussed in detail using a mathematical model in Section 7.2.

Another composite structure frequently used is schematically illustrated in Figure 7.5b, which also consists of a coating layer and a sublayer with highly porous connected skin. In this case, the skin layer is required to be permeable to have connected pores and therefore the selectivity of the composite membrane $\alpha_{\text{composite}}$ is mostly determined by the selectivity of the coating layer α_{coating}: $\alpha_{\text{composite}} \sim \alpha_{\text{coating}}$. Therefore, the coating layer is frequently called a selective layer. In this case, the separation performance is readily estimated from the material and thickness of the selective layer. Of course, the pore size and porosity of the sublayer also significantly influence the permeance of the composite membrane. Such composite membranes are mostly prepared for hydrogen separation and reverse osmosis (RO) applications. For RO membranes, the coating layer, prepared frequently by the interfacial polymerization, will serve as a selective layer while the microporous support only provides mechanical strength rather than selective transport.

The main difference between two composite structures in Figure 7.5a and b is the structure of the skin layer of the microporous sublayer. The skin layer porosity is required to be as small as possible for composite (a) whereas it should be as high as possible with small micropores and their connectivity for composite (b).

The effective thickness L of both composites (a) and (b) would not be readily estimated, resulting in difficulty in analyzing the separation performance in terms of the permeability P_i with Fick's law:

$$J_i = P_i \frac{\Delta p_i}{L} = Q_i \Delta p_i \tag{7.3}$$

Therefore, in the case of composite membranes, a new parameter, permeance, Q_i, is introduced, representing the pressure-normalized flux $(Q_i = J_i/\Delta p_i)$ or the thickness-normalized permeability $(Q_i = P_i/L)$. Therefore, the flux can be readily obtained by the simple product of the permeance Q_i and the pressure difference Δp_i across the membrane by following Equation (7.3). The unit of the permeance is (cm^3 (STP)/cm^2 sec cmHg) or gas permeation unit (GPU, 1 GPU $= 1 \times 10^{-6}$ cm^3 (STP)/(cm^2 sec cmHg).

7.2 Mathematical Model for Composite Membranes: Resistance Model

Henis and Tripodi have developed a mathematical model to explain the separation performance through composite membranes, which consist of a coating layer on the top of an asymmetrically microporous sublayer (the composite structures in Figure 7.5) [2]. An analogy between gas permeation and electric current is made for mathematical formulation in terms of the resistance to gas permeation. It is shown that skin layer porosity can significantly alter the separating properties of such a composite, and that sublayer and coating layer properties can be matched to optimize the permeability and selectivity. In this section, the original paper by Henis and Tripodi is solely introduced, and the results are summarized [2].

Ohm's law describes electron flow through a resistor:

$$I = \frac{E}{R} \tag{7.4}$$

where I, E, and R are electric current, potential difference, and resistance, respectively. Equation (7.4) is converted to:

$$j = \sigma\left(\frac{E}{L}\right) \quad \text{where} \quad \sigma = \frac{L}{AR} \tag{7.5}$$

Here, j is the electric current density with $j = I/A$. From conceptual equivalency between mass transport (Equation (7.2) for Fick's law) and electron transport (Equation (7.5) for Ohm's law), the mass permeability P_i can be equated with the electron conductivity σ. Therefore, the mass transport resistance, R_i, of component i is related with P_i as [2]:

$$R_i = \frac{L}{P_i A} \tag{7.6}$$

so that mass flux J_i can be expressed in terms of resistance R_i rather than P_i by combining Equations (7.2) and (7.6) as [2]:

$$J_i = \frac{\Delta p_i}{A R_i} \tag{7.7}$$

Therefore, the mass transport through the composite membrane can be described using the concept of resistance and is referred to as the resistance model [2].

We define four resistances in this composite membrane as shown in Figure 7.6: R_1 for the coating layer with surface area A_1 and thickness L_1; R_2 for "dense" portions of the skin layer with A_2 and L_2; R_3 for the pores in the skin layer with A_3 and $L_3(= L_2)$; and R_4 for the resistance of the porous matrix with $A_4(= A_1)$ and L_4, where the total surface area $A = A_1 = A_2 + A_3 = A_4$. Thus, A_3/A_2 is defined as a surface porosity of the skin layer [2].

The total resistance R_i of component i to permeate through the composite membrane is expressed as [2]:

$$R_i = R_{1,i} + \frac{R_{2,i} R_{3,i}}{R_{2,i} + R_{3,i}} + R_{4,i} \tag{7.8}$$

where $R_{1,i} = L_1/P_{1,i}A_1$, $R_{2,i} = L_2/P_{2,i}A_2$, $R_{3,i} = L_2/P_{3,i}A_3$, and $R_{4,i} = L_4/P_{4,i}A_4$.

From the standpoint of high flux and good separating properties, desirable sublayer structure will be very open and interconnectedly porous to have very small $R_{4,i}$, relative to $R_{1,i}$, $R_{2,i}$, and $R_{3,i}$. In such cases, Equation (7.8) simplifies to:

$$R_i = R_{1,i} + \frac{R_{2,i} R_{3,i}}{R_{2,i} + R_{3,i}} \tag{7.9}$$

The important aspect of such a composite membrane is that the resistance, R_3, of the pores to the passage of gases has been substantially increased by the coating material because the permeability $P_{1,i}$ of the coating material, which fills the pores, is so much lower than the permeability of an open channel. Even the highest permeability polymer is 10^5–10^6 times more resistive to gas flow than an open pore of equivalent area and depth [2].

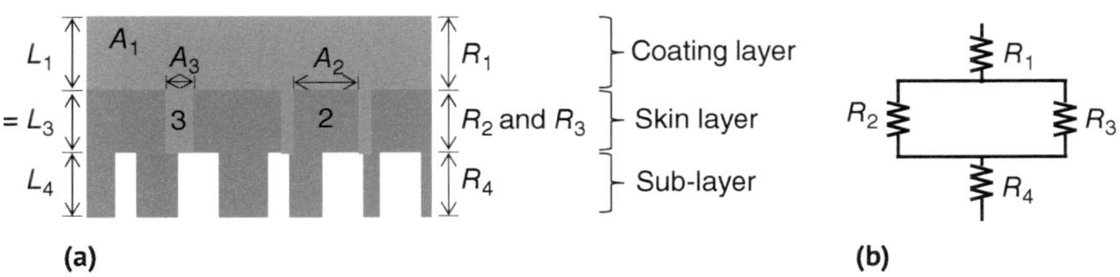

Figure 7.6 (a) Simplified schematic structure of composite membrane; and (b) corresponding equivalent circuit. *Source*: Adapted from [2].

If we make the assumption that the coating material has penetrated and filled the pores to a depth equal to the thickness of the skin layer, then $P_{3,i} = P_{1,i}$ and $L_3 = L_2$ and the expression for the resistance of the pores becomes:

$$R_{3,i} = \frac{L_2}{P_{1,i}A_3} \tag{7.10}$$

The total production rate N_i $(= AJ_i = A((\Delta p_i)/(AR_i)) = ((\Delta p_i)/R_i))$ for the composite membrane can be obtained from Equations (7.9) and (7.10) as [2]:

$$N_i = \Delta p_i \left(\frac{L_1}{P_{1,i}A_1} + \frac{L_2}{P_{2,i}A_2 + P_{3,i}A_3} \right)^{-1} \tag{7.11}$$

Practical membranes mostly have the surface porosity of the skin layer, defined as A_3/A_2, very small, typical value around 10^{-4} [2]. In this case, we may presumably assume $A_1 = A_2$. Equation (7.11) can then be rearranged to obtain the permeance, $Q_i = (J_i/\Delta p_i = P_i/L)$ [2]:

$$\frac{N_i}{A_1 \Delta p_i} = Q_i = \left(\frac{L_1}{P_{1,i}} + \frac{L_2}{P_{2,i} + P_{3,i}(A_3/A_2)} \right)^{-1} \tag{7.12}$$

The selectivity $\alpha_{i/j}$ through the composite membrane is just expressed as a permeance ratio of components i and j:

$$\alpha_{i/j} = \frac{Q_i}{Q_j} \tag{7.13}$$

Figure 7.7a shows how surface porosity (A_3/A_2) affects the H_2 fluxes from a H_2/CO gas mixture through a composite membrane made of microporous polysulfone hollow fiber support coated with silicone rubber [2]. It clearly indicates that changes in surface porosity have a much greater impact on the permeation properties of

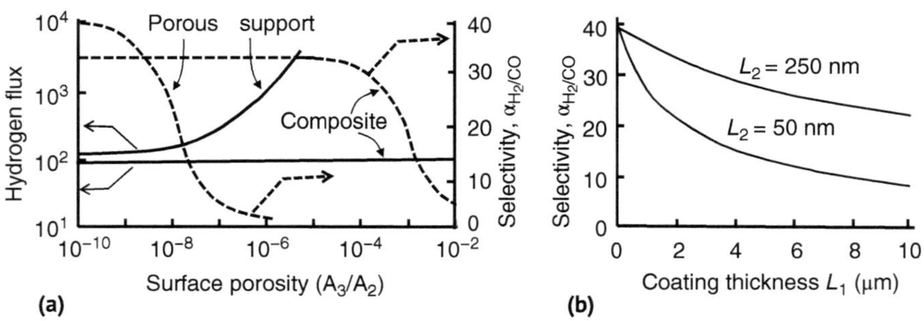

Figure 7.7 (a) The predicted effects of surface porosity on hydrogen flux and selectivity $\alpha_{H_2/CO}$ for a porous support membrane and a composite membrane. *Source*: Adapted from [2]; and (b) the predicted effects of coating thickness on selectivity $\alpha_{H_2/CO}$ for composite membranes at two effective skin layer thicknesses, $L_2 = 50$ and $L_2 = 250$ nm, with a fixed value of $A_3/A_2 = 1.0 \times 10^{-5}$ [2]. The information data used are summarized as: $P_1(H_2) = 520$ Barrer, $P_1(CO) = 250$ Barrer, $L_1 = 1 \times 10^{-4}$ cm, $P_2(H_2) = 12$ Barrer, $P_2(CO) = 0.3$ Barrer, and $L_2 = 1 \times 10^{-5}$ cm. Please refer to the original paper for the detailed values of the graphs. *Source*: Adapted from [2].

an uncoated microporous support compared to a coated composite membrane. Practically, this means that the surface porosity of the fiber in the composite membrane can vary more without significantly affecting the flux and separation factor, unlike in an uncoated microporous support membrane [2].

Figure 7.7b shows how changing the coating thickness affects the selectivity for an H_2/CO mixture in two composite fiber membranes with the same composition but different skin layer thicknesses L_2. The resistance model predicts that increasing the coating thickness L_1 above 1 mm has a much greater impact on hydrogen selectivity over CO, $a_{H_2/CO}$, in a fiber with a thin skin layer ($L_2 = 50$ nm), compared to one with a thicker skin layer ($L_2 = 250$ nm), where the skin layer mainly determines selectivity. Both fibers are silicone-coated polysulfone, but while the thinner-skinned fiber offers lower $a_{H_2/CO}$ values, it requires a carefully controlled, thinner coating to maintain these properties. Conversely, the thicker-skinned fiber can tolerate a thicker silicone coating without significantly reducing $a_{H_2/CO}$, but this comes at the cost of lower permeance (Q) values. This is because as L_2 decreases, the resistance of the skin layer to gas flow decreases. To ensure the coating does not significantly affect the membrane's separation properties, the coating's resistance (R_1) must also decrease, which is achieved by reducing the coating thickness (L_1). When L_1 is less than 10 times L_2, the silicone-coated polysulfone fibers will exhibit separation properties close to the intrinsic properties of polysulfone. This behavior applies to any coating/substrate combination with equivalent surface porosity [2].

In summary, the mathematical model predicts the separation performance of composite membranes. For these membranes, the combined resistance of the sublayer (R_2 and R_3) accounts for 90% or more of the total flow resistance for a specific gas at a given $A_3/A_2 = 1.0 \times 10^{-4}$ (neglecting R_4), even if the membranes cannot yet be fully characterized in terms of structure, porosity, and similar parameters. The skin layer's porosity and the nature of the sublayer matrix play critical roles in determining separation performance. The effects of surface pore size and the coating material's wetting properties also significantly impact separation performance and are areas for future research [2].

7.3 Fabrication of Composite Membranes

Membranes used in practical applications must combine excellent separation performance with the mechanical strength to withstand high-pressure or vacuum operations. This can be achieved by fabricating a composite structure, where an asymmetrically microporous support is coated with another polymer, as illustrated in Figure 7.5. The coating layer is essential because formation of micropores or defects (with porosity as low as 10^{-6}) on the skin layer of the asymmetrically microporous support is unavoidable, allowing gas to pass through and compromising separation. Thus, the structure of the asymmetrically microporous support is crucial for both separation efficiency and mechanical strength. A cross-section of a typical asymmetrically microporous support made from polysulfone is shown in Figure 7.3.

Controlling the structure of the asymmetrically microporous support and the subsequent polymer coating process is vital in fabricating composite membranes. The effects of these factors on separation performance were already discussed with the resistance model, in more detail, in Section 7.2. These supports are typically produced through phase separation processes, including thermally induced and nonsolvent-induced phase separation, which will be briefly introduced hereafter.

7.3.1 Thermally Induced Phase Separation

A homogeneous phase of polymer solution containing polymer 2 and solvent 1 at a temperature T_0 separates into two phases of a polymer-lean phase and a polymer-rich phase by decreasing the solution temperature to T_1 as shown in Figure 7.8. After removing the solvent 1, the polymer-lean phase forms pores, and the polymer-rich phase becomes a solid matrix. This phase separation occurs through two different mechanisms: the nucleation and growth (NG) mechanism and the spinodal decomposition (SD) mechanism.

When the temperature of a polymer solution at a given composition is reduced from T_0 to T_1, the solution first encounters the binodal curve, initiating phase separation via the NG mechanism until it reaches the spinodal curve. Below the spinodal temperature, the SD mechanism takes over. Therefore, during thermally induced phase separation (TIPS), phase separation typically occurs through NG first, followed by SD, except at the critical point. However, if T_1 is above the spinodal temperature, only the NG mechanism will operate, not SD. Therefore, the cooling rate or temperature profile significantly influences the final morphology of the asymmetrically microporous support and, consequently, its separation performance.

The phase separation by the NG mechanism prevails in the two metastable regions between $\phi^L_{2,B} < \phi_2 < \phi^L_{2,S}$ and $\phi^R_{2,S} < \phi_2 < \phi^R_{2,B}$ only when stable nuclei are formed. These nuclei grow over time leading to a phase-separated structure, as shown in Figure 7.9a. In contrast, SD occurs spontaneously within the concentration range $\phi^L_{2,S} < \phi_2 < \phi^R_{2,S}$, where the polymer solution becomes thermodynamically unstable at temperature T_1 (Figure 7.9b).

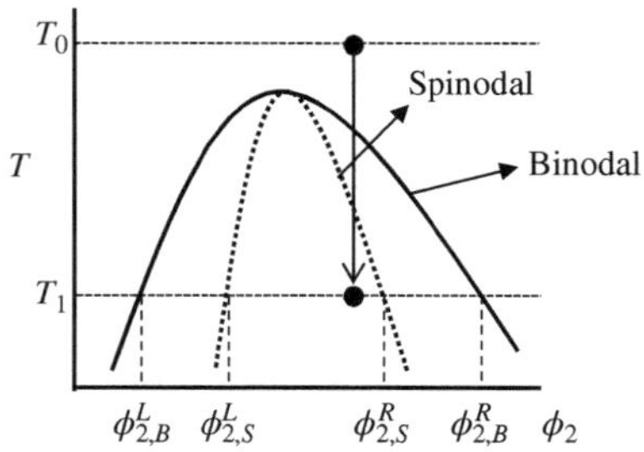

Figure 7.8 A phase diagram (T vs ϕ_2, the volume fraction of polymer 2) for two-component system of solvent 1 and polymer 2, and binodal and spinodal curves. *Source*: Y.S. Kang.

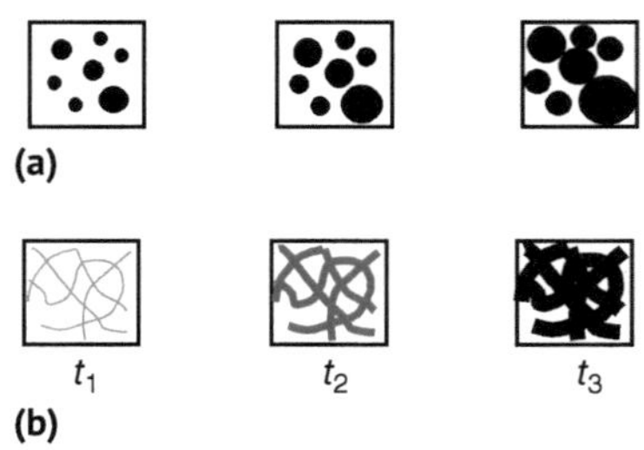

Figure 7.9 Phase-separated structure evolution with time ($t_1 < t_2 < t_3$) by (a) the nucleation and growth mechanism; and (b) spinodal decomposition mechanism. *Source*: Y.S. Kang.

The final structure resulting from phase-separated processes is determined mostly by the separation mechanism. The NG mechanism produces spherical particles dispersed within a continuous phase, as illustrated in Figure 7.9a. In contrast, the SD mechanism generates a lacy structure, as shown in Figure 7.9b. Therefore, controlling the phase separation mechanism is crucial for determining the final structure.

7.3.2 Nonsolvent-induced Phase Separation

Phase separation of polymer solutions consisting of polymers (P) dissolved in a solvent (S) can also be induced by incorporating a nonsolvent (NS) that is miscible with S but precipitates polymer chains at a given temperature. Water is the mostly used NS for preparation of asymmetrically microporous support. When a polymer solution (P + S) comes into contact with NS , NS diffuses into the polymer solution with a flux of J_{NS} whereas S diffuses out with a flux of J_S from the polymer solution as schematically drawn in Figure 7.10a. These fluxes, J_{NS} and J_S, mostly depend on factors such as viscosity of polymer solution, the affinity between S and NS, and the size of the diffusant at a given temperature.

Because of the exchange of S with NS at the interface between the polymer solution and the NS phase, the NS concentration in the polymer solution will be increased, leading to phase separation into polymer-lean and polymer-rich phases. The exchange ratio R, defined as $R = J_{NS}/J_S$, is a critical parameter in determining the resulting phase-separated structure. A homogeneous dense film is obtained with $R \ll 1$, following line 1 in Figure 7.10b, while a phase separation with large R values results in a microporous structure by following line 2 [3].

When the composition follows line 2, it first reaches the binodal line, initiating phase separation through the NG process and then continues to the spinodal curve. Afterward, the SD process takes place to finalize the support structure. In other words, the morphology of the microporous support is primarily determined by the R value. One simple way to decrease the R value is by increasing the NS size, for

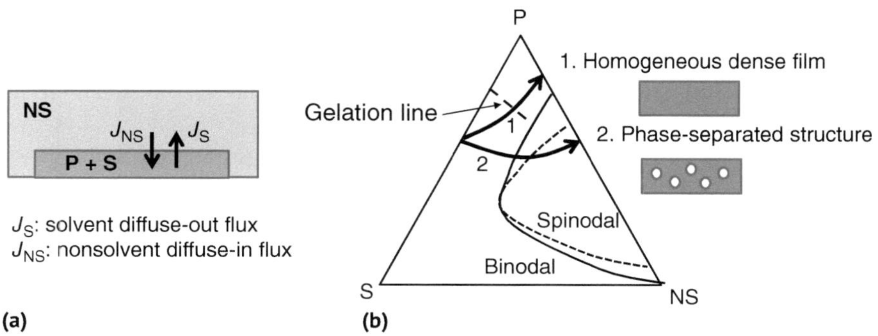

Figure 7.10 (a) Solvent (S) and nonsolvent (NS) exchange with NS flux-in J_{NS} and S flux-out J_S through the interface between polymer solution (P + S) and NS phases; and (b) ternary phase diagram of P/S/NS system at a given temperature T_1, where solid and dotted lines are binodal and spinodal curves, respectively. Arrows represent the composition changes by replacing S with NS during phase separation process with time at a given temperature. *Source*: Y.S. Kang.

example, by using larger molecules like isopropyl alcohol or poly(ethylene glycol). Therefore, understanding phase behavior, including phase separation kinetics and mechanisms, is key to controlling the final structure of asymmetrically microporous support [4].

7.4 Structures and Mass Transport in Modules

In separation applications, membranes are typically assembled into modules, which protect the delicate membranes during installation and maintenance, reducing the risk of damage. Modules also enhance scalability and allow for various configurations to maximize the effective membrane area per unit volume. The design of a module depends on the membrane geometry, whether flat sheet or hollow fiber. The characteristics and structures of membrane modules are discussed in detail in the book of *Membrane Technology and Applications* [1], and are briefly introduced here as follows.

7.4.1 Structures of Representative Membrane Modules

Membranes are generally produced in two forms: flat sheets or hollow fibers (including capillaries). Flat sheet membranes are commonly configured into plate-and-frame or spiral-wound modules, while hollow fiber membranes are assembled into hollow fiber or capillary modules. Different module designs and materials can be optimized to suit specific applications and operating environments.

7.4.1.1 Plate-and-frame Modules

Plate-and-frame modules are an early form of membrane assembly that sandwiches a single flat membrane sheet in a filter press. Still used for some small-scale or specialized applications in membrane bioreactors, electrodialysis, and pervaporation, but the configuration has been largely displaced by other membrane designs.

7.4.1.2 Spiral-wound Modules

Spiral-wound membranes consist of consecutive layers of membrane sheet and feed spacer wound around a perforated permeate collection tube at center. Spiral membranes offer lower energy costs due to reduced pumping requirements and higher packing density. Spiral construction also allows the module to be operated at elevated pressure and temperature. Spiral-wound modules range from laboratory-scale membranes to 12-inch diameter industrial modules up to 5 feet in length.

7.4.1.3 Hollow Fiber Modules

Hollow fiber membrane modules are made as small tubes, typically less than 1 mm in diameter. A single hollow fiber module contains tens of thousands of these fibers, a design that maximizes the membrane surface area. These modules are mainly used in crossflow applications where high permeate or recovery rates are needed, and membrane fouling is not a concern. While hollow fiber membranes are not suitable for wastewater treatment, where clogging can occur, they are ideal for high-pressure gas separation and RO polishing of drinking and ultrapure water. Although hollow fiber modules offer a large membrane surface area, they are costly to produce – often twice as expensive or more than spiral-wound modules – and not all membrane materials can be formed into fibers.

7.4.1.4 Capillary Modules

Capillary membrane modules are also manufactured as small tubes, though they are larger than hollow fiber membranes. The average diameter of a capillary fiber is 1–2 mm: hundreds to thousands of capillaries are assembled into housings to form membrane modules. Less prone to fouling, capillary membranes are used in crossflow methods for wastewater treatment, but are often being displaced by spiral-wound elements. The membranes are, however, finding a new niche in pervaporation processes.

7.4.2 Membrane Modules for Gas Separation

Gas separation membranes are typically configured as spiral-wound or hollow fiber modules. The choice of module design depends primarily on cost and membrane flux. High-pressure gas separation applications require hollow fiber membranes with very fine lumen diameters, typically between 50 and 200 μm. However, the small diameter of these fibers can cause significant pressure drops on the lumen side, which may negatively affect membrane performance.

For applications like nitrogen production from air, where membrane permeance is relatively low (1–2 GPU), pressure drops are generally not an issue. In contrast, for separating hydrogen from nitrogen or methane, and carbon dioxide from natural gas, membrane permeance is higher, and hollow fiber modules may experience excessive pressure drops. To address this, capillary fiber or spiral-wound modules are often preferred for these applications. Despite these limitations, hollow fiber membranes offer a cost advantage due to their lower production cost per square meter of membrane [1].

7.5 Applications of Gas Transport Membranes

Membrane gas separation systems offer numerous advantages, including compact size, high efficiency, low maintenance, low energy consumption, low capital investment, ease of operation, and exceptional flexibility. These systems are energy-efficient and environmentally friendly, making them well-suited for separating a wide variety of gas pairs. As a result, membrane gas separation is often favored over other processes such as pressure swing adsorption, absorption, and cryogenic distillation. Table 7.1 provides an overview of gas pairs where membrane separation has proven advantageous in both current and future applications. This section extensively cites two key references [1,5].

Two of the largest industrial gas companies now have membrane affiliates: Air Products (Permea) and Air Liquide (Medal). The affiliates focus mainly on producing membrane systems not only to separate nitrogen from air but also to produce

Table 7.1 Primary current and future applications of polymer membranes for industrial gas separations.

Current applications	Future applications
H_2/N_2	CO_2/N_2
H_2/CH_4	O_2/N_2
N_2/O_2	Ethylene/ethane
CO_2/CH_4	Propylene/propane
H_2O/Air	Ethanol/water
Olefin/N_2	CO_2/H_2

some hydrogen separation systems. Ube (Japan) and Aquillo (The Netherlands) are also active in these markets. Another group of companies – UOP (GMS/Separex), Cameron (Cynara), and MTR – produce membrane systems for natural gas separations. A third group of smaller independents is focusing on new applications, including vapor separation, air dehydration, and oxygen enrichment. Vapor/gas separation, such as propylene or ethylene/nitrogen, and dehydration of air or other gases also become important in near future but not covered here [1].

This section covers the major current applications, including hydrogen separation, oxygen/nitrogen separation, carbon dioxide capture, and olefin/paraffin separation.

7.5.1 Hydrogen Separation

Hydrogen recovery from nitrogen, methane, and argon in ammonia purge gas streams was the earliest industrial-scale application of gas separation membranes. Typical composition of the ammonia purge gas is 77% H_2, 17.5% N_2, and 5.5% CH_4. Membranes have been successful in this application since the gas is clean and free of condensable hydrocarbon vapors that might plasticize or foul the membrane. Another successful existing application is adjustment of the H_2/CO ratio in syngas plants.

A significant unmet application for hydrogen-permeable membranes lies in the recovery of hydrogen in refineries. The growing demand for hydrogen in refineries, driven by stricter environmental regulations and the processing of heavier crude oils, can be met cost-effectively by utilizing sources, such as refinery fuel gas streams, pressure swing adsorption tail gas, fluid catalytic cracking unit gas, and off-gases from hydrocrackers and hydrotreaters. These streams typically contain 30–80% hydrogen mixed with light hydrocarbons, making them ideal candidates for hydrogen recovery using membrane technology [5].

Membrane use on hydrogen-containing refinery streams has been limited by problems related to fouling, plasticization, and condensation of hydrocarbon vapors on the membrane surface. With the development of membrane materials that withstand the high hydrocarbon partial pressures and high temperatures of these streams, refinery applications are expected to grow.

7.5.2 Oxygen/Nitrogen Separation

Air can be separated by a membrane process into N_2-enriched air and O_2-enriched air. Currently, generating nitrogen from air is the largest volume end use of gas separation membranes. Natural gas treatment and hydrogen separation are the second and third largest markets, respectively [5].

7.5.2.1 Nitrogen-enriched Air

Producing enriched or pure nitrogen from air using membranes is relatively simple, since air contains ~80% nitrogen. Commercially available since the early 1980s, membrane technology has been widely applied in several industry sectors for generating moderate- to high-purity N_2 from air. Because air is ~80% nitrogen, the majority of air separation membranes operate to recover nitrogen. Combined applications in food processing, metal manufacturing and fabrication, and the electronics industry, account for an estimated 35–40% of global nitrogen use. In particular, membrane-produced N_2 is ideal for food industry, since it can be operated to deliver from 90 to 99 + % N_2 for different foods simply and demonstrates good economics for the 97–98% purities often desired.

In addition, "membrane-based inerting systems" utilizing polymer hollow fiber membranes for reducing the potentially hazardous flammable vapors in jet aircraft fuel tanks have been acquired by the Federal Aviation Administration (FAA) of USA. High-purity N_2 (O_2 content <0.1 %) also may be used to inert laser cutting of metals.

The first nitrogen production systems used membranes made from poly(methyl pentene) (TPX) with a selectivity of O_2/N_2 about 4. These membranes were incorporated into one-step designs to produce 95% nitrogen used to render flammable-liquid storage tanks inert. When we introduce the two- and three-step processes as shown in Figure 7.11, the reductions in both membrane area and pumping cost can be

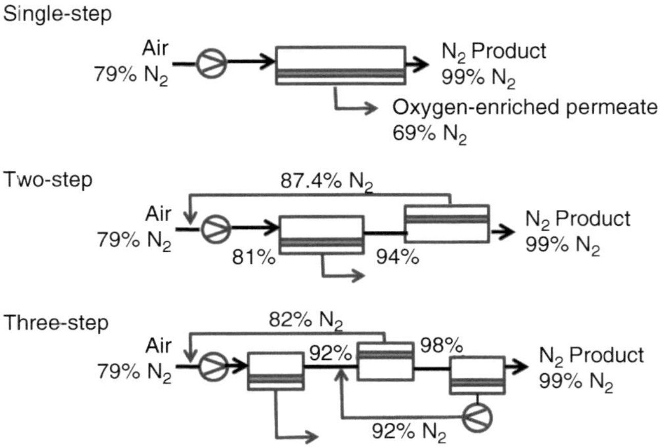

Figure 7.11 Single-, two-, and three-step designs for nitrogen production from air. *Source*: Adapted from [6].

achieved by 6 and 8%, respectively, demonstrating the importance of process design. However, the three-step process needs two compressors at the expense of 2% cost down and therefore the two-step process is generally adopted [6].

7.5.2.2 Oxygen-enriched Air

Oxygen is the third most widely used chemical globally, with an annual market exceeding $9 billion. Its applications range from medical breathing equipment to large-scale industrial processes, including chemical production, syngas generation, and coal or coke gasification [5].

For high-temperature furnaces, oxygen-enriched air (typically 25–35% oxygen) improves combustion efficiency and reduces fuel consumption. It is also used in cement kilns and refinery operations.

However, most industrial processes require pure oxygen (≥99%). Currently, pure oxygen is mainly produced by cryogenic distillation, a highly capital-, and energy-intensive process due to the close-boiling points of oxygen and nitrogen. This limits the broader use of oxygen-enriched combustion and oxygen-blown gasification.

Membrane-based separation offers a promising alternative. Advantages include lower energy consumption, modular scalability, and continuous operation. However, challenges remain, such as achieving high purity, optimizing permeability-selectivity trade-offs, and developing cost-effective membrane materials for large-scale applications [6].

In this regard, for years, researchers have actively investigated membranes to separate oxygen from the air, by selectively permeating oxygen and rejecting nitrogen. Producing pure oxygen from air is more difficult than producing nitrogen, since some nitrogen always permeates with oxygen. The separation typically results in oxygen-enriched air rather than in pure oxygen. Facilitated transport membranes also demonstrate excellent permeability and selectivity in the production of pure oxygen. If stability and other problems can be overcome, they hold great commercial promise in the application.

7.5.3 Carbon Dioxide Separation

Gas separation using membranes is emerging as an important tool for capturing CO_2 from gas streams since the approach offers significant advantages over conventional gas separation processes. Currently, there are no commercial operations separating CO_2 from flue gases, but membranes are employed in enhanced oil recovery operations to recycle CO_2 from associated gases and to remove CO_2 from natural gases and biogases or landfill gases [5].

Membranes with sufficient selectivities and fluxes already exist for this application. The drawback to commercial implementation is the amount of energy required to power the pumps or compressors for the separation. Additionally, membrane systems have not been tested on the scale needed to capture from utility power plants.

It has been identified membrane-based gas treatment as a leading candidate for pre-combustion and de-carbonization CO_2 capture, in which H_2 and CO_2 are separated from hydrocarbons. H_2 is used as fuel; the CO_2 is sequestered. Currently, most commercially viable membranes used for CO_2 removal are polymer-based products.

7.5.3.1 Natural Gas Purification or Sweetening

Purification of natural gas before it is sent through the pipeline is the second largest application of gas separation membranes. Contaminant concentrations in natural gas streams are controlled by strict regulations.

Amine absorption is the most widely used method for removing CO_2 from raw natural gas, although the technology is costly, complex, and labor-intensive. The process' drawbacks are especially acute for reserves with lower quality gas, smaller reserves, and remote locations where amine absorption may be too expensive or impractical [5].

Membrane systems have major advantages over more-traditional methods of CO_2 removal systems, including lower capital cost and lower operating costs. Membrane systems are skid-mounted, except for the larger pretreatment vessels, and so the scope, cost, and time taken for site preparation are minimal. Installation costs are significantly lower than alternative technologies, especially for remote areas.

Furthermore, membrane units do not require additional facilities, such as solvent storage and water treatment, needed by other processes. The only major operating cost for single-stage membrane systems is membrane replacement. This cost is significantly lower than the solvent replacement and energy costs associated with traditional technologies [6].

The designs of two typical carbon dioxide removal plants are illustrated in Figure 7.12. One-stage plants, which are simple, contain no rotating equipment, and require minimal maintenance, are preferred for small gas flows. In such plants, methane loss to the permeate is often 10–15%. The methane loss can be reduced to 1.9% by a two-stage process with one extra compressor. Two-stage processes are more expensive because a large compressor is required to compress the permeate gas.

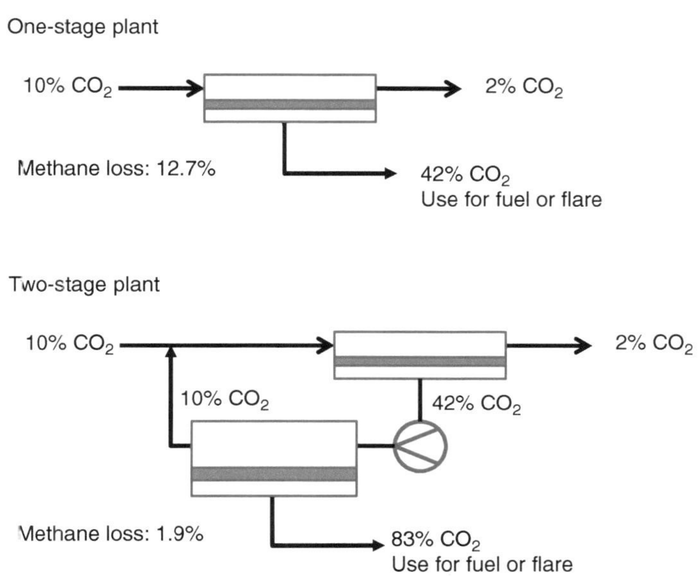

Figure 7.12 Flow scheme of one-stage and two-stage membrane separation plants to remove carbon dioxide from natural gas. *Source*: Adapted from [6].

7.5.3.2 CO_2 Separation from Flue Gases

The emergence of global warming as a significant environmental problem is likely to change the way the world produces and uses energy in the next decades. The costs involved in solving the problem are huge. One direct way of addressing the problem is to separate and sequester CO_2 generated in the production of electricity. Power production at electrical power plants is responsible for 35–40% of global CO_2 emissions. In addition, power stations are an attractive target for carbon capture, because they are large sources of CO_2 emissions; there are about 5,000 large power plants around the world. The amount of CO_2 emitted depends on the plant size and fuel feed, but an average (500 MWe) coal-fired power plant will emit about 10,000 tons of CO_2 per day. Separating the CO_2 from these emissions and then compressing the gas to pressure (80–100 bar) and injecting it deep underground would go a long way to mitigating the global warming problem.

Typical flue gases from natural gas-fired power plants may contain 8–10% CO_2, 18–20% H_2O, 2–3% O_2, and 67–72% N_2; flue gases from coal-fired boilers may contain 12–14% CO_2, 8–10% H_2O, 3–5% O_2 and 72–77% N_2. The furnace outlet temperature of flue gases is usually around 1,200 °C, which will decrease gradually along the pathway of heat transfer, while the temperature of the flue gases going to stack is around 150 °C. Pollution control technologies can remove the SO_x, NO_x, and particulate matter effectively, but CO_2 and H_2O as well as O_2 remain largely unchanged [7].

Amine absorption is again the most commonly considered technology to separate the CO_2 but would double the cost of the electricity produced. Membranes could be used instead, since several materials have the required CO_2/N_2 selectivities. However, for this application, extremely high permeance will be needed. With the currently highest permeance membrane, plant membrane areas will be in the 1–2 million m^2 range per power plant – as large as the largest RO plants.

If we develop membranes with CO_2 permeances of greater than 1,000 GPU and a CO_2/N_2 selectivity of 50, these membranes, in combination with a novel process design as shown in Figure 7.13 that uses incoming combustion air as a sweep gas

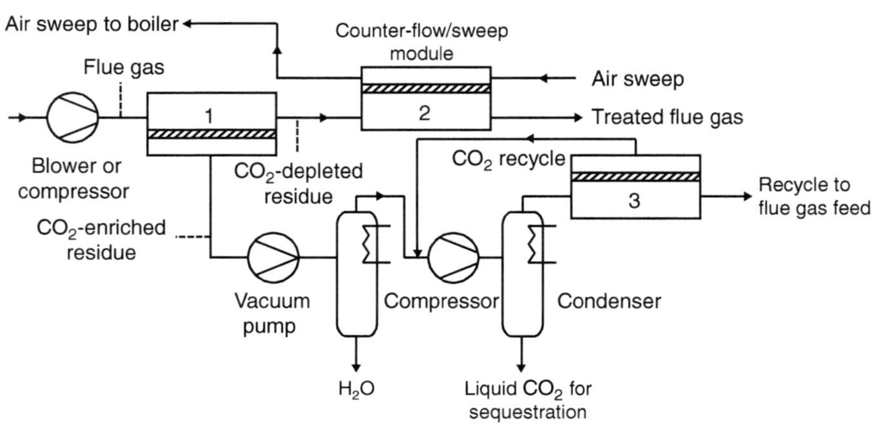

Figure 7.13 Simplified flow diagram of a two-step counter-flow/sweep membrane process to capture and sequester CO_2 in flue gas from a coal-fired power plant. The base-case membrane with a CO_2 permeance of 1,000 GPU and a CO_2/N_2 selectivity of 50 was used in the calculations. *Source*: Adapted from [8].

to generate driving force, show promise for meeting the US Department of Energy (DOE), CO_2 capture cost targets [8].

7.5.3.3 H_2/CO_2 Separation

Membrane separation of H_2/CO_2 is a promising technology for hydrogen purification and carbon capture, particularly in industrial processes like syngas production, natural gas reforming, and coal gasification. While CO_2-selective membranes often operate at temperatures below 40 °C, H_2-selective membranes operating at the syngas temperatures (100–300 °C) are desirable because they avoid the cooling of the syngas. In addition, increasing the temperature increases both H_2 permeability and H_2/CO_2 selectivity. Therefore, increasing the temperature moves up the permeability/selectivity upper bound. One effective design strategy to increase the H_2/CO_2 selectivity in polymers is to decrease free volume and increase size-sieving ability. One of the membrane process designs with N_2 sweep on the permeate suggests that membranes with H_2 permeance of 300 GPU or above and H_2/CO_2 selectivity of 30 or above provide high efficiency for membrane systems [9].

Polymeric membranes, typically made of materials like polybenzimidazoles, polyimides, polyamides, or polysulfones, are widely used due to their ease of fabrication and cost-effectiveness. However, they often suffer from trade-offs between permeability and selectivity, particularly under harsh conditions like high temperatures. Achieving both high permeability and high selectivity remains a major challenge. For instance, such polymers are frequently modified to improve the separation performance by cross-linking [10], blending [11], and rearranging or carbonizing by thermal treatment [12, 13].

7.5.4 Olefin/Paraffin Separation

A final group of separations likely to develop into a major application area for membranes is olefin/paraffin separations, such as ethylene (bp −93.9 °C) from ethane (bp −88.9 °C) and propylene (bp −47.2 °C) from propane (bp −42.8 °C). These close-boiling mixtures are separated on a very large scale in the production of ethylene and propylene, the two largest-volume organic chemical feedstocks. Because the mixtures are close-boiling, large towers and high reflux ratios are required to achieve good separations [6].

If membranes are to be used for these separations, highly selective materials must be developed. Simple membranes consisting of polymers only may not satisfy the separation performance for practical applications. Instead, composite membranes, including facilitated transport membranes, mixed matrix membranes, carbon molecular sieve membranes, and inorganic membranes, such as organosilica and ZIF-8 membranes, become a major trend to develop novel membranes having high separation performance [14].

Under common operating conditions of high pressure (8–10 bar) and temperatures sufficient to maintain the gas in the vapor phase, plasticization and loss of selectivity frequently occur for polymeric membranes. Reportedly, solid polymer electrolyte membranes also have shown good selectivities and stability in tests. In addition, composite membranes containing surface-activated silver nanoparticles dispersed in polymers result in extremely high propylene/propane separation performance of the mixed gas selectivity of 50 and the propylene permeance of 3.5 GPU and long-term operational stability [15]. In particular, 95 mole% of propylene feed is converted to 99.8% propylene with a single pass of membrane and operation is stable up to 450 hours (see Figure 5.22 in Section 5.8.2). Olefin/paraffin separation is especially attractive, although challenging, application for membranes likely to be commercialized soon.

7.6 Conclusions and Future Directions

The use of membranes for gas separation has grown rapidly since the first industrial plants were installed in the early 1980s. Overall, the future of membrane gas separation technology looks promising. Today, the current status of membrane gas separation processes falls into three categories: (1) Established, (2) Developing, and (3) Emerging [1, 6].

7.6.1 Established Processes

These include nitrogen production from air, hydrogen recovery, natural gas processing, petrochemical purge gas treatment, and air drying, representing over 90% of the current market. These processes have been in commercial use for over 10 years, with significant improvements in membrane selectivity, efficiency, and design. For example, modern nitrogen production modules generate more than 10 times the nitrogen with better quality and lower energy consumption than those from the 1980s. Most of these technologies have matured, with future improvements expected to be incremental. An exception is carbon dioxide removal from natural gas, where there is still room for significant advancement if more selective membranes can be developed.

7.6.2 Developing Processes

These include the recovery of light hydrocarbons from refinery and petrochemical plant purge gases, and the separation of C3+ hydrocarbons, hydrogen sulfide, nitrogen, and water from natural gas. These processes are already commercially viable, and ongoing development of better membranes and designs is driving their expansion.

7.6.3 Emerging Processes

These are potential future applications, such as oxygen-enriched air production, CO_2 separation from nitrogen or hydrogen, and organic vapor separation (e.g. olefin/paraffin mixtures). Research is active in these areas, particularly in developing facilitated transport membranes for separating oxygen, CO_2, and olefins. The success of these applications depends on significant improvements in membrane performance. For example, doubling the flux of oxygen/nitrogen separation could create a niche market, while a 5–10-fold improvement could make oxygen-enriched air economically viable for natural gas combustion. Similarly, CO_2 separation from nitrogen at power plants and hydrogen in coal gasification plants are being explored as environmental applications. The separation of organic vapors, especially olefin/paraffin mixtures, also holds promise for near-future application utilizing effective and sustainable olefin carriers such as surface-activated metallic nanoparticles for facilitated transport in the solid state.

References

1 R.W. Baker, Membrane Technology and Applications, John Wiley & Sons, 2023.
2 J.M.S. Henis, M.K. Tripodi, Journal of Membrane Science 8(3) (1981) 233–246.
3 S.K. Yong, J.K. Hyo, Y.K. Un, Journal of Membrane Science 60(2–3) (1991)219–232.
4 Y. Kang, J. Kim, H. Park, J. Won, H. Rhee, B. Jung, U. Kim, Korea Polymer Journal 5(3) (1997) 166–172.
5 S. Hanft, Membrane Technology for Liquid and Gas Separations, BBC Research, 2013.
6 R.W. Baker, Membrane Technology and Applications, 3 ed., John Wiley & Sons, 2012.
7 C.S. Song, W. Pan, S.T. Srimat, J. Zheng, Y. Li, Y.H. Wang, B.Q. Xu, Q.M. Zhu, Studies in Surface Science and Catalysis 153 (2004) 315–322.
8 T.C. Merkel, H.Q. Lin, X.T. Wei, R. Baker, Journal of Membrane Science 359(1–2) (2010) 126–139.
9 L.Q. Hu, S. Pal, H. Nguyen, V. Bui, H.Q. Lin, Journal of Polymer Science 58(18) (2020) 2467–2481.
10 Z. Ali, F. Pacheco, E. Litwiller, Y. Wang, Y. Han, I. Pinnau, Journal of Materials Chemistry A 6(1) (2018) 30–35.
11 N.P. Panapitiya, S.N. Wijenayake, D.D. Nguyen, Y. Huang, I.H. Musselman, K.J. Balkus, J.P. Ferraris, ACS Applied Materials & Interfaces 7(33) (2015) 18618–18627.
12 R.M. Joseph, M.M. Merrick, R. Liu, A.C. Fraser, J.D. Moon, S.R. Choudhury, J. Lesko, B.D. Freeman, J.S. Riffle, Journal of Membrane Science 564 (2018) 587–597.
13 V. Giel, Z. Morávková, J. Peter, M. Trchová, Journal of Membrane Science 537 (2017) 315–322.
14 M. Guo, M. Kanezashi, Membranes 11(5) (2021).
15 I.S. Chae, S.W. Kang, J.Y. Park, Y.G. Lee, J.H. Lee, J. Won, Y.S. Kang, Angewandte Chemie 123(13) (2011) 3038–3041.

Part II

Ion Transport Through Polymer Membranes

8

Basic Electrochemistry for Ion Transport

Electrochemistry studies chemical processes involving charge transfer, primarily through the movement of electrons and ions. These processes occur at interfaces, such as those between electrodes and electrolytes, where electrical and chemical interactions govern system behavior.

The electrochemical cell is at the core of electrochemistry, which enables charge transfer through electrochemical reactions. A typical cell consists of two electrodes – a cathode and an anode – separated by an electrolyte, which facilitates ion transport. Reactions at the electrode surfaces involve oxidation or reduction, generating or consuming charged species. This drives electron flow through an external circuit, leading to measurable changes in electrode potential.

Electrochemistry plays a crucial role in energy storage and conversion (e.g. batteries, fuel cells, water splitting, and supercapacitors), corrosion prevention, electroplating, and biosensors. Since charge conduction is significantly faster in metallic electrodes than in electrolytes, charge transport resistance through electrodes is typically negligible in electrochemical performance.

This section will introduce basic electrochemical concepts and terminology, followed by a discussion on electrode potential, reaction rates, and charge transport in electrolytes and charge transfer across interfaces. Key transport mechanisms, including diffusion, migration, and convection, will also be covered.

8.1 Electrochemical Devices and Key Terminologies

8.1.1 Reduction vs Oxidation

Electrochemical reduction–oxidation reactions, commonly known as redox reactions, involve the transfer of electrons from one species to another. The species that gains electrons is said to be reduced, while the species that loses electrons is said to be oxidized. Thus, the reducing agent is the electron donor, while the oxidizing agent is the electron acceptor [1]. In the reduction process, the oxidation number of a substance decreases by the gain of electrons, whereas, in the oxidation process, the oxidation number of a substance increases by the loss of electrons.

The overall equation of the redox half reaction for *Red*/*Ox* redox couple is

$$Ox + ne^- \underset{\text{oxidation}}{\overset{\text{reduction}}{\rightleftarrows}} Red$$

Oxidizing agent or oxidant *Ox* is the species that oxidizes another substance but reduces itself in reaction. While the reducing agent or reductant *Red* is the species that reduces other substances but oxidizes itself in the reaction. It is also necessary to keep in mind that *Ox* is both the oxidizing agent and the reduced form of the redox couple, whereas *Red* is both the reducing agent and the reduced form of the redox couple [2].

8.1.2 Galvanic Cells vs Electrolytic Cells

Electrochemical devices are conveniently categorized into two types depending on whether electricity is generated or consumed: galvanic cell and electrolytic cell, as shown in Figure 8.1. A galvanic cell produces electricity as a result of a spontaneous reduction and oxidation in the electrolyte solution, whereas an electrolytic cell uses electricity from the external power source to drive reduction and oxidation reactions, which are non-spontaneous (Figure 8.1a, b).

In a galvanic cell, oxidation and reduction occur spontaneously at two electrodes due to differences in the reduction potential of the respective half-cell reactions. The one with more positive reduction potential tends to undergo reduction, and the one with less positive (more negative) reduction potential tends to undergo oxidation. The electrode at which reduction takes place is called a cathode, and the other electrode where oxidation takes place is called an anode. For example, when H_2 and O_2 are fed into a fuel cell, H_2 is oxidized at the anode, and O_2 is reduced at the cathode. Specifically, H_2 is oxidized, releasing electrons by forming $H^+ (H_2 \rightarrow 2H^+ + 2e^-)$ at the anode [1]. These released excess electrons collected at the anode travel through the external circuit and reach the cathode, and eventually these electrons are consumed by the O_2 reduction taking place at the cathode $(1/2O_2 + 2H^+ + 2e^- \rightarrow H_2O)$ [1]. Thus, electrons flow from anode to cathode. In other view, the continuous collection of electrons at

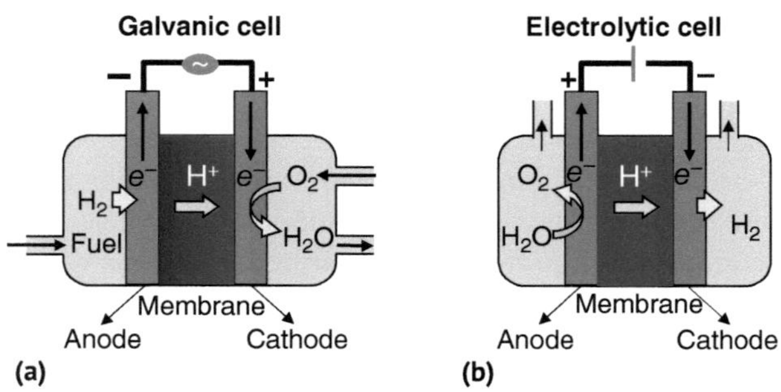

Figure 8.1 (a) Galvanic cell with spontaneous redox reactions (fuel cell: $H_2 \rightarrow 2H^+ + 2e^-$ for anode and $1/2O_2 + 2H^+ + 2e^- \rightarrow H_2O$ for cathode); and (b) electrolytic cell (water splitting: $H_2O \rightarrow 1/2O_2 + 2H^+ + 2e^-$ for anode and $2H^+ + 2e^- \rightarrow H_2$ for cathode). *Source*: Y.S. Kang [1].

the anode makes the electrode potential more negative, whereas continuous consumption of electrons at the cathode makes the electrode potential more positive. Therefore, the anode is negatively charged, and the cathode is positively charged, which leads electrical current to flow from the cathode (with more positive electrode potential) to the anode (with more negative electrode potential). Recall that by definition, electrical current flows from an area with high potential to low potential. The spontaneously generated electrical current in a galvanic cell can be used to do work. Fuel cells and batteries represent galvanic cells.

In an electrolytic cell, the situation is reversed due to the use of an external power source. Due to the external power source, electrons are withdrawn from a component (in its reduced form) through a reverse reaction ($H_2O \rightarrow 1/2O_2 + 2H^+ + 2e^-$) at an electrode with more positive electrode potential generated by the voltage applied from the external source. This positively charged electrode where oxidation occurs is the anode of the electrolyte cell. The as-withdrawn electrons are delivered through an external circuit and are provided to the other electrode with more negative electrode potential, where a component (in its oxidized form) is reduced ($2H^+ + 2e^- \rightarrow H_2$). This negatively charged electrode where reduction occurs is the cathode of the electrolyte cell. Note that despite the half-reactions are identical with those in the galvanic cell, the direction is reversed due to the use of external power source. The water splitting and production of caustic soda and chlorine are a typical application of electrolytic cell [1].

8.1.3 Cathode vs Anode and Positive Electrode vs Negative Electrode

A cathode is where reduction reaction undergoes regardless of the cell type, whereas an anode is where oxidation reaction undergoes. Positive and negative electrodes in a galvanic cell induce reduction and oxidation reactions, respectively, whereas those in electrolytic cells induce oxidation and reduction reactions, respectively. The results are summarized in Table 8.1.

8.1.4 Charge Transport vs Charge Transfer

Electrochemical performance results from the movement of charged species including electrons and ions in three different phases: electrodes with external circuit, electrolyte and their interface as shown in Figure 8.2. Electrons transport

Table 8.1 Comparison of characteristics of electrode reactions in different electrochemical cells.

Electrode	Cathode	Anode	Positive electrode	Negative electrode	Reaction
Galvanic cell	*Red*	*Ox*	*Red*	*Ox*	Spontaneous
Electrolytic cell	*Red*	*Ox*	*Ox*	*Red*	Non-spontaneous

Source: Adapted from [2].

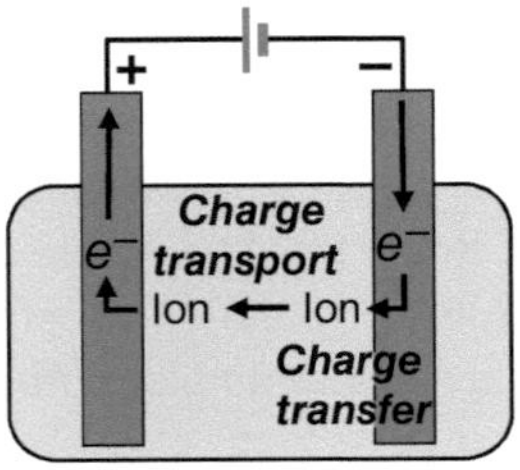

Figure 8.2 Charge transport and charge transfer in a common electrochemical cell. *Source*: Y.S. Kang.

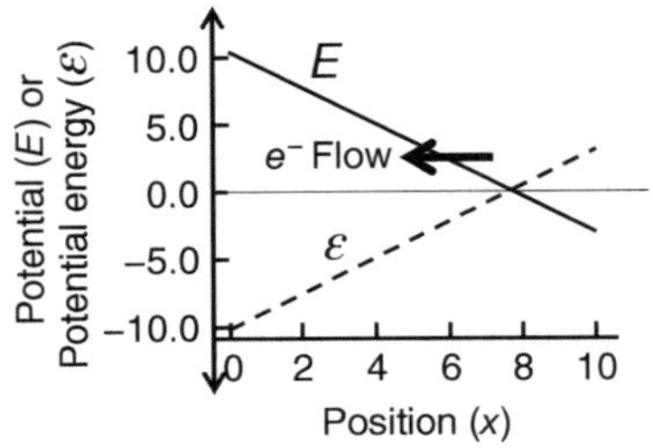

Figure 8.3 Electric potential and potential energy with electron transport direction. *Source*: Y.S. Kang.

through a single-phased electrode and external circuit, while ions transport through a single-phased electrolyte.

At the interface between the electrode and electrolyte, charge transfer occurs to convert electrons in electrode to ions in electrolyte (reduction reaction) or oppositely from ions to electrons (oxidation reaction), which is commonly termed as the charge transfer between two different phases. The charge transfer always involves the electrochemical redox reactions at the interfaces, the reduction reaction at the cathodic interface, and the oxidation reaction at the anodic interface. In this regard, electrochemical performance can be understood by both the transport and the transfer of the charged species, including electrons and ions involved.

8.1.5 Electric Potential vs Electric Energy

The electric potential energy ε (J) of a point charge Q (C) in the presence of an electric potential E (V = J/C) is defined as

$$\varepsilon = QE + \text{constant} \tag{8.1}$$

Electric potential E is, therefore, the electric potential energy ε per unit charge. Here, the term "electric potential" E is also often referred to as "the potential difference" because E represents the difference relative to a standard potential. The constant in Equation (8.1) depends on the choice of the standard reference electrode used.

Figure 8.3 schematically illustrates the relationship between electric potential, E, and potential energy, ε, as a function of position, x. Under a constant electric field, the electric potential, E, linearly decreases with position x, whereas electric potential energy ($\varepsilon = Q \cdot E$) increases with x. This behavior arises because electric potential is a measure of the work done per unit charge to bring a small positive test charge from infinity to a point x in an electric field. For a negatively charged electron ($Q < 0$), ε increases as E decreases. The arrow in Figure 8.3 indicates that the electron moves toward a region of higher electric potential (to the left) that has lower potential energy (it lowers its energy by moving to the left!). In simpler terms, electrons having negative charge move from lower electric potential to higher electric potential (opposite to the direction of the electric field), but toward lower potential energy, as the system naturally seeks to minimize energy.

8.2 Electric Potential

8.2.1 Electrode Potential

An electrode process involves the charge transfer across the interface between metallic electrode, *m*, and an electrolyte solution, *s*, as exemplified by the following reaction:

$$Fe^{3+}(s) + e^{-}(m) \rightarrow Fe^{2+}(s) \tag{8.2}$$

This reaction can only proceed once a suitable electrode is introduced into the solution, acting as a source or sink of electrons. Since the reaction involves the electron transfer between *m* and *s* phases, a net charge separation develops across the interface, leading to a potential difference, $\Delta\varphi_{m/s}$, between the electrode and the electrolyte as shown in Figure 8.4a [3].

$$\Delta\varphi_{m/s} = \varphi_m - \varphi_s \tag{8.3}$$

where φ_m and φ_s are electric potentials of metal and electrolyte solution, respectively. The charge on the metal resides in a very thin layer on the metal surface (<0.01 nm) [3].

Measuring $\Delta\varphi_{m/s}$ at a single interface is not experimentally feasible. To address this, a second electrode, known as a reference electrode, is introduced as shown in Figure 8.4b. Common reference electrodes include the standard or normal hydrogen electrode (SHE or NHE), the saturated calomel electrode (SCE, $Hg/Hg_2Cl_2/KCl$ saturated in water, 0.242 V vs NHE), and the silver–silver chloride electrode (Ag/AgCl/KCl saturated in water, 0.197 V vs NHE) [3].

When a voltage-measuring device is connected to the system, it records the difference in potential drops across the two electrode/solution interfaces. For example, in a simple electrochemical cell (Figure 8.4b), the cell potential, E_{cell}, is given by:

$$\begin{aligned} E_{cell} &= \Delta\varphi_{m/s}(\text{cathode}) - \Delta\varphi_{m/s}(\text{anode}) \\ &= \Delta\varphi_{m/s}(Fe^{3+}/Fe^{2+}) - \Delta\varphi_{m/s}(H_2/H^{+}) \end{aligned} \tag{8.4}$$

For convenience, normal hydrogen electrode (NHE, $Pt/H_2(a = 1)/H^{+}(a = 1,aq)$) comprising a platinum black electrode dipping into a hydrochloric acid solution is

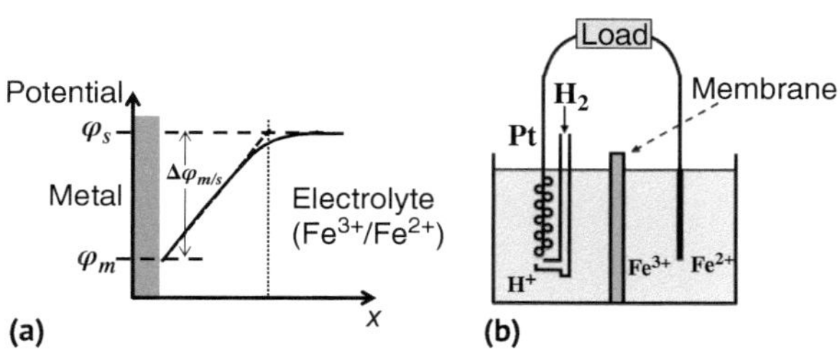

Figure 8.4 (a) The potential drop across the metal/electrolyte solution interface; and (b) a simple electrochemical cell with normal hydrogen electrode (NHE) as anode and metallic electrode (e.g. gold) in Fe^{3+}/Fe^{2+} solution as cathode. *Source*: Adapted from [3].

assigned an absolute potential of zero $(\Delta\varphi_{m/s}(H_2/H^+) = 0)$ at all temperatures [4]. For this special case, E_{cell} becomes:

$$E_{cell} = \Delta\varphi_{m/s}(Fe^{3+}/Fe^{2+}) \tag{8.5}$$

8.2.2 Standard Electrode Potential

The potentials of half-cells relative to NHE can be determined even though the NHE cannot exactly be realized in practice. The standard electrode potential of a half-cell is thus the potential of the cell made up with a half-cell (with all species in their standard states) and NHE.

In thermodynamics, a standard state refers to conditions where activities of all species are unity, elements and compounds are in their most stable forms at 298 K (25 °C). Similarly, in electrochemistry, fixing the concentrations of electroactive species is essential to establish a proper standard scale, as the cell potential depends on these concentrations. Selected standard reduction potentials at 298 K in aqueous solution are listed in Table 8.2.

Since $\Delta G^0 = -nFE^0$ (Equation (8.11) in Section 8.2.3), the standard potential is positive $(E^0 > 0)$ when the reaction proceeds in the spontaneous direction $(\Delta G^0 < 0)$ (galvanic cell). If $E^0 < 0$, the process is non-spontaneous $(\Delta G^0 > 0)$ (electrolytic cell).

8.2.3 Concentration-dependent Electrical Potential: Nernst Equation

Consider a cell in which the cathodic half-reaction for a reduction process:

$$n_O Ox + ne^- = n_R Red \quad \text{and} \quad K = a_R^{n_R}/a_O^{n_O} \tag{8.6}$$

Table 8.2 Standard electrode reduction potentials (E^0) in aqueous solutions at 25 °C vs NHE.

Reaction	E^0 (V)
$Li^+ + e^- \rightarrow Li$	−3.045
$K^+ + e^- \rightarrow K$	−2.925
$H_2O + e^- \rightarrow 1/2H_2 + OH^-$	−0.828
$Zn^{++} + 2e^- \rightarrow Zn$	−0.763
$Fe^{++} + 2e^- \rightarrow Fe$	−0.440
$1/2H_2 \rightarrow H^+ + e^-$	0.000
$AgCl + e^- \rightarrow Ag + Cl^-$	0.222
$Cu^{++} + 2e^- \rightarrow Cu$	0.340
$Fe^{+++} + e^- \rightarrow Fe^{++}$	0.771
$Ag^+ + e^- \rightarrow Ag$	0.799

Source: Adapted from [4].

where *Ox* and *Red* are the oxidant (or oxidizing agent, oxidized form of the redox couple) and the reductant (or reducing agent, reduced form of the redox couple), respectively. The equilibrium constant K is expressed as a function of the activities a_O and a_R of *Ox* and *Red*, respectively. The chemical activity, a_i, of component i is defined as $a_i = \gamma_i C_i$ with γ_i and C_i being the activity coefficient and the concentration, respectively.

In electrochemical systems, the Gibbs free energy change, dG, can be written as:

$$dG = -SdT + Vdp - dw_{\text{elec}} \tag{8.7}$$

where the electrical work, dw_{elec}, is included along with the conventional mechanical work, pdV, as:

$$dw = pdV + dw_{\text{elec}} \tag{8.8}$$

For constant pressure and temperature, Equation (8.7) simplifies to:

$$dG = -dw_{\text{elec}} \tag{8.9}$$

Electric work done by moving a charge Q (C) through an electrical potential difference E (V) is

$$w_{\text{elec}} = EQ \tag{8.10}$$

For charge transfer by n moles of electrons, $Q = nF$ and thus,

$$\Delta G = -nFE \tag{8.11}$$

Here, E is also known as the open circuit voltage or equilibrium potential, when no current is measured through the cell.

At equilibrium, E relates to the equilibrium constant K [4]:

$$E = E^0 - \frac{RT}{nF}\ln K \tag{8.12}$$

where E^0 is a standard electrode potential. Thus,

$$E = E^0 + \frac{RT}{nF}\ln\frac{a_O^{n_O}}{a_R^{n_R}} = E^0 + \frac{RT}{nF}\left(\ln\frac{\gamma_O^{n_O}}{\gamma_R^{n_R}} + \ln\frac{C_O^{n_O}}{C_R^{n_R}}\right) = E^{0'} + \frac{RT}{nF}\ln\frac{C_O^{n_O}}{C_R^{n_R}} \tag{8.13}$$

where E^0 and $E^{0'}$ are the standard potential and the formal potential, respectively, and they are related as:

$$E^{0'} = E^0 + \frac{RT}{nF}\ln\frac{\gamma_O}{\gamma_R} \tag{8.14}$$

when both n_O and n_R are the unity.

In the ideal solution $(E^0 = E^{0'})$ with $\gamma_i = 1$, Equation (8.13) becomes simplified as:

$$E = E^0 + \frac{RT}{nF}\ln\frac{C_O}{C_R} \tag{8.15}$$

This is the Nernst equation, which describes the potential of the cathode or anode relative to NHE as a function of the activities or concentrations of *Ox* and *Red*. The Nernst equation predicts how the electrical potential drop, $E\,(E = \varphi_m - \varphi_s)$, at the electrode/solution interfaces varies with the concentration.

In addition to the standard potential, E^0, reaction potentials are often reported using the convention of electromotive force, *emf*. This simplifies the calculation of cell potentials by combining two half-reactions. For reduction reactions, E^0 and *emf* have the same value, but for oxidation reactions, their signs are opposite. To calculate the cell potential using *emf*, the values for the half-reactions are simply added. When using standard potential E^0, the following convention applies [5]:

$$E^0{}_{\text{cell}} = E^0{}_{\text{cathode}} - E^0{}_{\text{anode}} \tag{8.16}$$

8.3 Electrochemical Redox Reactions and Charge Transfer Kinetics

The electric potential difference ($E = \Delta\varphi_{m/s}$) across the interface between the electrode and the electrolyte drives electrochemical redox reactions either spontaneously or with external input. These redox reactions determine the concentration of electrons, which transfer either to or from the electrode, generating an electric current.

Electrochemical redox reactions always involve electron transfer across the electrode/electrolyte interface. Consequently, the rate of the electrochemical reaction is directly or indirectly linked to the rate of electron transfer, which determines the electric current. A reduction reaction occurs when electrons transfer from the electrode (cathode) to the electrolyte, while an oxidation reaction occurs when electrons move from the electrolyte to the electrode (anode). Note that the current direction is opposite to the direction of electron flow.

Let us consider the following single electron transfer reaction:

$$Ox(s) + e^-(m) \longleftrightarrow Red(s) \tag{8.17}$$

Electrochemical reactions occur in the three consecutive steps, for instance, reduction reaction at cathode as shown in Figure 8.5. The first step is the transport of *Ox* by diffusion to the electrode surface and subsequent adsorption on the surface ($Ox \rightarrow Ox^*$), followed by the electrochemical reaction ($Ox^* + e^- \rightarrow Red^*$), and finally, the desorption of the product ($Red^* \rightarrow Red$) and the diffuse out from the electrode surface. Therefore, the electron transfer rate depends on both the mass transport, including adsorption and desorption, and electrochemical kinetics.

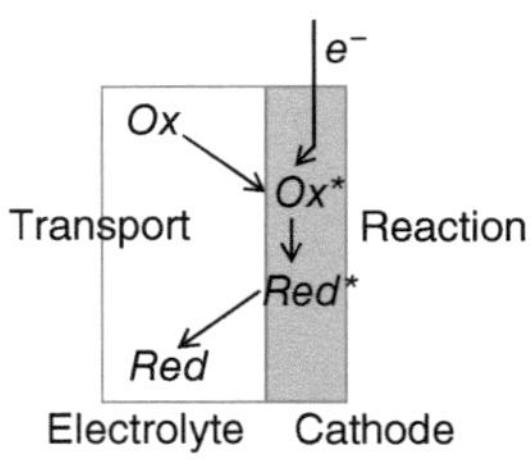

Figure 8.5 Electrochemical reaction steps: Transport and adsorption, electrochemical reaction, and desorption and transport for the reaction $Ox + e^- \rightarrow Red$. *Source*: Y.S. Kang.

8.3.1 Electrochemical Redox Reaction Kinetics

Within the electrode–electrolyte interface, the reduction and oxidation reactions described by Equation (8.17) take place at the electrode surface. Their respective reaction rates, v_c and v_a (mol/cm^2 sec), represent the rate of reaction per unit area of the interface. Assuming a first-order heterogeneous reaction, these rates can be expressed as

$$v_c = k_{Red}C_O^S \quad \text{and} \quad v_a = k_{Ox}C_R^S \tag{8.18}$$

where k_{Red} and k_{Ox} (1/sec) are the electrochemical reaction rate constants, and C_O^S and C_R^S (mol/cm^2) are the surface concentrations of *Ox* and *Red*, respectively. Since the electrochemical reaction occurs at the electrode surface, using surface concentration (mol/cm^2) is more appropriate for expressing reaction rates than the bulk concentration (mol/cm^3).

For reduction reaction at cathode, electrons transfer from electrode to electrolyte, whereas they transfer opposite direction for oxidation reaction as illustrated in Figure 8.6. Therefore, the net reaction rate v_{net} of *Ox* to *Red* at the cathode is then:

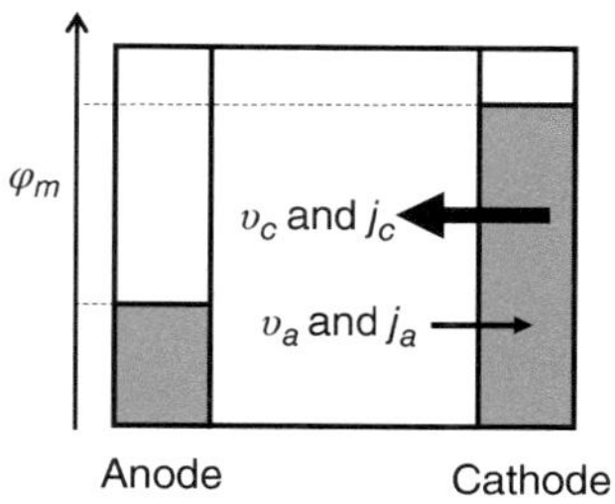

Figure 8.6 Charge transfer across the interface between cathode and electrolyte, where the reduction ($Ox + e^- \rightarrow Red$) is much faster than the oxidation ($Red \rightarrow Ox + e^-$). v_c and v_a are reduction and oxidation reaction rates, respectively. j_c and j_a are cathodic and anodic current densities. The arrow represents electron transfer direction. *Source:* Y.S. Kang.

$$v_{net} = v_c - v_a = k_{Red}C_O^S - k_{Ox}C_R^S \tag{8.19}$$

8.3.1.1 Free Energy of Activation

Transition state theory predicts the reaction rate constant of Equation (8.18):

$$k_{Red} = A_{Red}\exp\left(-\frac{\Delta G_{Red}^{+}}{RT}\right) \quad \text{and} \quad k_{Ox} = A_{Ox}\exp\left(-\frac{\Delta G_{Ox}^{+}}{RT}\right) \tag{8.20}$$

where ΔG_{Red}^{+} and ΔG_{Ox}^{+} are the free energy of activation for the reduction and oxidation reactions, respectively, with A_{Red} and A_{Ox} being the pre-exponential factors. Figure 8.7a and b illustrate (a) free energy plot for a simple one-electron reduction of an *Ox* (aq) to *Red* (aq) and (b) its potential energy changes depending on the α value.

8.3.1.2 Effects of Potential

The electrode potential change due to concentration variations, as expressed by the Nernst equation in Equation (8.15), leads to modifications in the free energies of activation, represented as:

$$\Delta G_{Red}^{+} = \Delta G_{Red,0}^{+} + \alpha F(E - E^{0'}) \tag{8.21}$$

and

$$\Delta G_{Ox}^{+} = \Delta G_{Ox,0}^{+} - (1 - \alpha)F(E - E^{0'}) \tag{8.22}$$

where α is the transfer coefficient, defined as $\alpha = -(1/F)(d\Delta G_{Red}^{+}/dE)$ at $E = E^{0'}$ [6], and $\Delta G_{Red,0}^{+}$ and $\Delta G_{Ox,0}^{+}$ are the free energy corresponding to standard equilibrium. The second term in Equations (8.21 and 8.22) represent the contribution of the concentration changes from the standard state, according to the Nernst equation. The value of α ranges from zero to unity ($0 \leq \alpha \leq 1$). If α is close to zero, the

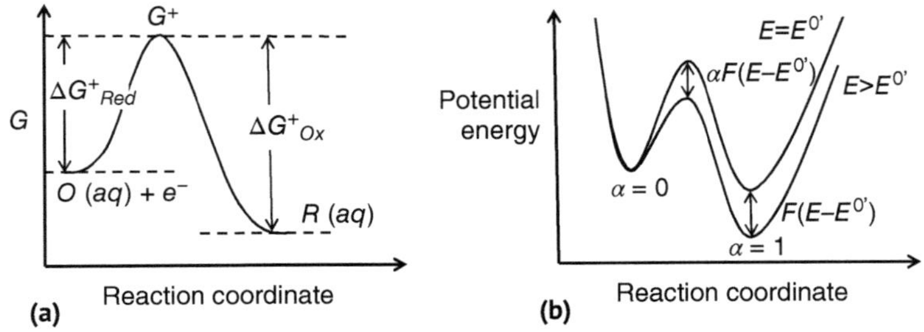

Figure 8.7 (a) Free energy plot for a simple one-electron reduction of a species *Ox* (aq) to *Red* (aq); and (b) potential energy curve for the standard equilibrium state $E = E^{0'}$ and for $E > E^{0'}$. *Source*: Adapted from [6].

transition state resembles the reactants in its potential dependence, whereas when it approaches unity, the transition state behaves in a product-like manner. In fact, α is typically found to be close to 0.5 for many reactions, suggesting that the transition state has intermediate behavior. The reaction rate constants thus become:

$$k_{Red} = A_{Red}\exp\left(-\Delta G^{+}_{Red,0}/RT\right)\exp\left(-\alpha F(E - E^{0'})/RT\right) \tag{8.23}$$

$$k_{Ox} = A_{Ox}\exp\left(-\Delta G^{+}_{Ox,0}/RT\right)\exp\left((1-\alpha)F(E - E^{0'})/RT\right) \tag{8.24}$$

Consider the special case in which the interface is in equilibrium with an electrolyte solution and also $C_R^s = C_O^s$. At equilibrium, the generation and consumption rates of electrons become equivalent to each other (the total current $j = 0$). In this case, $E = E^{0'}$ and $k_{Ox}C_R^s = k_{Red}C_O^s$ so that $k_{Red} = k_{Ox}$. Thus $E^{0'}$ is the standard potential where the reduction and oxidation reaction rate constants have the same value. That value is called the standard rate constant or intrinsic rate constant k^0 [4].

$$k_{Red} = k^0\exp\left\{\frac{-\alpha F(E - E^{0'})}{RT}\right\} \tag{8.25}$$

And

$$k_{Ox} = k^0\exp\left\{\frac{(1-\alpha)F(E - E^{0'})}{RT}\right\} \tag{8.26}$$

where $k^0 = A_{Red}\exp(-\Delta G^{+}_{Red}/RT) = A_{Ox}\exp(-\Delta G^{+}_{Ox}/RT)$. Note that $E^{0'} = E^0 + (RT/nF)\ln(\gamma_o/\gamma_R)$ and when an electrolyte is ideal ($\gamma_o = \gamma_R = 1$), $E^{0'} = E^0$. Finally, υ_{net} at the cathode can be expressed as:

$$\upsilon_{net} = k^0\left(\exp\left\{\frac{-\alpha F(E - E^{0'})}{RT}\right\} - \exp\left\{\frac{(1-\alpha)F(E - E^{0'})}{RT}\right\}\right) \tag{8.27}$$

Equation (8.27) suggests the relation between the net reaction rate υ_{net} and the electrode potential E.

8.3.2 Current-Voltage Relationship: The Butler–Volmer Equation

Electric current is the result of the flow of electrons generated or consumed by electrochemical oxidation and reduction reactions, respectively. Therefore, current density j (C/cm^2 sec), representing charge flux, is directly related to the electrochemical reaction rate υ:

$$j_c = F\upsilon_c \quad \text{and} \quad j_a = F\upsilon_a \tag{8.28}$$

where j_c and j_a are reductive and oxidative current densities, respectively, as shown in Figure 8.6. Then the net current density j becomes:

$$j = F\upsilon_{net} = j_c - j_a = F\left(k_{Red}C_O^s - k_{Ox}C_R^S\right) \tag{8.29}$$

This shows how the current density depends on the concentrations at the electrode surfaces (C_O^s and C_R^S) and also on k_{Red} and k_{Ox}, which are known to be potential-dependent as written in Equations (8.23) and (8.24).

It is known that no current is detected through the cell when the applied potential E is equal to the equilibrium potential E_{eq}. For other values of E, the electrodes are no longer in equilibrium, so redox reaction becomes thermodynamically viable. Whether current is detected or not depends on the kinetics of the particular electrode reaction under investigation. It has been well understood that a potential difference in value to that of E_{eq} must be applied to the working electrode to drive the electrode reaction. The deviation of E from E_{eq} is given the term "overpotential" η.

$$\eta = E - E_{eq} \tag{8.30}$$

where

$$E_{eq} = E^{0'} + \frac{RT}{nF}\ln\frac{C_O^b}{C_R^b}$$

The reaction rate constants of Equations (8.23) and (8.24) are substituted into Equation (8.29) and rearranged with Equation (8.30) give the net current density, j, flowing at the working electrode as:

$$j = j_0\left(\frac{C_O^S}{C_O^b}\exp\left\{\frac{-\alpha F\eta}{RT}\right\} - \frac{C_R^S}{C_R^b}\exp\left\{\frac{(1-\alpha)F\eta}{RT}\right\}\right) \tag{8.31}$$

where $j_0 = Fk^0[C_R^s]^{\alpha}[C_O^s]^{1-\alpha}$ is called the exchange current density. Equation (8.31) is the Butler–Volmer equation, which predicts how the observed current varies as a function of the overpotential η and transfer coefficient α.

Even though the net current is zero ($j = 0$) at equilibrium, we still envision balanced faradaic activity, representing $j_c = j_a$ from Equation (8.29), that can be expressed in terms of the exchange current density, j_0. For a particular case where $C_R^s = C_O^s = C^s, j_0 = Fk^0C^s$ [4].

When the solution under investigation is well stirred, $C_R^s = C_R^b,\ C_O^s = C_O^b$, the Butler–Volmer equation simplifies to: [4]

$$j = j_0\left(\exp\left\{\frac{-\alpha F\eta}{RT}\right\} - \exp\left\{\frac{(1-\alpha)F\eta}{RT}\right\}\right) \tag{8.32}$$

At the equilibrium potential ($j = 0$ with $\eta = 0$), both the anodic and cathodic currents have the same magnitude j_0 ($j_0 = j_c = j_a$), thus canceling each other.

Figure 8.8a depicts the variation of j_c, j_a, and j as a function of the overpotential η. When $\eta < 0$, j_c mostly determines j, whereas j_a mostly determines j when $\eta > 0$.

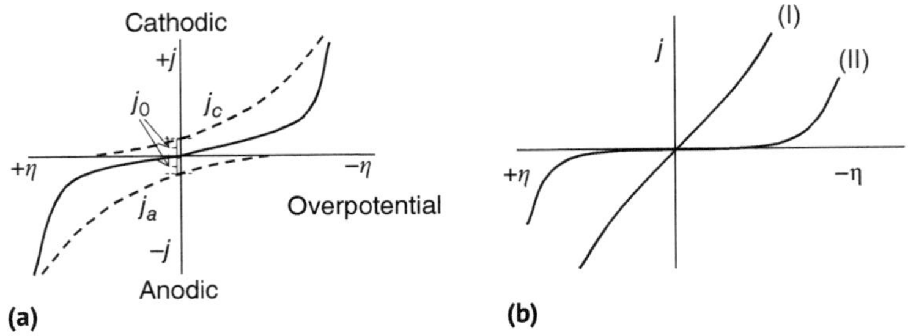

Figure 8.8 (a) The contributions of both the cathodic j_c and anodic j_a current densities to the total current density j; and (b) the variation of current density as a function of the overpotential for (I) a reversible electrode reaction with large j_0 and (II) an irreversible electrode reaction with small j_0. *Source*: Y.S. Kang.

Figure 8.8b illustrates two limiting cases with large and small values of j_0. When j_0 is large, called "reversible," little or no applied overpotential is required to drive the reaction and current flows readily in both anodic and cathodic directions according to the sign of the overpotential applied (Figure 8.8b (I)). For processes that have a small value of j_0, called "irreversible," a high overpotential is required to induce current flow (Figure 8.8b (II)) [3].

8.3.3 Tafel Plot

When $\eta \gg 0$, the Butler–Volmer equation, Equation (8.32), simplifies to:

$$\ln(j) = \ln(j_0) + \frac{(1-\alpha)F\eta}{RT} \tag{8.33}$$

whereas, when $\eta \ll 0$, it reduces to:

$$\ln(-j) = \ln(j_0) - \frac{\alpha F\eta}{RT} \tag{8.34}$$

A plot of $\ln(j)$ vs η known as the Tafel plot is useful for evaluating kinetic parameters such as the transfer coefficient α, the exchange current density j_0 and the limiting current density j_L (Figure 8.9). In general, the slope of $-\alpha F/RT$ is for cathodic Tafel zone and that of $(1-\alpha)F/RT$ is for anodic Tafel zone. Both linear segments extrapolate to an intercept of $\ln(j_0)$.

The plots deviate sharply from linear behavior as η approaches zero because the back reactions or oxidation reactions at cathode can no longer be regarded as negligible, called the polarization zone. In particular, for very small overpotential $\eta \ll 1$ or $E \to E_0$, the Butler–Volmer equation can be linearized by expanding the exponentials:

$$j = j_0 \frac{nF}{RT}\eta \tag{8.35}$$

Adapting Ohm's law at very small η, a new resistance R_{CT} is defined as $R_{CT} = \eta/j$ [6, 7]. Here, R_{CT} represents the resistance of ion transferring from a solvated ionic state in the electrolyte, crossing the electrode/electrolyte interface and inserting into

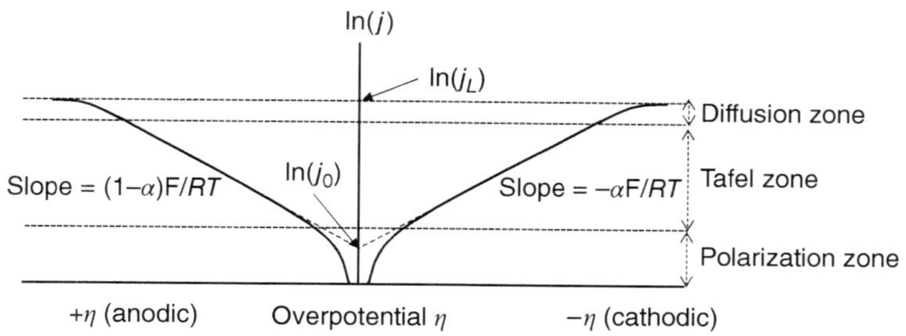

Figure 8.9 Tafel plots for cathodic and anodic branches of the current density-overpotential curves for $Ox + e^- \rightleftharpoons Red$ with $\alpha = 0.5$. *Source*: Y.S. Kang.

the electrodes, which causes the activation polarization. From Equation (8.35), R_{CT} can then be estimated:

$$R_{CT} = \frac{RT}{nF}\frac{1}{j_0} \tag{8.36}$$

On the other hand, when η increases beyond the Tafel zone, the curve deviates from the linear relation to reach the limiting current density j_L, called the diffusion zone, where the reaction becomes very fast and therefore the current is nearly completely determined by the mass transport. The diffusion coefficient D in the electrolyte is obtained from Equation (8.37) [8].

$$j_L = \frac{nFDC^b}{\delta} \tag{8.37}$$

where C^b is the bulk electrolyte concentration and δ is the thickness of diffusion layer [8, 9].

8.4 Interfacial Charge Transfer Through Electric Double Layer

The interfacial structure of an electrode and an electrolyte plays a crucial role in determining the performance of electrochemical devices, such as batteries and fuel cells. It influences the mass transport of ions, charge transfer kinetics, and overall device efficiency. The following is on the interfacial structure of electric double layer consisting of the Helmholtz layer and diffuse layer, and its relationship with charge transport.

8.4.1 Electric Double Layer

The interface between the electrode and electrolyte comprises the solid electrode surface, the electrolyte solution, and their interfacial layer, known as the electrical double layer. This double layer is further divided into two regions: the Helmholtz layer, where ions are tightly bound to the electrode surface, and the diffuse layer, where ion concentration gradually changes with distance from the electrode.

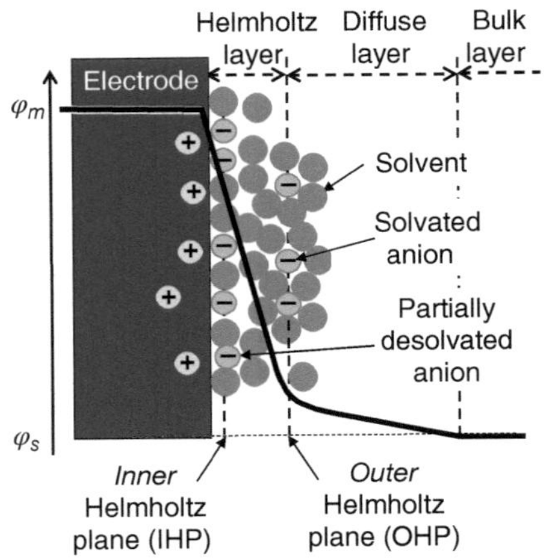

Figure 8.10 Schematic structure of electric double layer (EDL) and electrical potential profile, where φ_s and φ_m are the electric potentials of the electrolyte and the positive electrode material, respectively. The thicknesses of the Helmholtz and the diffuse layers are approximately in the range of 0.1–10 nm and 500 μm, respectively. *Source*: Y.S. Kang.

Consider an electrochemical cell with an electrolyte between two electrodes. When an electrode is positively charged, anions will be preferably adsorbed in the electrode surface, and corresponding electrical potential will be changed as illustrated in Figure 8.10.

EDL comprises the Helmholtz layer and the diffuse layer. The structure and electric potential profiles of these layers significantly influence the rate of charge transfer in redox reactions, depending on the strength of the applied potential. Therefore, the characteristics of both layers are briefly discussed as follows.

8.4.1.1 Helmholtz Layer

Microscopically, solvated or partly solvated anions, for instance, will be adsorbed at the positively charged surface because of both chemical and Coulombic interactions between the electrode surface and the corresponding ions. Therefore, the concentration profile of anions will be perturbed to have higher ion concentration at a certain layer of the interface than the bulk as shown in Figure 8.10. Such layer is called the Helmholtz layer, which therefore plays great roles in determining many electrochemical processes, including electrochemical reactions and charge transfer rates, and thereby becoming one interesting field for insight into the electrochemical fundamentals. In other words, the Helmholtz layer is the layer of ions that are strongly attracted to the charged surface. These ions thus form a compact, immobile layer.

Figure 8.10 also shows the *inner* Helmholtz plane corresponding to the position of specifically adsorbed and partially solvated ions, and the *outer* Helmholtz plane corresponding to the plane of the closest approach of the hydrated anions. The Helmholtz layer is essentially a molecular dielectric and stores charge electrostatically.

8.4.1.2 Diffuse Layer

When the electrochemical reactions start at both electrode surfaces, the reactant concentration, such as *Ox* at the cathode surface diminishes (Figure 8.11a) and, therefore, it should be replenished by ion transport from the bulk layer, across the

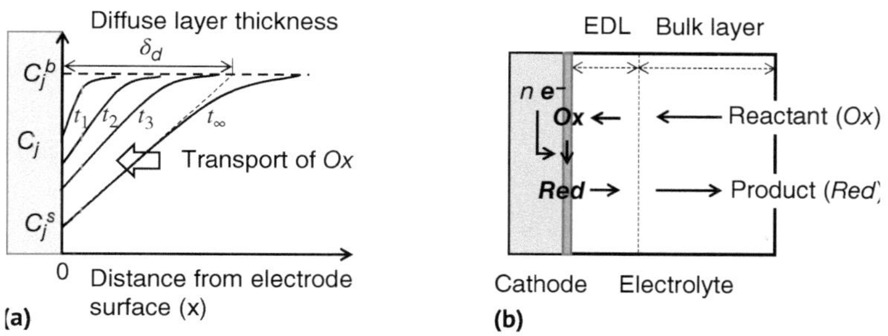

Figure 8.11 Schematic picture of (a) concentration profiles (*Ox*) and thickness δ_d of Nernst diffuse layer with increasing time *t* (from t_1 to t_∞); and (b) mass transport of *Ox* and *Red* across the diffuse layer and reduction reaction at cathode side. *Source*: Y.S. Kang.

electrical double layer, and finally to the electrode surface as shown in Figure 8.11b. Gradually, a "diffuse layer" is established close to the electrode in which the concentration of *Ox* differs from its value in the bulk solution. At the cathode surface, the electrochemical reaction of $Ox + e^- \rightarrow Red$ continues to consume *Ox* and generate *Red* and consequently the concentration C_j^s of *Ox* at the surface will be decreased (Figure 8.11a). The product *Red* will be then desorbed from the electrode surface and subsequently diffuse out from the surface to the bulk. At the anode surface, *Red* is adsorbed first and oxidized to form *Ox*, which will be released from the surface, followed by the diffuse out.

The thickness of the diffuse layer δ_d also grows as the reaction proceeds as shown in Figure 8.11a: It grows steadily ($\delta_d(t) = (\pi DT)^{1/2}$)) to a steady-state value where the reaction rate becomes equal to the mass transport rate. In practice, the bulk of the solution invariably experiences some mixing because of natural convection, which also limits the expansion of the diffuse layer. Under these steady-state conditions, the transport within the diffuse layer is pictured as occurring without natural convection, whilst outside the layer the concentration of *Ox* or *Red* is maintained at its bulk value through natural mixing [10]. As a result, δ_d is frequently estimated to be approximately 10–1,000 μm, whereas the Helmholtz layer thickness is 0.1–0.5 nm.

The concentration C_O^s of *Ox* at the electrode surface decreases continuously with the reaction time *t* and reaches a certain value at the steady state. When the reaction is very fast, compared to the mass transport rate of species *Ox*, C_O^s becomes zero, where the limiting current density j_L can be obtained.

8.4.2 Charge Transfer Through Electric Double Layer

The charge transfer at the electrode/electrolyte interface is accompanied by mass transport across EDL to the electrode surface, followed by electrochemical reactions. Under normal experimental conditions, for electrolytes with mm dimensions, the diffuse layer thickness, δ_d, is approximately in the range of 500 μm. The ionic mass transport is accompanied by the two different mass transport modes: Diffusion and migration fluxes.

The diffusional mass flux J_{jd} of species j to the electrode surface results in the current density j_{jd};

$$j_{jd} = FJ_{jd} = FD_j\frac{C_j^b - C_j^s}{\delta_d} \tag{8.38}$$

where C_j^b and C_j^s are the concentrations at bulk and surface, respectively, and D_j is the diffusion coefficient of species j. The maximum observable current is limited by the minimum surface concentration, i.e. C_j^s~0. Thus, a limiting current density, j_L, is reached:

$$j_L = FD_j\frac{C_j^b}{\delta_d} \tag{8.39}$$

Under these conditions, all the electroactive material transported to the electrode is converted to products. In other words, the electrochemical reaction is much faster than the ion transport rate. j_L is the limiting value of a Faradic current for which the charge transfer rate at the electrode is saturated. As the overpotential increases, the charge transfer rate at the electrode surface increases and then is limited by the deposition of electrolyte ions onto the electrode by diffusion. After the point, regardless of an increase in the overpotential, the current density is saturated to j_L.

The external electric field ($\xi = -\mathrm{d}\varphi/\mathrm{d}x$), developed across EDL as a result of the drop in electrical potential between the two phases ($\Delta\varphi_{m/s}$) is capable of exerting an electrostatic force on charged species present in the interfacial region and thereby of inducing the movement of ions to or from the electrode. The migratory flux, J_{jm} (mol/cm^2 sec), and the corresponding electric current j_{jm} (C/cm^2 sec) is proportional to the ionic mobility u_j (cm^2/secV) of species j, the ion concentration, C_j (mol/cm^3) and the electric potential gradient $\mathrm{d}\varphi/\mathrm{d}x$ (V/cm).

$$j_{jm} = FJ_{jm} = -Fu_jC_j\frac{\mathrm{d}\varphi}{\mathrm{d}x} \tag{8.40}$$

where u_j is directly related with the diffusion coefficient D_j (cm^2/sec), consequently depending on the ionic charge and size as well as on the solution viscosity, which will be discussed in more detail in Section 8.5.

To minimize charge transfer resistance and improve overall performance, various strategies can be employed, including optimizing the electrode and electrolyte materials, designing efficient electrode structures to increase surface area, and employing additives or membranes to enhance ion transport. Understanding and controlling the interfacial structure between the electrode and electrolyte is essential for developing high-performance electrochemical devices with improved charge transfer characteristics and enhanced energy storage or conversion capabilities.

8.5 Transport of Charged Species and Electric Current

Electric currents depend on the overall reaction rates because electrons are generated or consumed by oxidative and reductive electrochemical reactions (redox reactions), respectively. Such redox reactions are, in turn, determined by both mass transport from (or to) the bulk electrolyte to (or from) the electrode surface and electrochemical reactions at the interfaces as illustrated in Figure 8.11b. Three different ion transport processes in electrolyte can be observed including diffusion, migration, and convection. The significance of each of the three transport properties will now be considered in a little more detail.

8.5.1 Electrochemical Potential

According to Fick's law, diffusional mass transport of neutral species is driven by a concentration gradient. However, it can be more rigorously described using the chemical potential gradient, $\mathrm{d}\mu_j/\mathrm{d}x$, rather than the concentration gradient, $\mathrm{d}c_j/\mathrm{d}x$. These gradients are related by

$$\frac{\mathrm{d}\mu_j}{\mathrm{d}x} = \frac{RT}{c_j}\frac{\mathrm{d}c_j}{\mathrm{d}x} \tag{8.41}$$

where $\mu_j = \mu_j^0 + RT\ \ln a_j$ by assuming ideal conditions (activity $a_j = \gamma_j c_j$ with the activity coefficient $\gamma_j = 1$). Then, Fick's law can be expressed differently in terms of chemical potential gradient as:

$$J_j = -\left(\frac{c_j D_j}{RT}\right)\frac{\mathrm{d}\mu_j}{\mathrm{d}x} = -L_j\frac{\mathrm{d}\mu_j}{\mathrm{d}x} \tag{8.42}$$

where L_j is the phenomenological coefficient, $L_j = c_j D_j/RT$.

If the species j is charged, the transport must be described using the electrochemical potential, which combines chemical potential μ_j (J/mol) (related to concentration) and the electrical potential φ (V = J/C) (related to charge). Electrochemical potential $\bar{\mu}_j$ is mathematically expressed as:

$$\bar{\mu}_j = \mu_j + z_j F\varphi \tag{8.43}$$

where the charge number z_j is positive for cation and negative for anion. The gradient of the electrochemical potential thus drives ion transport, encompassing both chemical and electrical potential contributions.

Substituting $\mu_j = \mu_j^0 + RT\ln a_j$ into Equation (8.43) yields:

$$\bar{\mu}_j = \mu_j^0 + RT\ln a_j + z_j F\varphi \tag{8.44}$$

Equation (8.44) highlights the dependence of electrochemical potential $\bar{\mu}_j$ on a_j and φ.

8.5.2 Transport of Charged Species: Diffusion, Migration, and Convection

The mass flux $J_j\left(\mathrm{mol/cm^2\ sec}\right)$ of an ionic species j in the x-direction is driven by the electrochemical potential gradient ($\mathrm{d}\bar{\mu}_j/\mathrm{d}x$) as a driving force:

$$J_j = -\left(\frac{c_j D_j}{RT}\right)\left(\frac{\mathrm{d}\bar{\mu}_j}{\mathrm{d}x}\right) = -L_j\left(\frac{\mathrm{d}\bar{\mu}_j}{\mathrm{d}x}\right) \tag{8.45}$$

Using Equation (8.44), the electrochemical potential gradient $d\bar{\mu}_j/dx$ becomes:

$$\frac{d\bar{\mu}_j}{dx} = RT\left(\frac{1}{c_j}\frac{dc_j}{dx} + \frac{z_jF}{RT}\frac{d\varphi}{dx}\right) \tag{8.46}$$

Substituting Equation (8.46) into (8.45) yields the Nernst–Planck equation:

$$J_j = -D_jc_j\left(\frac{1}{c_j}\frac{dc_j}{dx} + \frac{z_jF}{RT}\frac{d\varphi}{dx}\right) \tag{8.47}$$

or

$$J_j = -D_j\frac{\partial c_j}{\partial x} - u_jc_j\frac{\partial \varphi}{\partial x} \tag{8.48}$$

where u_j is again the mobility of species j. According to the Einstein–Smoluchowski equation, u_j and D_j are related as:

$$u_j = D_j\frac{z_jF}{RT} \tag{8.49}$$

The mobility u_j is either positive or negative depending on the charge z_j; however, for practical convenience, it is typically expressed as a positive value.

The first term in Equations (8.47) and (8.48) is the diffusion term caused by concentration gradients while the second term is the migration term driven by the electric field. Therefore, the total mass flux J_j is the sum of the diffusional flux J_{jd} and the migrational flux J_{jm}:

$$J_j = J_{jd} + J_{jm}. \tag{8.50}$$

Diffusion arises from concentration gradients, aiming to equalize composition and maximize entropy. Migration, on the other hand, results from the electric field ($\xi = -d\varphi/dx$) at the electrode/solution interface, where a potential drop ($\Delta\varphi_{m/s}$) induces ion movement. These combined effects of diffusion and migration play a critical role in transport phenomena, particularly in electrochemical reactions.

Convection also contributes to the transport of ions in addition to diffusion and migration. There are two types of convection: natural and forced convections. The first, natural convection, can be present in any solution and arises from thermal gradients and/or density difference within the solution. Natural convection typically becomes a significant perturbation in electrolysis conducted with electrodes of conventional size (approximately mm or larger) on the time scale of 10–20 seconds and longer, and is generally undesirable since it is difficult to predict.

The second type of movement is forced convection, which is achieved by external forces such as gas bubbling through the solution, pumping, stirring, or rotating the electrode. This serves to overwhelm any contributions from natural convection, ensuring that reproducible experiments can be conducted over time scales beyond the 10–20 second limit.

Mathematically, concentration changes resulting from movement of solution with a velocity v (cm/sec) in the x-direction are given by

$$J_j = c_jv \tag{8.51}$$

Total mass transport equation, commonly termed as the Nernst–Planck equation, from three different modes of diffusion, migration, and convection can then be rewritten as:

$$J_j = -D_j\frac{\partial c_j}{\partial x} - u_j c_j\frac{\partial \varphi}{\partial x} + c_j v \quad \text{for cation and} \tag{8.52}$$

$$J_j = -D_j\frac{\partial c_j}{\partial x} + u_j c_j\frac{\partial \varphi}{\partial x} + c_j v \quad \text{for anion} \tag{8.53}$$

It must be noted that the Nernst–Planck equation is true only under specific physical conditions and is subject to a number of limitations. First, the diffusion coefficient characterizes the interaction between the dissolved electrolyte component and polymer matrix, but does not take into account the interaction between the diffusing species themselves. Second, it is the activity gradient or chemical potential gradient for the motive force of diffusion. However, the individual activities of the ions are thermodynamically difficult to determine and therefore, the averaged ion activity is used instead.

When three different mass transports, such as diffusion, migration, and convection, occur simultaneously in any liquid electrochemical system, the data interplay becomes extremely difficult. To simplify the problem, experiments are often performed under conditions where migration can be neglected. This is possible by means of a chemically and electrochemically inert "background electrolyte" added to the solution in high concentrations: levels around or more than 0.1 M may be commonly used. Note that the common concentration range of electrolytes is ~10^{-3} M. Calculations have shown that approximately 100 times the concentration of background electrolyte relative to that of the reactant is required. The addition of background electrolytes provides several benefits: (1) highly conductive, (2) compressed double-layer thickness ~1–2 nm, over which the full interfacial drop in potential ($\Delta\varphi_{m/s}$) occurs, and (3) ionic strength stays effectively constant.

8.5.3 Electric Current Density

The total electric current density j (A/cm^2) is the summation of each current density of the species j

$$j = \sum_{j=1}^{n} j_j \tag{8.54}$$

Since,

$$j_j = z_j F J_j \tag{8.55}$$

the electric current density j can be obtained from substituting Equation (8.47) into Equation (8.55) without considering the convection:

$$j = -\sum_j^n z_j F J_j = \sum_j^n z_j F D_j\left(\frac{\partial c_j}{\partial x}\right) + \sum_j^n \frac{z_j^2 F^2}{RT} D_j c_j\left(\frac{\partial \varphi}{\partial x}\right) \tag{8.56}$$

which suggests that, in the presence of the concentration gradient, the current density j is not proportional to the potential gradient $\partial\varphi/\partial x$ and the Ohm's law is not valid. Equation (8.56) also allows us to make an important conclusion that even when there is no current ($j = 0$) in a system, there is an electric potential gradient.

Finally, the electric neutrality condition is written as

$$\sum_{j=1}^{n} c_j z_j = 0 \tag{8.57}$$

8.5.4 Ion Conductivity, Transport, and Transference Numbers

In the bulk solution (away from the electrode), concentration gradients are generally small, and the total current is carried mainly by migration. For species j in the bulk region having a cross-sectional area A, $j_j = j_{jm}$ (migration current density of species j) [4]:

$$j_j = \frac{z_j^2 F^2 D_j c_j}{RT}\frac{\partial\varphi}{\partial x} = z_j F u_j c_j \frac{\partial\varphi}{\partial x} \tag{8.58}$$

For a linear electric field,

$$\frac{d\varphi}{dx} = \frac{\Delta E}{l} \tag{8.59}$$

where $\Delta E/l$ is the gradient of the change in potential ΔE over distance l. Thus,

$$j_j = |z_j| F u_j c_j \frac{\Delta E}{l} \tag{8.60}$$

Ohm's law ($i = \Delta E/R$) yields,

$$j = \frac{\Delta E}{AR} = \frac{l}{AR}\frac{\Delta E}{l} = \sigma\frac{\Delta E}{l} \tag{8.61}$$

and $j = \sum j_j$. Therefore, the ionic conductivity σ $[\Omega^{-1}\text{cm}^{-1}$ or $\text{Scm}^{-1}]$ is given by

$$\sigma = F\sum_{j}^{n} |z_j| u_j c_j = F^2 \sum_{j}^{n} |z_j|^2 \frac{D_j}{RT} c_j \tag{8.62}$$

This indicates that the ionic conductivity σ is proportional to the sum of the products of the mobility u_j and their concentration c_j of each charged species j.

In an electrolyte solution, the current is carried out by both cations and anions. However, cations and anions usually have different portions of the overall current. When many ionic species, at least, more than two are present, each ionic species contributes to the ionic current. For instance, in the case of lithium-ion battery of $LiPF_6$, both Li^+ and PF_6^- ions are dissolved in a solvent and can contribute the measured conductivity, σ. But Li^+ ion conductivity is much more important instead of the total ionic conductivity. Thus, it is necessary to obtain the contribution of each ionic species to the total current. To address this issue, the ion transport number T_j or transference number t_j is defined as given:

$$T_j = \frac{j_j}{j} = \frac{|z_j| u_j c_j}{\sum_{k=1}^{n} |z_k| u_k c_k} \tag{8.63}$$

and

$$t_j = \frac{T_j}{|z_j|} \tag{8.64}$$

The ion transport number T_j indicates that the fraction of the current carried out by the ion j, and the transference number t_j indicates the number of moles of ion j transported per mole of electrons, i.e. per Faraday [1]. The sum of the transport number or transference number of all ions in the solution becomes unity. In literature, the definition of the transport number or transference number is not consistently used and very often they are mixed up.

The membrane permselectivity describes the degree to which a membrane passes an ion of one charge and retains an ion of opposite charges: in other words, it means the ability of IEMs to permeate counter-ions (c) over co-ions (a). The permselectivity of cation and anion exchange membranes (CEMs and AEMs) is defined as:

$$P_c^m = \frac{T_c^m - T_c^s}{T_a^s} \quad \text{for CEM} \tag{8.65}$$

$$P_a^m = \frac{T_a^m - T_a^s}{T_c^s} \quad \text{for AEM} \tag{8.66}$$

where P_c^m and P_a^m are the permselectivity of CEM and AEM, and superscript m and s are for membrane and electrolyte solution, respectively [1, 11].

References

1 H. Strathmann, Ion-Exchange Membrane Separation Processes, Elsevier Science, 2004.
2 C. Lefrou, P. Fabry, J.-C. Poignet, Electrochemistry: The Basics, with Examples, Springer Science & Business Media, 2012.
3 A.C. Fisher, Electrode Dynamics, Oxford University Press, 1996.
4 A.J. Bard, L.R. Faulkner, Electrochemical Methods: Fundamentals and Applications, Wiley, 2000.
5 C.G. Zoski, Handbook of Electrochemistry, Elsevier, 2006.
6 W. Schmickler, E. Santos, Interfacial Electrochemistry, Springer Science & Business Media, 2010.
7 W. Plieth, Electrochemistry for Materials Science, Elsevier, 2008.
8 I. Hwang, K. Yong, ChemElectroChem 2(5) (2015) 634–653.
9 T.S. Sorenson, Surface Chemistry and Electrochemistry of Membranes, CRC Press, 1999.
10 J.-P. Diard, B. Le Gorrec, C. Montella, Journal of Electroanalytical Chemistry 584(2) (2005) 182–191.
11 J. Kamcev, Journal of Polymer Science 59(21) (2021) 2510–2520.

9

Polymer Electrolytes

Polymer electrolytes are ionically conductive because salts such as $LiClO_4$ are dissolved in solvents of polar polymers like poly(ethylene oxide) (PEO) to generate free ions, also called charge carriers. Because polymer electrolytes are in the solid state, they have many advantages over liquid electrolytes. For example, electrochemical devices using solid polymer electrolytes (SPEs) are much safer along with long-term stability in performance than those using liquid electrolytes, and additionally can be made them thinner, which are great advantages for practical applications. The safety in electrochemical energy conversion and storage devices for conventional consumer electronics (e.g. cell phones and laptop computers) and large-scale energy storage units for electric grids and vehicles has become a prioritized concern for the choice of power sources. Typical applications of polymer electrolytes include secondary batteries, sensitized solar cells, supercapacitors, and electrochromic displays.

On the other hand, challenges remain in the development of polymer electrolytes, especially (1) to increase ionic conductivity up to 10^{-4} S/cm, currently in the range of 10^{-8}–10^{-5} S/cm at room temperature, (2) to increase charge transport number, the fraction of the current transported by the specific ion, and (3) to decrease charge transfer resistance at the solid electrolyte/solid electrode interface. Therefore, many studies have been conducted to overcome those issues while at least maintaining the advantages unchanged. In this chapter, the generation of charge carriers of cations and anions by dissolving salts in polymeric solvents, the mobility of ionic species associated with the structure of polymer electrolytes, and finally ion transport behavior are introduced.

9.1 Definitions of Ion Conductivity, Mobility, and Transport Numbers

When salt dissolves in a solvent to form an electrolyte solution, it dissociates into free charge carriers – cations and anions – that make the solution ionically conductive. The ionic conductivity σ (S/cm) is proportional to the product of the

Molecule and Ion Transport through Polymer Membranes, First Edition. Yong Soo Kang.

concentration c_j (mol/cm^3) and the mobility u_j (cm^2/sec V) of each charged carrier j as described previously (Equation (8.62)) in Section 8.5:

$$\sigma = F\sum_{j}^{n}|z_j|u_jc_j = F^2\sum_{j}^{n}|z_j|^2\frac{D_j}{RT}c_j \tag{9.1}$$

The mobility u_j of each charged carrier j is related to its diffusion coefficient D_j (cm^2/sec) (Equation (8.49)):

$$u_j = D_j\frac{|z_j|F}{RT} \tag{9.2}$$

where F (C/mol) and z_j are the Faraday constant and the charge number or valence of species j, respectively. Therefore, σ can also be expressed by D_j as in Equation (9.1).

When multiple ionic species are present, each species contributes differently to the overall conductivity. To quantify the portion of the conductivity carried by a specific ionic species j to the total conductivity, we use the ion transport number T_j, given (Equation (8.63)) as:

$$T_j = \frac{\sigma_j}{\sigma} = \frac{|z_j|u_jc_j}{\sum_{k=1}^{n}|z_k|u_kc_k} \tag{9.3}$$

9.2 Formation of Polymer Electrolyte to Generate Charge Carriers

When a salt, MX, dissolves in a polymer solvent, P, containing electron-rich ligands Ls, free cations, M^+, and anions, X^-, are generated and stabilized as:

$$MX + P = [P\cdots M^+]X^- \text{ or } [P\cdots M^+]+[P\cdots X^-]$$

where $[\cdots]$ represents the interactions between free ion and polymeric chains. Typical structure of $LiPF_6$ solvated by PEO is schematically illustrated in Figure 9.1.

9.2.1 Thermodynamics for Formation of Polymer Electrolytes

The overall free energy change ΔG_m of the formation of polymer electrolytes should be negative. Therefore,

$$\Delta G_m = \Delta H_m - T\Delta S_m < 0 \tag{9.4}$$

Since the entropy change, ΔS_m, in mixing process is always positive ($\Delta S_m > 0$), the enthalpy change, ΔH_m, plays a major role in determining the value of ΔG_m, whether it is negative or positive. For instance, when ΔH_m is zero or negative, ΔG_m becomes always negative, which favors the mixing process to form

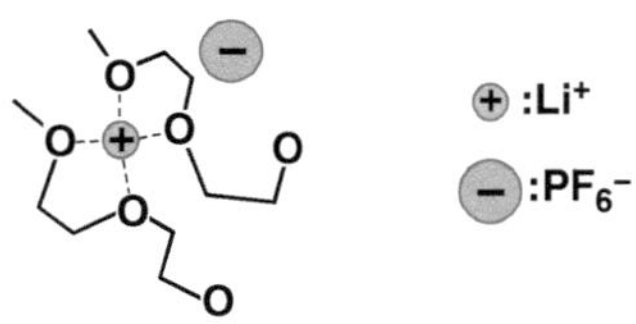

Figure 9.1 Schematic structure of polymer electrolyte containing $LiPF_6$ dissolved in poly(ethylene oxide). *Source:* Y.S. Kang.

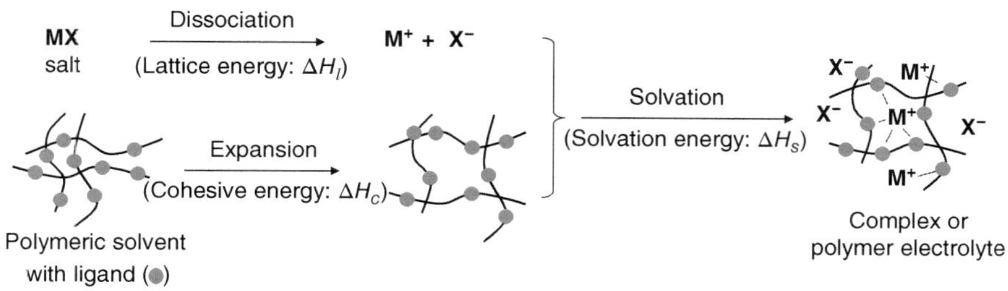

Figure 9.2 Schematic dissolution process of a salt (MX) into a polymeric solvent containing electron-rich ligands *Ls*. *Source*: Y.S. Kang.

polymer electrolytes. However, when ΔH_m is positive, whether ΔG_m is positive or negative depends on its magnitude relative to ΔS_m.

The dissolution process of salt in a polymer solvent is envisaged as three consecutive steps in a thermodynamic view: the dissociation of the salt into free ions, the conformational change or expansion of polymeric chains, and the solvation of the ions with polymer solvent as schematically drawn in Figure 9.2.

The dissociation of the salt into free cation and anion in the gaseous state requires the lattice energy, ΔH_l, the conformational change of polymeric chains needs an extra energy to overcome the cohesive energy for their expansion to accommodate ions, ΔH_c, and the solvation of the ions with a polymer solvent releases the solvation energy, ΔH_s. Thus, the total mixing enthalpy change, ΔH_m, is:

$$\Delta H_m = \Delta H_l + \Delta H_c + \Delta H_s \tag{9.5}$$

Commonly, both $\Delta H_l > 0$ and $\Delta H_c > 0$, and $\Delta H_s < 0$. Thus, ΔH_s must be negatively large enough to overcome the sum of both ΔH_l and ΔH_c to form polymer electrolyte. Accordingly, several criteria for the formation of polymer electrolytes have been put forward [1]:

1. The salt should have a small lattice energy, favorable when the anion is soft, large, and the conjugated base of a strong acid (negative effect of lattice energy of salt, $\Delta H_l > 0$).
2. The polymer should have low cohesive energy density (CED) (negative effect of the CED, $\Delta H_c > 0$).
3. The polymer should have good complexing properties, afforded by the base group for cations. Normally this requires a high concentration of electron-rich ligands such as oxygen, nitrogen, and sulfur atoms (positive effect of the solvation or complexation energy, $\Delta H_s < 0$).
4. The polymer chain should be flexible both to ensure effective solvation of cations and to provide favorable solvation entropy (positive effect of the solvation energy, $\Delta H_s < 0$).

The following sections describe the three energies required to dissolve a salt into a polymer solvent: the lattice energy of the salt, the cohesive energy of the polymer solvent, and the solvation energy of the ion into the polymer solvent.

9.2.1.1 Lattice Energy of Salt, ΔH_l

An ionic compound or salt is stable because of the electrostatic attraction between its positive and negative ions. ΔH_l is defined as the energy required to separate one mole of the solid salt into its component free ions in vapor state and is thus always positive ($\Delta H_l > 0$).

ΔH_l can be readily obtained from Equation (9.6), which is derived from the Coulomb's law, because the electrostatic interactions between M^+ and X^- in the vapor state is negligibly small due to $r \gg r_0$ as shown in Figure 9.3:

$$\Delta H_l = C \cdot z^+ \cdot z^- / r_0 \tag{9.6}$$

in which C is a constant that depends on the type of crystal structure; z^+ and z^- are the charges on the ions, and r_0 is the interionic distance between two charged ions in crystal state (the sum of the radii of the positive and negative ions). Representative ion sizes are listed in Table 9.1 [Ionic radius, Wikipedia]. Thus, the lattice energy of an ionic crystal decreases rapidly as the charges of the ions decrease and the size of the ions increases, which favors the formation of polymer electrolytes. For instance, ΔH_l = 787 and 648 (kJ/mol) for NaCl and $NaClO_4$, respectively [Wikipedia].

9.2.1.2 Cohesive Energy of Polymeric Solvents, ΔH_c

The cohesive energy ΔH_c is the interaction energy per unit mol of the polymeric molecules in the pure state, and consequently is the energy of all the interactions among polymeric molecules and their neighbors that have to be disrupted to remove the molecule from the pure state [2]. Therefore, the cohesive energy of polymeric chains is the minimum energy necessary to expand the polymeric chains large enough to accommodate ionic species as shown in Figure 9.4.

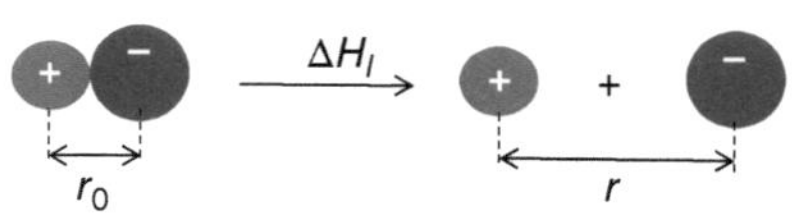

Figure 9.3 A salt MX in ambient condition is dissociated into M^+ and X^- in the vapor state, where $r \gg r_0$. *Source*: Y.S. Kang.

Table 9.1 Size of representative cations and anions [Wikipedia].

Cation	**H^+**	**Li^+**	**Na^+**	**K^+**	**Ca^{++}**	**Mg^{++}**	**Fe^{++}**	**Zn^{++}**	**Al^{+++}**
Radius (nm)		0.090	0.116	0.152	0.114	0.086	0.075	0.088	0.067
Anion	**OH^-**	**F^-**	**Cl^-**	**$CH_3CO_2^-$**	**PO_4^{3-}**	**SO_4^{2-}**	**CO_3^2**	**NO_3^-**	**ClO_4^-**
Radius (nm)	0.14	0.133	0.181	0.276	0.238	0.230	0.178	0.179	0.240

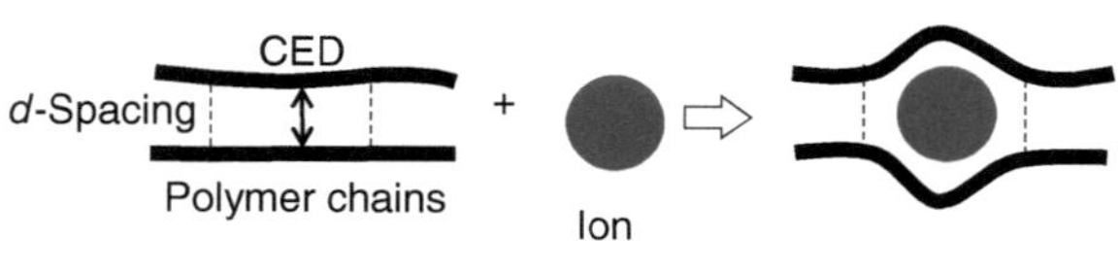

Figure 9.4 Schematic volume expansion against cohesive energy density (CED) required for accommodating ion between polymeric chains. *Source*: Y.S. Kang.

The cohesive energy ΔH_c of polymeric chains can be obtained from the energy of vaporization ΔE_v (J/mol). Here, ΔE_v cannot be directly measured, but readily estimated from $\Delta E_v = \delta^2 V$ where δ $(J/cm^3)^{1/2}$ and V (cm^3/mol) are the Hildebrand solubility parameter and the molar volume of the polymer repeating unit, respectively. The δ values are available in many textbooks and literatures, and can also be readily estimated from the chemical structure of the repeating unit of any polymer by the molar contribution method [3]. The δ values for poly(epichlorohydrine), PEO, and polyethylene are 26.8, 16.8, and 9.2 $(kJ/mol)^{1/2}$, respectively.

9.2.1.3 Solvation Energy of Ions with Solvents, ΔH_s

The solvation enthalpy ΔH_s of a salt in a polymer solvent depends on the cation–polymer and also anion–polymer interactions, and therefore it strongly influences the solubility of salts in polymeric solvents. The solubility can be frequently discussed in terms of the acid–base interactions between polymeric solvent and ions. Here, the solvent is commonly classified as *hard* or *soft,* based on the fact that the strongest interactions occur by matching *hard* with *hard* or *soft* with *soft*, called *hard–soft* acid and base principle (*HSAB* principle).

Hard Lewis acids are characterized by small ionic radii, high positive charge density, strongly solvated, empty orbitals in the valence shell, and high energy LUMOs (alkali, alkali earth ions), whereas soft Lewis acids are characterized by large ionic radii, low positive charge density, completely filled atomic orbitals, and low energy LUMOs (partially filled "*d*" orbitals and to less extent "*f*" orbitals) [4]. Hard Lewis bases are characterized by small ionic radii, strongly solvated, highly electronegative, weakly polarizable, and high energy HOMOs (oxygen in ether, thio group in thioether), whereas soft Lewis bases are characterized by large ionic radii, intermediate electronegativity, highly polarizable, and low energy HOMOs. It is worthy to note that H^+, Li^+, Na^+, and K^+ are strong acids, Fe^{2+}, Co^{2+}, Ni^{2+}, Cu^{2+}, and Zn^{2+} are intermediate, and Cu^+ and Ag^+ are weak acids, whereas H_2O, OH^-, F^-, Cl^-, ROH, R_2O, and RNH_2 are strong bases, aniline, pyridine, Br^-, NO_2^-, and SO_3^{2-} are intermediate, and RSH, RS^-, R_2S, and I^- are weak bases [4].

The large electronegativity differences between hard acids and hard bases give rise to strong ionic interactions, whereas the electronegativities of soft acids and soft bases are almost the same and hence have less ionic interactions. i.e. the interactions between them are more covalent. The interactions between hard acid-soft base or soft acid-hard base are mostly polar covalent and tend to be more reactive or less stable [4].

For neat polymer electrolytes without any additives, the repeating unit of the polymeric solvent commonly contains Lewis bases. For instance, a very common polymeric solvent is PEO, which has the strong Lewis base repeating group of ethylene oxide. Thus, the strongest solvation in PEO would be possible with a hard cation, e.g. Li^+, Na^+, Mg^{2+}, and Ca^{2+}. The best electron donors (bases) for hard Lewis acids or cations follow the relative values of the negative charge on the heteroatom [4]:

$$-O- > -NH- \gg -S-$$

For soft cations such as Ag^+; however, the relative stability is found to be:

$$—NH— > —S— > —O—$$

which is the inverse order to the hard cations, such as Li^+.

It has been suggested that Mg^{2+} (hard), for example, would be expected to form very stable complexes with PEO (hard), whereas the interaction of Hg^{2+} (soft) with PEO would be weak. Complexes of PEO with both cations of Mg^{2+} and Hg^{2+} are readily formed, whereas Mg^{2+} ions are immobile but Hg^{2+} ions are mobile, which highlights a possible inverse relationship between the promoting complex formation and the consequential effects on cation mobility or conductivity [5].

In polar solvents, such as water, alcohols, or H-bonding solvents, the anions are solvated and thereby stabilized by the hydrogen bonds. With less polar solvents such as THF, acetonitrile, or polymer solvent, the stability of the anion depends strongly on the charge dispersion. Large anions with delocalized charge require little solvation. These can be either soft (e.g. I^-) or hard (e.g. $CF_3SO_3^-$) bases. Thus, the most appropriate anions for polyether-type polymer electrolyte formation would be expected to follow the order [4]:

$$F^- \ll Cl^- < Br^- < I^- \sim SCN^- < ClO_4^- \sim CF_3SO_3^- < BF_4^- \sim AsF_6^-$$

This order parallels to that of the lattice energy of the appropriate salts. In this regard, the most suitable choices of anion for aprotic, low dielectric dipolar polymer-based polymer electrolytes are large anions with delocalized charge and very weak bases possessing low ion–dipole stabilization energies. In addition, their lattice energies are relatively low and they have little tendency to from tight ion pairs. These particular anions may be either soft (I^-) or hard (ClO_4^-) bases [4].

An interesting quantitative approach to evaluate the solvation energy ΔH_s (J/mol) is to utilize the salt/neutral polymer molecule interactions, which can be in turn estimated from three parameters: Donor number (DN), acceptor number (AN), and entropy term as listed in Table 9.2 [1, 6]. DN measures the ability of the solvent to donate electrons to solvate the cation, considered as a Lewis acid. On the other hand, AN quantifies the possibility of anion solvation by accepting electrons. The entropy term depends on the optimal spatial disposition of the solvating units. For example, 1,2-dimethyoxyethane and PEO, having the ethylene oxide unit, show the

Table 9.2 Donor (DN) and acceptor (AN) numbers of solvents [1, 6].

Solvents	DN	AN
Acetonitrile	14.1	19.9
Propylene carbonate	15.5	19.3
Methanol	19.1	41.5
1,2-dimethyoxyethane	22.0	10.2
THF	20.0	9.0
Water	16.4	54.8

most favorable spacing in addition to the highest DN, and thus PEO is most commonly used a polymeric solvent [1].

Ethers, such as 1,2-dimethyoxyethane are quite strong electron donors with DN of 20. They are, however, very poor acceptor with AN of 10.2 as they lack H-bonding ability for anion solvation. Therefore, salts complexing with PEO are those with a large delocalized anion, requiring little solvation, typical example being ClO_4^- and I^-.

9.2.2 Generation of Free Charge Carriers

Since ionic conductivity σ is determined by the product of the concentration c_j of free ions and their mobility u_j as described earlier (Equation (9.1)), increasing c_j enhances σ with a given u_j. Therefore, achieving a high concentration c_j of free ions is crucial for improving ionic conductivity σ by optimizing the choices of salts and polymer solvents [1].

9.2.2.1 Effects of Salts

When MX is dissolved in a polymeric solvent, the concentration c_j of free ionic species j is determined by the equilibrium constant, K as:

$$MX \leftrightarrow M^+ + X^- \tag{9.7}$$

with $K = [M^+][X^-]/[MX]$. When MX is highly soluble with large K, the salt added into a solvent readily dissolves intimately and therefore the free ion concentration c_j ($c_j = [M^+]$) becomes nearly equivalent to the salt concentration added, c_0:

$$c_j \cong c_0 \tag{9.8}$$

Therefore, with large K, Equation (9.8) will yield $\sigma \sim c_0$.

For weak dissociation with small K and $c_j \ll c_0$,

$$K = \frac{c_j^2}{c_0 - c_j} \cong \frac{c_j^2}{c_0} \tag{9.9}$$

Therefore,

$$c_j = \sqrt{Kc_0} \tag{9.10}$$

In the case of small K, we expect the ionic conductivity $\sigma \sim c_0^{1/2}$ from Equation (9.10). This suggests that the concentration of the charged species directly depends on the interaction between the ionic species and the polar ligands in the polymer chains.

When salt MX is dissolved in a solvent, only a portion of the added salt dissolves to form free ions (Figure 9.5a), but the rest remains undissolved. In this case, the remainder exists partially as contacting ion pairs or crystals, as shown in Figure 9.5b and c, respectively. In other words, MX can have three different states in solvent: free ions, contact ion pairs, and ion aggregate containing ion pairs and/or crystals. Among the three states in a solvent, the free ions mostly contribute to the ionic conductivity, while the contact ion pairs may also contribute to the ionic conductance to some extent [7].

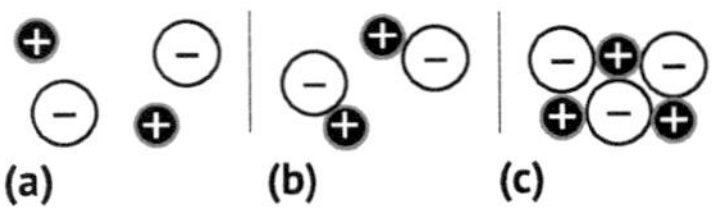

Figure 9.5 Three different states of a salt (MX) in solvent: (a) free ions, (b) contact ion pairs, and (c) crystal. *Source*: Y.S. Kang.

Forming polymer electrolytes via salt complexation or dissolution, it is preferable that the anion X^- would be a weak Lewis base (that is, its conjugate acid HX to be very strong) to increase the concentration of the free ionic species j, c_j. Otherwise, the complexed X^- ion will have a tendency to form ion pair with M^+, and such pairing is not favorable for conductivity [1].

9.2.2.2 Effects of Polymer Solvents

As discussed previously, in order to get high concentration of free ions, polymeric solvent has to have high interactions with free ions and high flexibility for increasing the solvation energy ΔH_S to get high energy of mixing ΔH_m. A polymer solvent that is capable of strong coordinative interactions with cations is necessary for electrolyte formation. Common polymer solvents to form polymer electrolytes contain a Lewis base ligands (Ls) such as O in PEO, poly(propylene oxide) (PPO), and poly(ethylene succinate), N in poly(ethylene imine), and S in poly(ethylene sulfide). Thus, polyethers, polyesters, polyimines, and polythiols may dissolve salts easily to form polymer electrolytes, whereas polyethylene, for example, having no coordinating group, does not.

The majority of polymer electrolyte studies so far concentrates on PEO, a polyether, and related polymers as a polymeric solvent for several salt species. The chemical structure of polyethers is $-[-(CH_2)_m-O-]_n-$, where m is 1, 2, 3, and 4 for poly(methylene oxide) (PMO), PEO, poly(trimethylene oxide), poly(tetramethylene oxide), respectively, along with poly(propylene oxide) (PPO) $-[-CH(CH_3)CH_2-O-]n-$. Polyethers are soft except PMO ($T_g = 180\,°C$) with $m = 1$, which is also called acetal resin as an engineering plastic.

Interestingly, PEO has regularity of the repeating unit, which allows a high degree of crystallinity of ~70–85%. It adopts a 7/2 helical configuration, that is, seven ethylene oxide repeat units with two turns in the fiber period of 1.93 nm. The melting point of the crystalline phase of PEO is $T_m = 65\,°C$ and the glass transition temperature of the amorphous phase is $T_g \sim -60\,°C$. The dipole–dipole interactions are probably responsible for a higher value of T_g compared with polyethylene ($-100\,°C$). However, the dielectric constant is still low (~5–8), and this strongly influences the behavior of PEO-based electrolytes.

Small ions such as Li^+ and Na^+ were suggested to lie inside the helix with oxygens as shown for the top view of Figure 9.6a, along with that of PMO in Figure 9.6b.

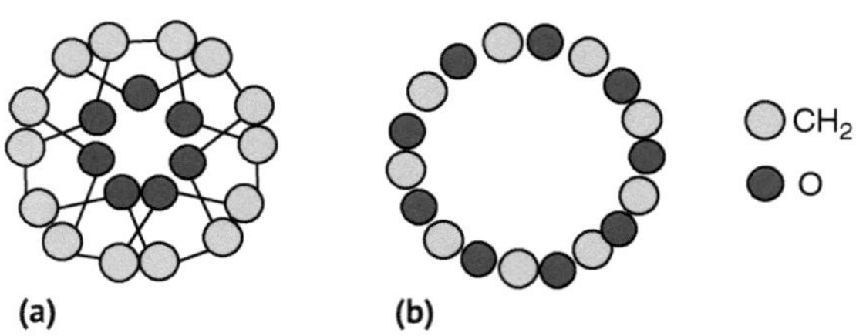

Figure 9.6 Schematic top view models of (a) poly(ethylene oxide) (PEO) and (b) poly(methylene oxide) (PMO). *Source*: Redrawn from [5, 8].

Some oversimplified ideas of Li^+ ion motion are developed, i.e. lithium cations move down through channels inside PEO helices with no anion mobility.

It has been reported that many MX/PEO complexes are partly crystalline and effective conduction occurs only though the amorphous elastomeric phase, but not through crystalline phase. Therefore, both crystalline PEO and salt-rich crystalline complex in the bulk sample inhibit conductivity [1]. This general behavior is not limited to alkali salts in PEO.

9.3 Polymeric Chain Mobility and Glass Transition Temperature

The chain mobility is another critical factor in determining the ionic conductivity in addition to the concentration of the free charge carrier because the ionic conductivity σ is linearly proportional to the product of the concentration c_j of free charged species j and the ionic mobility u_j as described previously (Equation (9.1)). It is understood that the polymeric chain mobility is directly related with the ionic mobility u_j according to the free volume theory.

Therefore, polymers having high chain mobility are required to achieve high ionic conductivity. One simple parameter to represent the chain mobility is the glass transition temperature T_g. In other words, T_g is directly related with the chain mobility: the lower T_g and the higher the chain mobility is, and vice versa. Therefore, low T_g polymers are required to achieve the high ionic conductivity σ. Table 9.3 shows T_gs of common polymers.

Upon the dissolution of salts into a polymer, cations such as Li^+ and Na^+ are coordinated by electron-rich ligands such as oxygen and nitrogen atoms in polymeric chains. When the coordination number (CN) of the dissolved cations is CN $\geq$ 2, effective cross-linking of polymeric chains would occur as shown in Figure 9.7. Therefore, the cross-linking density is directly related with the concentration of the charged species and consequently decreases the chain mobility, resulting in the lower conductivity.

Table 9.3 Glass transition temperatures of common polymers [Wikipedia].

Polymer	T_g (°C)	Polymer	T_g (°C)	Polymer	T_g (°C)
Poly(dimethyl siloxane)	−120	Poly(phosphazene) [9, 10]	−40	Poly(vinyl alcohol)	85
Poly(ethylene oxide)	−67	Poly(epichlorohydrin) [11]	−49	—	—

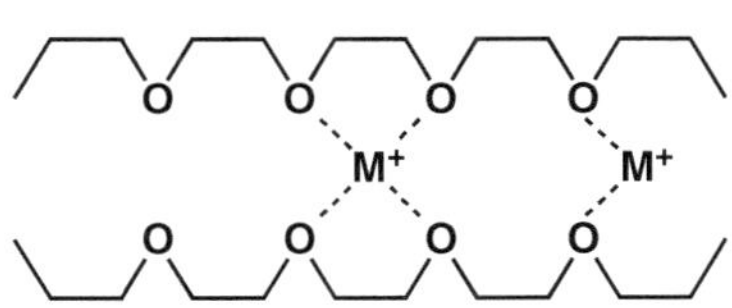

Figure 9.7 Schematic structure of cross-linking polymer chains by metal ion (M^+) coordination. *Source*: Y.S. Kang.

For instance, when silver salts AgX with X being ClO_4, BF_4, CF_3SO_3, and NO_3 are dissolved in poly(2-ethyl oxazoline) (POZ) to form silver polymer electrolytes, polymeric chains will be effectively cross-linked by Ag^+ when CN ≥ 2 as shown in Figure 9.8a.

In addition, heavy metallic ions will be dangled on polymeric chains with the coordination number of CN $= 1$ (Figure 9.8b), making polymeric chains less mobile to some extent. Therefore, the chain mobility will be decreased upon incorporation of silver salt because of the cross-linking and dangling effects of Ag^+.

As a measure of the chain mobility, glass transition temperature T_g was measured and illustrated in Figure 9.9. The increase in T_g of silver polymer electrolyte as a function of the silver salt concentration suggests a decrease in the polymeric chain mobility, mostly because of the transient cross-links.

The glass transition temperature increases with the mole fraction of the silver salt added at low salt concentrations, and reaches the maximum, followed by the decrease at higher salt concentrations. These results suggest that the POZ chains become stiffer because of the coordinative interactions between Ag^+ and the carbonyl oxygen atoms of the POZ chains as shown in Figure 9.8a. The increase in the glass transition temperature ΔT_g is nearly up to 40 °C, suggesting high degree of cross-linking. The change behavior in T_g by varying anion X is also noticeable

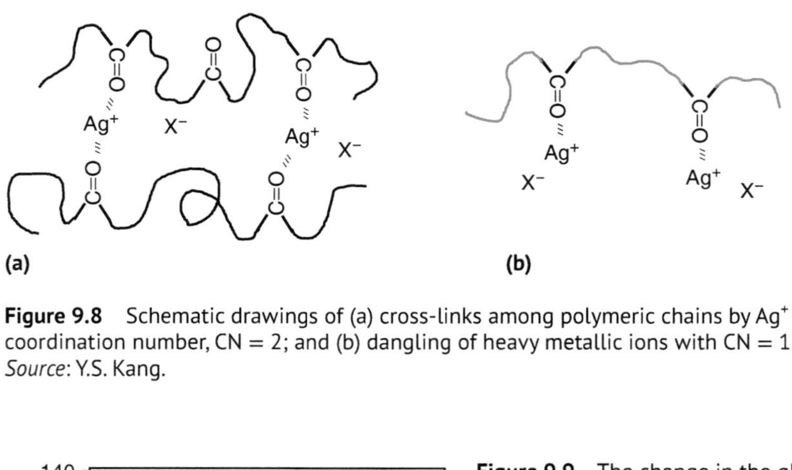

Figure 9.8 Schematic drawings of (a) cross-links among polymeric chains by Ag^+ with the coordination number, CN = 2; and (b) dangling of heavy metallic ions with CN = 1. *Source*: Y.S. Kang.

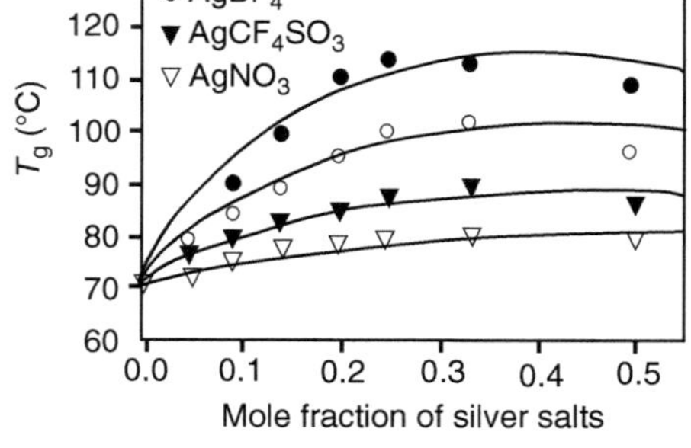

Figure 9.9 The change in the glass transition temperature T_g of silver polymer electrolytes containing various silver salts dissolved in poly(2-ethyl oxazoline) (POZ) and the solid lines calculated from the theoretical model (Equation (9.13)) based on the conformational entropy. *Source*: Replotted from [12–14].

depending on the types of AgX. The change is the highest for $AgClO_4$/POZ complex whereas the lowest for $AgNO_3$/POZ complex. These suggest that the coordinative interactions between Ag^+ and carbonyl oxygen are the strongest for $AgClO_4$/POZ and the weakest for $AgNO_3$/POZ complex in this experimental range [12].

In order to understand the change in the glass transition temperature T_g more quantitatively, mathematical models have been developed based on the conformational entropy theory. According to the viewpoint of Di Marzio and Gibbs, the glass formation is considered as a result of the system's loss of conformational entropy, S_c, where S_c is expressed as [15]:

$$S_c = S^{\text{liquid}} - S^{\text{glass}} \tag{9.11}$$

where S^{liquid} and S^{glass} are the conformational entropies of the liquid and glass states, respectively. Following Chow's assumption that $S^{\text{glass}}(n_1, 0, T) = S^{\text{glass}}(n_1, n_2, T) = 0$ for a mixture of 1 and 2 at glassy state, where n_1 and n_2 are numbers of species 1 and 2, respectively, the change in the glass transition temperature of polymer 1 from T_{g1} to T_{g12} for mixture of polymer 1 and diluent 2 can be described by [16]

$$\ln\left(\frac{T_{g12}}{T_{g1}}\right) = -\frac{1}{\Delta C_p}[S_c(n_1, n_2, T) - S_c(n_1, 0, T)] \tag{9.12}$$

where ΔC_p is the change in the heat capacity between the supercooled liquid and the glass, and assumed to be independent of temperature and composition. This concept was applied for polymer electrolyte containing metal salts 2 dissolved in a polymeric matrix 1, resulting in as [12]:

$$\ln\left(\frac{T_{g12}}{T_{g1}}\right) = \beta_{12}\left[\left(\frac{\phi_1}{x_1}\ln\phi_1 + \frac{\phi_2}{x_2}\ln\phi_2\right)\left(1 - \gamma_{\text{com}}\ln\left(\frac{z-1}{e}\right)\right)\right] \tag{9.13}$$

where ϕ_i and x_i are the volume fraction and the degree of polymerization of component i, respectively. Thus, $x_2 = 1$. $\beta_{12} = zR/M_u\Delta C_p$ with z, R, and M_u being the coordination number, the gas constant, and the molecular weight of repeating unit, respectively. γ_{com} is an adjustable parameter in Equation (9.13), representing the complexation interactions between polymer 1 and ions 2. The value of β_{12} for POZ/AgX was 0.68. The agreement between the experimental data and the theoretical predictions with $z = 12$, solid lines in Figure 9.9, is exceptional with one adjustable parameter γ_{com} [12].

However, in developing Equation (9.13), only polymer–salt interactions were considered, but the cation–anion interactions that are significant at high concentrations of silver salts were not. In this regard, the conformational entropy theory was extended to account for the cation–anion interaction through adopting the extended Debye–Hueckel theory modified by Guggenheim as [17]:

$$\ln\frac{T_{g12}}{T_{g1}} = \beta_{12}\left[\left(\frac{\phi_1}{x_1}\ln\phi_1 + \frac{\phi_2}{x_2}\ln\phi_2\right)\left(1 - \gamma_{\text{com}}\ln\left(\frac{z-1}{e}\right)\right) - \frac{\phi_2}{x_2\nu m}\frac{4}{3}A_{\text{diss}}I^{\frac{3}{2}}\tau(I^{\frac{1}{2}})\right] \tag{9.14}$$

Table 9.4 Interaction parameters γ_{com} between POZ/silver salts and A_{diss} between cation and anion [17].

Salt	$AgClO_4$	$AgBF_4$	$AgCF_3SO_3$	$AgNO_3$
γ_{com}	1.100	0.978	0.920	0.866
A_{diss}	0.156	0.103	0.212	0.272

Here, the ionic strength $I = 1/2m\nu|z_M z_X|$, where m is a solute molality (mol/kg polymer). ν ($= \nu_M + \nu_X$ where ν_M and ν_X are the number of M and X ions, respectively) is the number of ions per salt, and z_M and z_X are the valences of the M and X ions, respectively. $\tau(z) = 3/z^3[\ln(1+z) - z + z^2/2]$. A_{diss} is the Debye–Hueckel coefficient [17]. The theoretical predictions based on Equation (9.14) with two interaction parameters γ_{com} and A_{diss} as listed in Table 9.4 was improved to some extent relative to those based on Equation (9.13) [17].

Additionally, Equation (9.13) for binary system (polymer and salt) was successfully extended for two ternary systems of polymer blend with salt and polymer, diluent and salt [12].

9.4 Mechanism of Ionic Transport

Understanding the ion conduction mechanism in polymer electrolytes is essential for designing high-conductivity electrolytes that enable high performance. Ion transport in polymer electrolytes occurs even at sub-glass transition temperatures via the movement of polymer segments, primarily through crank-shaft motions around mainchain C—C and C—O bonds, as described in Section 2.4 (Figure 2.13a and b), although ion conductivity in glassy state is much lower than that in rubbery state. These segmental motions create free volume, facilitating ion diffusion within the polymer matrix [4]. Thus, ion transport in polymer electrolytes is generally viewed as a process where ion mobility is coupled with local polymer segment motion. However, cases of "decoupling" have also been observed, where high ionic conductivity is achieved through different mechanisms that will be discussed subsequently.

Figure 9.10 schematically illustrates various types of ion transport through polymers in molecular scale [4]. Figure 9.10a shows how a cation moves between coordinating sites on one chain (intra-chain hopping) or between sites on neighboring chains (inter-chain hopping), promoted by the segmental motion of the chains themselves. Figure 9.10b shows again two examples of how ionic clusters, including ion pairs and triple ions, may contribute to the ionic conduction via intra-chain hopping and inter-cluster hopping. Under an electrical field, continuous reversible coordination/decoordination behaviors result in long-range ion transport. Here, the polymer chains act solely as anchor points for ion conduction. All these mechanisms are feasible and may depend, to a large extent, on the concentration of salt in the matrix [4].

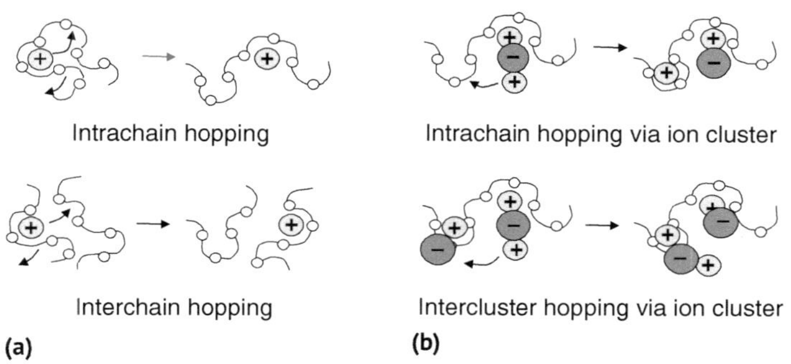

Figure 9.10 Representation of cation motion in a polymer electrolyte (a) assisted by polymer chain motion only; (b) taking account of ionic cluster contribution. *Source*: Adapted from [4].

An important unresolved area in the field of the ionic conduction through polymer electrolytes concerns the role of the anion in ion conductivity and the extent of ion association in these systems. It is generally accepted that the net anion mobility is small. However, in the case of PEO-based electrolytes, the Li^+ transport number T_{Li^+}, the fraction of the current transported by the Li^+ ion, was observed to be in the range of T_{Li^+}~0.3, indicating that small Li^+ ions in PEO move more slowly than much larger anions, such as PF_6^- or bis-(trifluoromethane)sulfonamide [Tf_2N]. This characteristic is ascribed to strong Li^+ coordination by several oxygen atoms of the PEO backbone [18, 19] as shown in Figure 9.1, which provides good ion solubility but restricts Li^+ mobility to some extent [20]. Based on these concepts on a molecular scale, four different mathematical models will be introduced hereafter: Free volume model, conformational entropy model, dynamic bond percolation (DBP) model, and Anderson and Stuart model.

9.4.1 Free Volume Model

The free volume approach [4] is one of the simplest ways of understanding polymer dynamics, which assumes that motion in liquid-like systems is not a thermally activated process but occurs as a result of the redistribution of free volume within the system by Cohen and Turnbull [21]. As the temperature increases, local voids or free volumes are produced by material expansion, allowing species such as polymer segments, or free ions to move into free volume, which is conceptually equivalent to the free volume theory for neutral molecules as described in Section 3.5 of Chapter 3. Cohen and Turnbull derived for diffusion coefficient with the hard sphere model [4], which is written again here:

$$D = D_d \exp\left(-\frac{\gamma v^*}{\langle v_f \rangle}\right) \tag{9.15}$$

where γ is a constant that allows for overlap of the free volume ($0.5 \le \gamma \le 1.0$), v^* is the minimum size of a free volume required for transport, and $\langle v_f \rangle$ is an average free volume per molecule or per repeating unit in the case of polymers ($\langle v_f \rangle = V_f/N$ with V_f being the total specific free volume and N being the number of molecules or repeating units in 1 g of polymer).

Since the free volume increases with temperature, it can be mathematically expressed around temperature at which the free volume disappears as [1]:

$$\langle v_f \rangle = \alpha_f v_m (T - T_0) \tag{9.16}$$

where α_f is the thermal expansion coefficient of free volume, which is equated with the difference in the thermal expansion coefficient in rubbery and glassy states, v_m is the mean molecular volume over the range ($T \sim T_0$), and T_0 is the equilibrium glass transition temperature. This implies that the free volume vanishes at $T = T_0$. Substituting Equation (9.16) into Equation (9.15) yields both the Vogel–Tamman–Fulcher (VTF) equation and Williams-Landal-Ferry (WLF) equation, highlighting their consistency [1].

Since diffusion processes have been known to be a thermally activated process, Miyamoto and Shibayama [22], Duda and Vrentas [23], for example, proposed a mathematical expression for diffusion coefficient D with an activation energy E^* and free volume $\langle v_f \rangle$ as:

$$D(0) = D_0 \exp\left(-\frac{\gamma v^*}{\langle v_f \rangle} \right) \exp\left(-\frac{E^*}{RT} \right) \tag{9.17}$$

In polymer electrolytes, ions are incorporated into a polymeric solvent, which may contribute to improving ion transport. To address this behavior, the original free volume theory has been extended to consider the salt dissociation into ions by the earlier work of Miyamoto and Shibayama [22], and by Cheradame [24]. This extended theory accounts for both ion-pairing (salt dissociation) and Arrhenius-type activation of the displacement:

$$\sigma = \sigma_0 \exp\left(-\frac{\gamma v^*}{\langle v_f \rangle} - \frac{E^*}{RT} \right) = \sigma_0 \exp\left(-\frac{\gamma v^*}{\langle v_f \rangle} - \frac{E_J + W/2\varepsilon}{RT} \right) \tag{9.18}$$

where the pre-exponential factor σ_0 depends on the electric field applied, and the apparent activation energy E^* for ion conduction is taken as

$$E^* = E_j + \frac{W}{2\varepsilon} \tag{9.19}$$

Here, E_j is the potential barrier for diffusional jump, W is the ionic dissociation energy of the salt, and ε is the dielectric constant. The introduction of the Arrhenius form in Equation (9.18) is one way of generalizing the free volume picture of the polymer solvent to account for the motion of the ionic guest.

Figure 9.11a is a plot of $\log(\sigma)$ as a function of $1/T$, demonstrating two different linear plots across the glass transition temperature of the polymer used. Interestingly, the activation energy calculated from the Arrhenius equation in the rubbery state is higher than that in the glassy state. Figure 9.11b shows the linear relationship between $\log(\sigma)+(E^*/2.303\ kT)$ and $1/\langle v_f \rangle$, demonstrating the validity of Equation (9.18) [22].

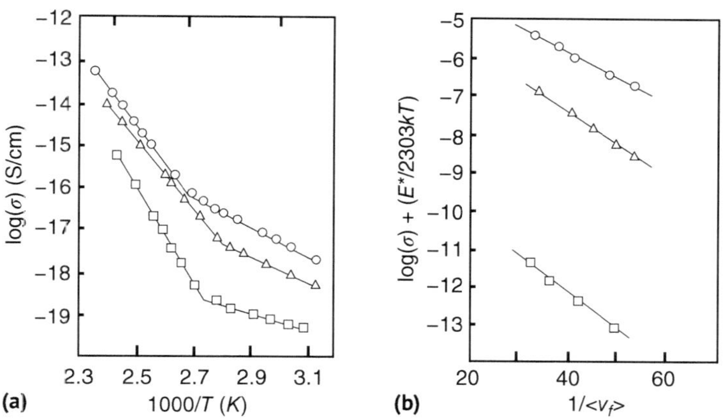

Figure 9.11 (a) Arrhenius plots of electrical conductivity; (b) $\log(\sigma)+(E^*/2.303\ kT)$ vs $1/\langle v_f \rangle$ for poly(methyl methacrylate) (○), unsaturated polyester (△) and polystyrene (□) where E^* is the apparent activation energy in the glassy state and $E^* = E_j + W/2\varepsilon$. *Source*: Ref. [22]/AIP Publishing.

The major weakness of the free volume approach is that it ignores the kinetic effects associated with macromolecular chain motion. In addition, the model does not relate directly to a macroscopic picture and therefore does not predict straightforwardly how such variables as ion size, polarizability, ion pairing, solvation strength, ion concentration, polymer structure, or chain length will affect the conduction process [4].

9.4.2 Conformational Entropy Model

An alternative approach to model polymer dynamics is based on the conformational entropy model [4] of Gibbs et al. [25, 26] and goes some way to overcoming the free volume model deficiencies. Again, conformational entropy model describes only the properties of the polymer, but the approach is preferable since it is based on entropy, not volume, with transport modeled on group cooperative rearrangements of the polymer chains rather than a void-to-void jumping mechanism. The average probability of a mass-transporting cooperative rearrangement P is given as [1]:

$$P = A\exp\left[-\frac{\Delta G}{kT}\frac{S_c^*}{S_c}\right] \tag{9.20}$$

where ΔG is the free energy barrier per mole that opposes the rearrangement and S_c is the conformational entropy at temperature, T. s_c^* is the minimum conformational entropy required for rearrangement and is often taken as $s_c^* = k\ln 2$. Using the thermodynamic definition of entropy, we can write [1]:

$$S_c(T) - S_c(T_0) = \int_{T_0}^{T} \frac{\Delta C_p}{T}\mathrm{d}T \tag{9.21}$$

with ΔC_p the heat capacity difference between liquid and glass. Note that $S_c(T_0)$ vanishes since T_0 is the equilibrium T_g. When we assume that ΔC_p is independent of T, it follows:

$$P = A\exp\left[-\frac{\Delta G}{kT}\frac{s_c^*}{\Delta C_p \ln(T/T_0)}\right] \tag{9.22}$$

As the glass transition temperature approaches, molecular relaxation times increase, preventing the system from maintaining equilibrium. Eventually, the dynamic conformational entropy reaches zero at a temperature T_0. Equation (9.22) serves as the basis for deriving both the WLF equation, by relating P to the inverse relaxation time, and the VTF equation, using $\ln(T/T_0) \sim (T - T_0)/T_0$ and $T/T_0 \sim 1$, thus confirming its validity [1].

The model shows promise, as analyses of various polymer electrolyte systems yield reasonable activation energy values and $(T_g - T_0)$ values close to 50 K. Unlike the simpler free volume theory, the conformational entropy model is based on realistic considerations of relaxation processes, state availability, and incorporates kinetic data. When kinetic effects are included in free volume-based transport models, the results closely align with the conformational entropy model [4, 27]. The success of free volume concepts in polymer electrolytes may stem from the strong correlation between volume and enthalpy or entropy [27].

9.4.3 Dynamic Bond Percolation Model

Mobility of ions in polymer electrolytes is strongly dependent on segmental motion of the polymer solvent. Based on this, and assuming a weak dependence of the conductivity on interionic interactions, Ratner and Shriverz [28, 29] proposed a microscopic model to describe long-range ion transport in polymer electrolytes, known as the DBP theory. Conductivity is visualized as being due to a combination of ion/polymer cooperative motion with the occasional independent ion movement; the timescale for the latter is much shorter than that for polymer relaxation.

Cation and anion motions are considered to be fundamentally different. The former is visualized as the making and breaking of coordinate bonds with motion between coordinating sites, while anion motion is regarded as a hopping between an occupied site and a void that is large enough to contain the ion. Such an assumption is logical, considering the largely different interactions of the polymer with anions (weak dipolar) and cations (strong solvation) [4].

Pulsed-field gradient (PFG)-NMR measurements on *a*PEO-$LiPF_6$ have shown that, for this system at least, the anion and cation diffusion have different mechanisms. Diffusion of the Li^+- and PF_6^--based species has different temperature dependencies, at least at high salt concentrations. Figure 9.12 shows the temperature variation of the diffusion coefficients of ^{31}P for the anion and ^{7}Li for the cation. The ^{31}P diffusion coefficients follow a VTF-type dependence at all concentrations and are always significantly faster than those of Li^+-based species. Not only is the anion more mobile than the cation, it must also be much more mobile than any ion pairs. Furthermore, if ion pairs have significant mobility, they may be more mobile

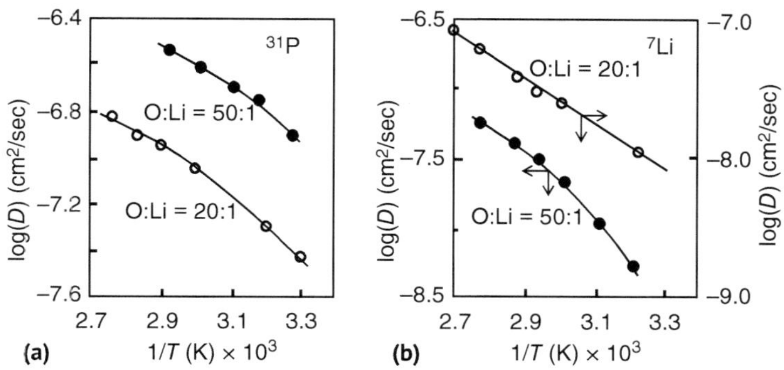

Figure 9.12 Diffusion coefficients of (a) ^{31}P and (b) ^{7}Li as a function of temperature for PEO-LiPF$_6$ electrolytes with an O:Li ratio of (●) 50:1 and (○) 20:1. *Source*: Replotted from [4, 30].

than the Li$^+$ species. Anions do not form strong bonds with any of the polymer hosts commonly studied, so their transport is likely to depend principally on the rate of polymer rearrangements. Such a mechanism may be described in terms of conformational entropy or free volume theories, both of which predict a VTF-like temperature dependence [4].

In contrast to ^{31}P diffusion, ^{7}Li diffusion shows a change in the ion diffusion mechanism from a process controlled by Vogel–Tamman–Fulcher (VTF) kinetics to a thermally activated mechanism as the salt concentration is increased. The diffusion coefficients in the high concentration region are not greatly affected by concentration, unlike with ^{31}P diffusion. Thus, a mechanism has been proposed [4, 31], which involves ions hopping between clusters such as:

$$M^+X^-M^+ + X^-M^+ \leftrightarrow M^+X^- + M^+X^-M^+ \tag{9.23}$$

Cation transport at high salt concentration appears to be largely independent of polymer segmental mobility, and therefore it seems likely that the ionic conductivity is an average of two distinct processes.

The DBP theory provides the simplest model that includes information on local mechanistic processes, and involves ion hopping between sites on a continually renewing lattice (not static as for a solid electrolyte), with the ions obeying a hopping-type equation:

$$P_i = \sum_j \left[P_j W_{ji} - P_i W_{ij}\right] \tag{9.24}$$

P_i is the probability of finding a mobile ion at site i, W_{ij} is the probability per unit time that an ion will hop from site i to site j and is equal to zero except between neighboring sites. It is then suggested that W_{ij} can be related to system parameters such as ion size, free volume, temperature, and pressure [4].

9.4.4 Anderson and Stuart Model

Many experimental results demonstrate a significant decoupling of ion transport from segmental relaxation in some polymer electrolytes [32–34] particularly for temperatures lower than their glass transition temperatures. Anderson and Stuart proposed a mathematical model to describe the activation energy E_σ for ion transport in superionic glasses including poly(ionic liquid)s to incorporate the electrostatic interactions of mobile ion with matrix ion and elastic force [35] as:

$$E_\sigma = \frac{\beta q_1 q_2}{4\pi\varepsilon_0\varepsilon(R_1 + R_2)} + G4\pi R_D(R_1 - R_D)^2 \tag{9.25}$$

The first term describes the electrostatic interaction of a mobile ion with charge q_1 and radius R_1 at contact with the matrix ion with charge q_2 and radius R_2; ε is the dielectric constant, and β is the "Madelung" constant, which indicates charge neutralization between the ion and its surroundings. The second term describes the elastic force contribution and assumes that it can be approximated by an opening of a "doorway" the size of the mobile ions. Here, G is the high-frequency shear modulus, and R_D is an effective radius of an existing open doorway allowing mobile ions to move from one site to another. R_D in this model can be considered a kind of "free volume" measure in polymers. In particular, it can represent a frustration in chain packing providing extra space for ion motions [36].

This model suggests that the diffusion of small ions such as Li^+ and Na^+ is controlled mostly by electrostatic interactions, whereas the diffusion of large ions, such as e.g. Tf_2N^-, is controlled mostly by the elastic contribution. Thus, to achieve a higher decoupled ionic conductivity for small cations (Li^+ or Na^+), the electrostatic interactions should be reduced significantly. This can be achieved by either developing a polymer with a large dielectric constant or employing anions with a strongly delocalized charge. Indeed, it has been emphasized that conductivity in single-Li^+-ion conducting polymers depends stronger on anion charge delocalization than on the structure of the polymer itself, confirming the dominant role of electrostatic interactions [36, 37].

9.5 Ionic Conduction and Transport Number Through Polymer Electrolytes

In solid polymer electrolytes (SPEs), ion conduction is widely understood to rely on the segmental motion of polymer chains, which facilitates ion hopping between coordination sites within the polymer matrix. This implies a strong coupling between ionic conductivity and polymer chain mobility. To enhance segmental mobility, various strategies have been explored, including modifications to key polymer properties such as crystallinity, physical aging, and glass transition temperature. However, a promising recent approach to improving ionic conductivity involves decoupling segmental mobility from ion transport, as demonstrated in polymer-in-salt (PIS) systems.

This section will first introduce PEO-based polymer electrolytes, then discuss emerging polymer solvents, PIS systems, and finally nanocomposites designed to improve ionic conductivity, transport number, and the electrochemical stability window.

9.5.1 Solid Polymer Electrolytes Based on PEO

Ether-based polymer electrolytes and their derivatives are some of the earliest and most promising systems due to their excellent ability to dissociate and coordinate lithium salts and their interfacial stability with lithium metal. PEO is a widely studied example of an ether-based polymer electrolyte.

Figure 9.13a illustrates the ionic conductivity, σ, of amorphous oxymethylene-linked PEO as a function of $LiClO_4$ concentration [5, 38]. At low salt concentrations, ionic conductivity rises as salt added due to increasing the concentration, c_j, of free ions. However, beyond a certain concentration, conductivity peaks and then declines.

This maximum results from two competing factors: free ion concentration and chain mobility. Initially, as salt concentration increases, so does the free ion concentration, improving conductivity. However, at higher salt levels, the free ion concentration plateaus due to solubility limits. Meanwhile, metal cations like Li^+ form coordination bonds with polymer oxygen atoms, creating transient cross-links that stiffen the polymer chains, raise the glass transition temperature, T_g, and ultimately reduce conductivity, as shown in Figure 9.13a [5, 38].

Diffusion coefficient, D, is also strongly influenced by the molecular weight, MW, of the matrix polymer, which reflects the mobility of the polymeric chains. For instance, as shown in Figure 9.13b, D is initially high but decreases sharply with increasing MW [39].

PEO typically exhibits a low lithium-ion transport number, T_{Li^+}, around 0.3, meaning that lithium ions, Li^+, move more slowly than larger anions, such as PF_6^- or Tf_2N^-. This is due to Li^+ ions being tightly coordinated by the PEO backbone's oxygen atoms, which enhances ion solubility but restricts Li^+ mobility [18, 19]. Despite these limitations, PEO-based electrolytes offer a relatively wide electrochemical stability window, up to ~4.5–5 V, extendable to 5.4 V in PEO composites [20, 40, 41].

However, achieving high ionic conductivity and a satisfactory Li^+ transport number at room temperature (25 °C) remains challenging, and PEO electrolytes often

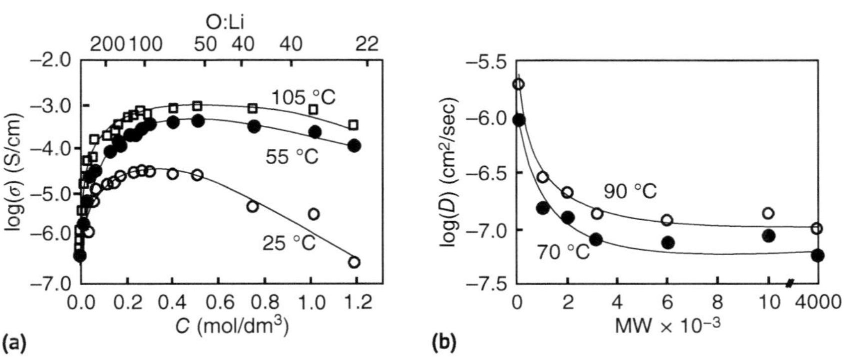

Figure 9.13 (a) Variation in conductivity σ with $LiClO_4$ concentration C for an amorphous oxymethylene-linked poly(ethylene oxide). *Source*: Adapted from [38]; and (b) variation of log(D) for ^7Li in PEO-$LiCF_3SO_3$ with an O:Li ratio of 20:1 at 70 °C (●) and 90 °C (○) as a function of polymer molecular weight, *MW*. *Source*: Adapted from [39].

suffer from limited interfacial compatibility with electrodes. Even in the amorphous phase, polymer electrolytes have ionic conductivities several orders of magnitude below the required levels at ambient temperatures [20]. For instance, PEO-based electrolytes typically achieve only around 1.0×10^{-5} S/cm, whereas inorganic solid electrolytes like $Li_{10}GeP_2S_{12}$ and $La_{0.51}Li_{0.34}TiO_{2.94}$ exhibit 2.0×10^{-2} and 1.0×10^{-2} S/cm, respectively, at 25 °C. Additionally, PEO is prone to oxidation at high charge potentials, limiting its interfacial compatibility and application potential in solid-state lithium-ion batteries [42, 43].

To address these challenges, several strategies have been explored to improve PEO's ionic conductivity and transport number. These include reducing PEO crystallinity (typically 60–80%) to increase ion mobility, leveraging its high dielectric constant ($\varepsilon_r \sim 5$–8) to enhance salt solubility, and incorporating additives such as nanoparticles and plasticizers. Both physical modifications (e.g. blending) and chemical modifications (e.g. copolymerization, cross-linking) have been widely applied to suppress crystallization, thereby improving ionic conductivity.

9.5.2 Emerging Polymeric Solvents

The dielectric constant of PEO is generally too low to fully separate ions, leading to their aggregation into clusters. The dielectric constant measures the ability of materials to polarize in response to an electric field, which affects their ability to dissolve salts. To improve ion solvation and enhance ionic conductivity, researchers have explored increasing the dielectric constant by incorporating high-dielectric polymers.

Poly(acrylonitrile) (PAN), poly(ethylene carbonate) (PEC), and poly(vinylidene fluoride) (PVdF) are commonly used to dissolve large amounts of salt for polymer electrolytes. Nitrile-containing polymers, such as PAN, have a high dielectric constant ($\varepsilon_r \sim 30$) due to the electron-withdrawing nitrile group (—CN) with a low-lying LUMO. This provides a broad electrochemical stability window, making nitrile-based polymer electrolytes suitable for solid-state lithium-ion batteries with high-voltage cathodes. Due to their strong ion-coordinating ability and favorable electrochemical properties, nitrile-based electrolytes, including PAN, its copolymers [44, 45], and blends [46], have been extensively studied.

However, nitrile-based polymers suffer from brittleness, high crystallinity, and poor electrode compatibility, which limit their electrochemical performance. To address these challenges, researchers have explored alternative polar polymers, such as carbonate-based materials [47, 48] and PVdF ($\varepsilon_r \sim 6-12$) [49–51].

9.5.3 Polymer-in-Salts: Novel Approach

In the 1990s, Angell et al. proposed a novel approach for achieving decoupled superionic systems (i.e. the ion transport which is decoupled from the mobility of polymeric chain segments or apparent viscosity of system) via a so-called polymer-in-salt (PIS) SPEs (Figure 9.14) [52, 53]. In this work, a high content (90 wt%) of a superionic four-salt melt (i.e. LiI/lithium acetate (LiOAc)/lithium chlorate ($LiClO_3$)/$LiClO_4$) was mixed with a small amount (10 wt%) of PPO with $MW = 4 \times 10^3$ g/mol, achieving a room temperature (e.g. 20–30 °C) ionic conductivity of about 10^{-4} S/cm.

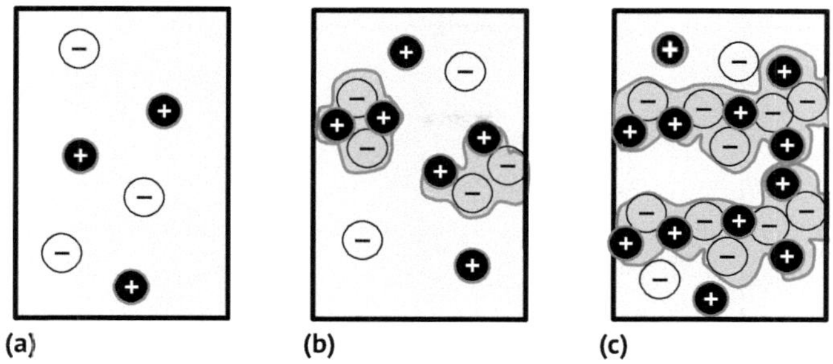

Figure 9.14 Salt status in polymer electrolytes at: (a) low (all salts are dissolved, $C < C^t$); (b) medium (aggregates start to be formed, $C \geq C^t$); and (c) high concentration (interconnected aggregates are formed, $C \gg C^t$) of salt, where C^t is threshold concentration. *Source*: Y.S. Kang.

The salts provide ionic conductivity and the role of polymer here is to provide mechanical strength, where their functions are decoupled [43, 54].

The concept of decoupling in such a system is formulated on a hypothesis that when the salt concentration keeps increasing to a threshold (C^t) (Figure 9.14a), the free ion concentration will reach a maximum, then further increasing the salt concentration $(C \geq C^t)$ will lead to the formation of salt aggregates as shown in Figure 9.14b. Those aggregates become more interconnected at higher salt concentrations (>50 wt% of salt in SPEs or $C \gg C^t$), enhancing the decoupling of the small metal ion aggregates from the polymer backbone, PIS-type polymer electrolyte (Figure 9.14c), and consequently promoting the metal ions to diffuse through this second conduction path [53].

In this case, the ionicity of lithium salt in PIS-type polymer electrolytes is of utter importance, where the ionicity is defined as a measure of the degree of ion dissociation, commonly referring to the effective fraction of ionic species being able to participate in ionic conduction [54, 55]. For PIS-type polymer electrolytes, the lithium salt should possess sufficient ionicity to ensure the high conductivity. In addition, the strategy of using salt mixtures to maintain their liquid state by achieving eutectic point at room temperatures to achieve high ionic conductivities is highly recommended [56].

Hereafter, there has been a large body of research to improve the ionic conductivity along with transport number for PIS-type heterophase polymer electrolytes, including primarily polycarbonate- and polyacrylonitrile-based PIS systems.

Similar to the Li^+-coordinating ether oxygens of PEO and other polyether electrolyte hosts, polycarbonates can coordinate to Li^+ ions by means of the carbonyl group oxygen – although some coordination by the ether oxygens adjacent to the carbonyl group has also been implicated in carbonate solvents [20]. Since the carbonyl symmetric stretch is a prominent feature in the FT-IR spectrum of polycarbonates, the interaction with Li^+ can be conveniently visualized as a shift of this band to lower wavenumbers on coordination [48].

Typically, polycarbonates with aromatic groups in the main chain, such as the well-known polycarbonate material based on bisphenol-A, are too rigid with high T_g. For higher molecular flexibility with concomitant improvements in ionic

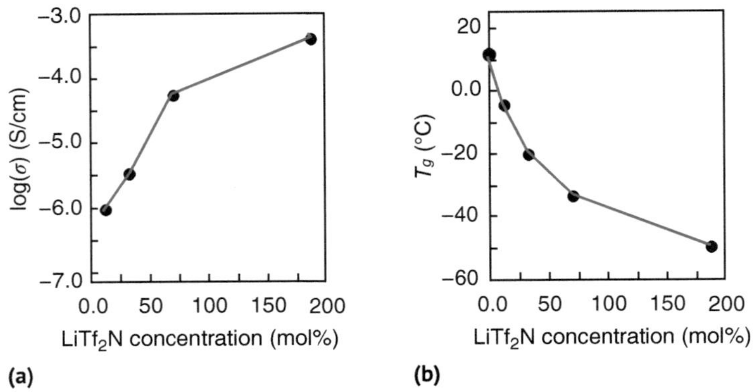

Figure 9.15 (a) Ionic conductivity and (b) glass transition temperature versus salt concentration for PEC:LiFSI electrolytes. *Source*: Adapted from [57].

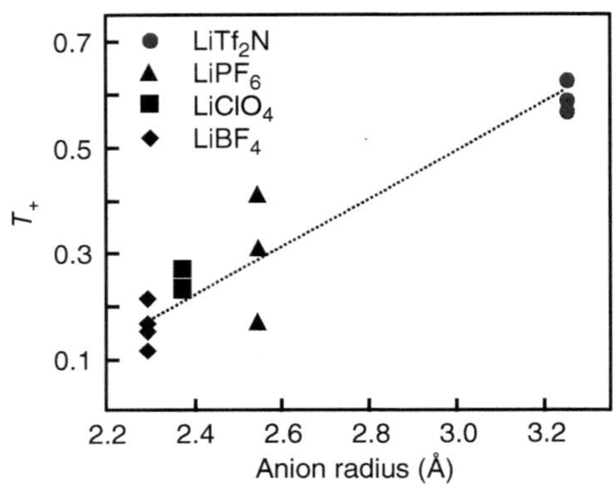

Figure 9.16 Variation of cation transport number T_+ with anion size for poly(ethylene carbonate) (PEC)-based electrolytes with different Li salts at 80 °C. *Source*: Adapted from [57].

conductivity, polycarbonates with aliphatic backbones are preferred. The simplest of these is poly(ethylene carbonate) (PEC). Thus, PEC and its derivatives have been frequently used [20].

Figure 9.15 illustrates that ionic conductivity of PEC/LiFSI electrolytes continuously increases with the salt concentration up to 200 mol%, resulting in Li^+ conductivity higher than 1.0×10^{-4} S/cm (Figure 9.15a), whereas the glass transition temperature continuously decreases with the salt concentration from ~10 to −50 °C (Figure 9.15b) [20, 57].

In addition to high ionic conductivity, PEC-based electrolytes have also shown impressive Li^+ transport numbers T_+, with values as high as 0.83 having been measured for PEC/$LiTf_2N$ [57]. As illustrated in Figure 9.16, this has been shown to be highly dependent on the size of the anion. The highest values of T^+ are seen in $LiTf_2N$ electrolytes and in contrast to other electrolyte salts, in the PEC : $LiTf_2N$ system, T_+ is seen to increase with increasing salt concentration. However, even with smaller-anion salts such as $LiBF_4$ and lithium bis(fluorosulfonyl)imide, LiFSI, as well as with $LiTf_2N$ at lower salt concentrations, PEC-based electrolytes clearly outperform polyether electrolytes, typically showing T_+ values close to 0.5 as opposed to closer to 0.3 for PEO electrolytes [20]. From these data it is apparent that PEC does not coordinate to and bind the Li^+ cations as tightly as ethylene oxide-based polyethers. Whether this is due to intrinsically weaker interactions or the absence of the chelating effects seen in ethylene oxide-based polyethers, leading to the formation of less well-defined complexes, remains to be definitely determined.

In addition to solvation by O atoms, as in the materials discussed thus far, N-containing groups may also constitute excellent Lewis bases that can coordinate to and solvate Li^+ in a polymeric matrix [20]. This can be exemplified by the nitrile (cyano) group, as found in, e.g. acetonitrile, succinonitrile, and adiponitrile, that have found use in liquid electrolyte formulations, in ion-conducting molecular plastic crystals, and as polymer electrolyte additives. Naturally, SPEs based on the Li^+-coordinating nitrile motif are also possible. The most typical nitrile-functional polymer is polyacrylonitrile, PAN, a semicrystalline material with a T_g of 80 °C and high stability toward electrochemical oxidation.

Given the thermal properties of PAN, i.e. semicrystallinity and T_g well above room temperature, fast ion dynamics should not be anticipated based on the conventional transport mechanism. However, Figure 9.17 shows that conductivity through PAN/lithium triflate system continuously increases with the salt concentration of lithium triflate up to 75 wt% for rigorously dried samples at 65 °C [55]. Ion conductivity reaches 2×10^{-6} S/cm at 50 °C and 10^{-5} S/cm at 75 °C for samples with 75 wt% salt. This behavior can be contrasted with that of more classical salt-in-polymer electrolytes, where a maximum in conductivity is typically found at much lower salt concentrations. Additionally, the salt was seen to act as a plasticizer, lowering the T_g from the initial 80 °C for pure PAN to an estimated 50–65 °C for the PIS electrolytes. This means that these high-salt electrolytes have appreciable conductivity at or below T_g, indicating a high level of decoupling of the ion transport from the polymer chain dynamics.

In summary, at low salt concentrations, ionic conductivity increases with salt concentration, reaching a maximum. In this region, known as the salt-in-polymer, SIP, phase, chain mobility and ionic conductivity are strongly coupled, as the salt dissolves in the polymer matrix to form a polymer electrolyte.

As the salt concentration increases beyond a threshold, salt aggregates begin to form and interconnect, leading to a polymer-in-salt, PIS, system, and a second rise in ionic conductivity, which reaches a much higher maximum than the

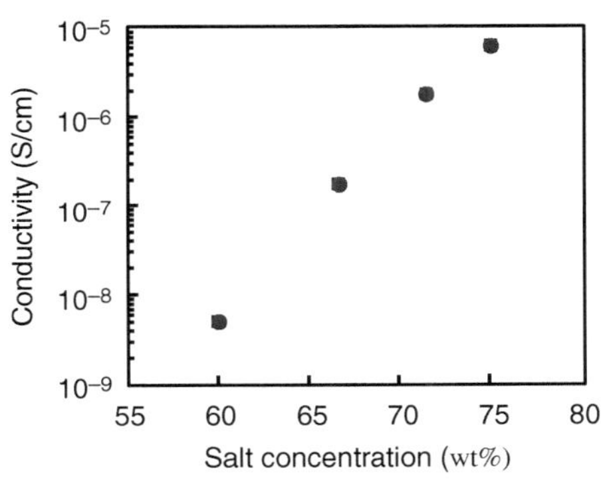

Figure 9.17 Conductivity as a function of salt concentration (between 60 and 75 wt% lithium triflate) for rigorously dried poly(acrylonitrile) (PAN) samples at 65 °C. This temperature is near the expected T_g for these samples. *Source:* Adapted from [55].

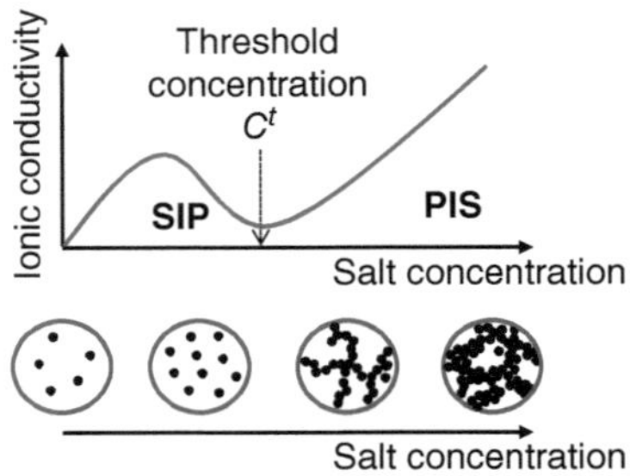

Figure 9.18 Variation of the ionic conductivity for the morphology change by increasing salt concentration from salt-in-polymer, SIP, to polymer-in-salt, PIS, electrolytes beyond the percolation threshold. *Source*: Adapted from [55].

first (Figure 9.18). This phenomenon presents a promising strategy for further enhancing ionic conductivity.

In the PIS phase, chain mobility and ionic conductivity become decoupled, allowing the independent design of polymers and salts. However, maximizing ionicity is crucial for improving conductivity in this regime, making it a key challenge in polymer electrolyte design.

9.5.4 Nanocomposites

Nanocomposites consisting of nano-sized inorganic fillers as active sites incorporated in a polymeric matrix have been successfully utilized to improve the ionic conductivity. For example, porous inorganic fillers, regardless of their electrochemical activity, could provide additional transport channels [58]. When inorganic fillers, in particular their surfaces, are electrochemically active with functional groups to reversibly interact with the metallic cations, their ionic conductivity will be increased, depending on the degree of the interactions and more importantly their kinetics, particularly the backward interaction rate as explained in Section 4.2.4.

In addition, to facilitate rapid lithium-ion transport, the morphology of inorganic fillers is expanded from nanoparticles (i.e. less than 100 nm for all three Cartesian dimensions of the nanoparticle) to nanowires (i.e. less than 100 nm for the diameter of the nanowire) [59, 60]. Functionalization of nanofiller is also suggested to improve the compatibility of organic and inorganic phases [61]. It was also reported that the transport along inorganic bulk phases is likely to only occur in inorganic rich (>50 vol%) SPEs or when particular morphologies (e.g. nanowire) of inorganic phases are used [54, 58]. Addition of nanofillers generally suppresses the crystallization of polymeric chains such as PEO chains, promoting the lithium-ion transport [62].

9.5.5 Transport Number

For liquid electrolyte, the absence of ion selectivity generally leads to comparable mobilities for the cation and anion, yielding the transport number $T_+ = T_-{\sim}0.5$, whereas solid-state electrolyte based on poly(ether)s gives mostly $T_+ < 0.5$. However, a high transport number is required for most applications, including solid-state Li^+ batteries, even to develop materials with unipolar conduction $T_+ = 1.0$ [1].

The low cation transport number T_+ of most polymer electrolytes based on polyether linkages often leads to a steep ion concentration gradient near the electrode surfaces that strongly reduces charge/discharge rate and favors Li metal dendrite growth.

Although the single-ion conducting polymer electrolytes (SIPEs) conceptually have the intrinsic potential to overcome this limitation, their very low ionic conductivities limit practical applications, especially under high current densities. Only a high cation transport number T_+ polymer electrolytes with simultaneously high ionic conductivity could provide a solution to enable high-performance solid-state batteries [63].

In the case of PEO-based polymer electrolytes with LiSCN, $LiClO_4$, and $LiBF_4$, T_+ is only in the range of 0.3–0.5, which should be better to be further improved for high performance. One simple way is to blend with polymers containing negatively charged groups. For instance, PEO blended with acrylate polymer consisting of pendant sulfonate or perfluorocarboxylate groups exhibited the unipolar conduction for the first time at the expense of the decrease in the ionic conductivity [64]. Since then, diverse anion centers, such as carboxylate ($-CO_2^-$), sulfonate ($-SO_3^-$), or sulfonylimide ($-SO_2N-SO_2^-$) have been investigated because they can be easily synthesized and covalently attached to the polymer chain or the surface of nanoparticle [63].

Based on the ionic transport mechanism, the high cation transport number polymer electrolyte could be divided into three different strategies: single lithium-ion conducting, hindered anion transport, and accelerated cation transport. First, most single-ion conducting polymer electrolytes have high T_+ ($T_+ = 1$) but low σ. Therefore, σ should be increased while maintaining the high value of T_+. Second, the strategy to hinder anion transport may include several approaches, such as the addition of anion acceptors that coordinate with anions selectively and slow down their migration, the incorporation of inorganic fillers that attract anions selectively via their surface functionalities, functionalization of inorganic ceramics to increase the interaction with anions. Third, the strategy to accelerate cation transport includes the use of polymer materials with a similar structure to carbonate solvents as solid electrolytes, in which physical interactions between Li^+ and polymer chain can effectively improve the cationic migration and hence higher T_+ [63].

Currently, conventional polymer electrolytes show low $\sigma \left(\sim 10^{-5}\ \text{S/cm}\right)$ as well as low T_+ (< 0.5), while SIPEs give σ in the range of 10^{-6}–10^{-5} S/cm with $T_+ > 0.85$. Therefore, both high $\sigma \left(10^{-4}\ \text{S/cm}\right)$ and high T_+ (> 0.8) polymer electrolytes should be developed for practical applications. In addition, cycling life, energy density, and operating conditions (current density, service temperatures, etc.) are the major concerns, and some high T_+ (> 0.8) polymer electrolytes should be designed with balanced ionic conductivity to match the requirements of future Li^+ batteries.

9.6 Temperature Dependence of Ionic Transport

Ionic transport increases with temperature as shown in Figure 9.19a. The plots of $\ln(\sigma)$ vs $1/T$ show deviation from the linearity suggesting that ionic transport show complex behavior and consequently that Arrhenius equation may not be always valid here.

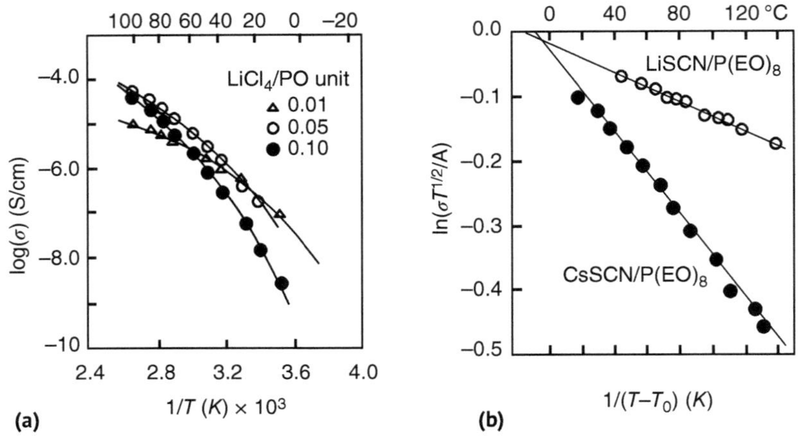

Figure 9.19 (a) Temperature dependence of ionic conductivity for polymer complexes consisting poly(propylene oxide) (PPO) from PPO triol (*MW* = 750) and $LiClO_4$. *Source*: Adapted from [1]. (b) Linearization of conductivity plots for LiSCN and CsSCN showing the determination of the Vogel–Tamman–Fulcher (VTF) parameters ($A = 2.57$, $B = 7.57 \times 10^{-2}$ eV, $T_0 = -15$ °C for LiSCN/P(EO)$_8$ and $A = 0.514$, $B = 2.71 \times 10^{-2}$ eV, $T_0 = -11$ °C for CsSCN/P(EO)$_8$. *Source*: Adapted from [1].

9.6.1 Vogel-Tamman-Fulcher Equation

The variation of ionic conductivity with temperature for a fully amorphous polymer electrolyte was observed to be more accurately represented by the Vogel-Tamman-Fulcher (VTF) equation rather than the simple Arrhenius function [4].

$$\sigma = \frac{A}{T^{1/2}} \exp\left[-\frac{B}{(T - T_0)}\right] = \sigma_0 \exp\left[-\frac{B}{(T - T_0)}\right] \tag{9.26}$$

where T_0 is the equilibrium glass transition temperature (commonly $T_0 = T_g - 51.6$), frequently used as a reference temperature, and A and B are pre-exponential factor and pseudo activation energy, respectively. Thus, whilst Arrhenius plots of $\ln(\sigma)$ vs $1/T$ show a typical curvature (Figure 9.19a), linear plots for $\ln(\sigma T^{1/2}/A)$ versus $(T - T_0)^{-1}$ are obtained as shown in Figure 9.19b. The physical meaning of T_0 is that, as T approaches to T_0, the viscosity goes infinitive and the mobility of the system nearly vanishes; the free volume becomes extremely small, representing equilibrium or true glass transition temperature.

Since T_0 and T_g represent a measure of the polymeric chain mobility, the VTF equation reveals that the ion transport is directly related to the segmental motion of the polymer chains. Polymers with lower T_g show higher segment mobility, which can promote the ion transport to increase the ionic conductivity [65]. One of the reasons that PEO is the most intensively investigated polymer conductor by far is the relatively low T_g of −60 to −50 °C compared to many other polymers in addition to its high salt solubility due to the hydrophilic ethylene oxide groups. The VTF equation really arises from considerations involving only the polymer host whereas the actual observations are made on a polymer/salt complex [1].

The $\sigma(T)$ curves for most polymer electrolytes show one of four patterns as shown in Figure 9.20 [20]. Each curve in Figure 9.20 represents a specific behavior as follows:

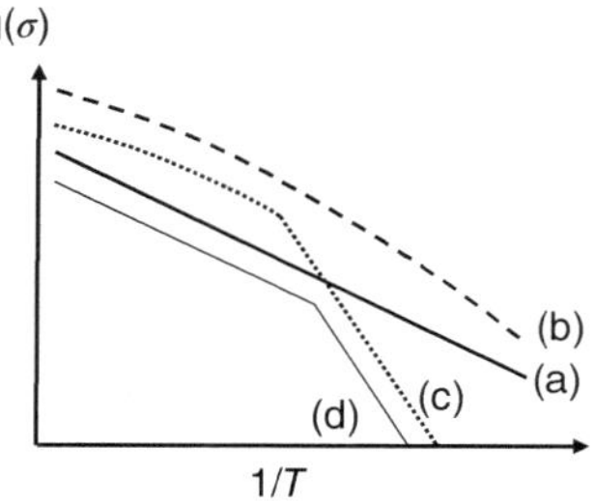

Figure 9.20 Idealized conductivity behavior of solid polymer electrolytes (SPEs). *Source*: Ref. [20]/with permission of Elsevier.

a. The Arrhenius behavior is typical for low-*MW* oligomer-salt solutions or rigid polymer systems. An example of the former is the crystalline complex $(PEO)_6/LiAsF_6$ [66].
b. The VFT behavior of completely amorphous high-*MW* SPEs, e.g. the poly(trimethylene carbonate) with $LiTf_2N$ [67].
c. For semicrystalline polymers (such as PEO-based systems), where a melting of the crystalline phases occurs after which the VFT behavior is displayed. A representative example is the poly(ε-caprolactone) with 28 wt% $LiTf_2N$ [68].
d. For crystalline systems where a solid–solid phase transition occurs, e.g. $(PEO)_8/NaAsF_6$ [66].

9.6.2 Williams-Landal-Ferry Equation and Master Curves

WLF equation extended the VTF treatment to characterize relaxation processes in amorphous systems in a general manner. Any temperature-dependent mechanical relaxation process can be expressed in terms of universal scaling law [4]:

$$\log(a_T) = -\frac{C_1(T - T_0)}{(C_2 + T - T_0)} \tag{9.27}$$

where a_T is the shift factor, defined as the ratio of any mechanical relaxation process at temperature T to its value at the reference temperature T_r, i.e. $a_T = \tau_T/\tau_r = \eta_T/\eta_r = \sigma_r/\sigma_T$ where τ_T and η_T represent a characteristic segmental relaxation time and viscosity, respectively, at temperature T. Here, T_0, known as the equilibrium T_g, is used as a reference temperature T_r. The WLF equation is empirical, but is consistent with the free volume theory as demonstrated in Section 3.5.

When the glass transition temperature T_g is taken as the reference temperature, it follows that C_1 and C_2 in Equation (9.27) are 17.44 and 51.6, which are universal being independent of the measured properties. Thus,

$$\log\left(\frac{\sigma_T}{\sigma_{T_g}}\right) = \frac{17.44(T - T_g)}{51.6 + (T - T_g)} \tag{9.28}$$

Figure 9.21 shows a master curve of the WLF plots of ionic conductivity for various PEO-alkali metal salt complexes. This is possible because the WLF equation is considered to be universal for all disordered, amorphous polymers.

Substituting Equation (9.16) for free volume theory into Equations (9.1) and (9.15) gives the VTF equation (Equation (9.26)) and WLF equation (Equation (9.28)), demonstrating the consistency among them.

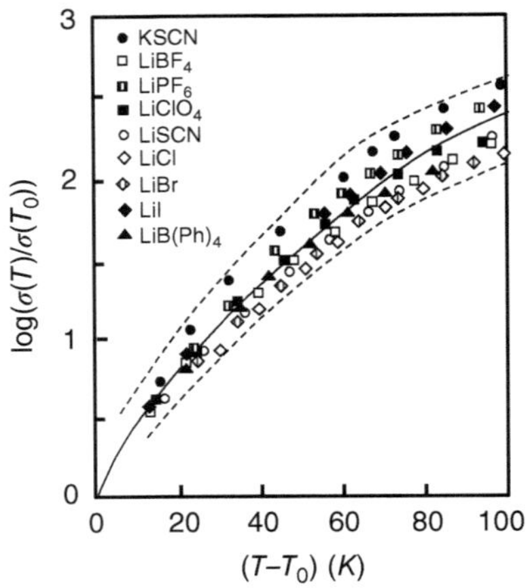

Figure 9.21 Master curve of the Williams-Landal-Ferry (WLF) plots of ionic conductivity $\sigma(T)$ for various poly(ethylene oxide) (PEO)-alkali metal salt complexes. The reference temperature is $T_0 = T_g - 50$ K. *Source*: Replotted from [4, 69].

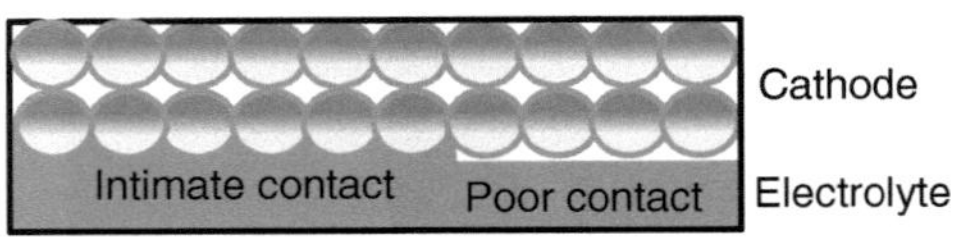

Figure 9.22 The schematic illustration of the interfaces in polymer electrolyte-based solid-state lithium-ion batteries. *Source*: Adapted from [43].

9.7 Interfacial Charge Transfer Between SPE and Electrode

When an electrolyte is incorporated between two electrodes in electrochemical devices, its interfacial layers with both anode and cathode significantly influence the charge transfer and consequently the device performance. SPEs offer a safer and more stable alternative to liquid electrolytes, with the potential for higher energy density. Recent reviews have examined interfacial phenomena in lithium-ion batteries [70, 71], while these issues are discussed in more detail at Section 12.1 of Chapter 12.

Unlike liquid electrolytes, SPEs cannot infiltrate electrode pores such as those in cathodes of secondary batteries, limiting interfacial contact and disrupting continuous ion transport. Ensuring intimate contact is crucial for reducing charge transfer resistance R_{CT}, yet this remains challenging due to solid–solid mechanical contact issues and chemical incompatibilities between the SPE and the electrode [72]. As a result, the interfacial layer typically consists of both intimate contact and poor contact regions, as illustrated in Figure 9.22.

To improve interfacial contact, various strategies have been explored, including thin coating layers, the incorporation of organic salts and inorganic ionic conductors to form solid electrolyte interface, SEI, or cathode electrolyte interface, CEI, layers [72], where CEI is formed by electrolyte decomposition at high cathode potentials during charge/discharge cycles. CEI typically consists of inorganic and organic decomposition products, including Li_2CO_3, LiF, transition metal fluorides, and polymerized organic compounds. For example, a PEO layer at the lithium metal anode enables dendrite-free plating, while a poly(*N*-methyl-malonic amide) layer at the cathode facilitates Li^+ extraction from an oxide host cathode without electrolyte oxidation, maintaining stable charge/discharge cycling in a 4 V cell at 65 °C [73].

9.8 Prospects of Polymer Electrolytes

Polymer electrolytes encounter several critical challenges that must be addressed to unlock their full potential in diverse electrochemical applications. The foremost issues include enhancing ionic conductivity and transport numbers to ensure efficient ion mobility and minimizing energy losses. Achieving improved interfacial contact and stability with electrodes is equally vital, as poor adhesion and chemical instability at interfaces can degrade device performance and longevity.

Another pressing concern is inhibiting lithium dendrite growth in batteries, which poses significant safety risks, such as short circuits and thermal runaway. Additionally, ensuring the chemical and mechanical robustness of polymer electrolytes under prolonged cycling conditions is crucial for maintaining reliability and performance.

Solving these challenges is pivotal for advancing technologies such as batteries, fuel cells, sensitized solar cells, and supercapacitors, paving the way for safer, more efficient, and higher-performing electrochemical devices.

References

1 J.R. MacCallum, C.A. Vincent, Polymer Electrolyte Reviews, Springer Netherlands, 1989.
2 M. Rubinstein, R.H. Colby, Polymer Physics, OUP Oxford, 2003.
3 D.W. Van Krevelen, K. Te Nijenhuis, Properties of Polymers: Their Correlation with Chemical Structure; Their Numerical Estimation and Prediction from Additive Group Contributions, Elsevier, 2009.
4 F.M. Gray, Polymer Electrolytes, Royal Society of Chemistry, 1997.
5 F.M. Gray, Solid Polymer Electrolytes: Fundamentals and Technological Applications, VCH, 1991.
6 V. Gutmann, The Donor-Acceptor Approach to Molecular Interactions, Springer US, 1978.
7 J. Popovic, C. Pfaffenhuber, J.P. Melchior, J. Maier, Electrochemistry Communications 60 (2015) 195–198.
8 Y. Takahashi, H. Tadokoro, Macromolecules 6(5) (1973) 672–675.
9 Z. Tian, A. Hess, C.R. Fellin, H. Nulwala, H.R. Allcock, Macromolecules 48(13) (2015) 4301–4311.
10 S. Rothemund, I. Teasdale, Chemical Society Reviews 45(19) (2016) 5200–5215.
11 S. Hayano, K. Ota, H.T. Ban, Polymer Chemistry 9(8) (2018) 948–960.
12 J.Y. Kim, S.U. Hong, J. Won, Y.S. Kang, Macromolecules 33 (2000) 3161–3165.
13 J.H. Kim, B.R. Min, C.K. Kim, J. Won, Y.S. Kang, The Journal of Physical Chemistry B 106 (2002) 2786–2790.

14 S. Choi, J.H. Kim, Y.S. Kang, Macromolecules 34 (2001) 9087–9092.
15 E.A. Dimarzio, J.H. Gibbs, Journal of Polymer Science Part A: General Papers 1(4) (1963) 1417–1428.
16 T.S. Chow, Macromolecules 13(2) (1980) 362–364.
17 J.H. Kim, B.R. Min, J. Won, Y.S. Kang, The Journal of Physical Chemistry B 107(24) (2003) 5901–5905.
18 Z. Xue, D. He, X. Xie, Journal of Materials Chemistry A 3(38) (2015) 19218–19253.
19 A. Bernson, J. Lindgren, W. Huang, R. Frech, Polymer 36(23) (1995) 4471–4478.
20 J. Mindemark, M.J. Lacey, T. Bowden, D. Brandell, Progress in Polymer Science 81 (2018) 114–143.
21 M.H. Cohen, D. Turnbull, The Journal of Chemical Physics 31(5) (1959) 1164–1169.
22 T. Miyamoto, K. Shibayama, Journal of Applied Physics 44(12) (1973) 5372–5376.
23 J. Vrentas, J. Duda, Journal of Polymer Science: Polymer Physics Edition 15(3) (1977) 403–416.
24 H. Benoît, P. Rempp, Macromolecules: Main Lectures Presented at the 27th International Symposium on Macromolecules, Strasbourg, France, 6–9 July 1981, Pergamon Press, 1982.
25 J.H. Gibbs, E.A. DiMarzio, The Journal of Chemical Physics 28(3) (1958) 373–383.
26 G. Adam, J.H. Gibbs, The Journal of Chemical Physics 43(1) (1965) 139–146.
27 E. Williams, C.A. Angell, The Journal of Physical Chemistry 81(3) (1977) 232–237.
28 M.A. Ratner, D.F. Shriver, Chemical Reviews 88(1) (1988) 109–124.
29 M.A. Webb, B.M. Savoie, Z.-G. Wang, T.F. Miller III, Macromolecules 48(19) (2015) 7346–7358.
30 S. Arumugam, J. Shi, D.P. Tunstall, C.A. Vincent, Journal of Physics: Condensed Matter 5 (1993) 153–160.
31 D. Teeters, R. Frech, Solid State Ionics 18 (1986) 271–276.
32 F. Fan, Y. Wang, T. Hong, M.F. Heres, T. Saito, A.P. Sokolov, Macromolecules 48(13) (2015) 4461–4470.
33 C. Imrie, M. Ingram, G. McHattie, The Journal of Physical Chemistry B 103(20) (1999) 4132–4138.
34 K. Nakamura, K. Fukao, Polymer 54(13) (2013) 3306–3313.
35 O.L. Anderson, D.A. Stuart, Journal of the American Ceramic Society 37(12) (1954) 573–580.
36 V. Bocharova, A.P. Sokolov, Macromolecules 53(11) (2020) 4141–4157.
37 H. Zhang, C. Li, M. Piszcz, E. Coya, T. Rojo, L.M. Rodriguez-Martinez, M. Armand, Z. Zhou, Chemical Society Reviews 46(3) (2017) 797–815.
38 F.M. Gray, Solid State Ionics 40–41 (1990) 637–640.
39 J. Shi, C.A. Vincent, Solid State Ionics 60(1–3) (1993) 11–17.
40 D.J. Brooks, B.V. Merinov, W.A. Goddard, B. Kozinsky, J. Mailoa, Macromolecules 51(21) (2018) 8987–8995.
41 Y. Zhu, J. Cao, H. Chen, Q. Yu, B. Li, Journal of Materials Chemistry A 7(12) (2019) 6832–6839.

42 X. Yang, M. Jiang, X. Gao, D. Bao, Q. Sun, N. Holmes, H. Duan, S. Mukherjee, K. Adair, C. Zhao, Energy & Environmental Science 13(5) (2020) 1318–1325.
43 P. Ding, Z. Lin, X. Guo, L. Wu, Y. Wang, H. Guo, L. Li, H. Yu, Materials Today 51 (2021) 449–474.
44 D. Zhou, Y.-B. He, R. Liu, M. Liu, H. Du, B. Li, Q. Cai, Q.-H. Yang, F. Kang, Advanced Energy Materials 5(15) (2015) 1500353.
45 P. Wang, J. Chai, Z. Zhang, H. Zhang, Y. Ma, G. Xu, H. Du, T. Liu, G. Li, G. Cui, Journal of Materials Chemistry A 7(10) (2019) 5295–5304.
46 B.K. Choi, Y.W. Kim, H.K. Shin, Electrochimica Acta 45(8) (2000) 1371–1374.
47 Y. Zheng, X. Li, C.Y. Li, Energy Storage Materials 29 (2020) 42–51.
48 A.M. Elmér, P. Jannasch, Journal of Polymer Science Part A: Polymer Chemistry 44(7) (2006) 2195–2205.
49 X. Zhang, S. Wang, C. Xue, C. Xin, Y. Lin, Y. Shen, L. Li, C.W. Nan, Advanced Materials 31(11) (2019) e1806082.
50 H. Wu, Y. Cao, H. Su, C. Wang, Angewandte Chemie International Edition 57(5) (2018) 1361–1365.
51 Q. Ma, H. Zhang, C. Zhou, L. Zheng, P. Cheng, J. Nie, W. Feng, Y.S. Hu, H. Li, X. Huang, L. Chen, M. Armand, Z. Zhou, Angewandte Chemie International Edition 55(7) (2016) 2521–2525.
52 C.A. Angell, C. Liu, E. Sanchez, Nature 362 (1993) 137–139.
53 H. Zhang, M. Armand, Israel Journal of Chemistry 61(1–2) (2021) 94–100.
54 Z. Song, F. Chen, M. Martinez-Ibanez, W. Feng, M. Forsyth, Z. Zhou, M. Armand, H. Zhang, Nature Communications 14(1) (2023) 4884.
55 M. Forsyth, J. Sun, D.R. Macfarlane, A.J. Hill, Journal of Polymer Science Part B: Polymer Physics 38(2) (2000) 341–350.
56 G. Guzmán-González, M. Alvarez-Tirado, J.L. Olmedo-Martínez, M.L. Picchio, N. Casado, M. Forsyth, D. Mecerreyes, Advanced Energy Materials 13(1) (2023) 2202974.
57 Y. Tominaga, K. Yamazaki, Chemical Communications 50(34) (2014) 4448–4450.
58 M. Keller, A. Varzi, S. Passerini, Journal of Power Sources 392 (2018) 206–225.
59 W. Liu, N. Liu, J. Sun, P.C. Hsu, Y. Li, H.W. Lee, Y. Cui, Nano Letters 15(4) (2015) 2740–2745.
60 W. Liu, S.W. Lee, D. Lin, F. Shi, S. Wang, A.D. Sendek, Y. Cui, Nature Energy 2(5) (2017).
61 L. Porcarelli, P. Sutton, V. Bocharova, R.H. Aguirresarobe, H. Zhu, N. Goujon, J.R. Leiza, A. Sokolov, M. Forsyth, D. Mecerreyes, ACS Applied Materials & Interfaces 13(45) (2021) 54354–54362.
62 F. Croce, G.B. Appetecchi, L. Persi, B. Scrosati, Nature 394 (1998) 456–458.
63 X. Shan, Z. Song, H. Ding, L. Li, Y. Tian, A.P. Sokolov, M. Tian, K. Xu, P.-F. Cao, Energy & Environmental Science 17(22) (2024) 8457–8481.
64 D.J. Bannister, G.R. Davies, I.M. Ward, J.E. McIntyre, Polymer 25(9) (1984) 1291–1296.
65 O. Borodin, G.D. Smith, Macromolecules 39 (2006) 1620–1629.
66 C. Zhang, S. Gamble, D. Ainsworth, A.M. Slawin, Y.G. Andreev, P.G. Bruce, Nature Materials 8(7) (2009) 580–584.

67 J. Mindemark, B. Sun, D. Brandell, Polymer Chemistry 6(26) (2015) 4766–4774.

68 J. Mindemark, B. Sun, E. Törmä, D. Brandell, Journal of Power Sources 298 (2015) 166–170.

69 M. Watanabe, M. Itoh, K. Sanui, N. Ogata, Macromolecules 20(3) (1987) 569–573.

70 S. Wu, C. Wang, S. Li, J. Weng, Materials Chemistry Frontiers 8(18) (2024) 2924–2943.

71 S.N. Patel, ACS Macro Letters 10(1) (2021) 141–153.

72 M. Geng, G. Su, S. Huang, S. Wang, M. Xiao, D. Han, Y. Meng, Materials Chemistry Frontiers (2023).

73 W. Zhou, Z. Wang, Y. Pu, Y. Li, S. Xin, X. Li, J. Chen, J.B. Goodenough, Advanced Materials 31(4) (2019) 1805574.

10

Ion-exchange Membranes

Ion-exchange membranes (IEMs) are selective barriers that allow specific ions to pass based on their charge, preventing others from crossing. In electrochemical systems, IEMs function as electrolytes for ion transport and as separation media between anode and cathode compartments, ensuring ion selectivity and mechanical stability. They contain ion-exchange groups, such as sulfonate or tetra-alkyl ammonium groups, which form microphase-separated structures consisting of hydrophilic ionic clusters or channels embedded within a hydrophobic matrix.

In recent decades, IEMs have drawn increasing interest, particularly for their role in energy conversion and storage technologies. Examples include fuel cells, water-splitting devices, and redox flow batteries for renewable energy storage. Traditional applications, such as electrodialysis for water treatment, also depend on IEM.

IEMs are classified into cation-exchange membranes (CEMs) with fixed negative charges, which allow cation transport, and anion-exchange membranes (AEMs) with fixed positive charges for anion transport. When a cation-exchange layer (CEL) and an anion-exchange layer (AEL) are paired, they form a bipolar membrane (BPM), sometimes with an optional catalyst layer at the interface for specific applications.

In this context, we will start by discussing ion transport in water, followed by the microstructures of IEMs and their structural impact on ion transport, particularly through ion channels, along with their transport mechanisms.

10.1 Definitions of Ion Conductivity, Transport Number, and Permselectivity

The ionic conductivity σ (S/cm), defined in Section 8.5.4, is written again as:

$$\sigma = F\sum_{j=1}^{n}|z_j|u_j c_j = F^2\sum_{j=1}^{n}|z_j|^2\frac{D_j}{RT}c_j \tag{10.1}$$

where F is the Faraday constant. u_j, D_j, and c_j, are the ion mobility, the diffusion coefficient, and the concentration of ion species j, respectively. Note that the

Molecule and Ion Transport through Polymer Membranes, First Edition. Yong Soo Kang.

mobility is related with the diffusion coefficient as $u_j = z_j FD_j/RT$ (Equation (8.49)). Equation (10.1) can be rearranged for one cation (+) and one anion (−) with monovalent system as:

$$\sigma = \frac{F^2}{RT}(D_+c_+ + D_-c_-) \tag{10.2}$$

Equation (10.2) indicates that the ionic conductivity σ is proportional to the sum of the products of the diffusion coefficient and the concentration of all ionic species in a system.

In an electrolyte solution, the current is carried out by both cations and anions. Depending on applications, the ionic conductivity σ_j of specific ion j plays important roles rather than the total ionic conductivity σ. For instance, in the case of the lithium-ion battery of $LiPF_6$, both Li^+ and PF_6^- ions are dissolved in a solvent and can contribute to the measured conductivity, σ. Thus, it is necessary to obtain the contribution of each ionic species to total conductivity. Thus, the fraction of the conductivity that is carried out by a specific ion species j is expressed by the ion transport number T_j (Equation (8.63)) or transference number t_j (Equation (8.64)) as given:

$$T_j = \frac{\sigma_j}{\sigma} = \frac{|z_j|u_jc_j}{\sum_{k=1}^{n}|z_k|u_kc_k} \tag{10.3}$$

And

$$t_j = \frac{T_j}{|z_j|} \tag{10.4}$$

The ion transport number T_j indicates that the fraction of the current carried out by the ion j, and the transference number t_j indicates the number of moles of ion j transported per mole of electrons [1]. The sum of the transport number or transference number of all ions in the solution becomes unity. In literature, the definition of the transport number or transference number is not consistently used, and very often they are mixed up [1].

The membrane permselectivity describes the degree to which a membrane passes an ion of one charge and retains an ion of opposite charges; in other words, it means the ability of IEMs to permeate counterions over co-ions. The permselectivity of CEMs and AEMs is defined as:

$$P_c^m = \frac{T_c^m - T_c^s}{T_a^s} \quad \text{for CEM} \tag{10.5}$$

$$P_a^m = \frac{T_a^m - T_a^s}{T_c^s} \quad \text{for AEM} \tag{10.6}$$

where P_c^m and P_a^m are the permselectivity of CEM and AEM, respectively. Superscripts m and s are for membrane and electrolyte solution, respectively, and subscripts c and a are for cation and anion, respectively [1, 2].

Table 10.1 Characteristics of ions in aqueous solution at 25 °C[a] [3].

Cation	r	r_s	u_j	D_j	Anion	r	r_s	u_j	D_j
H^+	—	0.28	36.23	10.31	OH^-	—	0.46	20.64	5.27
K^+	1.38	1.25	7.62	1.96	Cl^-	1.81	1.21	7.91	2.03
Li^+	0.59	2.38	4.01	1.03	F^-	1.33	1.66	5.70	1.46
Ca^{++}	1.00	3.10	6.17	0.793	SO_4^{-}	—	2.30	8.29	1.07

[a]The cations and anions are displayed in the ascending order of ionic Stokes radius r_s (Å); r (Å) is the ionic radius in salt crystals; u_j (10^8 m²/sV) and D_j (10^9 m²/s) are mobility and self-diffusion coefficient, respectively, of ions at infinite dilution in solution. u_j and D_j are correlated by $D_j = RTu_j/z_jF$ [3].
Source: Adapted from [3].

10.2 Ion Transport Through Water and Heterocycles

Ion transport such as proton and hydroxide ions through pure water is extremely fast with $D_{H^+} = 10.31 \times 10^{-9}$ m²/s at 25 °C. Table 10.1 lists the diffusion coefficient D_j and the mobility u_j of both representative cations and anions, respectively, along with ionic radius r in salt crystals and the Stokes radius, also called Stokes–Einstein radius, r_s in aqueous solution at 25 °C [3]. The Stokes radius r_s of an ion is the radius of a hard sphere that diffuses at the same rate as that ion. It is closely related to the ion mobility, factoring in not only size but also solvent effects. A smaller ion with stronger hydration, for example, may have a greater Stokes radius than a larger ion with weaker hydration ($r_s = 2.38$ Å for Li vs 1.25 Å for K). As can be seen, the mobility and diffusion coefficient decrease with the Stokes radius r_s except H^+ and OH^-.

The ionic mobilities of both H^+ and OH^- are much larger than the common cations and anions, respectively. Notably, the diffusion coefficient of H^+ in aqueous solution at 25 °C is reported to be 10.31×10^{-9} m²/sec whereas that of OH^- 5.27×10^{-9} m²/sec, nearly, the half of the protons. It also unexpectedly shows that small ions such as Li^+ or F^- have lower mobility than large ions such as Ca^{2+} or SO_4^{2-} [1], suggesting the importance of the size of hydrated ions, represented by the Stokes radius, rather than the neat size of atom, estimated from crystals. Therefore, the ionic mobility does not significantly change with common cations and anions except proton cation and hydroxide anion.

10.2.1 H^+ and OH^- Ion Transport Mechanisms

The exceptionally high mobility of the H^+-ion and OH^--ion is very complicated and thus difficult to elucidate the precise transport mechanism. However, it has been frequently explained by the two different transport mechanisms: vehicle and Grotthuss mechanisms.

Because of the molecular interactions of water dipoles with electrically charged protons, protons form hydronium ions (H_3O^+) with the water. The three attached

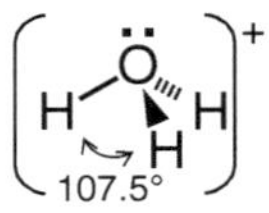

Figure 10.1 Structure of hydronium ion. *Source*: Y.S. Kang.

H-atoms in the hydronium ion are no longer distinguishable in terms of their electric charges. They form a tripod with three H-atoms equally apart, carrying the positive charge together as shown in Figure 10.1. The formation of the hydronium ions has consequences for the high mobility of the protons in aqueous media, called the vehicle mechanism [3].

Additionally, the proton can also be transported via a so-called Grotthuss mechanism from one hydronium ion to the next water molecule, as indicated in Figure 10.2. This explains not only the extraordinarily high mobility of protons, but it is also one of the reasons for the high proton conductivity even through ice water state. The same mechanism as the one described here for the transport of protons also holds true for the transport of hydroxide ions (Grotthuss mechanism), and thus the conductivity of hydroxide ions in an aqueous solution is much higher than that of other salt anions, which generally is very low [1, 3].

Figure 10.3a schematically illustrates the vehicle mechanism, and Figure 10.3b does the consecutive hydrogen bonding and debonding processes for Grotthuss mechanism differently [1, 3].

10.2.2 Proton Conduction Through Heterocycles

Interestingly, the proton conduction has been observed from heterocycle compounds without water. The proton conductivity is in the range of 1.0×10^{-3} S/cm at 130 °C through imidazole, benzimidazole, pyrazole, or 1,2,4-triazole having a five-membered ring with two or three nitrogen atoms (Figure 10.4), as listed in Table 10.2. The ionic conduction mechanism is the same Grotthuss mechanism observed with water, as shown in Figure 10.5. Therefore, heterocycles can be a

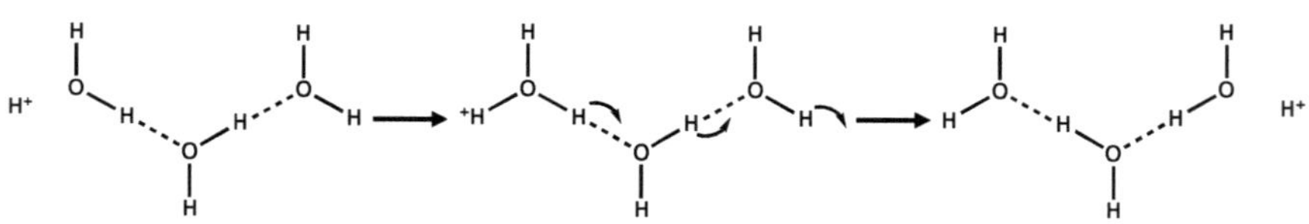

Figure 10.2 Grotthuss mechanism of H^+ in aqueous solution by multiple hydrogen bonding and debonding processes. *Source*: Adapted from [3].

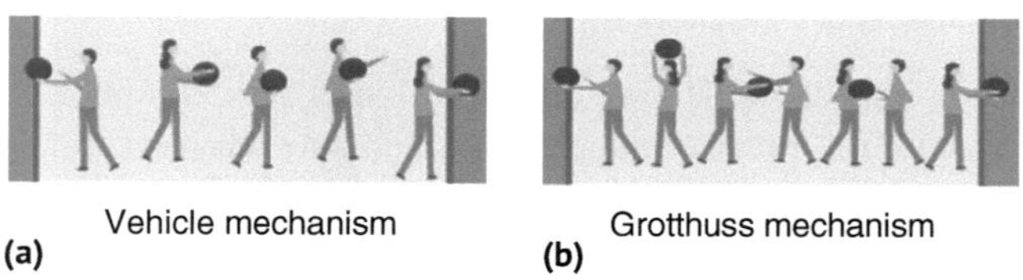

Figure 10.3 Schematic drawings of (a) the vehicle and (b) the Grotthuss mechanisms. *Source*: Y.S. Kang.

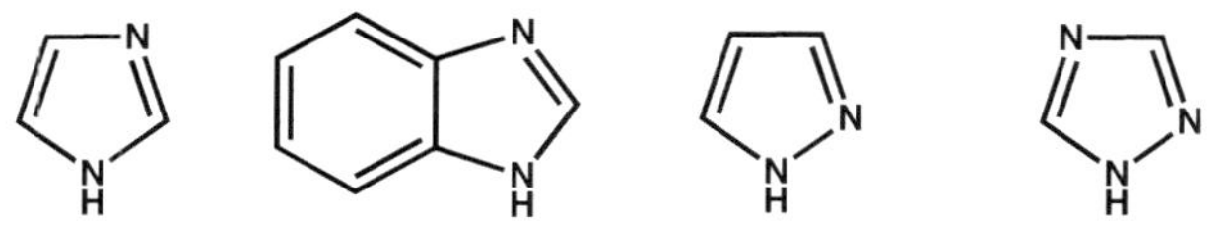

Figure 10.4 Chemical structures of representative heterocycle compounds. *Source*: Y.S. Kang.

Table 10.2 Proton conductivity of heterocyclic compounds without humidification [4].

Heterocycle	Melting temperature, T_m (°C)	Proton conductivity (S/cm)	Measured temperature (°C)
Imidazole	90	$\sim 6 \times 10^{-3}$	130
Benzimidazole	170	$\sim 3 \times 10^{-3}$	140
Pyrazole	70	$\sim 2 \times 10^{-3}$	130
1,2,4-Triazole	120	$\sim 1 \times 10^{-3}$	120

Source: Adapted from [4].

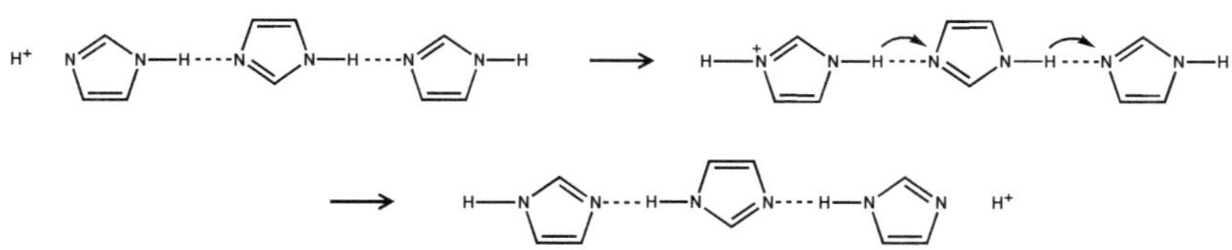

Figure 10.5 Grotthuss mechanisms of H^+ transport through a heterocycle. *Source*: Adapted from [4].

good candidate to replace water as a proton carrier, in particular at temperatures higher than 120 °C [4].

10.3 Structures of Ion Exchange Membranes: CEM, AEM, and BPM

A typical IEM consists of a polymer matrix embedded with fixed ionic groups, allowing selective ion transport. These fixed charge groups are covalently bound and determine ion selectivity, while the polymer matrix provides structural support and houses these functional groups.

Due to the incompatibility between the hydrophilic phase (containing fixed charge groups and counter ions) and the polymer matrix, IEMs naturally adopt a microphase-separated structure, as shown in Figure 10.6. In this structure, CEMs contain negatively charged fixed groups that attract cations and repel anions, while AEMs have positively charged fixed groups that attract anions and repel cations.

CEMs and AEMs are monopolar membranes, while a BPM consists of both a CEL and an AEL joined together, creating a space charge region at their interface

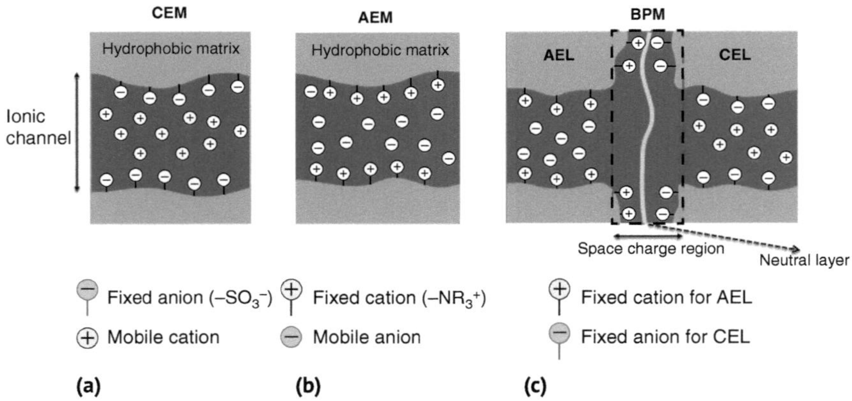

Figure 10.6 Schematic of microphase-separated structures: (a) cation-exchange membrane (CEM) with negatively fixed charge, attracting cations and repelling anions; (b) anion-exchange membrane (AEM) with positively fixed charge, attracting anions and repelling cations; and (c) bipolar membrane (BPM) consisting of a cation-exchange layer (CEL) and an anion-exchange layer (AEL), forming a space charge region at the bipolar junction. *Source*: Y.S. Kang.

(Figure 10.6c). This interface, known as the bipolar junction, generates a strong electric field, up to 10^8 V/m, which significantly enhances electrochemical reactions such as water dissociation ($H_2O \rightarrow H^+ + OH^-$), as discussed in Sections 10.4 and 12.4.

Ion exchange in IEMs is a process in which ions attached to polymer chains (fixed ions) are exchanged for other ions (counter ions) in solutions. The ion-exchange capacity (IEC) primarily determines this process, representing the total number of functional groups available for ion exchange. IEC is commonly expressed in terms of equivalent weight (EW), defined as the grams of polymer per mole of SO_3^- (EW = g polymer/mol SO_3^-). The relationship between IEC and EW is given by EW = 1,000/IEC. IEC is typically measured using a conventional acid-base titration method.

In this section, we first describe the chemical structure and morphology of monopolar IEMs, followed by a discussion on BPMs and their ability to generate strong electric fields.

10.3.1 Chemical Structure of Cation Exchange Membranes

CEMs possess fixed negatively charged groups covalently attached to hydrophobic polymer chains. Typical fixed negatively charged groups for CEMs are acidic func-

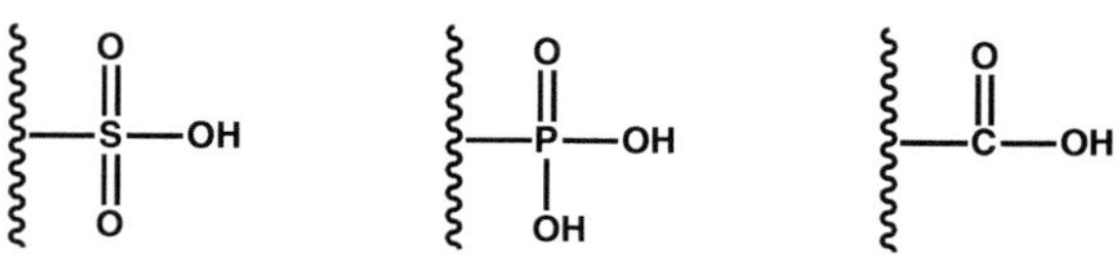

Figure 10.7 Representative fixed charged groups for cation-exchange membranes. *Source*: Y.S. Kang.

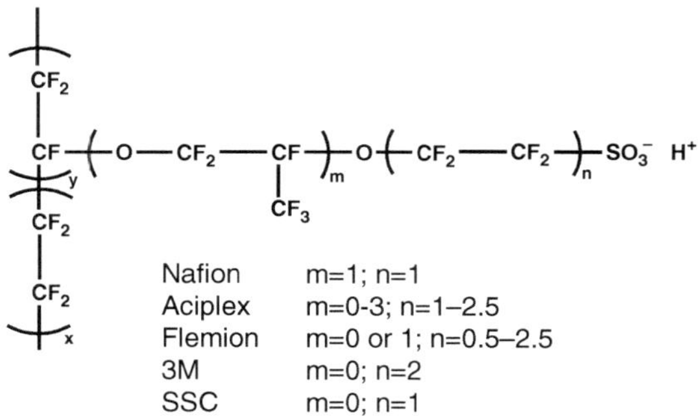

Figure 10.8 Chemical structure of perfluorosulfonic acid/tetrafluoroethylene copolymer, PFSA. Commonly $x \sim 5$ and $y \sim 1,000$. *Source*: Adapted from [5].

tional groups such as sulfonic, carboxylic, and phosphonic acids (Figure 10.7) having different equilibrium constants pK_a (−2.6 for CH_3SO_3H, 2.38 for $CH_3PO_3H_2$, and 4.8 for CH_3CO_2H). In fact, sulfonates have been much more frequently used for practical applications because sulfonic acids show high proton conductivity and good chemical stability against highly acidic environments (pH~0). However, the phosphonic acid also has a strong advantage of high proton conductivity at low humidity or low water-swollen state.

Representative CEMs are perfluorosulfonic acid/tetrafluoroethylene copolymer, often referred to as PFSA. Commercially available PFSAs are Nafion (DuPont), PFA (3M), Aciplex (Asahi Kasei), Flemion (Asahi Glass), and Hyflon (Solvay), and their linking group between the polymeric main chain and the sulfonic acid group is slightly different, as shown in Figure 10.8. Their structure and properties have been rigorously reviewed by Kusoglu and Weber [5].

PFSAs provide high proton conductivity and necessary mechanical strength for practical applications, which have been limited primarily due to their high cost. Therefore, new hydrophobic backbone polymers such as aromatic polymers like engineering plastics have been seriously investigated to replace the PFSAs. Common backbone polymers include polysulfone (PSF), poly(ether-sulfone) (PES), poly(ether-ether-ketone) (PEEK), polyimide, etc., which are tough chemically and mechanically, as shown in Figure 10.9.

10.3.2 Chemical Structure of Anion Exchange Membranes

Representative fixed cationic groups for AEMs include ammonium-, phosphonium-, and sulfonium-, as well as organometal-based cations (Figure 10.10). Among them, ammonium groups have been most extensively investigated in AEMs due to the good nucleophilic activity and tunability of amine groups. Quaternary phosphonium, tertiary sulfonium cations, and organometallic salts have also been recently explored and are briefly reviewed at the end of this subsection.

Despite their wide use, conventional quaternary ammonium (QA) and imidazolium (IM) cations are quite prone to degradation under alkaline conditions via

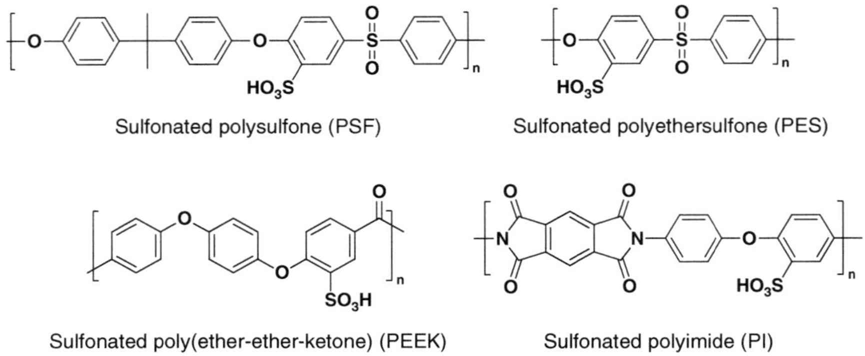

Figure 10.9 Chemical structure of cation-exchange membranes with sulfonic acid as fixed charge group. *Source*: Y.S. Kang.

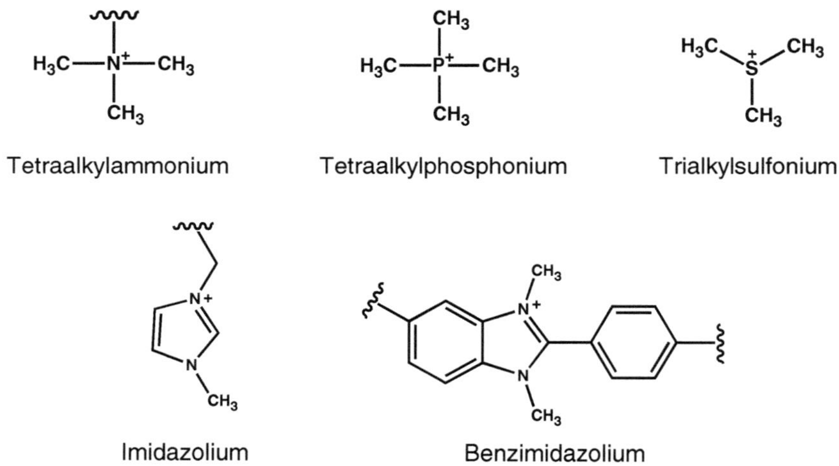

Figure 10.10 Chemical structures of fixed cationic groups for anion-exchange membranes. *Source*: Y.S. Kang

the Hofmann degradation (E2), nucleophilic substitution (S_N2), or ring-opening reaction, especially at elevated temperatures and high-pH conditions [6]. The Hofmann degradation is mainly triggered by the elimination of β-hydrogen via an anti-coplanar intermediate, while S_N2 degradation undergoes by a direct nucleophile attack, largely influenced by the steric hindrance. Therefore, highly sustainable AEMs under high pH and temperature conditions have been actively pursued along with improvements in ionic conductivity.

10.3.2.1 Tetramethyl Ammonium and Piperidinium Cations

Tetramethylammonium, TMA, exhibiting high OH^- conductivity up to 89 mS/cm at 60 °C, was considered to be the most stable QA among early QAs primarily due to the absence of the β-hydrogens [7]. Therefore, TMA-based AEM is frequently used as a reference material for the OH^- conductivity and durability. There has been a large body of research to overcome those two issues simultaneously using quaternary ammonium cations, among which two approaches are noticeable by utilizing piperidinium and imidazolium cations.

Recently, two noticeable cationic groups containing piperidinium, such as *N*-heterocyclic 6-azonia-spiro[5.5]undecanium (ASU) and *N,N*-dimethyl piperidinium (DMP) exhibit higher alkaline stabilities among the common quaternary ammonium groups [8]. In addition, poly(fluorenyl aryl piperidinium) (Figure 10.11a) demonstrates high OH^- conductivity of 208 mS/cm at 80 °C, low H_2 permeability, excellent mechanical properties (84.5 MPa), and 2,000 hours *ex situ* durability in 1 M NaOH at 80 °C along with high water vapor permeability and low phenyl adsorption [9].

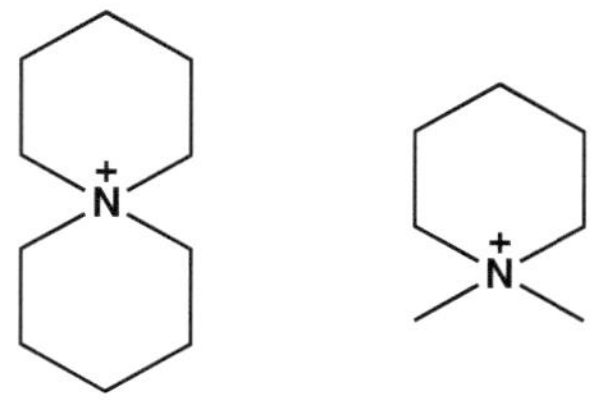

ASU DMP

A series of piperidinium side-chain-type poly(dibenzyl-*co*-terphenyl piperidinium) (s-PDTP) copolymers featuring aryl ether-free polymer backbone and pendent piperidinium cations also endow both high alkaline stability and excellent micromorphology (Figure 10.11b) [10].

10.3.2.2 Imidazolium Cations

Imidazolium-based AEMs have been paid much attention mostly because of their chemical stability in alkaline conditions when suitably chemically modified. Among diverse investigations on designing alkaline-resistant imidazole cations, a noticeable is that the substituent at each position, including C-4 and C-5 positions of the imidazolium ring, have a dramatic effect on the overall cation stability. In this regard, *N,N'*-dibutylimidazolium cation groups with alkyl groups substituted at C-4 and C-5 were very stable in 5 M KOH/CD_3OH at 80 °C for 30 days [11].

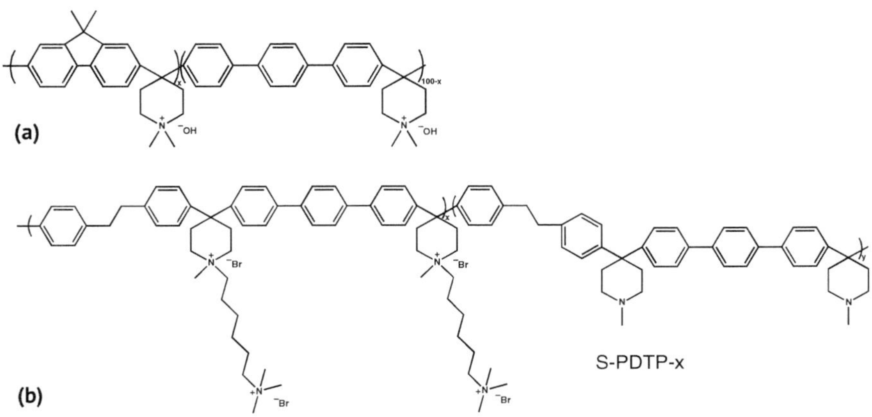

Figure 10.11 Chemical structures of (a) poly(fluorenyl aryl piperidinium). *Source*: Adapted from [9]; and (b) poly(dibenzyl-*co*-terphenyl piperidinium) (s-PDTP). *Source*: Adapted from [10].

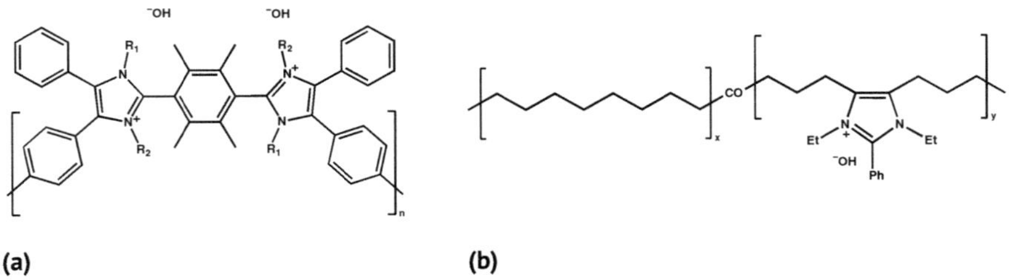

Figure 10.12 Chemical structure of (a) poly(aryl imidazolium)s. *Source*: Adapted from [12]; and (b) random copolymer of *trans*-cyclooctene and *trans*-cyclooctene–fused imidazolium derivative (HC-$[1]_{498}[2]_{200}$). *Source*: Adapted from [13].

AEMs with the sterically protected *N,N′*-dibutylimidazolium cations ($R_1 = R_2 = n$-butyl-) incorporated in poly(aryl imidazolium)s, PAIms, as shown in Figure 10.12a, exhibited both high ion conductivity and high stability. The OH^- ion conductivity was reached up to nearly 100 mS/cm at 80 °C [12]. Crosslinked random copolymer (Figure 10.12b) of *trans*-cyclooctene and *trans*-cyclooctene–fused imidazolium monomers also showed very promising results of the hydroxide conductivity about 50 mS/cm and the nearly unchanged conductivity at 1 M KOH solution at 22 °C for 30 days [13].

Recently, imidazolium compounds containing sterically hindered *N,N*-dibutylimidazolium groups show the increased half-life time $t_{1/2}$ up to almost five-fold longer than that of TMA benchmark [12]. However, their OH^- ion conductivity becomes relatively lower than quaternary ammonium-based AEMs [8].

Imidazolium model compounds 4, 7–10, and 12–14 possessing steric-protecting groups at the C-2 position and different alkyl side chains attached to the *N*-1/*N*-3 imidazole position have been investigated systematically with a reference compound 4 as shown in Figure 10.13. The half-life $t_{1/2}$ and IEC of these representative compounds in 3 M NaOD at 80 °C are listed in Table 10.3. The half-life of ammonium cation 4 was within 0.1 hour. The major degradation pathway evolved from ring-opening and demethylation. These results suggest that increased steric hindrance at the C-2 position raises the energy barrier for ring-opening degradation [12].

In order to elongate the half-life further by suppressing dealkylation of imidazolium, longer alkyl side chains were attached to the *N*-1/*N*-3 positions. Compound 9, with its longer *n*-butyl chains, was found to be significantly more stable ($t_{1/2} > 10,000$ hours) than the corresponding compound with short dimethyl groups. No degradation of compound 9 was observed by NMR spectroscopy after 10 days, which not only supports that dealkylation has been suppressed but also indicates that Hoffman elimination (potentially due to the presence of β-hydrogens) is absent. However, a consequence of attaching longer chains at the *N*-position is that IEC decreases, which may adversely affect ionic conductivity of membranes prepared therefrom. To counter this, bis-imidazolium cations 12–14 show IEC in the range 2.78–2.41 mmol/g, compared with 2.13 mmol/g for compound 9, resulting in the half-life longer than 10,000 hours, as indicated in Table 10.3. This result

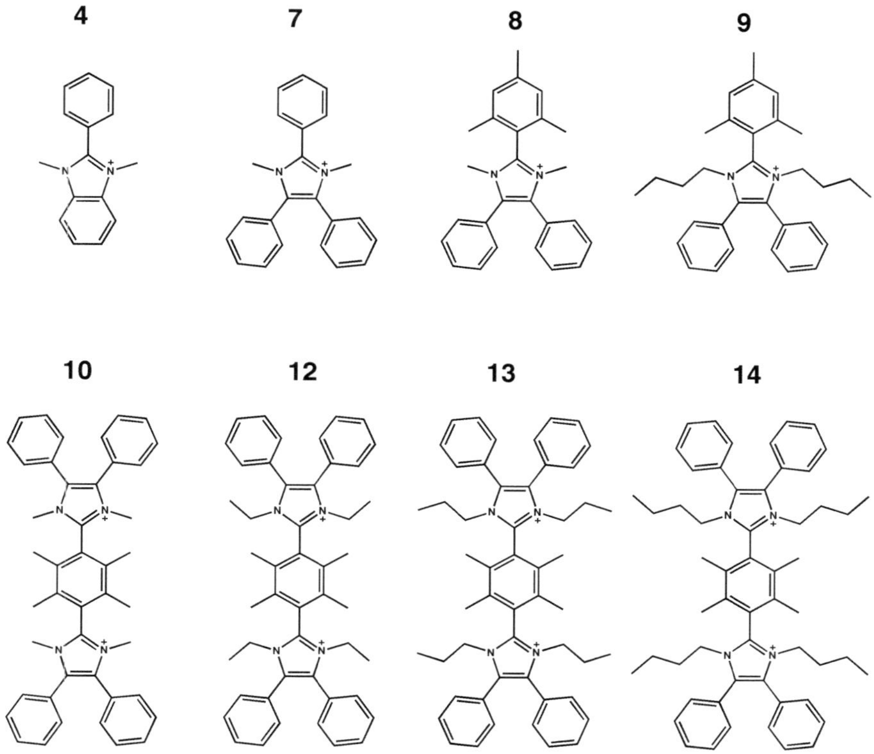

Figure 10.13 Chemical structures of imidazole compounds (numbers 4, 7–10, and 12–14). *Source*: Adapted from [12].

Table 10.3 Half-life $t_{1/2}$ and ion-exchange capacity (IEC) of imidazolium cations [12][a]: Chemical structures of compound numbers (4, 7–10, and 12–14) are in Figure 10.13.

Compound no.	4	7	8	9	10	12	13	14
Half-life (hours)	<0.1	1,370	7,790	>10,000	2330	>10,000	>10,000	>10,000
IEC (mmol/g)	4.16	2.92	2.60	2.13	3.02	2.78	2.58	2.41

[a]The half-life $t_{1/2}$ of imidazolium cations 4, 7–9, 10, and 12–14 was studied at 0.02 M solutions in 3 M NaOD containing 7:3 wt. CD_3OD : D_2O at 80 °C.

can be explained by the higher electrostatic potential (ESP) between the cation and hydroxide anion. The high ESP may lead to more rapid attack of hydroxide on the imidazolium according to the DFT calculations. Attaching longer alkyl chains, with their stronger electron-donating effect on the imidazolium ring, reduces the ESP values. Moreover, longer alkyl chains confer increased steric hindrance around the alkyl—N bond. Thus, compounds 12–14 exhibit exceptional stabilities of >10,000 hours in 3 M NaOD/D_2O/CD_3OD [12].

10.3.2.3 Quaternary Phosphonium and Tertiary Sulfonium Cations, and Organometallic Salt

Quaternary phosphonium and tertiary sulfonium cations are known to be less stable thermally and chemically than QA groups [14]. However, recently, the noticeable improvements in both OH^- conductivity and chemical stability against alkaline conditions have been reported [15, 16]. For instance, tris(2,4,6-trimethoxyphenyl) quaternary phosphonium-based PSF membrane exhibited higher ion conductivity (27 mS/cm at 20 °C) and stability in 1 M KOH for 30 days [17]. Organometallic salts have also been attempted to improve both the OH^- conductivity and also chemical stability, including Ru, Ni, Co, K, and Na ions [8, 18].

10.3.2.4 Backbone Polymers

Poly(aromatic ether)s like PEEK, PSF, and polyphenylene oxide (PPO) are commonly used as backbone polymers for AEMs due to their low cost, easy modification, and strong mechanical properties. Their industrial availability makes them viable for commercial applications such as fuel cells and water electrolysis. These polymers also allow for functionalization to enhance ion conduction in alkaline conditions.

However, their chemical stability is limited in high-pH environments. The presence of β-hydrogens in certain backbone structures also contributes to degradation via elimination reactions, particularly under alkaline conditions, further reducing polymer longevity. Studies by Fujimoto et al. [19] and Arges and Ramani [20] show that aryl ether bonds in these materials degrade under alkaline conditions, causing structural failure. The presence of electron-withdrawing groups like carbonyl and sulfuryl exacerbates this issue by increasing degradation susceptibility. This instability hinders the long-term performance of AEMs [21].

To counter this, researchers developed aryl ether-free polymers like styrene-ethylene-butylene-styrene (SEBS) block copolymer and poly(biphenyl alkylene), as illustrated in Figure 10.14, which exhibit better stability in alkaline environments [8]. These hydrocarbon-based backbones with carbon—carbon (C—C) and carbon—hydrogen (C—H) bonds resist nucleophilic attack. For example, poly(biphenyl alkylene), with stable biphenyl units, and SEBS demonstrate improved durability and versatility for AEM use.

Another approach uses fully aromatic or aliphatic hydrocarbon polymers, as shown in Figures 10.11 and 10.12. Fully aromatic types, like polyphenylene [22, 23], offer thermal and oxidative stability, while aliphatic ones [24], such

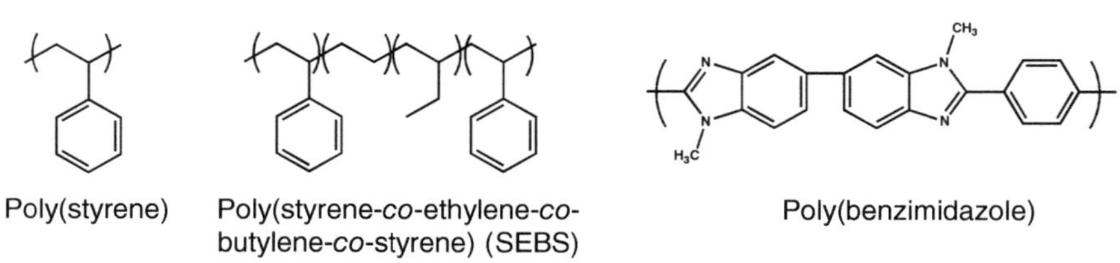

Figure 10.14 Chemical structures of polymer backbones of anion-exchange membranes: polystyrene, poly(styrene-*co*-ethylene-butylene-*co*-styrene) (SEBS), and poly(benzimidazole). *Source*: Y.S. Kang.

as polyethylene derivatives [25], resist oxidative and hydrolytic damage. Combining these in block copolymers balances mechanical and chemical properties. Molecular weight and crosslinking density are also critical in enhancing mechanical strength and stability.

In conclusion, the shift from poly(aromatic ether)s to aryl ether-free and fully aromatic or aliphatic polymers marks significant progress. Future research integrating advanced synthesis, modeling, and testing will help overcome challenges in hydroxide stability and durability, paving the way for next-generation AEMs.

10.3.3 Morphology of Ion Exchange Membranes

IEMs are typically required to meet several key criteria: high ionic conductivity for specific ions, selective barrier properties for others, mechanical strength for efficient module design, and electrochemical stability for long-term operation. One effective approach to achieving these properties is the formation of a microphase-separated structure, as exemplified by such as Nafion, as illustrated in Figure 10.15a, b [26]. This structure consists of interconnected hydrophilic ion channels, formed by ion clusters, embedded within a hydrophobic matrix. The hydrophilic channels provide pathways for ionic conduction, while the hydrophobic matrix ensures the necessary mechanical strength.

The microphase-separated structure is readily formed because of the incompatibility between the hydrophilic clusters and channels and the hydrophobic matrix. Increasing the content of the hydrophilic clusters, ionic clusters may interconnect together to form ion channels. Figure 10.15a is a classical two-phase model of the Nafion membrane [4, 27], assuming that the —SO_3H concentration in the hydrophobic matrix is nearly zero. The fixed site acidic or basic groups in hydrophilic clusters are concentrated on their interface, as shown in Figure 10.15b.

The morphological difference of PEMs between the perfluorinated polymer (Nafion) and sulfonated PEEK is illustrated in Figures 10.16a and b, respectively. The ion channel in Nafion is wider, less branched, and less tortuous than that in PEEK-based PEM. Therefore, Nafion has generally much higher ion conductivity than PEEK-based PEM [27].

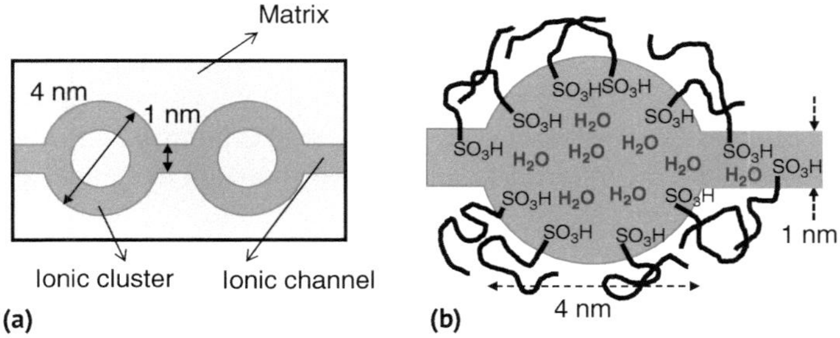

Figure 10.15 Schematics of (a) microphase-separated structure for Nafion configuring hydrophilically interconnected ion clusters with ion channels dispersed in a hydrophobic matrix; and (b) ion cluster and channel structure. *Source*: Adapted from [26].

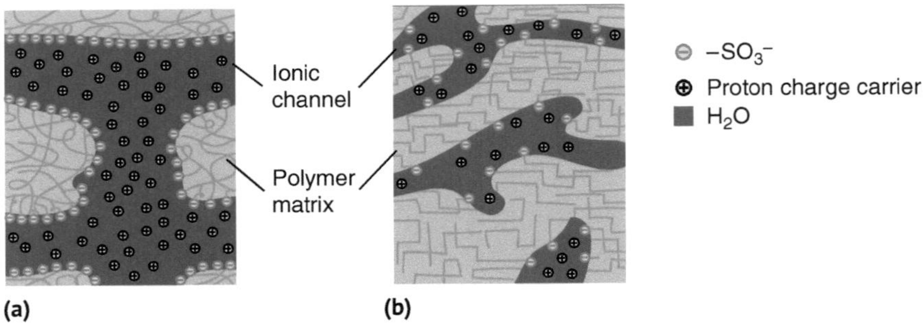

Figure 10.16 Schematic morphologies of cation-exchange membranes (CEMs) containing $-SO_3H$ with (a) perfluorinated polymers (Nafion) and (b) sulfonated poly(ether-ether-ketone) (PEEK). *Source*: Adapted from [27].

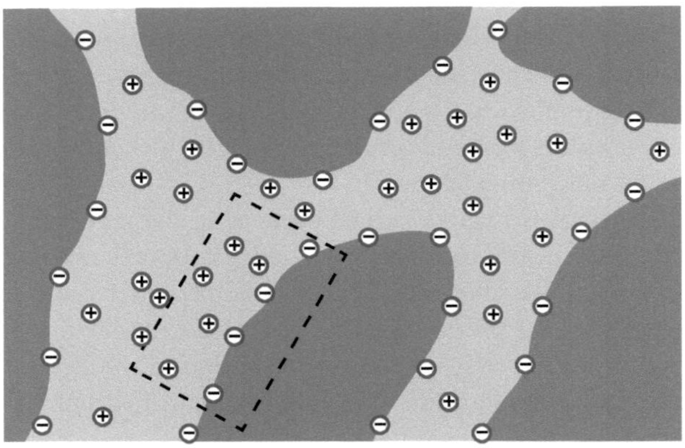

Figure 10.17 Two-phase structure model composed of interstitial (the gray part with ions) and gel phases (the dark part). *Source*: Adapted from [3].

A new, different two-phase model was proposed to have gel and interstitial phases [4]. The gel phase is composed of hydrophilic ion-exchange groups bound to the polymer main chains, and the interstitial phase is the void among the gel phases (Figure 10.17). Likewise, the interstitial phase is filled with electro-neutral solution when the membranes are hydrated with salt solutions, and responsible for the transport of co-ions through IEMs [3].

Another notable example of a CEM is poly(benzimidazole) doped with phosphoric acid. Unlike the phase-separated structure, this material exhibits a homogeneous polymer electrolyte structure, offering an alternative approach to achieving the desired membrane performance [28, 29].

10.3.4 Water Absorption and Structural Evolution

Hydrophilic ion clusters and channels are usually swollen by water solvent and interconnected together at high water concentrations. They provide pathways for

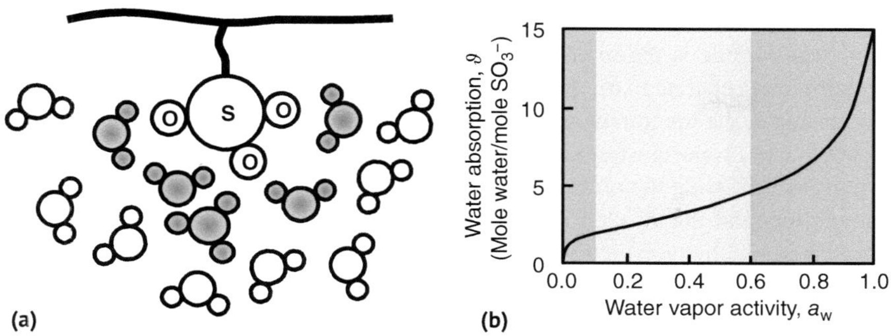

Figure 10.18 (a) The two types of absorbed water molecules in proton exchange membrane (PEM): five strongly bound water molecules ϑ^C in the primary hydration shell, akin to chemisorption, and eight physically bound water molecules, akin to physisorption ϑ^P. Adapted from [30]. (b) The water sorption amount ϑ, in Nafion (EW = 1100) as a function of water vapor activity. *Source*: Adapted from [30].

ion transport, and consequently their size and connectivity with others will mostly determine the ion conductivity.

In an aqueous environment, water is absorbed into IEMs, particularly through ionic clusters and channels, and their structure changes to a large extent depending on the amount of water absorbed. The amount of water absorbed in IEMs is frequently characterized by ϑ, the mole number of water absorbed per one mole of ion-exchange site. The absorbed water consists of surface or bound water ϑ_B and bulk or free water ϑ_F, where ϑ_B in turn is assumed to be two types: (1) chemically or strongly bound waters (akin to chemisorption), represented by ϑ^C; and (2) "free" to physically equilibrated waters (akin to physisorption) in the fluid phase, ϑ^P. Thus, $\vartheta = \vartheta_B + \vartheta_F$ where $\vartheta_B = \vartheta^C + \vartheta^P$. A schematic of these two different types of water molecules in Nafion is shown in Figure 10.18a. In the case of Nafion, ϑ^c was reported to be ~5 and ϑ^P was about 8 [30].

Figure 10.18b illustrates a typical example of the total water ϑ absorbed as a function of water vapor activity, a_w. In the case of Nafion (EW = 1100), the water uptake, expressed as the ϑ value, shows distinct trends: a sharp increase at low water activities ($a_w < 0.1$), a nearly linear rise in the intermediate range ($0.1 < a_w < 0.6$), and a steeper increase again at higher water activities ($a_w > 0.6$) [30].

The structural change depending on the ϑ value is schematically represented in Figure 10.19. The pictures are cross-sectional representations of the membrane

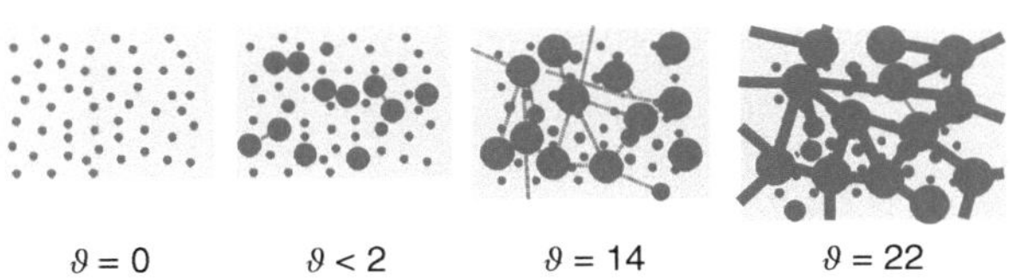

Figure 10.19 Evolution of the membrane structure as a function of water content, ϑ, moles of water per mole of sulfonic acid. *Source*: Adapted from [31].

where the gray area is the hydrocarbon matrix, the black circle is the hydrophilic cluster, and the line is the interconnected ion channel. When $\vartheta = 0$, Nafion shows the microphase-separated structure with extremely small sized hydrophilic clusters. As increasing ϑ, the membrane structure evolves to the interconnected structure. For instance, at $\vartheta = 14$, the clusters start to interconnect together through slightly swollen ion channels, initiating the proton conductivity increase. When $\vartheta > 22$, both the sizes of the clusters and the swollen ion channels increase to a large extent, from which the proton conductivity increases sharply. This issue will be discussed in more detail in the next section to explain the relationship between ϑ and ionic conductivity σ. These indicate that the morphology of the microphase-separated structure, which depends on the amount of water absorbed, strongly affects the ion conductivity [31].

10.4 Electric Field Generation at Bipolar Membrane Junction

A BPM is a layered structure composed of one CEL and the other AEL joined together in series, as shown in Figure 10.6c. The most important characteristic of these BPMs is to build an electric field up to $\xi = 10^8$ V/m at the bipolar junction between CEL and AEL, called the space charge region or depletion region, as shown in Figure 10.20 [32]. The high electric field built at the space charge region markedly enhances, for instance, water dissociation rate to generate H^+ and OH^- ($H_2O \rightarrow H^+ + OH^-$) up to 20–$10^6$ fold, compared to that for without the electric field, according to the theoretical models [32, 33]. In this section, the generation of high electric fields at the bipolar junction from the structural models of the space charge region will be focused, while its effects on the water dissociation will be discussed in Section 12.4 of Chapter 12.

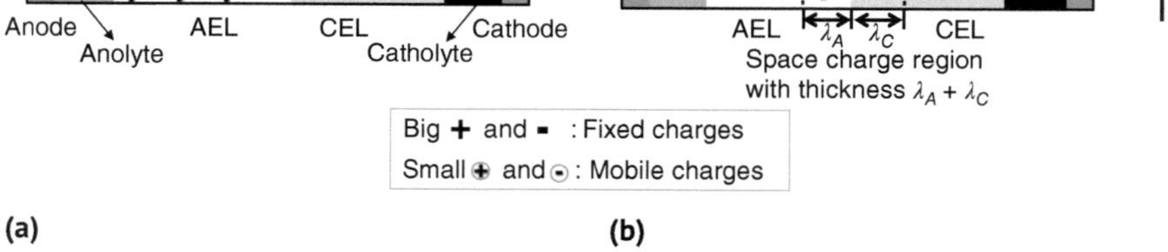

Figure 10.20 Schematic illustration of the formation of the space charge region after contacting an anion-exchange layer (AEL) and cation-exchange layer (CEL) to make a bipolar membrane (BPM): (a) before contact; and (b) after contact. The electrical potential difference ($\Delta\varphi^{sc}$) is generated across the space charge region with its thickness ($\lambda_A + \lambda_C$). *Source*: Y.S. Kang.

When a CEL and an AEL contact together, mobile ions of H^+ and OH^- in CEL and AEL near the bipolar junction diffuse to AEL and CEL, respectively, meet together to form water, as shown in Figure 10.20. Therefore, both the mobile ions will be depleted in each layer at the bipolar junction while the fixed ions maintain their initial sites and charges without change, so that the large electric potential difference ($\Delta\varphi^{sc}$) will be developed to generate an electrical field ($\xi = \Delta\varphi^{sc}/(\lambda_A + \lambda_C)$) across the space charge region with thickness ($\lambda_A + \lambda_C$). The electric field can thus act as a driving force of ion transport, called migration, with opposite direction to diffusion. After a while, the diffusion and migration rates eventually become equivalent to each other with opposite direction, resulting in a steady state and a constant thickness of the space charge region. These results suggest that the space charge region generates an electric potential difference across the narrow bipolar junction, yielding a huge electric field, as shown in Figure 10.20.

There are three different structural models for the space charge region developed so far: (1) abrupt bipolar junction, (2) smooth bipolar junction, and (3) neutral layer at the bipolar junction models. In this section, the abrupt bipolar junction and neutral layer at the bipolar junction models, as illustrated in Figure 10.21, will be introduced to characterize the electrical potential, thickness, and electric field of the space charge region [32]. For the smooth bipolar junction model, you may refer to ref. [33].

10.4.1 Abrupt Bipolar Junction Model

A CEL and an AEL contain a fixed negative charge concentration X^{CEL} and a fixed positive charge concentration, X^{AEL}, respectively, throughout whole BPM. The concentrations of mobile cations and anions, denoted by c_C and c_A at CEL and AEL, respectively, change with position x accordingly, as illustrated in Figure 10.21a.

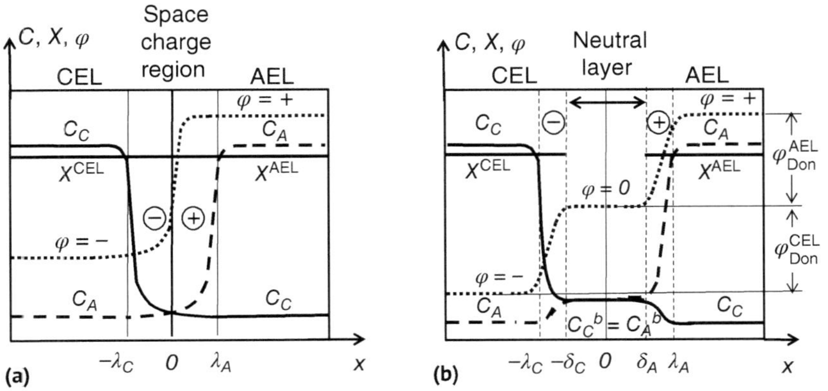

Figure 10.21 Schematic structures of (a) abrupt bipolar junction; and (b) neutral layer at the bipolar junction models, where c_j, φ, and λ are the concentration of charged mobile species j, electric potential, and the thickness of the space charge layer, respectively. The cation and anion-exchange layers contain a fixed charge concentration X^{CEL} and X^{AEL}, respectively. Subscript C and A are for cation and anion, and superscript b is for bulk. *Source*: Adapted from [32].

When it is assumed that the electric potential, φ, satisfies the Poisson equation, which is given by [32]:

$$\frac{\partial^2 \varphi}{\partial x^2} = -\frac{\rho_e}{\varepsilon_0 \varepsilon_r} \tag{10.7}$$

Here, ε_0 is the permittivity of free space (vacuum) and ε_r the relative permittivity or dielectric constant; ρ_e is the space charge density for the abrupt bipolar junction model which can be calculated from [32]:

$$\rho_e = F\left(\sum z_j c_j + \sum z_j X^m\right) \tag{10.8}$$

Here, subscript j is either c for mobile cation and a for mobile anion and superscript m is for either CEL or AEL. X is the fixed charge density, and z_j is the electrochemical valence. In the case of the abrupt junction model, $c_j \sim 0$, so that the mobile charge effects are commonly nullified. Therefore, ρ_e can be expressed as:

$$\rho_e = FX^{\mathrm{CEL}} \quad \text{at} - \lambda_C \leq x \leq 0 \tag{10.9}$$

$$\rho_e = -FX^{\mathrm{AEL}} \quad \text{at } 0 \leq x \leq \lambda_A \tag{10.10}$$

Poisson's equation (Equation (10.7)) is solved using two boundary conditions given in Equations (10.9) and (10.10). In the symmetrical case, where identical fixed charge densities are equal ($X^{\mathrm{CEL}} = X^{\mathrm{AEL}} = X$) and thicknesses of the CELs and AELs are the same ($\lambda_C = \lambda_A = \lambda$) the total thickness 2λ of the space charge or depletion layer can be calculated by [32]:

$$2\lambda = 2\sqrt{\frac{\Delta\varphi^{\mathrm{sc}} \varepsilon_0 \varepsilon_r}{XF}} \tag{10.11}$$

where $\Delta\varphi^{sc}$ is the potential difference across the space charge layer, which is given by:

$$\Delta\varphi^{\mathrm{sc}} = \varphi_{\mathrm{Don}}^{\mathrm{CEL}} + \varphi_{\mathrm{Don}}^{\mathrm{AEL}} = 2\varphi_{\mathrm{Don}}^{\mathrm{sc}} \tag{10.12}$$

with an assumption of $\varphi_{\mathrm{Don}}^{\mathrm{CEL}} = \varphi_{\mathrm{Don}}^{\mathrm{AEL}} = \varphi_{\mathrm{Don}}^{\mathrm{sc}}$ where $\varphi_{\mathrm{Don}}^{\mathrm{CEL}}$ and $\varphi_{\mathrm{Don}}^{\mathrm{AEL}}$ are Donnan potentials developed in corresponding ion-exchange layers (IELs), as shown in Figure 10.21a and 10.22. If we assume that the system contains only monovalent ions and that the concentration c_j^b of charged mobile species j at $x = 0$ is much smaller than X ($c_j^b \ll X$), the potential difference $\Delta\varphi^{\mathrm{sc}}$ across the space charge region can then be determined by [32]:

$$\Delta\varphi^{\mathrm{sc}} = 2\left|\frac{RT}{z_j F}\ln\frac{c_j^b}{X}\right| = 2\frac{RT}{F}\ln\frac{X}{c_j^b} \tag{10.13}$$

In the case of water medium, $c_{\mathrm{H}^+}^b = c_{\mathrm{OH}^-}^b = 10^{-7}$ mol/l at the center of the space charge region ($x = 0$).

With a fixed charge density $X = 1.0$ mol/l and a relative permittivity of water $\varepsilon_r = 78$, the calculated thickness 2λ of the space charge layer using Equations (10.11) is $2\lambda = 4.84$ nm. Equation (10.13) gives that the calculated potential difference $\Delta\varphi^{\mathrm{sc}}$ across the space charge region of BPM in water at $T = 20\,°\mathrm{C}$ is $\Delta\varphi^{\mathrm{sc}} = 0.814$ V.

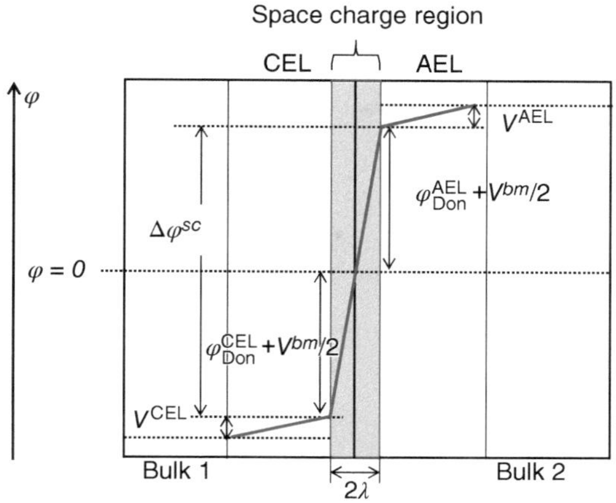

Figure 10.22 Simplified potential profile for the abrupt bipolar junction model across the bipolar membrane. *Source*: Adapted from [32].

Equation (10.11) describes the thickness 2λ of the space charge layer when no external electrical field is applied. When an external voltage V^{bm} is established across the BPM, the voltage drops across the space charge layer will be changed accordingly as illustrated in Figure 10.22. For small current densities, the voltage drops across the monopolar membranes V^{CEL} and V^{AEL} can also be neglected and $\Delta\varphi^{sc}$ can be described by [32]:

$$\Delta\varphi^{sc} = \varphi^{CEL}_{Don} + \varphi^{AEL}_{Don} + V^{bm} = 2\varphi^{sc}_{Don} + V^{bm} \tag{10.14}$$

In Equation (10.14), V^{bm} is positive, the thickness of the space charge region 2λ is then [32]:

$$2\lambda = 2\sqrt{\frac{\left(2|\varphi^{sc}_{Don}| + V^{bm}\right)\varepsilon_0\,\varepsilon_r}{XF}} \tag{10.15}$$

Equation (10.15) shows that, with increasing applied reverse bias, the thickness of the depletion layer increases from 4.8 to 7 nm with increasing the applied external voltage from 0 to 1.0 V.

10.4.2 Neutral Layer at Bipolar Junction Model

A neutral water layer between the two IELs may form in some BPMs [34], representing a significant departure from the idealized, abrupt bipolar junction model discussed earlier. For simplicity, consider a symmetric BPM consisting of two ion-exchange layers with equal fixed charge concentrations but opposite signs (i.e., $X^{CEL} = X^{AEL}$). These layers are separated by a transition zone that contains no fixed charge groups, referred to as the neutral layer, spanning the region $-\delta_C < x < \delta_A$ (see Figure 10.21b).

The neutral layer is a thin, uncharged region, that typically forms under open-circuit conditions or when no significant external electric field is applied across the bipolar membrane. This layer may arise from water trapped during the lamination of the BPM or from an inserted neutral porous membrane between the two charged layers. For simplicity in analysis, we assume that this zone has the same electrical permittivity ε as the IELs.

The concentration of charged mobile species c_j in an electric field with an electrical potential of $\varphi(x)$ can be calculated by the Maxwell–Boltzmann relation [32]:

$$c_j^{\varphi(x)} = c_j^{\varphi=0} e^{-z_j \varphi(x) e_0/kT} \tag{10.16}$$

where e_0 and k are the elementary charge and the Boltzmann constant, respectively.

From Equation (10.16), the Donnan potential φ_{Don} for IELs in equilibrium with a surrounding solution can be calculated. It is assumed that in the bulk phase, i.e. at a large distance from the membrane, $\varphi(x) = 0$ and $c_j^{\varphi=0} = c_j^{b}$. In the membrane, the potential is assumed to be $\varphi(x) \neq 0$. With Equation (10.16), this results in [32]:

$$c_j^m = c_j^b e^{-z_j(\varphi^m - \varphi^b) e_0/kT} \tag{10.17}$$

where superscript m represents either CEL or AEL membrane. The Donnan potential can then be calculated by [32]:

$$\varphi_{\text{Don}} = \varphi^m - \varphi^b = \frac{RT}{z_j F} \ln \frac{c_j^b}{c_j^m} \tag{10.18}$$

The space charge density ρ_e for the Poisson's equation (Equation (10.7)) for the neutral layer at bipolar junction model as shown in Figure 10.21b, can also be obtained from Equation (10.8).

Figure 10.21b illustrates the concentration and potential profiles of a CEL and AEL separated by an electrolyte solution and a strong electric field forms across the space charge region. The figure indicates that the electroneutrality at the interfaces is disturbed due to changes in the concentration of counterions and co-ions in the membranes and the surrounding solution. As a result, a space charge region remains at the interfaces [32].

10.5 Transport Models of Ions in Ion Exchange Membranes

IEMs possess a microphase-separated structure, as illustrated in Figure 10.15. This structure comprises hydrophilic ion clusters and interconnected water-filled ion channels embedded within a hydrophobic polymer matrix. Ion transport occurs exclusively through these water-filled ion clusters and channels.

Ion transport in IEMs is described using the dual-mode transport concept. The first model, known as the facilitated transport model, explains ion movement through two distinct mechanisms: bulk transport and interfacial transport. Bulk transport refers to the movement of ions through the water-filled cores of the ion channels. Interfacial transport, also called carrier-mediated transport, involves fixed charges located along the channel interface, which act as immobile carriers for counter ions. Reversible interactions between these fixed charges and mobile counter ions facilitate ion movement along the interface.

The second model, the condensed counterion model, distinguishes between two types of counterions. Condensed counterions are tightly bound to the fixed charges within the membrane, while uncondensed counterions move more freely. Recent findings have shown that condensed counterions contribute more than twice as much to overall ion transport as uncondensed counterions [37].

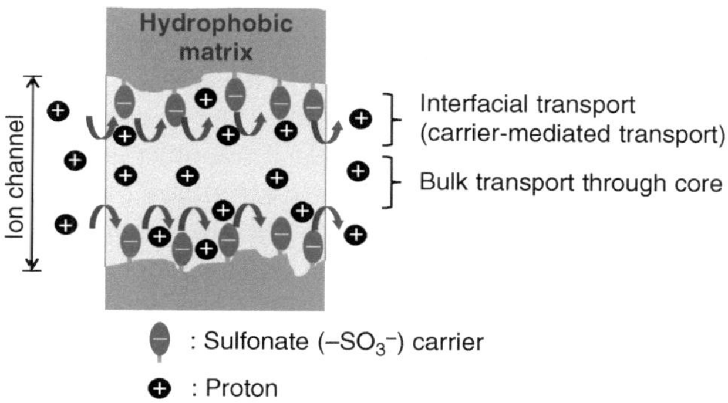

Figure 10.23 Schematic microphase-separated structure of proton-exchange membrane containing ion channels dispersed in hydrophobic matrix and two different proton conduction modes of bulk and interfacial transports. *Source*: Y.S. Kang.

The following discussion first examines the facilitated transport model, followed by an overview of the condensed counterion model.

10.5.1 Facilitated Ion Transport Model Through Ion Channels

An ion channel consists of two different phases depending on whether the effects of fixed charge groups bound to the interface persist or not, as shown in Figure 10.23. The bulk core phase may not be influenced by the fixed charge, whereas the interfacial layer consisting of fixed charge significantly affects the ion transport by reversible interactions of mobile ions such as H^+ and OH^- with fixed charge groups as carriers for facilitated ion transport.

Ion transport through the cores of ion channels is nearly equivalent to that in bulk water, as ions in these cores are hardly affected by fixed charges on the interface. Consequently, they behave like free ions. For instance, H^+ and OH^- ions follow their typical transport mechanisms found in bulk water: the vehicle mechanism and the Grotthuss mechanism, as described previously in Section 10.2.

Fixed ions such as sulfonate groups at ion channel interfaces of CEMs including Nafion, *reversibly* interact with proton H^+, and therefore they can play as a proton carrier for facilitated proton transport.

$$\Big\}\text{--}\mathbf{SO_3{-}} + \mathbf{H^+} \underset{k_2}{\overset{k_1}{\rightleftharpoons}} \Big\}\text{--}\mathbf{SO_3H} \qquad K = k_1/k_2 \tag{10.19}$$

where k_1 and k_2 are the forward and backward reaction rate constants, respectively, and $K(= k_1/k_2)$ is the equilibrium constant of the carrier-solute reaction. Cations in AEMs, including quaternary ammonium ions, for instance, also reversibly react with free anion X^- such as OH^-, thus acting as OH^- carrier to facilitate the OH^- transport.

$$\Big\}\text{---}\mathbf{N}^{+} + \mathbf{X^-} \underset{k_2}{\overset{k_1}{\rightleftharpoons}} \Big\}\text{---}\mathbf{N}^{+}\mathbf{X^-} \qquad K = k_1/k_2 \tag{10.20}$$

Because of the reversible reactions of the fixed ions with counterions, the transport of the counterions will be facilitated by the carrier action of the fixed ions attached to polymeric chains at the interface. The carrier-mediated transport occurs

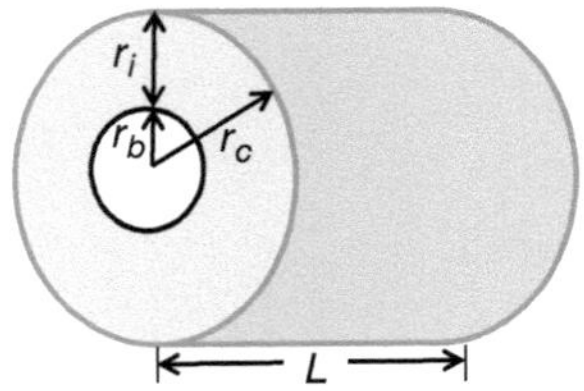

Figure 10.24 Schematic structure of an idealized ion channel in ion-exchange membranes (IEMs) with r_c and r_b being the radii of the ion channel and the bulk phase, respectively, and r_i the interfacial layer thickness. r_i ($r_i = r_c - r_b$) can be represented by the Bjerrum length λ_B ($r_i = \lambda_B$). L is the film thickness or the length of the channel. *Source*: Y.S. Kang.

only through the interface layer with the thickness r_i of ion channels (Figure 10.24) because the fixed ions acting as carriers are present there, but nowhere else. The carrier-mediated transport has been called differently in literatures, one common example is the "hopping transport."

For the case of PEMs, the fixed sulfonate anions confined to the interface are ionized in the presence of water and are also surrounded by water (forming a hydration shell), as are the mobile ions in solution. The strong coordination of the water molecules around the fixed ions reduces the Coulombic interactions between the mobile counter ions like H^+ and fixed groups. Quantum-mechanical calculations predict a substantial weakening of the Coulombic interactions when water molecules are present between them [3, 35]. Because of the weakened Coulombic interactions, the backward reaction rate will be enhanced to give a higher rate constant k_2, resulting in the facilitated proton transport by the carrier action of the sulfonate anions according to the concentration fluctuation model described in Section 4.2.4 (Equations (4.11) and (4.12)) of Chapter 4.

Adopting the analogy between the permeability P_j of a neutral molecule through facilitated transport membranes in the solid state and the conductivity σ_j of an ionic species j (as described in Equations (4.11) and (4.12) in Chapter 4), the mathematical equation for facilitated transport with fixed site carriers in the solid state will be converted into for ion transport without considering the tortuosity effects through the interfacial layer of the ion channel structure:

$$\frac{\sigma_f}{\sigma_b} = 1 + \left(\frac{V_f}{V}\right)\sqrt{n^2 + \left\{\frac{2\pi k_2 L^2 C_C}{\sigma_b}\frac{\ln(1+KV)}{V}\right\}^2} \tag{10.21}$$

In this equation, σ_f and σ_b represent the ionic conductivity through the interfacial layer of ion channels with and without fixed charge carriers, respectively. C_C, k_2, and K represent the carrier (or fixed ion) concentration, the backward reaction rate constant, and the equilibrium constant for the reversible reactions of Equations (10.19) and (10.20), respectively, with applied voltage V. V_f is the potential fluctuation due to the concentration change associated with the reversible reaction of Equations (10.19) or (10.20). This formulation provides a framework for understanding ion transport in fixed site carrier membranes under applied voltage.

Equation (10.21) is only for the interfacial layer in ion channels, suggesting the importance of the backward reaction rate constant k_2 on the final conductivity. When we consider an idealized ion channel with a radius of r_c where r_i and r_b are the thickness of the interfacial layer and radius of the bulk phase ($r_c = r_b + r_i$), respectively (Figure 10.24), the total ionic conductivity σ(S/cm) through the ion channel can be modeled as:

$$\sigma = \frac{\sigma_b r_b^2 + \sigma_f (r_c^2 - r_b^2)}{r_c^2} \tag{10.22}$$

where σ_f can be obtained from Equation (10.21).

If we assume $r_i = \lambda_B$, known as the Bjerrum length, Equation (10.22) is readily converted into:

$$\sigma = \frac{\sigma_b (r_c - \lambda_B)^2 + \sigma_f (r_c^2 - (r_c - \lambda_B)^2)}{r_c^2} \tag{10.23}$$

where λ_B, is defined as $\lambda_B = e^2/4\pi\varepsilon_0 \varepsilon kT$. For water at room temperature ($T = 298$ K), $\varepsilon_r = 80$, so that $\lambda_B = 0.71$ nm [36]. The Bjerrum length λ_B is the separation at which the electrostatic interaction between two elementary charges is comparable in magnitude to the thermal energy scale, kT. Since parameters like σ_b, r_c, and λ_B are readily obtainable, σ_f can be estimated from Equation (10.23) with one adjustable parameter V_f, the degree of potential fluctuation.

10.5.2 Condensed Counterion Model

A recent approach uses the effects of condensed counterions on understanding ion transport through ion channels [37]. When fixed ions are densely concentrated near the channel interface, nearby counterions readily interact with them, while those farther away in the bulk phase interact less. Counterions close to the interface are termed "condensed counterions," while those farther away from it are "uncondensed counterions." This distinction affects overall ionic conductivity, similar to how the interfacial layer and bulk core phase contribute differently: the interfacial layer with condensed counterion phase and the bulk core with uncondensed counterion phase. Consequently, the interfacial layer thickness r_i in Figure 10.24 can be represented by the condensed layer thickness, known as the Bjerrum length λ_B.

The ratios of diffusion coefficients for condensed counterions, D_c^m, to uncondensed counterions, D_u^m, were measured as a function of external NaCl concentration in IEMs, including CR61 (a CEM based on crosslinked polystyrene), AR103 (an AEM based on crosslinked polystyrene), and AR204 (an AEM based on crosslinked poly(ethyl methacrylate)) from GE Power and Water. The results revealed that D_c^m/D_u^m consistently ranged between 2.0 and 2.5, as shown in Figure 10.25 [37]. Additionally, the diffusion coefficients for counterions were experimentally found to be higher than those for co-ions.

A new model based on counterion condensation model was proposed by Freeman et al. to explain these results. This model classifies condensed and uncondensed counterions using the dimensionless linear charge density, ξ, as defined in [37].

$$\xi = \frac{e^2}{4\pi\varepsilon_0 \varepsilon kTb} = \frac{\lambda_B}{b} \tag{10.24}$$

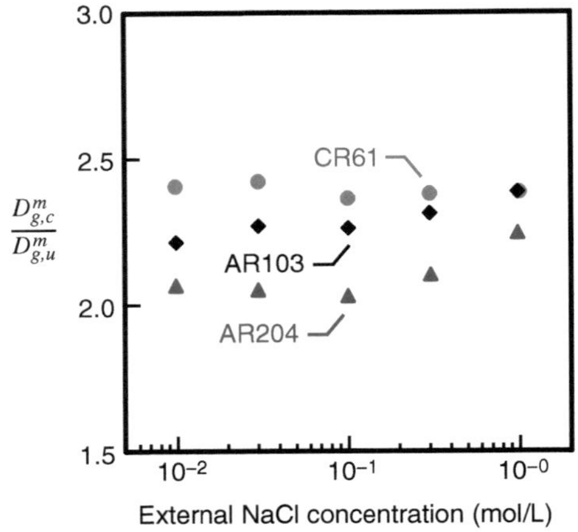

Figure 10.25 The ratio of diffusion coefficients of condensed counterion, $D^m_{g,c}$, to that of an uncondensed counterion, $D^m_{g,u}$, with subscript g being for counterion, as a function of external NaCl concentration in three ion-exchange membranes (IEMs): CR61 (CEM based on crosslinked polystyrene), AR103 (AEM based on crosslinked polystyrene), and AR204 (AEM based on crosslinked poly(ethyl methacrylate)) from GE Power and Water. *Source*: Adapted from [37].

where e is the elementary electron charge, ε_0 the vacuum permittivity, ε the dielectric constant, k the Boltzmann's constant, T the absolute temperature, b the average distance between the fixed ions on the polymer chain, and λ_B the Bjerrum length, defined previously.

For a monovalent system, when $\xi > 1$ and therefore $\lambda_B > b$, the counterions will be in condensed state, while when $\xi < 1$ and therefore $\lambda_B < b$, the counterions will be in uncondensed state, as shown in Figure 10.26.

Manning initially assumed that the condensed counterions were immobile, and therefore did not contribute to the ionic conductivity, but only the uncondensed counterions were mobile to contribute the ionic conductivity [38, 39]. However, Freeman et al. [37] proposed a new mathematical model to estimate the diffusion

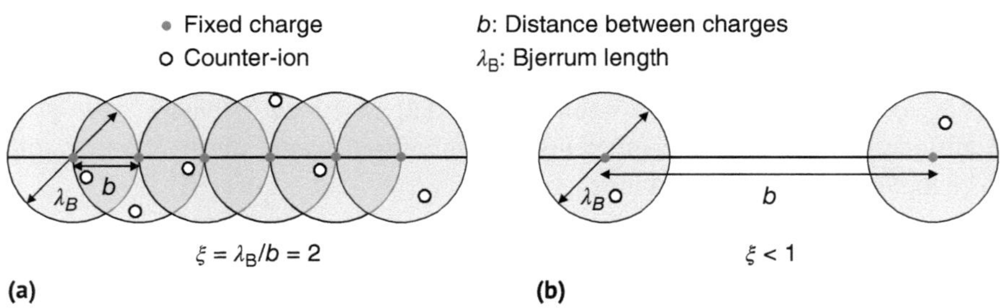

Figure 10.26 The illustration of polyelectrolyte chains in Manning's model for the case of monovalent fixed charges and counterions when (a) counterion condensation occurs (i.e. $\xi = 2$); and (b) counterion condensation does not occur (i.e. $\xi < 1$). When $\xi = 2$, 50% of the counterions are "condensed." *Source*: Adapted from [36].

coefficients of counterions and co-ions through IEMs by considering the effect of electrostatic interactions between the mobile ions (both counterion and co-ions) and the fixed charge groups of the membrane, along with tortuosity of ion diffusion paths in membranes.

The expression for counterion diffusion coefficients, D_g^m, with subscript g being for counterion in IEMs is given by

$$D_g^m = f_u D_{g,u}^m + f_c D_{g,c}^m \tag{10.25}$$

where f_u and f_c are the fractions of uncondensed and condensed counterions, respectively, and $D_{g,u}^m$ and $D_{g,c}^m$ are the diffusion coefficients of the uncondensed and condensed counterions, respectively, in the membrane. With the help of the expression for uncondensed counterion diffusion coefficients $D_{g,u}^m$ obtained from the Manning and Mackie/Meares models [39–41], the expression for condensed counterion diffusion coefficients, $D_{g,c}^m$, is given by [37]:

$$D_{g,c}^m = \frac{1}{3}\alpha D_g^s \tag{10.26}$$

where D_g^s is diffusion coefficient of counterion in aqueous solution. The factor α accounts for the molecular architecture of the polymer and the structured water on or near it. It is expected to be less than unity and dependent on the type of counterion and polymer. An expression for this factor is currently not available, so it will be treated as an adjustable constant. The final expressions for counterion diffusion coefficients in IEMs are given by:

$$\frac{D_g^m}{D_g^s} = f_u\left(1 - \frac{1}{3}A\left(1;\frac{x}{\xi}\right)\right)\left(\frac{\phi_w}{2-\phi_w}\right) + f_c\frac{1}{3}\alpha \tag{10.27}$$

where $A\left(1;\frac{x}{\xi}\right) = \sum_{m_1=-\infty}^{\infty}\sum_{m_2=-\infty}^{\infty}\left(\pi(m_1^2+m_2^2)+1+(2\xi/x)\right)^{-2}$ and $(m_1,m_2)\neq(0,0)$ with $x = c_f^m/c_s^m$ where c_f^m and c_s^m are the fixed charge group concentration and the mobile salt concentration, respectively, and $f_u = (X/\xi + 1)/(1+X)$.

The theoretical model (Equations (10.26) and (10.27)) identifies three key parameters influencing salt permeability coefficients in IEMs: the dimensionless linear charge density (ξ, Equation (10.24)), the concentration of fixed charge groups (X), and the water volume fraction (ϕ_w). According to this model, the diffusion coefficients of condensed counterions under an externally applied electric field are approximately 2–2.5 times higher than those of uncondensed counterions, as shown in Figure 10.25 [37]. The strong agreement between theoretical predictions and experimental data validates the condensed counterion model, incorporating the Bjerrum length (λ_B).

Condensed counterions are primarily located near the interfaces of ion channels, while uncondensed counterions are typically farther from the interfaces or in the bulk phase of the channels. Accordingly, the transport of condensed counterions corresponds to interfacial transport, whereas uncondensed counterions primarily contribute to bulk core transport. Figure 10.23 illustrates this distinction in the context of facilitated transport in the solid state. These findings highlight that interfacial transport plays a critical role in determining ionic conductivity through ion channels in IEMs. This observation aligns with the predictions of the condensed counterion model, emphasizing the importance of accounting for interfacial dynamics.

10.6 Ion Transport Through Ion Exchange Membranes

Here, we will first introduce the fundamentals of ion flux using the Nernst–Planck equation. This will be followed by a discussion of ion conduction in CEMs and AEMs, focusing on how structural changes, particularly due to water content, affect ion transport.

10.6.1 Nernst–Planck Equation for Ion Transport

Ion transport can be usually described by the extended Nernst–Planck equation, which constitutes the three different contributions of diffusion, migration, and convection as [3]:

$$J_j = -D_j \frac{\partial c_j}{\partial x} - u_j c_j \frac{\partial \varphi}{\partial x} + c_j v \quad \text{for cation} \tag{10.28}$$

$$J_j = -D_j \frac{\partial c_j}{\partial x} + u_j c_j \frac{\partial \varphi}{\partial x} + c_j v \quad \text{for anion} \tag{10.29}$$

where D_j, u_j, and c_j are the diffusion coefficient, the mobility, and the concentration of ion j, respectively. φ, v, and x are the electric potential, the convective velocity of the solvent (water), and the distance coordinate across the membrane, respectively. It is worthy to note that D_j and u_j are directly interrelated: $u_j = RTD_j/z_jF$. The first term in Equations (10.28) and (10.29) represents ion diffusion due to the concentration gradient $(\mathrm{d}c_j/\mathrm{d}x)$, and the second represents ion migration as a result of the electric potential gradient ($\mathrm{d}\varphi/\mathrm{d}x$). The third term represents convective transport of ions associated with the solvent (water) transport, which can be commonly nullified in common IEMs [3]. Therefore, both diffusion and migration terms mostly contribute to the ion conductivity.

Despite the limitations of the Nernst–Planck equation (without the convective term) in describing ion transport through inhomogeneous media, the wide applicability of the equation has been proven for the measured ionic conductivity regardless of any contribution of the interfacial transport, thereby frequently serving as a basis of quantitative treatment today [3].

In the Nernst–Planck equations, the diffusion term is described by the Fick's law. However, this law is limited in describing the diffusion of multicomponent systems, especially the diffusion of individual ionic species. In this case, the Maxwell–Stefan approach is more rigorous and suitable [42]. If you are interested in more detail about it, you may refer to the reference of Prof. A. Heintz [42]. Another interesting approach is based on irreversible thermodynamics, which is more rigorous and realistic. Such approaches have been discussed and reviewed in great detail by Lakshminarayanaiah [3, 43].

10.6.2 Proton Conductivity and Diffusion Coefficient in Cation Exchange Membranes

Since ions transport exclusively through ion channels of microphase-separated structures, but not through hydrophobic polymer matrix of IEMs, their ion conductivity is primarily determined by the size, density and straightness (or tortuosity) of these ion channels. These structural features, in turn, depend largely on the water content within the membrane. The water content is primarily influenced by the external water environment, specifically its chemical activity and physical state, as well as the membrane's ion exchange capacity (IEC) or equivalent weight (EW), which reflects the concentration of fixed ionic groups.

For instance, when PEMs are exposed to an aqueous environment, the size, amount and straightness of the ion channels will be significantly increased to improve the ionic conductivity, depending on the extent of water absorbed in PEMs. Figure 10.27 schematically demonstrates the structural changes depending on the amount of water: narrow channels with vapor water while widely interconnected water channels among water clusters in liquid water [44].

The effects of EW, where EW = 1/IEC, on proton conductivity are demonstrated in Figure 10.28a. It illustrates the decrease in the ion conductivity with increasing the EW of PFSA of 3M Co. As the EW value increases, the size of the

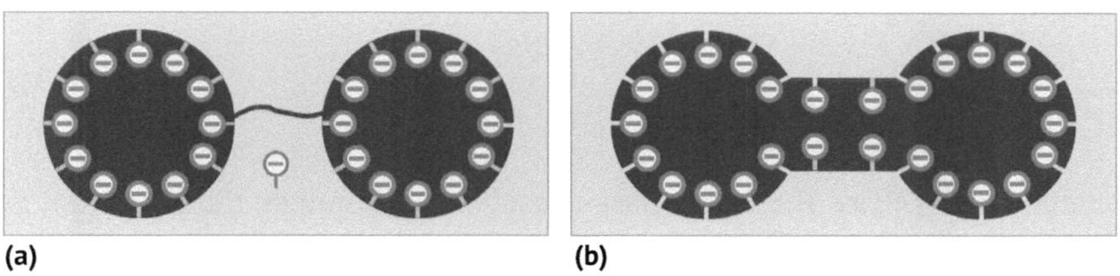

Figure 10.27 (a) Vapor water- and (b) liquid water-equilibrated membrane structures. *Source*: Adapted from [44].

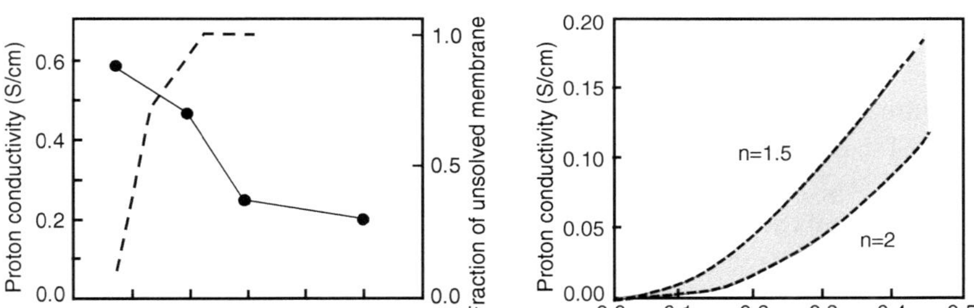

Figure 10.28 (a) Conductivity-stability correlation as a function of equivalent weight (EW) for 3M PFSA ionomers, where the stability was quantified by the membrane's ability to withstand water-boiling process; and (b) hydration effect of conductivity in Nafion membranes as a function of water volume fraction, ϕ_w, at 25–30 °C. The dotted lines are obtained from Equation (10.30) with $n = 1.5$ and 2.0. *Source*: Adapted from [5].

ion channel becomes smaller and its tortuosity higher, resulting in the decrease in the ion conductivity.

Figure 10.28b shows that the proton conductivity, σ, in Nafion increases with increasing the water volume fraction, ϕ_w. The relationship between σ and ϕ_w has been proposed as [5]:

$$\sigma(\phi_w, T) = \sigma_0(T)(\phi_w - \phi_0)^n \tag{10.30}$$

where $\sigma_0(T)$ is a material parameter that can include temperature effects through an Arrhenius expression, and n is empirically determined to be between 1.0 and 1.95. ϕ_0 is the water volume fraction at a percolation threshold (ϕ_0~0.1), i.e. where an interconnected network of conductive water clusters forms as shown in Figure 10.27. The experimental data show some scattered results primarily because of the difficulties maintaining the same experimental conditions to measure the proton conductivity (the data not shown in the figure). The most data are inside of the two lines obtained from Equation (10.30) with $n = 1.5$ and 2. Again, higher ϕ_w results in the ion channel larger and straighter (less tortuous) and consequently in the increase in the proton conductivity.

Mackie and Meares proposed a different mathematical model to predict ion diffusion coefficients in IEMs [41, 45], suggesting that a polymer with higher water content exerts smaller tortuosity effects to enhance ion diffusion coefficient. The model is expressed as:

$$\frac{D_s^m}{D_s^0} = \left(\frac{\phi_w}{2 - \phi_w}\right)^2 \tag{10.31}$$

where D_s^m and D_s^0 are the salt diffusion coefficients through IEMs and dilute solution, respectively. The D_s^0 value was around 1.3–2.0 × 10^{-5} cm^2/sec at 25 °C for LiCl, NaCl, and KCl, respectively [41, 46]. Equation (10.31) provides a rather close fit with experimental results only for moderate electrolyte concentrations and high ϕ_w.

10.6.3 Hydroxide Ion Conductivity in Anion Exchange Membranes

The hydroxide ion diffusivity in pure water (5.27 × 10^{-5} cm^2/sec) is quite high, compared to those (2.0–0.96 × 10^{-5} cm^2/sec) of other common anions such as Br^-, NO_3^-, and CO_3^- as listed in Table 10.1. However, it is much smaller than the proton diffusivity in bulk water (10.31 × 10^{-5} cm^2/sec). Therefore, the hydroxide ion conductivity through common AEMs is intrinsically lower than the proton conductivity through common PEMs. However, the overall views of hydroxide ion conduction behaviors in AEMs are very similar to those of the proton conduction in PEMs. For instance, the OH^- conductivity increases with the IEC, the amount of water, ϕ_w or ϑ, and temperature T.

Another serious issue is that durable hydroxide-conducting AEMs in high hydroxide concentration media (high pH~14 like 1 M NaOH solution) have proved elusive, because polymers bearing cations are inherently unstable under high pHs, i.e. weak

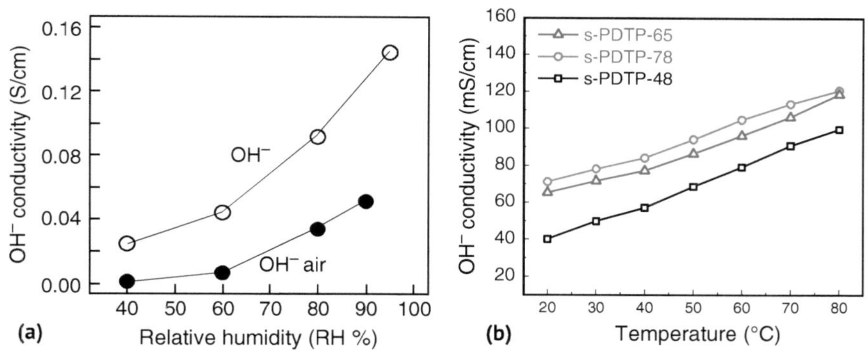

Figure 10.29 (a) OH^- (○) vs air-exposed OH^- (●) conductivity for anion exchange membranes-3.2 (AEM-3.2) with ion-exchange capacity (IEC) of 3.2 mmol/g at 60 °C as a function of relative humidity. Replotted from [47]; and (b) the OH^- conductivity as a function of temperature by varying IEC of poly(dibenzyl-*co*-terphenyl piperidinium) (s-PDTP). *Source*: Adapted from [10].

chemical or electrochemical stability of common AEMs at high pHs. Therefore, there has been a large body of studies to develop new AEM materials demonstrating high OH^- conductivity as well as high chemical stability [8]. Issues on chemical stabilities of varying AEM structures, particularly at high pHs, are discussed in Section 10.3.2 of this chapter.

Figure 10.29a shows the OH^- conductivity as a function of relative humidity, RH, for AEM-3.2 with IEC of 3.2 mmol/g based on poly(2,6-dimethyl-1,4-phenylene oxide)-*b*-poly(vinyl benzyl trimethyl ammonium) (PPO-*b*-PVBTMA) and for the same film exposed to air. The OH^- conductivity of the membrane increased when RH increased from 80 to 95% by 91–138 mS/cm and is still at a practical value of 44 mS/cm at a relatively low 60% RH [47]. The OH^- conductivity of the air-exposed membrane goes up with the increasing RH with less extent. The OH^- conductivity in AEMs without air exposure is always higher than that with air-exposed ones, which could be due to the decrease in the OH^- concentration by its reaction with dissolved CO_2 from air exposure ($OH^- + CO_2 \rightarrow HCO_3^-$). Furthermore, the OH^- conductivity of AEM-3.2 was stably maintained for 6 days with $\sigma = 120$ mS/cm at 60 °C and 95% RH even though AEM-3.2 contains aryl ether, known to be labile with OH^-. Such stable OH^- conductivity >50 mS/cm is indicated for practical applications [47].

A series of side-chain-type poly(dibenzyl-*co*-terphenyl piperidinium) (s-PDTP-*x*) copolymers featuring aryl ether-free polymer backbone and pendent piperidinium cations, s-PDTP-*x* ($x = 48$, 65, and 78), thus demonstrating high alkaline stability, were prepared by controlling IEC where their chemical structures are shown in Figure 10.11b [10]. Figure 10.29b shows that the OH^- conductivity increases with temperature as usual at a given IEC. The OH^- conductivity also increases with IEC at a given temperature where representative data are listed in Table 10.4. These IEC-controlled s-PDTP membranes possess high OH^- conductivity >120 mS/cm at 80 °C, reasonable molecular weight (MW = 150–290 kDa), and excellent *ex situ* durability in 1 M NaOH at 80 °C for 1,000 hours (~90% conductivity remaining) [10].

Table 10.4 Characteristics of side-chain-type poly (dibenzyl-co-terphenyl piperidinium) (s-PDTP-*x*) membranes at 80°C [10].

x	48	54	60	65	78
IEC (mmol/g)	2.4	2.6	2.9	3.0	3.4
σ_{OH^-}(mS/cm)	100	101	117	119	121

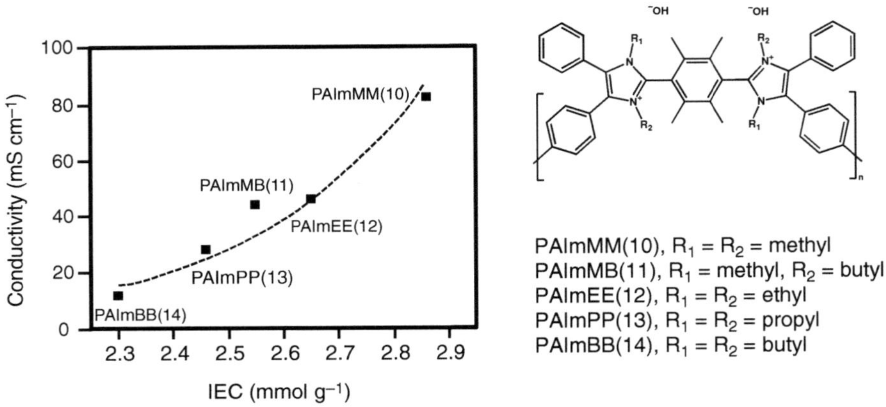

Figure 10.30 Ionic conductivity at 80 °C, 95% relative humidity (RH) vs ion-exchange capacity (IEC) of PAImXY(**#**). *Source*: [12]/Springer Nature.

Figure 10.30 demonstrates the effects of IEC on the OH^- conductivity of poly(bis-arylimidazoliums)s, PAIms by varying the alkyl group length, where PAImMM(10) is the smallest and PAImBB(14) the largest. PAImMM(10)Cl^- exhibits a swellability in water up to 45 wt%, which is low for the membrane's high IEC 2.86 mmol/g [12]. Polymers prepared with longer alkyl side chains correspondingly possess lower IEC, and consequently the swellability of PAImEE(12), PAImPP(13), and PAImBB(14) decreases from 26.1 to 20.3, to 14.0%, with the ϑ-values of 6, 5, and 3, respectively. Compared with PAImEE(12), the unsymmetrically alkylated polymer PAImMB(11) has a higher water uptake and hydration level, although it possesses a lower IEC and swelling ratio than PAImEE(12). Ionic conductivities of 82 and 12 mS/cm were observed at 80 °C under 95% RH for PAImMM(10) and PAImBB(14), respectively. The large difference in conductivity is due to the 0.4 mmol/g difference in IEC and subsequent decrease in water content, demonstrating a clear relationship between conductivity and IEC.

However, imidazolium units of 66.0, 93.6, 93.9, and 97.7% in PAImMM(10), PAImEE(12), PAImPP(13), and PAImBB(14), respectively, remained intact after 240 hours for chemical stability test in 10 M KOH at 80 °C, which is reverse order to the OH^- conductivity [12].

10.6.4 Concentration Polarization

In electrolytic membrane cells including chemical water splitting, ED, and other electrochemical processes, polarization refers to phenomena occurring at the surface of

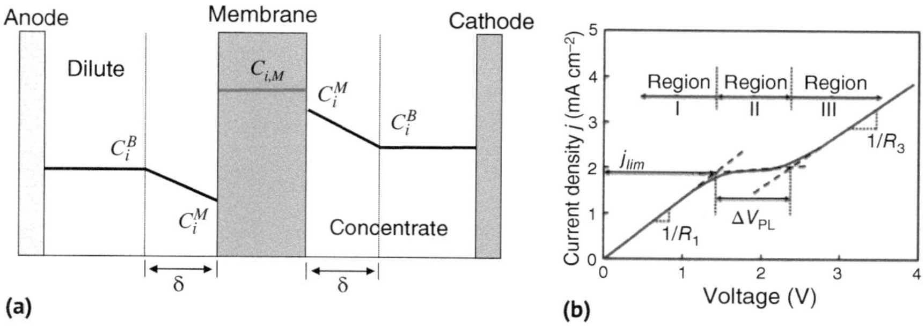

Figure 10.31 (a) Schematic drawing of concentration profiles of a salt in boundary layer on both sides of a cation-exchange membrane (CEM); and (b) typical current–voltage (*I-V*) curve of a CEM. *Source*: Adapted from [1].

an electrode or a membrane during mass transfer. These phenomena limit the rate of mass transfer.

Concentration polarization arises in electrolytic membrane cells when a current flows across an interface at a current density greater than the interface's capacity to transmit ions. During this process, the current in the solution phase is carried by both positive and negative ions with nearly the same transport numbers (~0.5) for both cations and anions, whereas, inside the membrane, the current is primarily carried by counterions, high transport number for counterions [1].

This difference in transport numbers depletes salt ions near the membrane surface on the dilute side [48, 49], while increasing concentration near the membrane on the concentrate side. As a result, a concentration gradient forms across the diffusion boundary layer (DBL). Figure 10.31a illustrates the concentration profile across a CEM during current flow [1].

The current–voltage (*I-V*) response is a useful tool for investigating ion transport and concentration polarization phenomena in IEMs. When an IEM is placed between two electrolyte solutions, the *I–V* response typically shows three distinct regions [50, 51]: regions I, II, and III, as shown in Figure 10.31b.

10.6.4.1 Under-limiting Current Density (Region I)

At low voltage, the current density increases linearly with voltage, displaying quasi-Ohmic behavior. This region assumes a quasi-equilibrium state at the membrane/solution interface. For ion transport analysis in IEMs, local electroneutrality is assumed, and electro-osmotic flow is neglected.

The fluxes J_i^M and J_i^B of counterions through the membrane and the boundary layer, respectively, at steady state are:

$$J_i^M = T_i^M \frac{j}{|z_i|F} \tag{10.32}$$

$$J_i^B = T_i^B \frac{j}{|z_i|F} - D_i \frac{\mathrm{d}C_i}{\mathrm{d}x} \tag{10.33}$$

where T_i^M and T_i^B are transport numbers of counterions through membranes and boundary layer, respectively, and D_i and C_i are the diffusion coefficient and concentration of counterion at the boundary layer, respectively, with a thickness of δ.

At steady state, fluxes in the membrane and boundary layers are equal $\left(J_i^M = J_i^B\right)$. Combining Equations (10.32) and (10.33) yields:

$$J_i^M = J_i^B = T_i^M \frac{j}{|z_i|F} = T_i^B \frac{j}{|z_i|F} - D_i \frac{\mathrm{d}C_i}{\mathrm{d}x} \tag{10.34}$$

By integrating Equation (10.34), the concentration profiles in the boundary layer can be expressed as:

$$C_i(x) = C_i^B - \left(T_i^M - T_i^B\right) \frac{j}{|z_i|FD} x \tag{10.35}$$

in the dilute side of membrane in Figure 10.31a and

$$C_i(x) = C_i^B + \left(T_i^M - T_i^B\right) \frac{j}{|z_i|FD} x \tag{10.36}$$

in the concentrate side of membrane in Figure 10.31a.

From Equation (10.35), the current density j^B through the boundary layer can be rearranged as:

$$j^B = \frac{|z_i|F}{\left(T_i^M - T_i^B\right)} D_i \frac{\left(C_i^B - C_i^M\right)}{\delta} \tag{10.37}$$

Equation (10.37) describes the current density through an electrolytic membrane cell as a function of the transport number difference $\left(T_i^M - T_i^B\right)$ of ions across the boundary layer and the concentration difference $\left(C_i^B - C_i^M\right)$ with thickness δ and D_i. This suggests that the change in current density will result in the concentration change in the boundary layer and vice versa.

10.6.4.2 Limiting Current Density (Region II)

When, due to the concentration polarization, in the dilute side, the salt concentration at the membrane surface is reduced to zero, reaching a limiting current density j_{lim}. The limiting current density is the current required to transfer all available ions and depends on bulk electrolyte concentration and hydrodynamic conditions near the membrane [1]. Thus, the voltage drop across the boundary layer increases drastically, resulting in a higher energy consumption and in electric field-enhanced water dissociation. Consequently, water dissociation is a loss of current utilization and drastic pH value change with an increase in the pH value at the surface of AEMs and with a decrease in CEMs.

Mathematically, limiting current density j_{lim}, is directly obtained with $C_i^M = 0$ from Equation (10.37) as [51, 52]:

$$j_{\text{lim}} = \frac{z_i F}{\left(T_i^M - T_i^B\right)} D_i \frac{C_i^B}{\delta} \tag{10.38}$$

10.6.4.3 Overlimiting Current Density (Region III)

At higher applied volts, the electric current starts to increase again, and electrical noise is observed. Overlimiting current density is linked to chaotic mechanical mixing in the depletion layer adjacent to the IEM [53, 54]. Under extreme concentration polarization, space charges form and interact with the electric field, creating spatially inhomogeneous forces that drive fluid motion in the depletion layer [55, 56]. Overlimiting conduction occurs when electro-convection dominates molecular diffusion [51].

10.6.4.4 Water-splitting Phenomena

Water splitting may occur in Region III but is not always significant. The extent of water splitting depends on the membrane type: in AEMs, water splitting is catalyzed by tertiary alkyl amines and strong local electric fields [57], whereas, for CEMs, water splitting typically occurs only in the presence of impurities and is associated with sulfonic acid groups.

A chemical reaction model explains these behaviors, highlighting the intrinsic properties of AEMs and the specific conditions required for water splitting in CEMs, which will be described in more detail in Section 12.4 of Chapter 12.

10.7 Selective Transport Through Ion Exchange Membranes: Permselectivity

Ion channels in IEMs, in ideal cases, selectively allow the passage of oppositely charged ions (counterions), while obstructing equally charged ions (co-ions). However, practically in most cases, the co-ion transport also occurs, even in the minor extent, through the ion channels. Therefore, the selective transport property of counterion toward co-ion through IEMs, frequently called the permselectivity, becomes a key characteristic of IEMs for both conventional and emerging applications, including modern energy conversion and storage devices such as water-splitting devices and redox flow batteries, in addition to electrodialysis, diffusion dialysis, and electrolysis [3].

10.7.1 Donnan Effects for Permselectivity of Counterion over Co-ion

When a charged membrane is equilibrated with an electrolyte, an unequal distribution of ions arises between phases, generating the so-called Donnan electrical potential at the electrolyte solution/membrane interface. The Donnan potential results in the exclusion of co-ion, providing the basis of permselectivity [58].

The typical concentration profiles of counterion and co-ions are illustrated in Figure 10.32 at the interface between CEM and electrolyte solution. In general, counterions are predominantly absorbed in IEM owing to the mutual Coulombic interactions with fixed charges, while co-ions are subjected to repulsive forces toward opposite directions. The resultant ion imbalance generates an electric

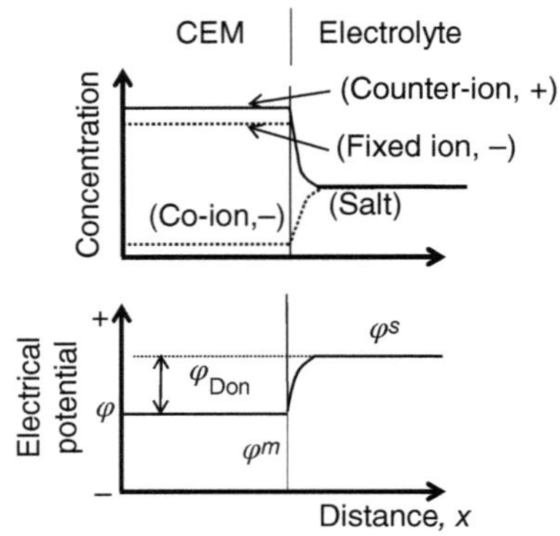

Figure 10.32 Schematic drawings illustrating the ion concentration and corresponding electrical potential in cation-exchange membrane (CEM) and bulk electrolyte solution, and the electrical potential difference between the membrane and bulk electrolyte, defined as the Donnan potential $\varphi_{Don} = \varphi^m - \varphi^s$ where φ^m and φ^s are CEM and electrolyte solution potentials, respectively. *Source*: Adapted from [1].

potential difference (Donnan potential, also called membrane potential) that prohibits the undesirable co-ion crossover during practical operation. This phenomenon is referred to as Donnan exclusion, which is a fundamental principle for the high permselectivity of IEM [1].

The Donnan exclusion depends on three major issues: (1) fixed charge density of IEM, (2) specific ion effects of counterion and co-ions, and (3) bulk electrolyte concentration. These parameters physically strengthen or weaken the Donnan potential at the IEM surface according to the structure of electrical double layer at the junction of IEM and the electrolyte [3].

The high fixed charge density in IEM makes much more selective counterions sorption and co-ions exclusion at the IEM surface by elevating the Donnan potential. It is conventional but the most powerful way to improve permselectivity of IEM. However, excessive amounts of fixed charges cause the unintended membrane swelling, followed by a considerable co-ion sorption. This trade-off phenomenon can be alleviated to some extent by employing inorganic nanomaterials in IEMs or coating IEMs with a durable porous matrix.

Types of counterion and co-ions dissolved in the bulk electrolyte change the equilibrated potential at the space charge region on IEMs, which is strongly associated with the Donnan potential. For instance, strong interactions of counterions with fixed charge in IEM critically diminish the local electric fields generated by the fixed charge [3].

Specific co-ion effects are much more complicated than those of counterions. Several parameters, such as size, charge valence, and chemical structure of co-ions, affect its sorption behavior. Among them, the polarizability of co-ions was identified as a primary factor for co-ion effects on permselectivity. Highly polarizable co-ions tend to interact strongly with solvent molecules and IEM matrix with electrostatic and dispersion forces, reducing the activation barrier of co-ion sorption and permselectivity of IEM.

The concentration of bulk electrolyte also apparently affects Donnan potential. Excessive amounts of ions in the bulk electrolyte compared to the ion-exchange capacity of the membrane makes the perfect screening of fixed charges, which decreases the Donnan potential and permselectivity of IEM as mentioned above. However, a high concentration of electrolytes improves the ionic conductivity and limiting current density. Therefore, it is highly important to find the optimal environment to obtain the decent conductivity and permselectivity applicable to the practical operation.

The surface hydrophilicity or hydrophobicity of IEMs is also considered to explain the alteration in the permselectivity of IEM. Recently, the hydrophobic functionality in AEM could help exclude co-ions, resulting in the improvement of the permselectivity. This result is quite interesting; however, the effect is relatively small compared with those mentioned above to govern the overall permselectivity trend. Further investigation is required to demonstrate that the hydrophobization of fixed charge group is effective to the permselectivity of IEM [3, 59].

Although there are well-established ways to indirectly estimate the Donnan potential, it has been widely reported that it cannot be measured directly. Recently, the first direct measurement of the Donnan potential of an IEM equilibrated with salt solutions was proposed. The results highlight the dependence of the Donnan potential on external salt concentration and counterion valence, and show a reasonable agreement with current theoretical models of IEMs, which incorporate ion activity coefficients [58].

10.7.1.1 The Electrochemical Equilibrium and the Donnan Potential

The change of the chemical potential of a component i in a real mixture at isothermal conditions is obtained by introducing the activity:

$$d\mu_i = V_i\,dp + RT\,d\ln(a_i) \tag{10.39}$$

Here μ_i, V_i, p, and a_i are the chemical potential, the partial molar volume, pressure, and the chemical activity of the component i. Integration of Equation (10.39) leads to:

$$\Delta\mu_i = \mu_i^m - \mu_i^s = V_i(p^m - p^s) + RT\ln\left(\frac{a_i^m}{a_i^s}\right) \tag{10.40}$$

with superscript m and s stands for membrane and electrolyte solution, respectively.

An IEM and an adjacent electrolyte solution will be in equilibrium if the electrochemical potentials of all ions in the two phases are equal. Thus, for each ion in equilibrium is:

$$\bar{\mu}_i^m = \bar{\mu}_i^s = \mu_i^m + z_i F\varphi^m = \mu_i^s + z_i F\varphi^s \tag{10.41}$$

Here, $\bar{\mu}_i$ and φ_i are the electrochemical potential and electrical potential of the component i. Thus, the electrochemical potential of an ion is composed of two additive terms: the first is the chemical potential, and the second is the electrical potential multiplied by the Faraday constant F and the valence z_j of ion j.

The Donnan potential, φ_{Don}, associated with the concentration change, is defined as the electrical potential difference between membrane and electrolyte solution [1]:

$$\varphi_{\text{Don}} = \varphi^m - \varphi^s \tag{10.42}$$

where φ^m and φ^s are the electrical potentials at the membrane and the electrolyte solution, respectively.

Rearranging Equations (10.40), (10.41), and (10.42) gives the electrical potential difference between the membrane and the adjacent solution to:

$$\varphi^m - \varphi^s = \varphi_{\text{Don}} = \left(\frac{1}{z_i F}\right)\left[RT\ln\left(\frac{a_i^s}{a_i^m}\right) + V_i(p^s - p^m)\right] \tag{10.43}$$

Here, the pressure difference can be converted into the osmotic pressure $\Delta\pi$ as [1]:

$$\Delta\pi = p^s - p^m = \left(\frac{RT}{V_w}\right)\ln\left(\frac{a_w^m}{a_w^s}\right) \tag{10.44}$$

where V_w and a_w are the partial molar volume and the chemical activity of water, respectively. From Equations (10.43) and (10.44), the Donnan potential developed at the interface between electrolyte solution/IEM can be readily obtained, which will influence the co-ion transport and consequently the permselectivity.

The Donnan model has thus been frequently adopted to explain the ion sorption phenomena in IEMs so far. However, it is only valid in some limited environments because of the serious assumptions made to derive the model. Recently, the assumption has been relaxed by introducing the effects of electrostatic interactions between polymeric segments and the ions on the change in the activity coefficient by Manning et al. [37, 60].

10.7.2 Permselectivity of the Same Charged Ions

Up to now, factors affecting the permselectivity between counterion and co-ion have been briefly introduced. In this subsection, the transport selectivity among the same charges, cation/cation and anion/anion selectivity, will be addressed. Originally, the Hofmeister series or lyotropic series, indicating the ability to salting-out (also known as salt-induced precipitation) or salting-in (the solubility increases by adding salt) in a protein solution, has been adopted to explain the difference in the transport rate of the same charge ions. Hofmeister discovered a series of salts that have consistent effects on the solubility of proteins. The precise mechanism of the effect has yet to be understood clearly, but it is thought to stem from the change in bulk water structure by ions. In other words, it is relevant to the specific interaction between hydrated ions in microscopic view, influencing the ion permselectivity in IEMs.

The difference in the anion transport rates through IEMs almost follows the Hofmeister series, while that in the cation transport rates does not correlates with it well. The reason for the difference is not clear yet. However, the difference in the transport rate could be explained according to the fact that the interactions between ion pairs are much stronger when they have similar hydration behavior (the Collins

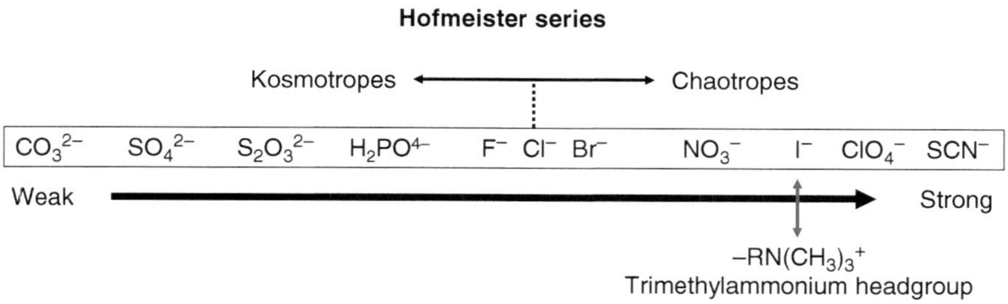

Figure 10.33 Hofmeister series (for anions) with the classification of kosmotropes and chaotropes. The anions are ordered with respect to their affinity for the trimethylammonium group in surfactants. The ordering is based on Collins' empirical rule that ions of similar water affinity exhibit strong interactions. *Source*: Adapted from [3].

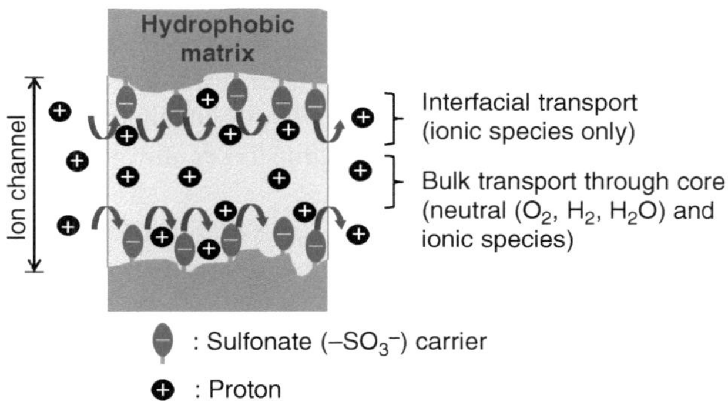

Figure 10.34 Schematic structure of ion channel and transports of ions and neutral molecules including water, fuel, and oxidant. *Source*: Y.S. Kang.

rule: e.g. chaotropes with chaotropes or kosmotropes with kosmotropes, as shown in Figure 10.33). Here, the chaotropes help inhibit the interactions among the water molecules, whereas the kosmotropes help promote the interactions. AEM having the quaternary ammonium ion (a chaotrope) shows the higher permselectivity in ClO_4^- compared to $H_2PO_4^-$, which is consistent with the Hofmeister series in Figure 10.33.

10.7.3 Transport of Water and Gas

Ion channels in IEMs provide a pathway for neutral components such as water, fuels (hydrogen, methanol), and oxidants (oxygen, air). Water transport helps ions like H^+ and OH^- flow through ion channels in IEMs, whereas fuel and oxidant transport should be minimized to improve the efficiency and safety of electrochemical devices like fuel cells. As shown in Figure 10.34, ions such as H^+ and OH^- permeate primarily by interfacial transport via facilitated transport through narrow ion channels and also through the bulk core of the channels by Grotthuss mechanism and vehicle transport. In the meantime, the water transports mostly through the core and partially through interface. Fuels (methanol and hydrogen) and oxidants (oxygen and air) also flow primarily through the bulk core of the ion channels.

In this regard, the size of the ion channels plays major roles in selective transport of ion and water over neutral fuels and oxidants. Here, water transport will be introduced briefly, followed by fuels like methanol, hydrogen, and oxygen transports.

10.7.3.1 Water Transport

Water transport also plays important roles in controlling sustainable operation. In proton exchange membrane fuel cells (PEMFCs), water diffuses from anode to cathode along with H^+ as vehicles so that anode is dried out while cathode is sufficiently filled with water or sometimes flooding. For anion exchange membrane fuel cells (AEMFCs), water generated at cathode must move to anode, to maintain the water balance. Therefore, water transport should be well balanced for efficient cell operation [1].

Water transport through IEMs occurs via three main mechanisms: (1) electroosmotic flux, where solvent moves along with ions under an electrical potential gradient, (2) flux coupled to ion transport, where water in the hydration shell of ions moves with them, and (3) osmotic flux, where water moves due to a chemical potential gradient across the membrane [1].

Water permeation is accompanied by electroosmosis, when an electric potential is applied across an ion channel in a CEM. This causes excess positively charged ions in the electrical double layer to migrate toward the cathode (−), generating an electric current and dragging water along with them. The permeation rates of ion and water are determined by the strength of the electrical field in electroosmosis [1].

The water diffusion coefficients D_w measured through many commercialized PFSAs typically falls within the range of 1–10 × 10^{-6} cm^2/sec, and increases with the water content ϑ inside, as shown in Figure 10.35 [61]. As observed, D_w increases

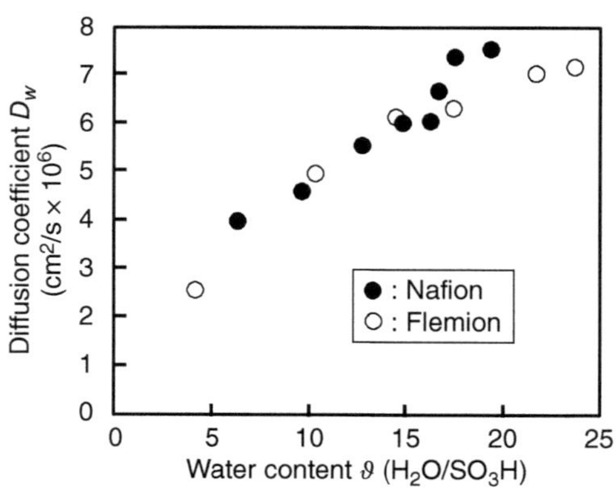

Figure 10.35 Measured self-diffusion coefficient D_w of water in solid polymer electrolytes (Nafion and Flemion) with various water contents. *Source*: Ref. [61]/with permission of Elsevier.

nearly linearly with the water content ϑ. This dependence of D_w on the water volume fraction, ϕ_w, or water content, ϑ, has been often expressed as [62]:

$$D_w(\phi_w) = \frac{\phi_w}{\tau} D_{w,aq} \tag{10.45}$$

where $D_{w,aq}$ is the diffusion coefficient of water in neat water, and τ is the tortuosity factor. This is consistent with the structure evolution of ion channels depending on the amount of water inside.

10.7.3.2 Methanol Transport

In the case of direct methanol fuel cells (DMFCs), methanol is used directly as a fuel mixture with water. Methanol is miscible with water so that it also penetrates together with water through ion channels, which represents the loss of methanol fuel. Therefore, methanol transport across the ion channel must be minimized to improve the efficiency of DMFCs while increasing the proton transport as much as possible.

Consider wide- and narrow-sized ion channels; the only difference between the two ion channels is the portion of the bulk core with the same interfacial area. It has been widely accepted that proton transports dominantly through the interface by the facilitated transport as described in Section 10.5 and also bulk core of ion channels by Grotthuss and vehicle mechanisms, as illustrated in Figure 10.23. However, methanol transports mostly through the bulk core of the ion channels, but hardly through the interface because it is neutral but does not have an electric charge. Then it can be readily anticipated that the proton conductivity difference between the two differently sized ion channels will be small, whereas the methanol transport will be reduced significantly through the narrow channels in CEMs, which is extremely important in enhancing the DMFC performance. In other words, the selective transport of H^+ over MeOH is low in wide channels, whereas narrow channels result in a higher selective one.

CEMs with wide ion channels are obtained from the sulfonated poly(styrene-*b*-(ethylene-*co*-butylene)-*b*-styrene), sSEBS, which were swollen in liquid water to make wide ion channels, while CEMs representing narrow ion channels are formed by crosslinking and sulfonating poly(styrene-*b*-butadiene-*b*-styrene), csSBS, where narrow channels are maintained even after immersion in water because of highly crosslinked structure of the base polymer matrix SBS. Their proton conductivity and methanol permeability through sSEBS, csSBS, and Nafion as a reference are listed in Table 10.5. The methanol permeability was reduced markedly from 1.7×10^{-6} to 8.1×10^{-8} cm^2/sec by crosslinking while small decrease in the proton conductivity from 3.8×10^{-2} to 2.3×10^{-2} cm^2/sec. These results demonstrate the importance of the channel size in reducing methanol permeability with minimal decrease in the proton conductivity [63].

10.7.3.3 Gas Permeation

In fuel cell applications, H_2 and O_2 are introduced to anode and cathode sides, respectively, at high pressures, leading to high driving force for their diffusion across IEMs. Since they are small in size, they permeate across ion channels in IEMs

Table 10.5 Ion conductivity, MeOH permeability, H_2O swellability, and thickness of sulfonated poly(styrene-*b*-(ethylene-*co*-butylene)-*b*-styrene), sSEBS, crosslinked and sulfonated poly(styrene-*b*-butadiene-*b*-styrene), csSBS, by controlling the degree of sulfonation, and Nafion.

	Ion conductivity (S/cm)	MeOH permeability (cm^2/sec)	H_2O swellability (%)	Thickness (μm)
sSEBS	3.8×10^{-2}	1.7×10^{-6}	75	90
csSBS	2.3×10^{-2}	8.1×10^{-8}	42.1	76
Nafion	6.3×10^{-2}	2.9×10^{-6}		170

Source: Adapted from [63].

rapidly to meet and react together to generate heat, compared to that through polymeric matrix, losing fuel and oxidant, and reducing the cell efficiency. Therefore, their permeabilities through ion channels should be as small as possible for efficient cell operation.

Permeation of gases through polymeric membranes have been described in detail in Section 3.7. The permeability, P_j, of a gas j can be described as a product of its diffusion coefficient, D_j, and its solubility coefficient, S_j.

$$P_j = D_j \times S_j \tag{10.46}$$

While D_{O_2}, for instance, increases with temperature and humidity, its solubility S_{O_2} decreases. These trade-offs often result in the oxygen permeability P_{O_2} being less sensitive to various parameters than either the diffusivity or solubility. Generally speaking, gas permeability in PFSAs like Nafion 117 increases with relative humidity and temperature, as shown in Figure 10.36.

Figure 10.36a and b illustrates that the oxygen and hydrogen permeabilities through the Nafion 117 membrane increase with gas relative humidity [64]. Gas

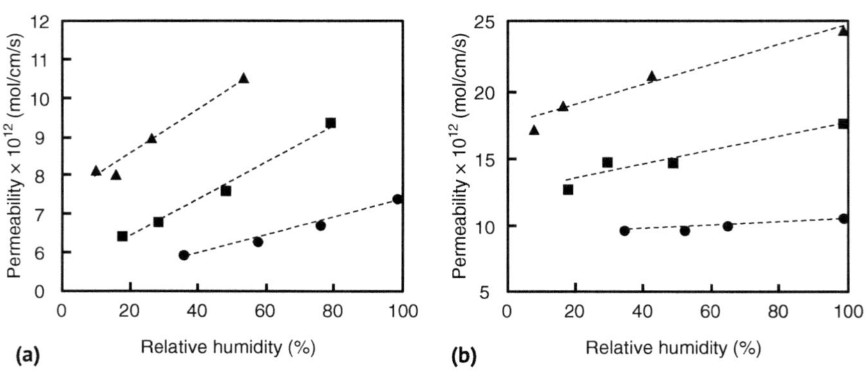

Figure 10.36 (a) Oxygen and (b) hydrogen permeabilities through Nafion 117 membrane at different values of gas relative humidity: (▲) at 80 °C; (■) 60 °C, and (●) 40 °C. *Source*: Ref. [64]/Springer Nature.

permeability also increases with temperature, although this temperature dependence is not particularly significant because of the trade-off between increasing diffusion and decreasing solubility coefficients by varying temperature. Within the experimental range, the oxygen permeability is nearly half that of hydrogen permeability.

The increase in gas permeability as a function of relative humidity or water content within the membranes is largely attributed to the diverse roles of water. One contributing factor is the increase in free volume associated with water uptake. Additionally, water swelling and plasticization effects in the fluorinated matrix are evident, although these effects are relatively modest. At high water content, water forms interconnected narrow channels within the hydrophobic fluorinated matrix, as shown in Figure 10.19.

The ionic channels play a key role in gas transport, as the permeability of common gases in water is 10 to 100 times higher than that in a dry fluorinated matrix. Therefore, the high gas permeability can be attributed to the formation of continuous narrow water channels, which act as pathways for gas transport [65]. Interestingly, Ogumi et al. [66] suggested that gas permeability primarily occurs in the amorphous regions of the membrane, with minimal permeation through ionic clusters. This contrasting view highlights the complexity of gas transport mechanisms within the Nafion 117 membrane.

10.8 Interrelationship Between Conductivity and Selectivity

IEMs are used to mediate the transport of various ionic and neutral solutes. In general, IEMs are required to efficiently permeate counterions (i.e. ions with a charge opposite to that of the fixed charge groups) while preventing the passage of other solutes [2]. For example, IEMs used in ED, reverse electrodialysis (RED), and redox flow battery (RFB) must prevent the transport of co-ions (ions with a same charge to that of the fixed charge groups). In this regard, high-performance membranes for ED, RED, and RFB should have both high ionic conductivity and high permselectivity. However, currently available IEMs are bound by a trade-off between membrane permselectivity and ionic conductivity [67]. That is, membranes that have high ionic conductivity generally have low permselectivity and *vice versa*. A similar trade-off between membrane permeability and selectivity is observed for virtually all synthetic polymer membranes, regardless of applications [68].

Even though permselectivity-conductivity plots are commonly used to evaluate the performance of IEMs [69, 70], they may not give sufficient information on their materials properties and morphology. This is mostly due to the fact that both permselectivity and conductivity used may not depend solely on the material properties and morphology of IEM, but also on other measurement environments. For instance, permselectivity is not an intrinsic membrane property since it depends on the properties of the solution contacting the membrane. Taken together, these complications make it difficult to conclude that the observed trade-off behavior is due to intrinsic membrane properties. Further, the use of membrane permselectivity restricts to IEMs for applications such as ED, RED, and RFB, in which counterion/co-ion selectivity is important [2].

In this regard, Kamcev recently reformulated the permeability-selectivity relations to aid us to understand wholly materials properties in designing better materials with two parameters β and λ:

$$\alpha_{i/j} = \beta/(C_i^m D_i^m)^\lambda \tag{10.47}$$

where the membrane selectivity, $\alpha_{i/j}$, is defined as:

$$\alpha_{i/j} = \frac{z_i^2 C_i^m D_i^m}{z_j^2 C_j^m D_j^m} \tag{10.48}$$

where z_i, C_i^m, and D_i^m are the valence, the concentration, and diffusivity of ion i in membrane m, respectively. Ion i is the more permeable species of the i/j pair. The reformulated trade-off relation is generalizable to applications including redox flow batteries other than ED/RED and is more consistent with the related literature on the permeability-selectivity trade-off relations for gas separation and reverse osmosis membranes [2].

For example, Figure 10.37 illustrates the applicability of Equations (10.47) and (10.48) to trade-off relation between the proton permeability ($C_H^m D_H^m$) and proton selectivity over vanadium ion ($\alpha_{H/V}$) observed in vanadium redox flow batteries (VRFBs). Here, IEMs are used to separate the catholyte, containing the V^{5+}/V^{4+} redox couple, from the anolyte solution, which contains the V^{2+}/V^{3+} redox couple. IEMs must allow transport of the charge-balancing ions – such as protons for CEMs and sulfate/bisulfate for AEMs – to complete the electrical circuit. At the same time,

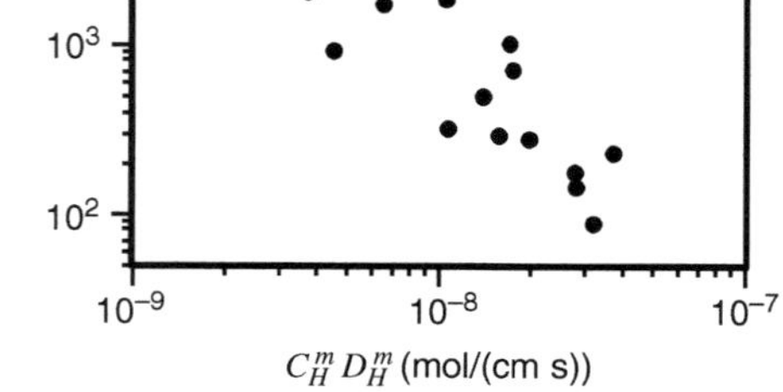

Figure 10.37 Reformulated permeability-selectivity trade-off relation for cation-exchange membranes (CEMs) used in vanadium redox flow batteries (VRFBs). The solid line represents the empirically constructed upper-bound relation $\alpha_{H/V} = \beta/(C_H^m D_H^m)^\lambda$, where $\beta = 2.82 \times 10^{-9}$ and $\lambda = 1.65$. The subscripts H and V are for proton and vanadium ion, respectively. *Source*: Adapted from [2].

they must suppress crossover of the vanadium ions between two compartments, as such crossover leads to self-discharge and reduced Coulombic efficiency [2].

References

1 H. Strathmann, Ion-Exchange Membrane Separation Processes, Elsevier Science, 2004.
2 J. Kamcev, Journal of Polymer Science 59(21) (2021) 2510–2520.
3 T. Luo, S. Abdu, M. Wessling, Journal of Membrane Science 555 (2018) 429–454.
4 K.-D. Kreuer, S.J. Paddison, E. Spohr, M. Schuster, Chemical Reviews 104(10) (2004) 4637–4678.
5 A. Kusoglu, A.Z. Weber, Chemical Reviews 117(3) (2017) 987–1104
6 B. Lee, D. Yun, J.-S. Lee, C.H. Park, T.-H. Kim, The Journal of Physical Chemistry C 123(22) (2019) 13508–13518.
7 M. Marino, K. Kreuer, ChemSusChem 8(3) (2015) 513–523.
8 N. Chen, Y.M. Lee, Progress in Polymer Science 113 (2021) 101345.
9 N. Chen, H.H. Wang, S.P. Kim, H.M. Kim, W.H. Lee, C. Hu, J.Y. Bae, E.S. Sim, Y.-C. Chung, J.-H. Jang, S.J. Yoo, Y. Zhuang, Y.M. Lee, Nature Communications 12(1) (2021) 2367.
10 H.M. Kim, C. Hu, H.H. Wang, J.H. Park, N. Chen, Y.M. Lee, Journal of Membrane Science 644 (2022) 120109.
11 K.M. Hugar, H.A.I.V. Kostalik, G.W. Coates, Journal of the American Chemical Society 137(27) (2015) 8730–8737.
12 J. Fan, S. Willdorf-Cohen, E.M. Schibli, Z. Paula, W. Li, T.J.G. Skalski, A.T. Sergeenko, A. Hohenadel, B.J. Frisken, E. Magliocca, W.E. Mustain, C.E. Diesendruck, D.R. Dekel, S. Holdcroft, Nature Communications 10(1) (2019) 2306.
13 W. You, E. Padgett, S.N. MacMillan, D.A. Muller, G.W. Coates, Proceedings of the National Academy of Sciences 116(20) (2019) 9729–9734.
14 B. Bauer, H. Strathmann, F. Effenberger, Desalination 79(2–3) (1990) 125–144.
15 H. Han, H. Ma, J. Yu, H. Zhu, Z. Wang, European Polymer Journal 114 (2019) 109–117.
16 B. Zhang, R.B. Kaspar, S. Gu, J. Wang, Z. Zhuang, Y. Yan, ChemSusChem 9(17) (2016) 2374–2379.
17 S. Gu, R. Cai, T. Luo, Z. Chen, M. Sun, Y. Liu, G. He, Y. Yan, Angewandte Chemie International Edition 48(35) (2009) 6499–6502.
18 M. Treichel, J.C. Gaitor, C. Birch, J.L. Vinskus, K.J. Noonan, Polymer 249 (2022) 124811.
19 C. Fujimoto, D.-S. Kim, M. Hibbs, D. Wrobleski, Y.S. Kim, Journal of Membrane Science 423 (2012) 438–449.
20 C.G. Arges, V. Ramani, Proceedings of the National Academy of Sciences 110(7) (2013) 2490–2495.
21 A.D. Mohanty, S.E. Tignor, J.A. Krause, Y.-K. Choe, C. Bae, Macromolecules 49(9) (2016) 3361–3372.
22 X. Chu, H. Zhang, C. Zhang, R. Shao, Z. Huang, H. Lv, S. Liu, L. Liu, N. Li, S. Zhao, Journal of Membrane Science 715 (2025) 123455.

23 F. Liu, K. Miyatake, A.M.A. Mahmoud, V. Yadav, F. Xian, L. Guo, C.Y. Wong, T. Iwataki, Y. Shirase, K. Kakinuma, Advanced Energy Materials (2024) 2404089.
24 Z. Wang, J. Parrondo, S. Sankarasubramanian, K. Bhattacharyya, M. Ghosh, V. Ramani, Journal of the Electrochemical Society 167(12) (2020) 124504.
25 C.R. Peltier, W. You, D. Fackovic Volcanjk, Q. Li, A.J. Macbeth, H.c.D. Abruña, G.W. Coates, ACS Energy Letters 8(5) (2023) 2365–2372.
26 T.D. Gierke, G. Munn, F. Wilson, Journal of Polymer Science: Polymer Physics Edition 19(11) (1981) 1687–1704.
27 K.D. Kreuer, Journal of Membrane Science 185(1) (2001) 29–39.
28 X. Glipa, B. Bonnet, B. Mula, D.J. Jones, J. Rozière, Journal of Materials Chemistry 9(12) (1999) 3045–3049.
29 L. Qingfeng, H.A. Hjuler, N. Bjerrum, Journal of Applied Electrochemistry 31 (2001) 773–779.
30 P. Choi, R. Datta, Journal of the Electrochemical Society 150(12) (2003) E601.
31 I.M. Model, Journal of The Electrochemical Society 151(2) (2004) A311–A325.
32 H. Strathmann, J.J. Krol, H.-J. Rapp, G. Eigenberger, Journal of Membrane Science 125 (1997) 123–142.
33 S. Mafé, P. Ramírez, Acta Polymerica 48(7) (1997) 234–250.
34 S. Mafe, J.A. Manzanares, P. Ramrez, Physical Review A 42(10) (1990) 6245–6248.
35 A.B. Yaroslavtsev, V.V. Nikonenko, V.I. Zabolotsky, Russian Chemical Reviews 72(5) (2003) 393–421.
36 J. Kamcev, D.R. Paul, B.D. Freeman, Macromolecules 48(21) (2015) 8011–8024.
37 J. Kamcev, D.R. Paul, G.S. Manning, B.D. Freeman, Macromolecules 51(15) (2018) 5519–5529.
38 G. Manning, The European Physical Journal E 34 (2011) 1–18.
39 G.S. Manning, The Journal of Chemical Physics 51(3) (1969) 924–933.
40 J. Kamcev, D.R. Paul, G.S. Manning, B.D. Freeman, ACS Applied Materials & Interfaces 9(4) (2017) 4044–4056.
41 J. Mackie, P. Meares, Proceedings of the Royal Society of London. Series A. Mathematical and Physical Sciences 232(1191) (1955) 498–509.
42 A. Heintz, E. Wiedemann, J. Ziegler, Journal of Membrane Science 137(1–2) (1997) 121–132.
43 N. Lakshminarayanaiah, Chemical Reviews 65(5) (1965) 491–565.
44 A.Z. Weber, J. Newman, Journal of the Electrochemical Society 150(7) (2003) A1008.
45 T. Sata, Ion Exchange Membranes: Preparation, Characterization, Modification and Application, RSC, 2004.
46 E.-S. Jang, J. Kamcev, K. Kobayashi, N. Yan, R. Sujanani, T.J. Dilenschneider, H.B. Park, D.R. Paul, B.D. Freeman, Polymer 192 (2020) 122316.
47 T.P. Pandey, H.N. Sarode, Y. Yang, Y. Yang, K. Vezzu, V. Di Noto, S. Seifert, D.M. Knauss, M.W. Liberatore, A.M. Herring, Journal of the Electrochemical Society 163(7) (2016) H513.
48 M. Block, J. Kitchener, Journal of The Electrochemical Society 113(9) (1966) 947.
49 B. Gavish, S. Lifson, Journal of the Chemical Society, Faraday Transactions 1: Physical Chemistry in Condensed Phases 75 (1979) 463–472.

50 J.-H. Choi, H.-J. Lee, S.-H. Moon, Journal of Colloid and Interface Science 238(1) (2001) 188–195.

51 M. Taky, G. Pourcelly, F. Lebon, C. Gavach, Journal of Electroanalytical Chemistry 336(1–2) (1992) 171–194.

52 M. Mulder, Basic principles of membrane technology, Springer Science & Business Media, 2012.

53 J. Krol, M. Wessling, H. Strathmann, Journal of Membrane Science 162(1–2) (1999) 155–164.

54 F. Maletzki, H.-W. Rösler, E. Staude, Journal of Membrane Science 71(1–2) (1992) 105–116.

55 I. Rubinstein, E. Staude, O. Kedem, Desalination 69(2) (1988) 101–114.

56 I. Rubinstein, F. Maletzki, Journal of the Chemical Society, Faraday Transactions 87(13) (1991) 2079–2087.

57 R. Simons, Nature 280(5725) (1979) 824–826.

58 P. Aydogan Gokturk, R. Sujanani, J. Qian, Y. Wang, L.E. Katz, B.D. Freeman, E.J. Crumlin, Nature Communications 13(1) (2022) 5880.

59 Y. Ji, H. Luo, G.M. Geise, Physical Chemistry Chemical Physics 22(14) (2020) 7283–7293.

60 G.S. Manning, Accounts of Chemical Research 12(12) (1979) 443–449.

61 S. Tsushima, K. Teranishi, Hirai, Energy 30(2–4) (2005) 235-245.

62 Q. Zhao, P. Majsztrik, J. Benziger, The Journal of Physical Chemistry B 115(12) (2011) 2717–2727.

63 J. Won, H.H. Park, Y.J. Kim, S.W. Choi, H.Y. Ha, I.-H. Oh, H.S. Kim, Y.S. Kang, K.J. Ihn, Macromolecules 36(9) (2003) 3228–3234.

64 K. Broka, P. Ekdunge, Journal of Applied Electrochemistry 27(2) (1997) 117–123.

65 T. Sakai, H. Takenaka, E. Torikai, Journal of The Electrochemical Society 133(1) (1986) 88.

66 Z. Ogumi, T. Kuroe, Z.i. Takehara, Journal of The Electrochemical Society 132(11) (1985) 2601.

67 G.M. Geise, M.A. Hickner, B.E. Logan, ACS Applied Materials & Interfaces 5(20) (2013) 10294–10301.

68 H.B. Park, J. Kamcev, L.M. Robeson, M. Elimelech, B.D. Freeman, Science 356(6343) (2017) eaab0530.

69 H. Fan, Y. Huang, N.Y. Yip, Journal of Membrane Science 610 (2020) 118259.

70 D.H. Cho, K.H. Lee, Y.M. Kim, S.H. Park, W.H. Lee, S.M. Lee, Y.M. Lee, Chemical Communications 53(15) (2017) 2323–2326.

11

Measurement of Ion Transport Properties

Characterizing ion transport properties in polymer electrolytes and ion exchange membranes is essential for understanding their performance in applications, such as batteries, fuel cells, sensitized solar cells, water-splitting electrolyzers, and supercapacitors. Key parameters of ion transport include ionic conductivity, σ, diffusion coefficient, D, ion mobility, μ, transport number, T_i, charge transfer resistance, R_{CT}, and electrical double-layer capacitance, C_D. Additionally, the forward, k_1, and backward, k_2, reaction rate constants for redox reactions ($Ox + ne^- \rightleftharpoons Red$) and their equilibrium constant, K, provide fundamental insights into electrochemical processes. Definitions and backgrounds of key parameters are described in Chapter 8. Here, experimental techniques used to measure key ion transport parameters will be briefly introduced.

11.1 Experimental Techniques

Electrochemical systems are evaluated by applying either a constant voltage or current and observing the resulting current or voltage response. Typically, to investigate ion transport, both direct current, DC, and alternating current, AC, methods are employed. For example, chronoamperometry as a DC method examines transient current responses under a constant voltage, while galvanostatic techniques provide insights by maintaining a constant current and measuring the resulting voltage changes. For a comprehensive understanding of ion transport properties in polymer electrolytes and ion exchange membranes, foundational texts, such as Bard and Faulkner [1], MacCallum and Vincent [2], and Strathmann [3] are recommended.

This chapter introduces the measurement techniques for parameters associated with ion transport, starting their representative DC and AC techniques. In addition, they will be applied to measure transport number.

11.2 Direct Current Techniques

DC techniques provide essential information, including diffusion coefficients, interfacial reaction kinetics, and exchange current densities. Here, three representative techniques, including linear sweep voltammetry (LSV), cyclic voltammetry (CV), and potentiostatic intermittent titration technique (PITT) will be introduced.

Molecule and Ion Transport through Polymer Membranes, First Edition. Yong Soo Kang.

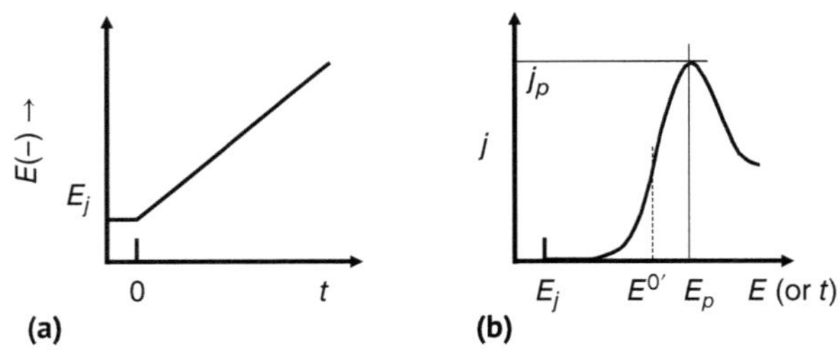

Figure 11.1 A schematic linear sweep voltammetry (LSV) curve highlighting the peak and its relationship to mass transport and electron transfer processes: (a) linear potential sweep or ramp starting at E_i; and (b) resulting j–E curve. *Source*: Y.S. Kang.

Among various DC methods, LSV and CV are particularly noteworthy for their diagnostic power and versatility. While CV is applicable to most cases where LSV is used, there are some instances where LSV is more useful. In cases where the reaction is irreversible, CV will not give any additional data that LSV would give us.

Both LSV and CV are a method of voltammetry where the current at a working electrode (WE) is measured while the potential between the WE and a reference electrode (RE) is swept linearly in time commonly with three-electrode setups. The three-electrode setup is typical for both LSV and CV, which represent a WE, counter electrode (CE), and RE. While the current flows between the WE and CE, the RE is used to accurately measure the applied potential relative to a stable reference reaction.

11.2.1 Linear Sweep Voltammetry (LSV)

In LSV, the electrode potential is swept linearly with time while the resulting current is measured as a function of potential. This produce an i–E (or a j–V) curve that reflects key aspects of mass transport and reaction kinetics. As the potential reaches the redox region, the system transitions from non-faradaic to faradaic behavior, leading to a characteristic peak-shaped curve (Figure 11.1), which will be discussed further in Section 11.2.2. Initially, only capacitive (non-faradaic) currents are observed. As the potential approaches the redox potential, faradaic reactions begin, and the current rises in response to the combined effects of mass transport and electron transfer.

The peak current density, j_p, and peak potential, E_p, are central to analyzing LSV data. These values depend on factors, such as the scan rate, r (mV/sec), and diffusion coefficient, D (cm^2/sec). Notably, for reversible systems, j_p is proportional to $r^{1/2}$, and E_p remains independent of r. This property distinguishes reversible processes from those with kinetic limitations or irreversibility.

11.2.2 Cyclic Voltammetry (CV)

CV extends LSV by reversing of the potential sweep direction as shown in Figure 11.2. This reversal allows the observation of both reduction and oxidation processes within a single experiment. CV is particularly valuable for its ability to provide insights into reaction mechanisms, reversibility, and intermediate species.

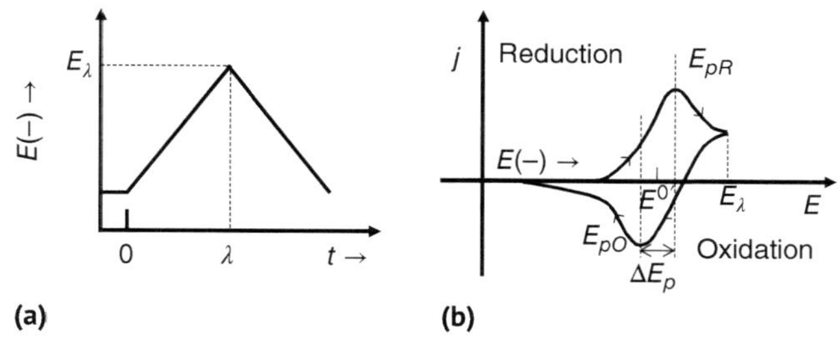

Figure 11.2 Example of a cyclic voltammogram for a reversible system, showing the forward and reverse peaks: (a) cyclic potential sweep; and (b) resulting cyclic voltammogram. *Source*: Adapted from [1].

The resulting cyclic voltammogram features a forward peak corresponding to reduction and a reverse peak for oxidation. The peak separation, ΔE_p, offers a diagnostic criterion for reaction reversibility. For reversible reactions, ΔE_p is approximately $59/n$ mV at room temperature, where n is the number of electrons involved.

11.2.2.1 *i–E* Curves and Concentration Changes

Consider a representative single electron transfer reaction at the equilibrium between ferrocenium, Fc^+, and ferrocene, Fc, as:

$$Fc^+ + e^- = Fc \quad (\text{or } Ox + e^- = Red) \tag{11.1}$$

When a solution containing Fc^+ is scanned toward negative potentials, Fc^+ is reduced to Fc at the electrode surface, generating a measurable current and depleting Fc^+ at the electrode surface. The resulting cyclic voltammogram is presented in Figure 11.3 accompanied by the corresponding concentration–distance profiles in Figure 11.4 for Fc^+ (solid line) and Fc (dotted line) at various points in the voltammogram.

Experimentally, as the potential is scanned negatively (cathodically) from point I to point IV in Figure 11.3, the concentration of Fc^+ (or Ox) near the electrode steadily decreases as it is reduced to Fc (or Red). At point III, where the cathodic peak current $i_{p,c}$ appears, the current is controlled by the diffusion of additional Fc^+ from

Figure 11.3 Voltagram of the reversible reaction for $Fc^+ + e^- = Fc$ at a scan rate of 100 mV/sec, where Fc^+ and Fc are ferrocenium and ferrocene, respectively. *Source*: Adapted from [4].

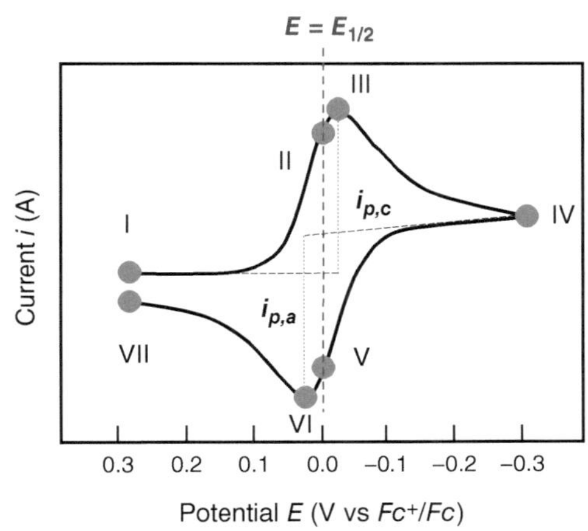

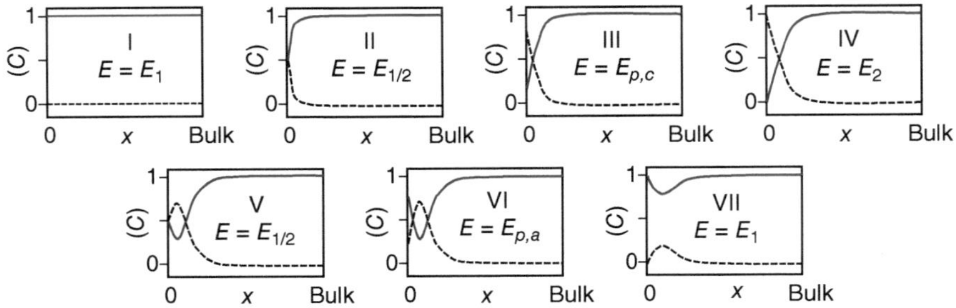

Figure 11.4 Concentration profiles [C] (mM) for Fc^+ (solid) and Fc (dotted) by varying applied potential E from I to VII, plotted as a function of the distance x from the electrode to the bulk solution. *Source*: Adapted from [4].

the bulk solution, through the diffuse layer, to the electrode surface. As the scan continues, the diffuse layer thickens, slowing Fc^+ transport and causing the current to fall at more negative potentials (III → IV) Fc^+(or Ox).

When the scan reaches the switching potential at point IV, the direction is reversed and the potential is swept in the positive (anodic) direction, allowing observation of the oxidation process in CV [4]. At points II and V, the surface concentrations of Fc^+ and Fc become equal. For a reversible redox couple, this satisfies the Nernst equation ($E = E^{o\prime}$), and corresponds to the midpoint between the cathodic and anodic peak potentials (points III and VI). The separation between these peaks, ΔE_p, is due to the time-dependent diffusion of the analyte to and from the electrode. For a chemically and electrochemically reversible system, ΔE_p is 57 mV at 25 °C (equal to 2.22 RT/F), and the width at half max on the forward scan of the peak is 59 mV [4].

11.2.2.2 Characterization of *i–E* Curves

Up to now, *i–E* curves are explained qualitatively. In order to get quantitative information, the changes of the analyte concentrations by electrochemical reactions are necessary to obtain electric potential E by the Nernst equation. The Nernst equation for $Ox + ne^- = Red$ at equilibrium is described as:

$$E = E^0 + \frac{RT}{nF}\ln\left(\frac{a_{Ox}}{a_{Red}}\right) \tag{11.2}$$

where E^0 is the standard potential, a_i is the chemical activity of species i, and F is the Faraday's constant. In application of the Nernst equation (Equation (11.2)) to the one-electron reduction of Fc^+ to Fc, the activities are replaced with their concentrations ($a_i = \gamma_i c_i$ with γ_i and c_i being the activity coefficient and the concentration of species i, respectively), which are more experimentally accessible, the standard potential E^0 is replaced with the formal potential $E^{0\prime}$. Equation (11.2) is readily converted into:

$$E = E^{0\prime} + \frac{RT}{nF}\ln\left(\frac{c_{Ox}}{c_{Red}}\right) \tag{11.3}$$

where $E^{0\prime} = E^0 + (RT)/(nF)\ln(\gamma_{Ox}/\gamma_{Red})$.

The formal potential $E^{0\prime}$ is specific to the experimental conditions employed and is often estimated with the experimentally determined $E_{1/2}$ value, the half-wave potential, defined as the potential at which the current is half of the peak current density, j_p, during a redox process. It can be mathematically obtained from Equation (11.3) when $c_{Ox} = c_{Red}$ at point II in Figure 11.4. In other words, $E_{1/2}$ can often approximate $E^{0\prime}$, especially for simple, reversible, one-electron transfer reactions, but not for irreversible, non-ideal, and coupled chemical reactions. The Nernst equation provides a powerful way to predict how a system will respond to a change of concentration of species in solution or a change in the electrode potential. To illustrate, if a potential of $E = E^{0\prime} \approx E^{1/2}$ is applied to the example of Fc^+ solution, the Nernst equation predicts that Fc^+ will be reduced to Fc until $[Fc^+] = [Fc]$, and equilibrium is achieved [4]. While $E^{1/2}$ is frequently used as a practical approximation for $E^{0\prime}$ in CV, it is not a strict replacement. Additional considerations, such as symmetry of the redox peaks and reversibility of the system, are necessary to determine whether $E^{1/2}$ closely reflects $E^{0\prime}$.

11.2.2.3 Scan Rate Effects

The scan rate r (V/sec) of the experiment controls how fast the applied potential is scanned. Faster scan rates lead to a decrease in the thickness of the diffuse layer, as a consequence, higher currents are observed.

We consider again the reaction $Ox + ne^- = Red$, assuming semi-infinite linear diffusion and a solution initially containing only species Ox, with the electrode held initially at a potential E_i, where no electrode reaction occurs. The potential is swept linearly with r so that the potential at any time is [1].

$$E(t) = E_i - rt \tag{11.4}$$

Then, substituting Equation (11.4) into Equation (11.3) yields:

$$\frac{c_{Ox}(0,t)}{c_{Red}(0,t)} = \exp\left[\frac{nF}{RT}(E_i - rt - E^{0\prime})\right] \tag{11.5}$$

where n is the number of electrons transferred in the redox event. Equation (11.5) indicates the r-dependent concentration changes. Since the mass flux J_{Ox} at the electrode surface is proportional to the current density $j(t)$, $J_{Ox} = -i(t)/nFA = -D_{Ox}[\mathrm{d}c_{Ox}(x,t)/\mathrm{d}x]$ at $x = 0$ with the electrode surface area A (cm^2) (usually treated as the geometric surface area). The peak current density j_p can be obtained as [1]:

$$j_p = 0.4463\left(\frac{F^3}{RT}\right)^{1/2} n^{3/2} D_{Ox}^{1/2} c_{Ox}^{b} r^{1/2} \tag{11.6}$$

with the diffusion coefficient D_{Ox} and bulk concentration c_{Ox}^{b} of O_x.

For electrochemically reversible electron transfer processes involving freely diffusing redox species, the Randles–Sevcik equation (Equation (11.6)) describes how the peak current density j_p increases linearly with $r^{1/2}$. In addition to verifying that the analyte is freely diffusing, the Randles–Sevcik equation may be used to calculate diffusion coefficients.

The peak potential, E_P, is also found to be [1]:

$$E_p = E_{1/2} - 1.109\frac{RT}{nF} \tag{11.7}$$

Because the peak is somewhat broad, so that the peak potential may be difficult to determine, it is sometimes convenient to report the potential at $i_p/2$, called the half-peak potential, $E_{p/2}$ [1], which is:

$$E_{p/2} = E_{1/2} + 1.09\frac{RT}{nF} \tag{11.8}$$

Note that $E_{1/2}$ is located just about midway between E_p and $E_{p/2}$, and that a convenient diagnostic for a Nernstian wave is:

$$|E_p - E_{p/2}| = 2.20\frac{RT}{nF} \tag{11.9}$$

Thus, for a reversible wave, E_p is independent of r, and j_p is proportional to $r^{1/2}$. The latter property indicates diffusion control and is analogous to the variation of the current density j_d due to the diffusive flux with $t^{-l/2}$ in chronoamperometry (Equation (11.6)). A convenient constant in LSV is $j_p/r^{1/2}C_0^b$ (sometimes called the current function), which depends on $n^{3/2}$ and $D_{Ox}^{1/2}$. This constant can be used to estimate n for an electrode reaction, if a value of D_{Ox} can be estimated, for example, from the LSV of a compound of similar size or structure that undergoes an electrode reaction with a known n value [1].

11.2.2.4 Adsorption at Surface

The Randles–Sevcik equation can give indications as to whether an analyte is freely diffusing in solution. As analytes can sometimes adsorb to the electrode surface, it is essential to assess whether an analyte remains homogeneous in solution prior to analyzing its reactivity.

When the species are adsorbed on the electrode surface, the current density response is described by [4]:

$$j_p = \frac{n^2F^2}{4RT}r\Gamma^* \tag{11.10}$$

where Γ^* (mol/cm^2) represents the surface coverage of the adsorbed species in moles per square centimeter. For species adsorbed on the electrode, the current response is expected to vary linearly with r (the scan rate, not its square root). If the analyte is suspected to be a freely diffusing species, deviations from linearity in the j_p vs $r^{1/2}$ plot may indicate one of two possibilities: The reaction is quasi-reversible (meaning the electron transfer kinetics are not fast enough to maintain equilibrium). The reaction involves surface-adsorbed species instead of freely diffusing species. To identify the cause of these deviations, the peak-to-peak separation, ΔE_p, can be examined: For quasi-reversible processes, ΔE_p increases with the scan rate. For surface-adsorbed species, no peak-to-peak separation is observed [4].

Potential sweep methods are further influenced by non-ideal factors such as double-layer capacitance, C_D, which generates a baseline current proportional to the scan rate, and uncompensated resistance, which distorts peak shapes. These effects are more pronounced at higher scan rates or lower concentrations, requiring careful experimental design [1].

Ultramicroelectrodes, UMEs, are particularly useful in addressing these challenges. Their small size minimizes resistance and capacitance effects, enabling high-speed sweeps while maintaining accurate current–potential relationships. The unique behavior of UMEs, including their steady-state response, further expands the applicability of potential sweep methods [1].

If you are interested in the effects of irreversible and quasi-reversibly processes as well as non-ideal factors such as double-layer capacitance and uncompensated resistance in more detail, you may refer to as Chapter 6 of the book by Bard and Faulkner [1].

Potential sweep methods, especially LSV and CV, are indispensable tools for electrochemical analysis. Their ability to elucidate reaction kinetics, transport phenomena, and system dynamics makes them foundational in both research and practical applications. By combining diagnostic clarity with flexibility, these techniques continue to shape our understanding of electrochemical systems.

11.2.3 Potentiostatic Intermittent Titration Technique (PITT)

PITT is an electrochemical method used to investigate the diffusion of ions or charge carriers within materials, particularly in batteries and fuel cells. It is conducted by applying a constant potential (potentiostatic control) to an electrochemical cell and intermittently switching the potential to analyze the time-dependent current response [5, 6].

In PITT, a constant potential is applied across an electrochemical system for a specific time period, allowing the system to equilibrate. The resulting current (which is initially high and decays over time) is measured as a function of time. After reaching equilibrium (or nearly so), the applied potential is incrementally changed, and the process is repeated. The key information including the material's ionic mobility or diffusion rates is obtained from analyzing the decay of the time-dependent current, which relates to diffusion-controlled processes, such as ion migration in solid-state materials.

PITT is widely used to study the diffusion of lithium ions in battery materials, such as electrodes and solid-state electrolytes. The method helps in studying the ion dynamics within electrode materials, particularly for supercapacitors, where fast ion transport is critical for achieving high power densities. PITT can be used to explore ion transport properties in proton-exchange membrane (PEM) fuel cells, providing insights into diffusion kinetics within the membrane material. In corrosion science, PITT is used to study how ions diffuse in corrosion layers or films on metals, providing insights into the protective properties of coatings [5, 6].

PITT is sensitive to small changes in the electrochemical system, making it ideal for studying slow ion diffusion processes. Since each potential step requires equilibrium, the technique can be slow. Extracting meaningful diffusion parameters from the data involves complex modeling, requiring expertise in electrochemical analysis.

Although conventional PITT is widely used to characterize diffusion coefficient in electrochemical materials, it assumes a diffusion-controlled system and neglects interfacial resistance. However, in real electrode systems, both interfacial reaction and diffusion can limit the rate. Recently, PITT has been extended to account for finite surface reaction rates, allowing simultaneous determination of the diffusion coefficient, surface reaction rate, and exchange current density by fitting analytical solutions to the measured transient current [7].

In summary, PITT is powerful for studying diffusion-controlled processes and also interfacial reactions in materials, with widespread applications in battery technology, supercapacitors, fuel cells, and corrosion science.

11.3 Alternating Current Techniques

AC techniques are favored for their ability to measure both bulk and interfacial properties using simple cells with inert, blocking electrodes. However, unlike DC methods, they require more advanced instrumentation and more complex theoretical analysis.

AC methods are particularly valued for their ability to provide insights not only into the long-range transport of ions but also into polarization phenomena, such as the relaxation of trapped ions. This section provides a brief introduction to the fundamental principles and measurement techniques, primarily based on a tutorial paper by Lazanas and Prodromidis [8], and two textbooks by Bard and Faulkner [1] and by MacCallum and Vincent [9].

The discussion begins with Nyquist and Bode plots, which are central to electrochemical impedence spectroscopy (EIS). EIS is then used to extract key ion transport properties, such as ion conductivity and charge transfer numbers.

11.3.1 Nyquist and Bode Plots

In AC circuits, impedance, Z, represents the total opposition to the current flow in an electrical circuit composed of resistors, capacitors, and inductors. These oppositions are called as resistance, capacitive reactance, and inductive reactance, denoted as R, X_C, and X_L, respectively.

When a low amplitude alternating voltage, $v(t)$, at a particular angular frequency ω ($\omega = 2\pi f$ with f being conventional frequency):

$$v(t) = V_0 \sin(\omega t) \tag{11.11}$$

is applied to an electrical circuit and the resulting AC at the same frequency

$$i(t) = I_0 \sin(\omega t + \varphi) \tag{11.12}$$

is measured where φ stands for phase angle as shown in Figure 11.5a. In this case, the impedance of the circuit can be more conveniently expressed by complex number as:

$$Z = Z' - jZ'' \tag{11.13}$$

where $j = (-1)^{1/2}$. Figure 11.5b illustrates the impedance, Z, of a cell on a vector or Argand diagram, where Z' and Z'' are respectively the real and imaginary components of the complex impedance [9].

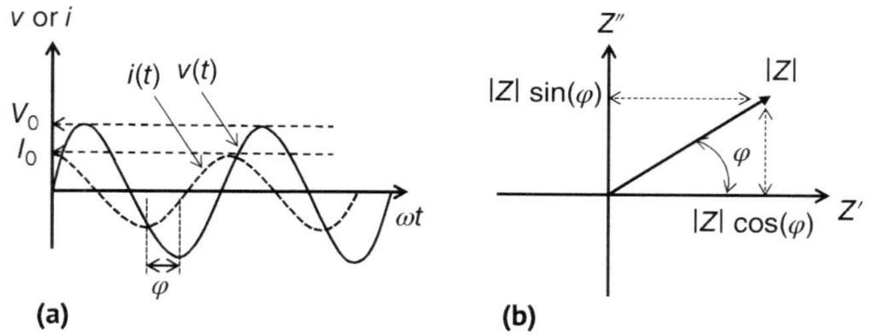

Figure 11.5 (a) The sinusoidal waveforms $v(t)$ and $i(t)$ with a phase angle φ.
(b) Representation of the impedance, Z, of a cell on a vector or Argand diagram. Z' and Z'' are respectively the real and imaginary components of the complex impedance $Z = Z' - jZ''$. *Source*: Y.S. Kang.

The amplitude of Z, $|Z|$, representing the opposition to the flow of charge, can be obtained from $|Z| = V_0/I_0$. Therefore, we can obtain that:

$$|Z|^2 = (Z')^2 + (Z'')^2 \quad \text{where} \quad Z' = |Z|\cos(\varphi) \quad \text{and} \quad Z'' = |Z|\sin(\varphi) \tag{11.14}$$

Thus,

$$Z(\omega) = |Z|(\cos(\varphi) - j\sin(\varphi)) \tag{11.15}$$

and

$$\tan(\varphi) = \frac{Z''}{Z'} \tag{11.16}$$

Therefore, the impedance is characterized by both amplitude $|Z|$ and phase angle φ at a given frequency ω. Generally, for electrochemical cells, both $|Z|$ and φ are functions of ω [9].

11.3.1.1 Impedance Through a Resistor and a Capacitor

When an alternating voltage $v(t)$ is applied to a resistor with resistance R, the resulting alternating current i(t) follows the voltage without phase delay. The sinusoidal waveforms $v(t)$ and $i(t)$ are in-phase ($\varphi = 0$), and thus $\sin(\varphi) = 0$. As a result, the impedance Z_R of the resistor in this case is independent of the frequency and contains only the real component:

$$Z_R = R \tag{11.17}$$

In the case of a capacitor, the resulting alternating current $i(t)$ flowing in the circuit, as a result of the fluctuation of the electrical charge q stored in the capacitor is given as:

$$i(t) = \frac{dq}{dt} = C\left(\frac{dv}{dt}\right) \tag{11.18}$$

since $q = Cv$ with C being the capacitance. Combining Equations (11.11) and (11.18) gives:

$$i(t) = V_0 \omega C \cos(\omega t) \tag{11.19}$$

For $\omega t = \pi/2$, the magnitude of the current, I_0, is:

$$I_0 = V_0 \omega C \tag{11.20}$$

and capacitive reactance, X_C, is given as:

$$X_C = \frac{V_0}{I_0} = \frac{V_0}{V_0 \omega C} = \frac{1}{\omega C} = \frac{1}{2\pi f C} \tag{11.21}$$

Equation (11.21) shows that X_C is inversely proportional to ω and C. Equation (11.19) can also be written as:

$$i(t) = V_0\ \omega C \sin\left(\omega t + \frac{\pi}{2}\right) \tag{11.22}$$

showing that the voltage applied to the capacitor lags the current by $\pi/2$. According to Equation (11.15), for $\varphi = \pi/2$, the real part of impedance is zero.

11.3.1.2 Nyquist and Bode Plots for a Resistor and a Capacitor

A typical AC experiment consists of determining the complex impedance Z of a cell as a function of the signal frequency ω and presenting the results in the form of a complex impedance plot including the Nyquist and Bode plots. Thus, the Nyquist and Bode plots for various simple electrical circuits containing only a single passive element such as resistor and capacitor will be first introduced here, followed by their combinations in different arrangements in parallel are illustrated in Figures 11.6 and 11.7, respectively.

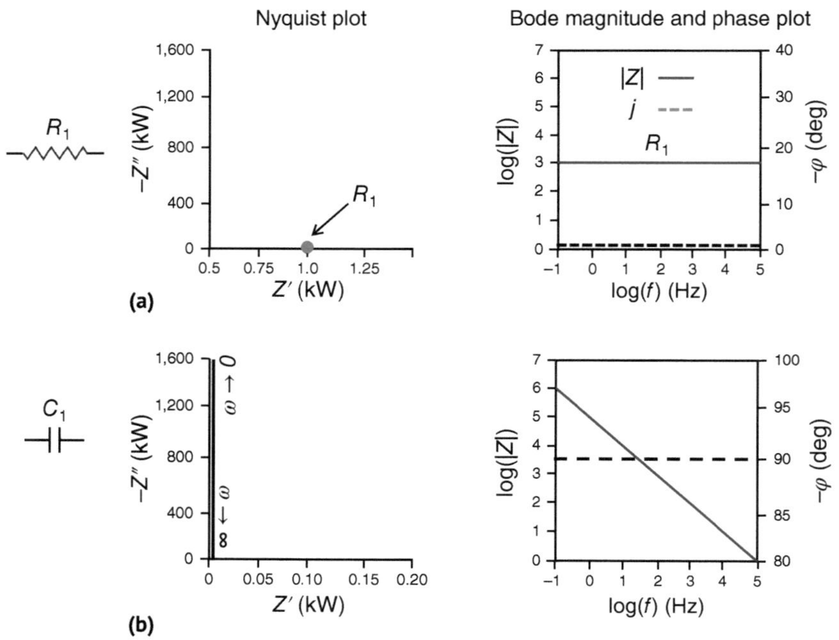

Figure 11.6 Nyquist and Bode plots of (a) a resistor and (b) a capacitor. *Source*: Adapted from [8].

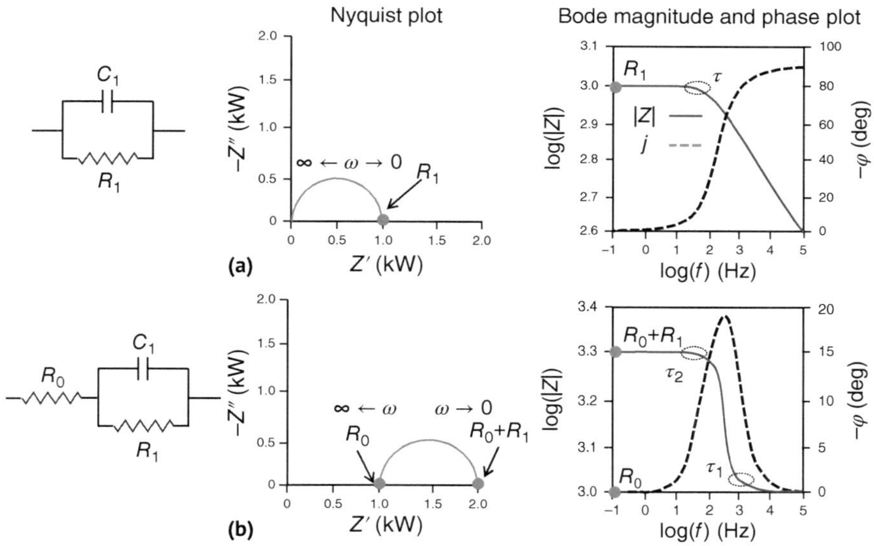

Figure 11.7 Nyquist and Bode plots of (a) parallel combination model circuit of a resistor and a capacitor and (b) additional resistor in series with the parallel combination of a resistor and a capacitor. $R_0 = R_1 = 1$ kW. *Source*: Adapted from [8].

When the circuit contains only a resistor, Figure 11.6a, $Z' = R_1$ and $Z'' = 0$. As a result, the Nyquist plot shows a single point lying in the real axis. That is, the impedance values at all the excitation frequencies are exactly the same as the resistance R_1 of the resistor. As a result, the Bode magnitude plot shows a straight line that crosses the left axis at $|Z| = R_1$, while the Bode phase angle φ plot shows a straight line that crosses the right axis at $\varphi = 0$, representing that the voltage and current waveforms are "in-phase" ($\varphi = 0$).

In the case of a capacitor only, Figure 11.6b, the equation of the impedance with help of Equation (11.21) becomes:

$$Z = 0 + \frac{1}{j\omega C} = 0 - j\left(\frac{1}{\omega C}\right) \tag{11.23}$$

because the phase angle is constant at for $\varphi = \pi/2$ regardless of the frequency ω. Thus, the real part is zero ($Z' = 0$), while the imaginary part Z'' is reversely proportional to C and ω (Equation (11.23)). As a result, the Nyquist plot shows a straight line lying along the y-axis (the real impedance is zero). Values close to zero corresponds to high frequencies, while at lower frequencies the impedance values are higher. The Bode magnitude plot shows a straight line with slope -1, while the Bode phase angle plot shows a straight line that crosses the right axis at $\varphi = \pi/2$, as in accordance with Figure 11.6b, through a capacitor, the voltage and current waveforms are "out-of-phase" with $\pi/2$. Note that in real electrochemical cells the phase between voltage and current due to a "capacitive" element is typically lower than $\pi/2$.

11.3.1.3 Nyquist and Bode Plots for a Parallel Combination of a Resistor and a Capacitor

There are diverse combinations of resistors, capacitors and inductors, and corresponding equivalent circuits, for specific devices. Here, only two examples are introduced: parallel combination of a resistor and a capacitor, and an additional resistor in series with the parallel combination of a resistor and a capacitor, as shown in Figure 11.7a and b.

When the circuit contains a resistor and a capacitor connected in parallel (Figure 11.7a), the equation of the impedance is:

$$Z(\omega) = \frac{1}{1/R_1 + j\omega C_1} = \frac{R_1}{1+(\omega R_1 C_1)^2} - j\frac{\omega R_1^2 C_1}{1+(\omega R_1 C_1)^2} \tag{11.24}$$

and the Nyquist plot corresponds to a semicircle. In this case, at very high frequencies the capacitive reactance, X_C, tends to zero ($\omega \to \infty, X_C \to 0$), and thus, all the current passes through the capacitor. The circuit acts as a short circuit and the impedance is zero. At very low frequencies, the capacitive reactance tends to infinity ($\omega \to 0,\ X_C \to \infty$), and all the current passes through the resistor. The impedance contains only real part and $Z' = R_1$. In response to the discussion above, when the $\omega \to 0$, the current is constant. Since the constant current cannot flow through the capacitor, it flows only though the resistor.

At intermediate frequencies, the current passes at the same time through the capacitor and the resistor. Traversing from high to medium frequencies, the capacitive reactance becomes larger (see Equation (11.23), $X_C = 1/\omega C$), but remains still lower than the ohmic resistance ($X_C < R_1$), which causes more AC to go through the capacitor while less constant current goes through the resistor. However, there is a single characteristic frequency corresponding to equal values of reactance and resistance ($X_C = R_1$). At this frequency, the imaginary part of the impedance is maximum ($\omega_{Z''\max}$). By substituting X_C with R_1, in Equation (11.21), we get that $R_1 = 1/(\omega_{Z''\max} C_1)$, and consequently

$$\omega_{Z''\max} = \frac{1}{\tau} = \frac{1}{R_1 C_1} \tag{11.25}$$

where $\tau = R_1 C_1$ is the characteristic time constant of the system. Note that by substituting $R_1 = 1/(\omega_{Z''\max} C_1)$ to Equation (11.24), it becomes $Z(\omega) = R_1(1-j)/2$, thus manifesting that, in an RC parallel circuit, τ is found when the imaginary and real parts of the impedance are equal.

Looking at the Bode magnitude plot at very low frequencies, according to the Equation (11.24), $|Z| = R_1$, while at very high frequencies $|Z| = 1/\omega C_1$. The slope of the curve is changing (this breaking point is highlighted with a dotted-line circle (τ) in Figure 11.7a) around the frequency that corresponds to the time constant of the system.

If a resistor R_0 is connected in series with the circuit in Figure 11.7a to form the configuration in Figure 11.7b, $Z(\omega)$ will be obtained from:

$$Z(\omega) = R_0 + \frac{R_1}{1+(\omega R_1 C_1)^2} - j\frac{\omega R_1^2 C_1}{1+(\omega R_1 C_1)^2} \tag{11.26}$$

At both very high and very low frequencies, the behavior of the circuit is resistive and thus the semicircle is shifted along the real axis to a value equal to the ohmic resistance of R_0, as indicated in Figure 11.7b. At very high frequencies ($\omega \to \infty$; $X_C \to 0$), $Z = R_0$, while at very low frequencies ($\omega \to 0$; $X_C \to \infty$), $Z = R_0 + R_1$. These boundaries conditions are illustrated in the Nyquist plot as the first and the second crossing points of the semicircle on the axis of the real impedance, respectively.

In contrast with the previous circuit (Figure 11.7a), this circuit of the Bode magnitude plot in Figure 11.7b shows two time constants indicating the two breaking points. The τ_1 at the high frequency domain is influenced by R_0 and can be calculated as $\tau_1 = [(R_0 R_1)/(R_0 + R_1)]C_1$. The τ_2, at lower frequencies, which as the slowest one is considered as the characteristic time constant of the system, can be calculated from the Nyquist plot as in the case of the simple RC parallel circuit, as $\omega_{Z''\max} = 1/\tau_2 = 1/R_1C_1$. Looking at the Bode plot of this circuit, the change of $|Z|$ and phase over a wide frequency range is described by an S-shaped and a bell-shaped curve, respectively.

11.3.2 Electrochemical Impedance Spectroscopy (EIS)

The basic principles of the AC techniques involving resistors, capacitors, and their simple parallel and/or series combinations are introduced in Section 11.3.1 using Nyquist and Bode plots. These concepts are then extended to more complex or practical devices through EIS [8, 10].

EIS measures the response of the electrochemical devices to an applied alternating voltage over a range of frequencies. It provides information on the impedance of the system, which can be used to calculate and analyze ionic conductivity, contributed from different processes like bulk ion conductivity, electrode polarization, and interfacial resistance (charge transfer resistance). From Nyquist plots (real vs imaginary impedance) and Bode plots (magnitude and phase vs frequency), the ion transport parameters described previously can be estimated.

The most common electrochemical setup refers to a three-electrode configuration consisting of WE, RE, and CE, in connection with a potentiostat as shown in Figure 11.8. WE offers the electro-catalytic surface on which a redox reaction takes place under a desirable potential, or in other words, under a desirable potential difference with respect to the (stable) potential of RE.

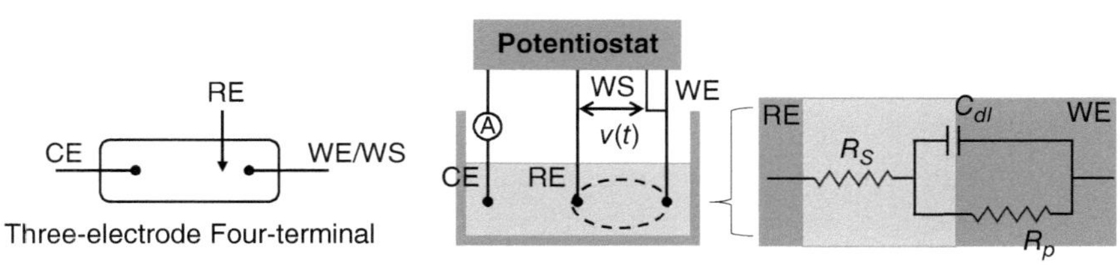

Figure 11.8 Electrochemical cell of three electrodes with the working (WE), working sensor (WS), counter (CE), and reference (RE) electrodes. *Source*: Adapted from [8].

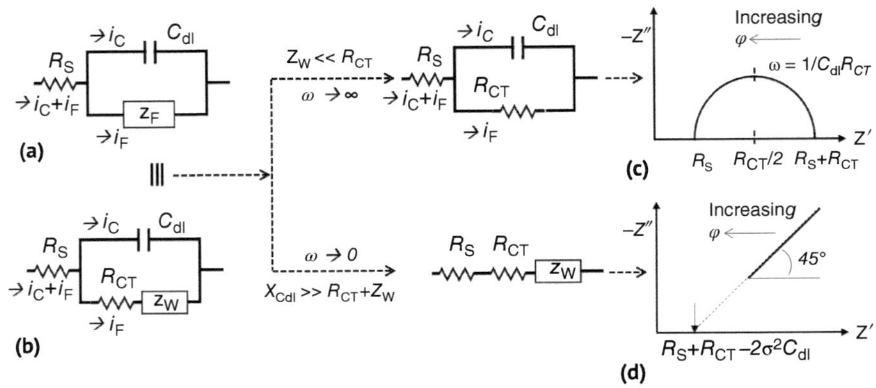

Figure 11.9 The Randles equivalent circuit (a, b) and their behavior at high (c) and low (d) frequencies. *Source*: Replotted from [8].

The term "faradaic EIS" refers to impedance measurements of an electrochemical cell – such as a three-electrode, four-terminal setup in potentiostatic mode with a working sensor (WS), as shown in Figure 11.8 – in the presence of a redox couple such as:

$$Ox + ne^- = Red \tag{11.27}$$

In this case, the total current that passes through an electrolyte resistor, represented by R_S, is divided into the current related to the charging/discharging of the electrical double layer, i_C, and the current related to the faradaic process, i_F. As illustrated in Figure 11.9a, the faradaic process is represented as a general impedance, Z_F, which accounts for both the kinetics of the redox reaction and the diffusion of the redox species to the surface of the WE. In this regard, Z_F can be divided into two components:

$$Z_F = R_{CT} + Z_W \tag{11.28}$$

where R_{CT} and Z_W are the charge-transfer resistance and the Warburg impedance [11], respectively. R_{CT} is related with the kinetics of the heterogeneous electrochemical process assuming that the redox species are not adsorbed on the electrode surface, as per equation:

$$R_{CT} = \frac{RT}{k^0 n^2 F^2 Ac} \tag{11.29}$$

where k_0 (cm/sec) is the heterogeneous electron transfer rate, n is the number of electrons transferred in the electrochemical reaction, F is Faraday's constant ($F = 96,485$ C/mol), R is the gas constant ($R = 8.314$ J/mol K), (K) is the temperature, A (cm^2) is the electroactive surface area of the WE, and c is the concentration of the redox species assuming that this concentration is the same as the one in the bulk solution and that $c_{Ox} = c_{Red} = c$.

The Warburg impedance, Z_W, expresses the difficulty of mass transport of the redox species to the electrode surface, considering a semi-infinite linear diffusion. Z_W behaves as an R_W and C_W circuit in series, where both R_W and C_W are frequency-dependent, and thus, the Z_W can be written as:

$$Z_W = R_W + C_W = \left[\sigma\omega^{-1/2} - j\sigma\omega^{-1/2}\right] \tag{11.30}$$

where $\sigma = 2RT/n^2F^2\sqrt{2}\sqrt{Dc}$ and D is the diffusion coefficient of the redox couple, assuming that $D_{Ox} = D_{Red} = D$, while the rest of the symbols have their aforementioned meaning [8].

The equivalent electrical circuit involving the charge-transfer resistance, R_{CT} and the Warburg impedance, Z_W (Figure 11.9) is the most common representation of the so-called Randles circuit [12]. Based on the above analysis, the impedance values of the Randles equivalent circuit over a wide range of frequencies can be separated into the real (in-phase), Z', and imaginary (out of phase), Z'', given by the equations [13, 14] as:

$$Z' = R_S + \frac{R_{CT} + \sigma\omega^{-1/2}}{(\sigma\omega^{1/2}C_{dl} + 1)^2 + \omega^2 C_{dl}^2 (R_{CT} + \sigma\omega^{-1/2})^2} \tag{11.31}$$

$$-Z'' = \frac{\omega C_{dl} (R_{CT} + \sigma\omega^{-1/2})^2 + \sigma^2 C_{dl} + \sigma\omega^{-1/2}}{(\sigma\omega^{1/2}C_{dl} + 1)^2 + \omega^2 C_{dl}^2 (R_{CT} + \sigma\omega^{-1/2})^2} \tag{11.32}$$

The physical meaning for the combination of the individual components (R_S, C_{dl}, R_{CT} and Z_W) is the following: R_S is connected in series with the parallel combination of C_{dl} and Z_F (or according to Equation (11.28), R_{CT} and Z_W) indicating that all the current will pass through the electrolyte. The parallel combination of C_{dl} with R_{CT} and Z_W indicates that the total current, $i_t = i_C + i_F$, will be split to the current for charging of C_{dl} (i_C) and the faradaic current (i_F) for the redox reaction depending on the relative values of the components C_{dl}, R_{CT}, and Z_W at different frequency. Note that information on kinetics such as k_0 and mass transport including D for electrochemical reactions can be obtained from R_{CT} (Equation (11.29)) and Z_w (Equation (11.30)), respectively [8].

As a summary of AC measurements, advantages are: (1) simply constructed two-terminal blocking electrode cells may be used to determine the total DC conductivity of the electrolyte; (2) blocking electrodes, e.g. Au and Pt, are frequently stable over wide temperature ranges and therefore the temperature dependence of the electrolyte conductivity may be obtained very easily; and (3) information on charge migration and polarization within the electrolyte and at the electrodes may be obtained, permitting a very full characterization of the entire cell.

Disadvantages are: (1) relatively sophisticated equipment is required; (2) care must be taken to eliminate lead impedances, etc., otherwise false interpretation of the data is a significant possibility; and (3) interpretation of the data requires some understanding of the theory of AC measurements of polymer electrolyte cells. In other words, we must construct possible equivalent circuit models for the AC response of the cell and consider which of these fits the data [9].

11.4 Transport Number Measurements

When a polymer electrolyte contains more than one mobile-charged species, it is essential to determine not only the total conductivity σ but also the contribution of each species to the overall current [15]. This contribution is expressed by the transport number T_j, defined in Equation (8.63) in Section 8.5.4 for any mobile species j:

$$T_j = \frac{j_j}{j} = \frac{\sigma_j}{\sigma} \tag{11.33}$$

Both ions and electrons can contribute to total conductivity. Here, we focus on the transport numbers, where electronic conduction is negligible. Various techniques have been developed to measure transport numbers in ionically conducting electrolytes, such as polymer electrolytes and ion exchange membranes. Sørensen and Jacobsen [16] formulated an expression for T_j by analyzing a system with symmetric, non-blocking electrodes and a thin electrolyte layer:

$$T_1 = \frac{1}{1 + Z_d(0)/R_b} = \frac{R_b}{R_b + Z_d(0)} \tag{11.34}$$

where R_b is the electrolyte resistance, and $Z_d(0)$ represents the low-frequency impedance due to Warburg diffusion [17].

Since then, several methods have been introduced. Watanabe et al. developed the isothermal transient ionic current (ITIC) method, which combines DC and AC measurements [18, 19]. Later, Vincent and Bruce proposed a similar technique that operates at lower voltages [20, 21]. Wiemhöfer et al. also introduced an approach where voltage is applied to symmetric cells and stopped once the current stabilizes. The subsequent voltage decay back to the open-circuit potential, governed by the diffusion of charge carriers, is then used to determine the transference number [17, 22].

Each method has its advantages in terms of ease of measurement, sensitivity, repeatability, and long-term stability. Among them, the approach by Sørensen and Jacobsen stands out as the simplest, most reliable, and least prone to experimental errors [17]. It produces T_j values with the lowest standard deviation, making it a preferred choice for accurate transport number measurements.

References

1 A.J. Bard, L.R. Faulkner, Electrochemical Methods: Fundamentals and Applications, Wiley, 2000.
2 J. R. MacCallum, C.A. Vincent, Polymer Electrolyte Reviews, Elsevier Applied Science, 1989.
3 H. Strathmann, Ion-exchange Membrane Separation Processes, Elsevier Science, 2004.
4 N. Elgrishi, K.J. Rountree, B.D. McCarthy, E.S. Rountree, T.T. Eisenhart, J.L. Dempsey, Journal of Chemical Education 95(2) (2018) 197–206.

5 C. Wen, C. Ho, B. Boukamp, I. Raistrick, W. Weppner, R. Huggins, International Metals Reviews 26(1) (1981) 253–268.
6 S. Malifarge, B. Delobel, C. Delacourt, Journal of the Electrochemical Society 164(14) (2017) A3925–A3932.
7 J. Li, X. Xiao, F. Yang, M.W. Verbrugge, Y.-T. Cheng, The Journal of Physical Chemistry C 116(1) (20 11) 1472–1478.
8 A.C. Lazanas, M.I. Prodromidis, ACS Measurement Science Au 3(3) (2023) 162–193.
9 J.R. MacCallum, C.A. Vincent, Polymer Electrolyte Reviews, Springer Netherlands, 1989.
10 S. Wang, J. Zhang, O. Gharbi, V. Vivier, M. Gao, M.E. Orazem, Nature Reviews Methods Primers 1(1) (2021) 41.
11 W. Plieth, Electrochemistry for materials science, Elsevier, 2008.
12 J.E.B. Randles, Discussions of the Faraday Society 1 (1947) 11–19.
13 A.J. Bard, L.R. Faulkner, H.S. White, Electrochemical Methods: Fundamentals and Applications, John Wiley & Sons, 2022.
14 C.M. Brett, O. Brett, Electrochemistry 67(2) (1993) 444.
15 C. Tian, K.R. Kristiansen, S. Kjelstrup, V.M. Barragán, International Journal of Thermophysics 43(1) (2021).
16 P.R. Sørensen, T. Jacobsen, Electrochimica Acta 27(12) (1982) 1671–1675.
17 S. Bublil, G. Peta, M. Fayena-Greenstein, H. Alon-Yehezkel, O. Raskin, Y. Elias, D. Aurbach, Journal of the Electrochemical Society 169(11) (2022) 110523.
18 M. Watanabe, M. Rikukawa, K. Sanui, N. Ogata, Journal of Applied Physics 58(2) (1985) 736–740.
19 M. Watanabe, M. Rikukawa, K. Sanui, N. Ogata, Macromolecules 19(1) (1986) 188–192.
20 P.G. Bruce, C.A. Vincent, Journal of electroanalytical chemistry and interfacial electrochemistry 225(1–2) (1987) 1–17.
21 J. Evans, C.A. Vincent, P.G. Bruce, Polymer 28(13) (1987) 2324–2328.
22 M. Hiller, M. Joost, H. Gores, S. Passerini, H.-D. Wiemhöfer, Electrochimica Acta 114 (2013) 21–29.

12

Applications of Ion Transport Membranes

Polymer electrolytes and ion-exchange membranes (IEMs) are essential components in electrochemical energy conversion and storage devices, such as batteries, solar cells, fuel cells, and hydrogen production by water splitting. Polymer electrolytes, which conduct ions while insulating electronically, enable efficient ion transport between electrodes. They are flexible, lightweight, and safe, making them ideal for both portable and large-scale energy applications. IEMs, on the other hand, selectively allow the passage of specific ions while blocking others, essential for maintaining charge balance in devices such as fuel cells and water splitting. This selectivity is vital for preventing reactant crossover, which enhances device performance and longevity. Together, these materials drive advancements in high-efficiency, next-generation energy technologies.

12.1 Polymer Electrolyte for Secondary Batteries

Polymer electrolytes are a versatile class of materials known for their ion-transporting ability by dissolving relatively high concentrations of salts in a polymeric solvent. These materials have been of interest for a long time because of their potential in various current and emerging technologies, such as electrochemical energy storage (e.g. secondary and flow batteries) and energy conversion (e.g. solar cells and fuel cells) technologies, electrochromic devices, sensors, actuators, and even separation membranes (e.g. for gas separation). In this context, we will introduce a representative application of polymer electrolytes in secondary batteries.

12.1.1 Basic Chemistry and Configuration of Secondary Batteries

In secondary batteries, electricity is saved by two reversible electrochemical reactions during the charging and discharging processes. Charging (forward) and discharging (backward) reactions at the cathode and anode for lithium-ion batteries are:

$$\mathrm{LiCo^{3+}O_2 = Li_{1-n}Co^{3+}_{1-n}Co^{4+}_{n}O_2} + n\mathrm{Li^+} + ne^- \tag{12.1}$$

$$n\mathrm{Li^+} + n\mathrm{e^-} = n\mathrm{Li} \tag{12.2}$$

Molecule and Ion Transport through Polymer Membranes, First Edition. Yong Soo Kang.

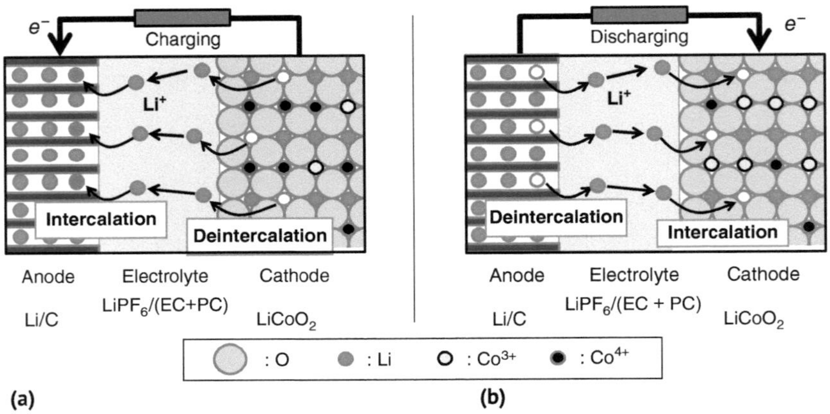

Figure 12.1 Schematic structure of Li^+ secondary batteries with $LiPF_6$ electrolyte positioned between anode of Li/carbon and cathode of $LiCoO_2$ for (a) charging process; and (b) discharging process. *Source*: Y.S. Kang.

During charging process, lithium ions are deintercalated from cathodic materials like $LiCoO_2$ by Equation (12.1) and diffuse across the cathode layer, through the electrolyte layer, and finally reach the anode layer (Figure 12.1a), where they are intercalated into a graphite layer by Equation (12.2). Meanwhile, electrons generated by Equation (12.1) transport through an external circuit as partial Co^{3+} is oxidized to Co^{4+}. During discharging, these reactions reverse, as illustrated in Figure 12.1b.

Two mobile ions mostly studied for secondary batteries are Li^+ and Na^+. Currently, lithium-ion batteries are much more widely used than sodium-ion batteries due to several advantages: higher energy density, operating voltage, efficiency, and longer cycle life. However, lithium resources are geographically limited, which poses challenges for large-scale, sustainable use. Sodium-ion batteries have slightly lower ion conductivity and charge density, but sodium resources are more abundant, making them attractive for sustainable energy applications. Developing cathode materials for sodium-ion batteries remains a significant technical challenge.

A typical Li^+ battery structure is shown schematically in Figure 12.1, with an electrolyte layer sandwiched between the anode and cathode. The anode usually consists of lithium-intercalated graphite or bulk lithium metal to enhance energy density, while the cathode is based on $LiCoO_2$, often combined with elements like nickel, manganese, or iron to partially replace cobalt. The electrolyte contains lithium salts such as $LiPF_6$, $LiBF_4$, or $LiClO_4$ dissolved in a polar solvent mixture, including cyclic carbonates (e.g. ethylene carbonate, propylene carbonate) and linear carbonates (e.g. dimethyl carbonate, diethyl carbonate). The electrolyte layer commonly includes a microporous polymer separator to prevent contact between the electrodes for safety, though this separator is omitted in Figure 12.1 for clarity.

When an electrolyte is incorporated between two electrodes for electrochemical devices, its interfacial layers with both anode and cathode significantly influence the charge transfer and, consequently, the device performance. Solid polymer electrolytes (SPEs) are an alternative to liquid electrolytes and offer advantages such as improved safety, stability, and potential for high energy density. Recent reviews on the interfacial phenome are written for anode [1], cathode [2], and interfaces [3, 4] for lithium-ion batteries. However, different from the liquid electrolyte, SPE cannot infiltrate into the pores of cathode, and could not make sufficient interfacial contact and offer a continuous ions transport pathway. Then the available capacity and energy density of solid-state lithium-ion batteries are not as high as that of the liquid electrolyte-based batteries. In this regard, an overview of the interfacial structure and its impact on charge transfer in electrode-SPE systems is addressed here briefly.

12.1.2 Issues of Liquid Electrolytes

Most current batteries rely on organic liquid electrolytes contained in a microporous polymer separator positioned between the anode and cathode. The organic liquid electrolytes containing, for instance, $LiPF_6$ dissolved in a mixture of cyclic and linear carbonates, generally provide the capacity and power needed for devices like mobile phones. The bulk lithium metal anode delivers higher capacity with less occupying volume than the Li/C anode. Therefore, it becomes very attractive for practical applications. However, they have significant limitations: the flammability and toxicity of the materials raise safety concerns, and they lack both a broad electrochemical stability window and resistance to dendrite growth [5, 6].

As a result, challenges remain in further improving battery performance, especially in terms of capacity and power as well as chemical and electrochemical stability and safety. Solid electrolytes offer promising solutions to many of these issues and open doors to new high-energy batteries, such as lithium–sulfur systems, which are not feasible with traditional liquid electrolytes [7].

12.1.3 Advantages and Issues of Polymer Electrolytes

Replacing the liquid electrolyte in Figure 12.1 with a SPE could resolve many issues, especially those related to safety, such as flammability and toxicity of organic solvents, as well as dendrite formation. However, SPEs introduce new challenges: they typically have low ionic conductivity and a low lithium-ion transport number, T_{Li^+}, which can reduce battery capacity. Additionally, poor contact between the electrode and SPE, due to intrinsic incompatibility in forming a solid–solid interface, can also be problematic. To enhance the performance and safety of secondary batteries using polymer electrolytes, two main issues must be addressed: (1) improving the inherent properties of SPEs, including ionic conductivity and lithium-ion transport number; and (2) resolving interfacial issues between the SPE and solid electrodes.

12.1.4 Challenges for Interfaces with Polymer Electrolytes

Two interfacial layers are formed where the SPE meets each of the two electrodes. These layers may undergo slight chemical modification – either intentionally or unintentionally – due to chemical or electrochemical reactions, interdiffusion,

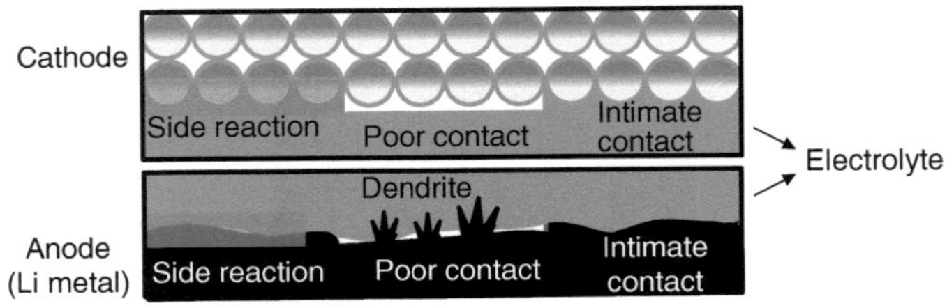

Figure 12.2 The schematic illustration of the interfaces in polymer electrolyte-based solid-state lithium-ion batteries. *Source*: Adapted from [9].

surface coatings, or additives. Ion transport across these interfacial layers affects the device's performance, which is often characterized by the charge transfer resistance, $R_{CT.}$ Reducing R_{CT} is therefore essential to enhance device performance [8], which was also described in Section 9.7 of Chapter 9.

In addition, electrochemical stability at the interface is also essential for stable operation. Potential issues include side reactions such as oxidation and decomposition of the polymer electrolyte, dissolution of transition metal ions, and interfacial aging during long-term cycling [9–11].

Issues associated with improvements in charge transfer resistance R_{CT} and electrochemical stability arise mostly from side reactions, poor contact, and dendrite formation and growth, as shown in Figure 12.2.

12.1.4.1 Interfacial Charge Transfer Resistance

Interfacial charge transfer resistance, R_{CT}, can be reduced from the intimate contact between SPE and the solid electrode. Achieving this contact is challenging due to the intrinsic incompatibility at the solid–solid interface, exacerbated by volume changes during charging and discharging cycles. Potential side reactions and poor interfacial contact lead to high R_{CT}, reducing capacity and energy density compared to liquid electrolytes. Strategies to mitigate these issues include forming a stable solid electrolyte interphase (SEI) or cathode electrolyte interphase (CEI) layer, applying surface coatings, and optimizing polymer formulations [12–14].

12.1.4.2 Side Reactions

Interfacial side reactions at cathode/SPE interfaces in lithium secondary batteries primarily involve the decomposition of the electrolyte and degradation of the cathode surface. These reactions are driven by factors like high voltage, temperature, and poor interfacial contact. SPE can decompose, forming by-products such as lithium carbonate (Li_2CO_3), lithium fluoride (LiF), or other inorganic species. Additionally, cathode dissolution and transition metal ion migration can occur, adversely affecting performance. Efforts to mitigate these issues include using interfacial coatings, enhancing electrolyte stability, and optimizing the electrode–electrolyte compatibility [14].

At the anode/SPE interface, several issues need resolution. First, maintaining chemical stability of the lithium anode/SPE interface is essential for stable operation of solid-state lithium metal batteries. When the polymer electrolyte contacts lithium metal, chemical reaction can occur, causing reduction and structural changes in SPE. Enhancing the anti-reduction ability of SPEs through adjustments of its functional units is necessary [15].

12.1.4.3 Poor Contact

Poor contact and compatibility between the lithium anode and SPE can increase R_{CT} especially due to mechanical and structural instability from thickness changes in the lithium anode during prolonged cycling. Minimizing R_{CT} through improved compatibility is necessary to maintain efficient charge transfer.

The compatibility can be improved by constructing a thin but stable SEIs by surface modification technologies. For instance, (1) adding lithium salts [16, 17]; (2) applying inorganic material [18, 19]; and (3) using polymer coatings, such as multi-layered coatings, optimized polymer electrolytes, and *in situ* polymerization techniques [20–23] have been frequently used.

12.1.4.4 Dendrite Formation and Growth

Finally, lithium dendrite growth is unavoidable with metallic anodes during charge/discharge cycles as they increase capacity. Lithium dendrite can penetrate the thin SPE layer, potentially causing internal short circuits and serious safety issues. Achieving a high shear modulus, higher than 10^6 Pa or self-healing properties of the polymer electrolyte could suppress dendrite formation and growth [9, 24]. Additionally, designing polymers that interact favorably with lithium metal surfaces to inhibit dendrite growth is a key research area [24, 25].

12.1.4.5 Applications and Opportunities

SPEs are well-suited for flexible and wearable electronics due to their inherent flexibility and ability to form thin films [26]. They are also critical for advancing solid-state batteries, offering higher energy density and improved safety. The integration of polymer electrolytes into next-generation batteries for electric vehicles, grid storage, and medical devices highlights their vast potential. While polymer electrolytes face significant challenges, such as low ionic conductivity, low cation transport number, dendrite growth, and interfacial resistance, ongoing research is unlocking their potential. Innovations in material design, interface engineering, and processing techniques will lead to safer, more efficient, and more sustainable secondary batteries, paving the way for their widespread adoption in diverse applications.

12.2 Polymer Electrolytes for Sensitized Solar Cells

Reusable energy resources, like solar energy, are essential for reducing our dependence on fossil fuels, mitigating climate change, and ensuring long-term energy sustainability. Solar cells, which directly convert sunlight into electricity, play a vital role in capturing this clean energy. They offer a scalable, low-maintenance solution that can be implemented globally, contributing to energy security and reducing carbon emissions. Expanding the use of solar cells also promotes technological innovation and creates economic growth in the green energy sector.

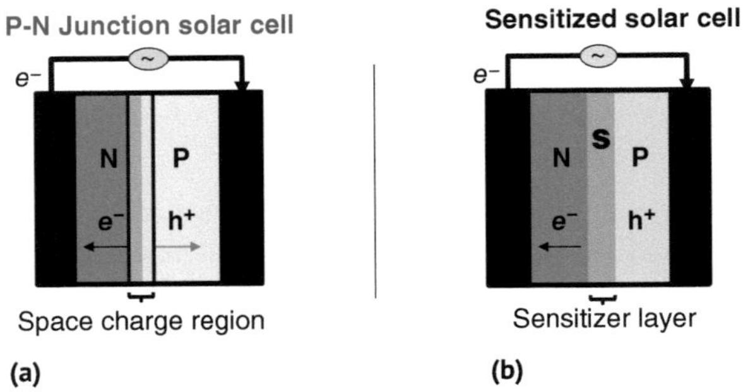

Figure 12.3 Schematic configuration of (a) *p-n* junction and (b) sensitized solar cells, where *p*-type semiconductor is frequently replaced by electrolyte containing redox couples. *Source*: Y.S. Kang.

Solar cells can be broadly categorized into *p-n* junction solar cells and sensitized solar cells, depending on materials used for light absorption. The former, the most commonly used, includes silicon semiconductor-based cells (monocrystalline, polycrystalline) and thin-film solar cells primarily with compound semiconductors or amorphous silicon. These cells operate based on a semiconductor *p-n* junction by forming a space charge region (SCR) or a depletion layer (Figure 12.3a). Here, excitons constituting electron-hole pairs generated upon sunlight absorption are then separated into free charge carriers (electrons and holes) due to the electric potential difference generated across the SCR that subsequently move opposite directions to each other to produce electricity.

The second type, sensitized solar cells, includes dye-sensitized solar cells (DSSCs), perovskite solar cells, and quantum dot-sensitized solar cells. In these cells, photosensitive materials, called sensitizers, such as dyes, perovskites, or quantum dots, are placed between *n*- and *p*-type semiconductors, as shown in Figure 12.3b. The sensitizer absorbs solar light and generates excited electrons, which are then transferred to a *n*-type semiconductor, typically titanium dioxide TiO_2, to produce electricity. Here, the *p*-type semiconductor can be frequently replaced by an electrolyte containing redox couples dissolved in a solvent.

12.2.1 Basic Chemistry and Configuration of Sensitized Solar Cells

Sensitizers are attached to the surface of *n*-type semiconductors for efficient injection of electrons generated by light absorption to the *n*-type semiconductors. Since only sensitizers generate electrons, the cell performance is directly related to the concentration of sensitizers. Therefore, a large amount of sensitizers should be attached onto the surface to improve the cell performance, requiring a large surface area. One simple way to do it is to make an interconnected nanoparticle layer with *n*-type semiconductor like TiO_2.

Figure 12.4a shows the typical interconnected structure of TiO_2 nanoparticles (diameter 20–50 nm) for the *n*-type semiconductor layer with enlarged surface area

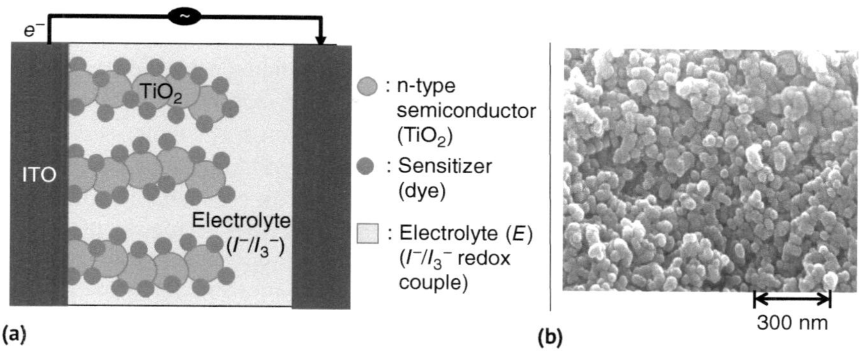

Figure 12.4 (a) Simplified dye-sensitized solar cell with an electrolyte solution containing the redox couple of $I^-$and I_3^- instead of *p*-type semiconductor; and (b) SEM image of the TiO_2 interconnected nanoporous layer. *Source*: Y.S. Kang.

at a given thickness of 10–20 mm. The average pore diameter of the TiO_2 layer is in the range of 10–20 nm, as shown in Figure 12.4b.

When sensitizers, Ss, absorb solar light $h\nu$, they will be excited to generate free electrons e^- (Equation (12.3)), which will be injected into the n-type TiO_2 semiconductor layer to generate electricity, and concomitantly oxidized to form S^+ as:

$$S = S^+ + e^- \tag{12.3}$$

Subsequently, S^+ should react with electrons to be regenerated to convert to S by the backward reaction of Equation (12.3), where the electrons are supplied by the oxidation (forward) reaction of I^- by Equation (12.4) at the anode surface as:

$$3I^- = I_3^- + 2e^-(Red = Ox + e^-) \tag{12.4}$$

where I^- and I_3^- are a redox couple most commonly used. Thus, the overall regeneration reaction of the oxidized sensitizer becomes as:

$$2S^+ + 3I^- = 2S + I_3^- \tag{12.5}$$

Equation (12.5) emphasizes the critical role of the contact between S^+ and I^- in maintaining solar cell performance. The concentration of I^- is regulated by three key processes: (1) the reduction reaction at the cathode ($I_3^- + 2e^- \rightarrow 3I^-$, the backward reaction of Equation (12.4)); (2) the transport of I^- through the bulk electrolyte; and (3) its diffusion through the interconnected nanopores (10–20 nm in diameter) of the TiO_2 layer to reach the oxidized sensitizer (S^+) for dye regeneration (Equation (12.5)).

These processes highlight two critical factors for efficient dye regeneration and enhanced cell performance: (1) the conductivity of I^- (and also I_3^-) in the bulk electrolyte; and (2) the diffusion of I^- (and I_3^-) through the interconnected nanopores to meet oxidized sensitizers for their regenerations. For performance improvement, the nanopores must be optimized to be large enough and well-connected to ensure effective transport of redox couples.

12.2.2 Issues of Liquid Electrolytes

An electrolyte can be used instead of *p*-type semiconductor, which contains the redox couples such as I^-/I_3^- dissolved in a liquid solvent including acetonitrile. In liquid electrolytes, redox couples of I^- and I_3^- transport across the nanopores in the TiO_2 layer rapidly without serious transport resistance, resulting in high power conversion efficiency, up to 13% efficiency [27]. In other words, cells with a liquid electrolyte exhibit very high performance and become attractive. However, they face serious problems associated with the liquid state electrolyte, as expected, such as safety, solvent-diffuse out, and long-term operational stability. Therefore, solid-state electrolyte has been attempted to solve those problems.

12.2.3 Advantages and Issues of Polymer Electrolytes

The discovery of high ionic conductivity (5×10^{-5} S/cm) at room temperature in the polymer electrolyte based on polyethylene oxide (PEO) with NaSCN salt by Fenton et al. (1973) [28] marked a breakthrough in developing solid-state electrochemical devices for energy storage and conversion such as rechargeable batteries, sensitized solar cells, supercapacitors, and electrochromic windows [29].

Since then, efforts have been focused on developing polymer electrolytes with even higher ionic conductivity (higher than 1×10^{-4} S/cm) to enhance energy conversion efficiency. For instance, a pioneering work, done by Nogueira et al. [30], demonstrated the successful use of polymer electrolyte based on an elastomeric copolymer of ethylene glycol and epichlorohydrin, showing 1.6–2.6% energy conversion efficiency at early 2000s. The key feature of the copolymer is its ability to penetrate the pores of the TiO_2 film in spite of its high molecular weight [31].

Another important challenge to be solved is the problem associated with the shallow penetration of redox couples into interconnected nanopores, leading to poor contact with sensitizers. For instance, PEO with a molecular weight of 1 million is used as a polymer matrix to dissolve redox couples. The flexible PEO chains form random coils with an approximate diameter of 80–120 nm. These large polymer chains struggle to penetrate into nanopores of the TiO_2 layer with the average diameter of 10–20 nm, as shown in Figure 12.5. This limits contact and impedes the regeneration of oxidized sensitizers within the nanopores, making it challenging for them to repeatedly absorb solar light to continuously produce electricity.

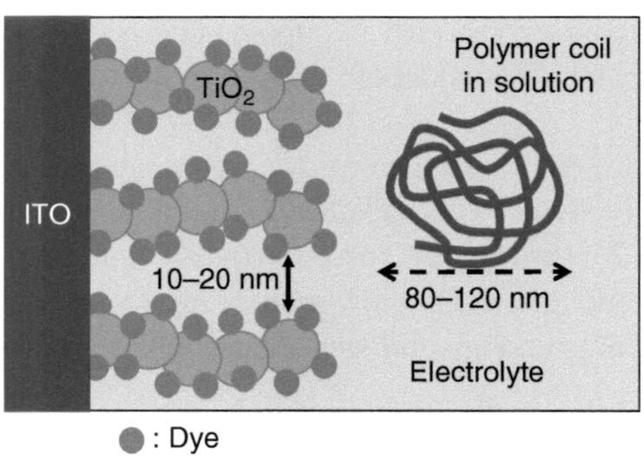

Figure 12.5 Schematic image of the anode side for comparison between pore (10–20 nm) and polymer coil sizes (80–120 nm) dissolved in a solvent. *Source*: Y.S. Kang.

Therefore, challenges for utilizing solid electrolytes are to solve both the low ionic conductivity and the shallow penetration of solid electrolyte into the interconnected nanopores in *n*-type semiconductor layer.

12.2.4 Oligomer Approach

To overcome challenges with SPEs, liquid oligomers with molecular weights around 1,000 and coil sizes of 2–4 nm in diameter have been effectively used to create solid-state polymer electrolytes. This approach, called as the "oligomer approach," takes advantage of liquid oligomers' ability to deeply penetrate into nanopores (10–20 nm in diameter) during cell fabrication. After penetration, the solvent is evaporated to transit the liquid electrolyte to a solid state using one of three methods.

The liquid oligomer electrolytes are solidified into SPEs by (1) incorporating polymers [32], (2) adding nanoparticles [33], or (3) inducing H-bondings [34], as illustrated in Figure 12.6. As a result, both the contact and the ionic conductivity have been significantly improved, yielding the increase in power conversion efficiency approximately from around 1.9% to nearly 9% [35, 36]. Additionally, PEO-based electrolytes are crosslinked to enhance mechanical strength, making them suitable for practical applications [37].

In addition, in a log–log plot of photocurrent density versus ionic conductivity, there is a nearly linear relationship at low ionic conductivities, which levels off to a saturated photocurrent density at higher ionic conductivities (Figure 12.7). This indicates a threshold ionic conductivity of 1×10^{-4} S/cm. Below this threshold, ionic conductivity significantly influences photocurrent density and, therefore, cell performance. To enhance cell efficiency, the ionic conductivity should exceed this threshold value of 1×10^{-4} S/cm.

This approach leverages the liquid properties of oligomers during cell fabrication, which then transition to a solid state upon solvent evaporation, allowing for good ionic conductivity, deep penetration into nanopores, and efficient contact with sensitizing dyes. This method has achieved power conversion efficiencies nearing 8% [39, 40].

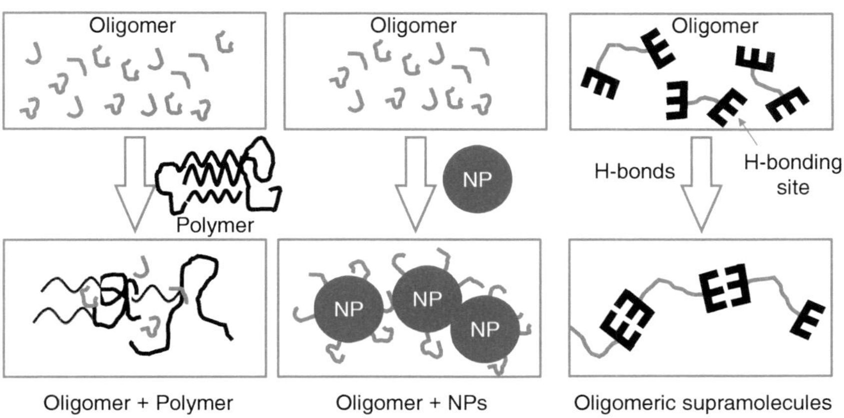

Figure 12.6 Three different concepts of utilizing liquid oligomers to transform into solid polymer electrolytes. *Source*: Adapted from [38].

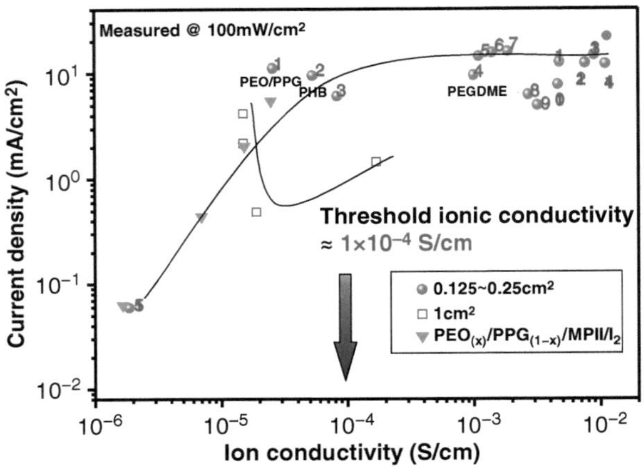

Figure 12.7 Current densities of dye-sensitized solar cells as a function of ion conductivity, proposing the minimum ion conductivity to improve the cell efficiency. *Source*: Adapted from [38].

12.2.5 Future and Challenges of Polymer Electrolytes for Sensitized Solar Cells

To further enhance ionic conductivity in polymer electrolytes, researchers have explored new polymers and innovative concepts, including poly(ionic liquid)s like poly(imidazolium ion), polymeric solvents such as polyurethanes (PUs), poly(vinylidene fluoride-co-hexafluoropropylene (PVdF-HFP), bio-based polymers, and composites.

Poly(ionic liquid)-based gel electrolytes, particularly those with imidazolium and pyridinium cations, have shown promising results. For example, poly((1-(4-ethenylphenyl)methyl)-3-butyl-imidazolium iodide) (PEBII) has achieved power conversion efficiencies up to 8.2% [41–43]. Pyridinium-based poly(ionic liquids) have also yielded significant results, with efficiencies reaching 5.64% in all-solid-state DSSCs [44].

PU is another promising material, formed by polymerization of liquid-state monomers or oligomers at room temperature without generating any side products. Therefore, solid-state polymer electrolytes are readily formed even at ambient temperatures, taking advantage of liquid-state fabrication to convert into solid-state in the final device. PU-based polymer electrolytes also offer high ion conductivity but yet to achieve high power conversion efficiency [45].

In addition, bio-based polymer solvents are gaining interest due to environmentally benign properties. Examples include chitosan and modified polysaccharides like carboxymethyl (CM)-κ-carrageenan and CM-cellulose. CM-κ-carrageenan/NaI/I_2/EC, for instance, has demonstrated an ionic conductivity of 5.53×10^{-2} S/cm at room temperature and power conversion efficiency of 2.06%. Notably, after 250 hours of thermal stress at 60°C for stability test, the cell retained nearly 94% of its initial efficiency [46]. Blends of CM-κ-carrageenan and CM cellulose-NH_4I/I_2

biopolymer electrolytes have achieved an ionic conductivity of 2.32×10^{-3} S/cm at room temperature [47].

Finally, gel or quasi-solid polymers are receiving attention for their high ion conductivity and photoconversion efficiency, comparable to liquid electrolytes, along with enhanced mechanical strength [31, 48]. PVdF-HFP, in particular, has shown potential as a gel medium with relatively high ionic conductivity [49, 50].

12.3 Ion Exchange Membranes for Fuel Cells

A fuel cell is an electrochemical energy conversion device from chemical energy of fuel to electric energy, which is highly energy-efficient and a clean alternative to traditional combustion-based power sources. Therefore, fuel cells are increasingly used in transportation, stationary power generation, and portable power systems. Their benefits include low emissions, quiet operation, and the potential for high energy density, making them promise for reducing carbon footprints in diverse sectors.

There have been various kinds of fuel cells, utilizing diverse fuels like hydrogen, methanol, and hydrocarbons, as summarized in Table 12.1. Representative reactions for H_2 and CH_3OH fuels are presented in Equations (12.6) and (12.7).

$$H_2 + 1/2O_2 = H_2O \qquad \Delta G^0 = -237\,\text{kJ/mol}, E^0 = 1.23\,\text{V} \tag{12.6}$$

$$CH_3OH + 3/2O_2 = 2H_2O + CO_2 \qquad \Delta G^0 = -702\,\text{kJ/mol}, E^0 = 1.25\,\text{V} \tag{12.7}$$

Depending on the charge carrier used, fuel cells are divided into phosphoric acid fuel cells (PAFCs), alkaline fuel cells (AFCs), proton exchange membrane fuel cells (PEMFCs), anion-exchange membrane fuel cells (AEMFCs), direct methanol fuel cells (DMFCs), molten carbonate fuel cells (MCFCs), and solid oxide fuel cells (SOFCs), each optimized for specific applications [51].

Table 12.1 Fuel cells and their representative characteristics.

	PAFC	PEMFC	AEMFC	DMFC	MCFC	SOFC
Fuel	H_2	H_2	H_2	CH_3OH	H_2, CH_4, CH_3OH	H_2, CH_4, CH_3OH
Electrolyte	Phosphoric acid	Proton exchange membrane	Anion Exchange Membrane	Proton exchange membrane	Alkali carbonate mixture	Yttria stabilized zirconia
Temperature(°C)	160–200	80–120	6–250	80–110	600–800	800–1,000
Charge Carrier	H^+	H^+	OH−	H^+	CO_3^{2-}	O^{2-}
Catalyst	Pt	Pt	Pt or Ni	Pt	Ni	Perovskite
Efficiency	40–45	40–60	35–55	~40	50–60	50–60

Source: Adapted from [51].

IEMs are used in PEMFCs, AEMFCs and DMFCs, where IEMs are ionically and selectively conductive and mechanically strong enough to separate two electrodes. IEMs include cation-exchange membranes, CEMs, anion-exchange membranes, AEMs, and bipolar membranes, BPMs, consisting of cation and anion exchange layers in series, as illustrated in Figure 10.6.

In this section, applications of CEMs and AEMs for fuel cells will be briefly introduced, whereas those of BPMs for electrochemical water splitting will be addressed in Section 12.4.

12.3.1 Basic Chemistry and Configuration of Fuel Cells

The basic electrochemical reactions in a hydrogen fuel cell involve the oxidation of hydrogen at the anode and the reduction of oxygen at the cathode, which performs by either acidic electrolyte with PEMFCs or alkaline electrolyte for AEMFCs. These electrochemical reactions for H_2 fuel occur as follows:

For hydrogen in acidic electrolyte (pH~0), electrochemical reactions occur:

$$\text{Anode: } H_2 = 2H^+ + 2e^- \qquad E^0 = 0\ \text{V} \tag{12.8}$$

$$\text{Cathode: } 1/2O_2 + 2H^+ + 2e^- = H_2O \qquad E^0 = 1.23\ \text{V} \tag{12.9}$$

$$\text{Overall: } H_2 + 1/2O_2 = H_2O \qquad E^0\ \text{cell} = 1.23\ \text{V} \tag{12.10}$$

For hydrogen in alkaline electrolyte (pH~14), electrochemical reactions occur:

$$\text{Anode: } H_2 + 2OH^- = 2H_2O + 2e^- \qquad E^0 = -0.83\ \text{V} \tag{12.11}$$

$$\text{Cathode: } 1/2O_2 + H_2O + 2e^- = 2OH^- \qquad E^0 = 0.40\ \text{V} \tag{12.12}$$

$$\text{Overall: } H_2 + 1/2O_2 = H_2O \qquad E^0_{\text{cell}} = 1.23\ \text{V} \tag{12.13}$$

The standard potential E^0_{cell} of the fuel cell ($E^0_{\text{cell}} = E^0_{\text{cathode}} - E_{\text{anode}}$) is $E^0_{\text{cell}} = 1.23$ V at $T = 25°\text{C}$ at both pH = 0 and pH = 14.

For methanol fuel,

$$\text{Anode: } CH_3OH + H_2O = CO_2 + 6H^+ + 6e^- \qquad E^0 = -0.02\ \text{V} \tag{12.14}$$

$$\text{Cathode: } 3/2O_2 + 6H^+ + 6e^- = 3H_2O \qquad E^0 = 1.23\ \text{V} \tag{12.15}$$

$$\text{Overall: } CH_3OH + 3/2O_2 = CO_2 + 2H_2O \qquad E^0_{\text{cell}} = 1.25\ \text{V} \tag{12.16}$$

In order to utilize the electrochemical reactions mentioned above, fuel cells are made of anode and cathode with optional catalysts, and IEM with electrolyte, as illustrated in Figure 12.8a and b.

The anode in PEMFCs is where the fuel, such as hydrogen, is fed, and is made of a porous conducting material like carbons and catalysts such as platinum or nickel that facilitates the oxidation reaction of hydrogen (Equation (12.8)) into protons and electrons. Oxygen is delivered to the cathode, which is also made from porous, conductive carbons with catalysts. The oxygen reacts with protons and electrons to produce water (Equation (12.9)).

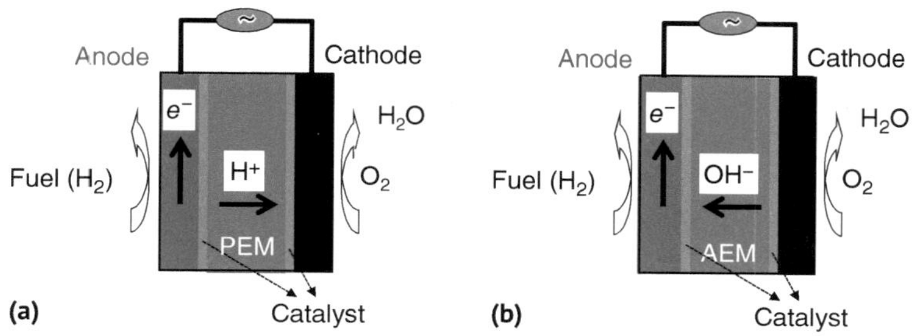

Figure 12.8 Configurations of (a) proton-exchange membrane fuel cell (PEMFC) and (b) alkaline fuel cell (AEMFC). *Source*: Y.S. Kang.

The electrolyte is ionically conductive with either acidic (most commonly H_2SO_4 solution) or alkaline (KOH solution) solution depending on PEMFC or AEMFC, respectively. CEM allows only H^+ to pass through from the anode to the cathode while blocking the electrons and other ions. On the other hand, AEM must be conductive for OH^-, but insulating to electrons and other ions.

12.3.2 Roles and Requirements of Ion Exchange Membranes

IEMs must exhibit high proton or hydroxide ion conductivity in PEMFCs or AEMFCs, respectively, across the membrane with minimal resistance, critical for high-efficiency energy conversion. Ions and neutral molecules such as fuels and oxidants transport through ion channels within a microphase-separated structure, as shown previously in Figure 10.15. To improve performance, these channels should be designed to promote ion transport while limiting the passage of fuels and oxidants. Because ion channels are naturally tortuous, reducing their tortuosity is important for enhancing ion conductivity. Rather than using large channels, it is more effective to have many small, straight channels with a given surface fraction. Ion exchange membranes (IEMs) with such structures can achieve high ion transport while minimizing fuel crossover, thereby improving both cell performance and safety.

Membranes must resist degradation under acidic or alkaline environments for PEMFCs and AEMFCs, respectively. It has been quite well accepted that CEMs based on perfluorosulfonic acid are quite stable even in strongly acidic electrolytes. In the case of AEMs, they are generally prone to chemical reactions to degrade polymeric chains and fixed ions in strong alkaline solutions (pH~14). Therefore, AEMs should be chemically stable against the very high alkaline electrolytes and withstand high operating temperatures without significant loss of functionality.

In PEMFCs, effective water retention at anode is critical for maintaining membrane hydration and sustainable operation. For this purpose, PEMs play a key role in managing water transport, essential for maintaining high ionic conductivity while preventing excessive flooding at cathode or drying at anode, whereas, for AEMFCs, controlled water transport is required to avoid ion dilution while supporting ionic movement.

Fuel cells undergo cyclic stresses, and IEMs must withstand mechanical stresses, swelling, and shrinkage over time to maintain structural integrity. Finally, IEMs should be affordable and scalable for large-scale production to enable commercial feasibility of fuel cell technologies.

A well-designed IEM ensures enhanced efficiency, durability, and reliability of fuel cells, paving the way for sustainable and high-performance energy solutions.

12.3.3 Performance of PEMFCs and AEMFCs

The performance of both PEMFCs and AEMFCs is primarily influenced by factors such as ion-exchange capacity (IEC), membrane conductivity, catalyst efficiency, temperature, pressure, and humidity control. The performance comparison between PEMFCs and AEMFCs highlights their distinct operating characteristics, efficiency, durability, and cost-effectiveness. AEMFCs have long been valued for their high efficiency and ability to operate with non-platinum catalysts, making them a potentially lower-cost alternative to other fuel cell types. Below is a performance comparison between PEMFCs and AEMFCs depending on the factors described above.

Figure 12.9 shows the *J–V* and power density curves of (a) representative PEMFC; and (b) AEMFC, where the amount of Pt catalyst and IEC of AEM vary, respectively. For PEMFC, as can be seen in Figure 12.9a, the performance sharply increases, as expected, with increasing the amount of Pt catalyst used from 50 to 125 mg/cm^2. Notably, the power density achieved with a catalyst loading of 125 mg/cm^2 is nearly equivalent to that obtained with a much higher loading of 400 mg/cm^2, simply by switching the Pt catalyst preparation method to pulsed laser deposition [52]. In the case of AEMFCs, by varying IEC of s-PDTP-*x* copolymers (Figure 12.9b), the maximum power density of 1.5 W/cm^2 at 80 °C was obtained with the medium IEC value of ~3.0 mmol/g, even though the OH^- conductivity increases nearly linear with the IEC value, as shown in Figure 10.29b in Section 10.6 [53].

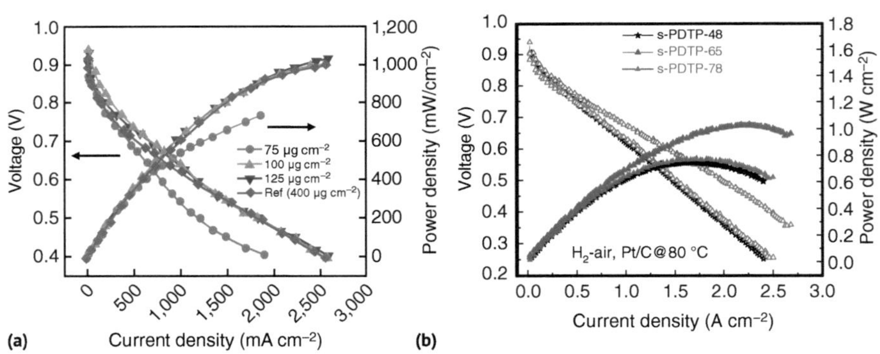

Figure 12.9 *J–V* and power density curves of (a) PEMFCs by varying Pt loadings deposited from pulsed laser deposition (PLD) with Nafion XL membrane. The performance of a PEMFC using E-TEK® Pt/C for the cathode with a Pt loading of 400 (μg/cm^2) is shown for comparison as a reference. All data were measured at an outlet pressure of 100 kPa. *Source*: [52]/MDPI/Public Domain. (b) AEMFC performance with different ion-exchange capacity of side-chain-type poly (dibenzyl-co-terphenyl piperidinium) (s-PDTP)-*x* ionomers (*x* = 48, 65, 78), based on A/C Pt/C catalyst. *Source*: Adapted from [53].

PEMFCs generally achieve higher power densities than AEMFCs, especially at lower operating temperatures (around 60–80 °C). The high proton conductivity of perfluorinated membranes, like Nafion, contributes to superior power output. AEMFCs have lower power density due to the low intrinsic hydroxide ion conductivities (60–130 mS/cm at 60–80 °C), which is even further reduced by decreasing the OH^- concentration through the chemical reaction with CO_2 ($OH^- + CO_2 \rightarrow HCO_3^-$) when exposed to CO_2 (Figure 10.29a in Section 10.6) [53, 54]. However, they can still reach reasonably high efficiencies under controlled conditions.

PEMFCs rely heavily on platinum-based catalysts, which are efficient but expensive. Platinum's scarcity and cost present a significant barrier to large-scale PEMFC deployment. AEMFCs, in contrast, can utilize non-precious metals, such as nickel or silver, which reduces the activation energy for the oxygen evolution reaction (OER). This lowers catalyst costs and makes AEMFCs potentially more economical [55, 56].

PEMFCs are less sensitive to CO_2 contamination, which does not react with the acidic membrane but only slightly affects the cathode over time. AEMFCs are, however, highly susceptible to CO_2, because it forms carbonates by reacting with OH^- in the KOH electrolyte, reducing ionic conductivity as described previously. This may necessitate purified air to improve operational stability [57].

In summary, PEMFCs offer high power density and durability, but their cost remains high due to reliance on platinum catalysts, PFSA membranes, and the need for precise water management. AEMFCs present a cost-effective alternative, with simpler catalyst requirements and easier scalability, though challenges with CO_2 sensitivity and durability remain. This comparison illustrates the potential of both types in different applications, from stationary power generation to portable energy sources (Table 12.2).

Table 12.2 Comparison between cation exchange membrane (CEM) and anion exchange membrane (AEM) fuel cells.

Property	**CEMs**	**AEMs**
Mobile charge carrier	Cation (H^+, Na^+)	Anion (OH^-, Cl^-)
Operating environment	Acidic	Alkaline
Ionic conductivity	Higher proton conductivity	Lower hydroxide conductivity
Catalyst requirements	Noble metals (e.g. platinum)	Non-noble metals (e.g. nickel, cobalt)
Fuel crossover	Higher (e.g. in methanol fuel cells)	Lower
Cost	Higher due to expensive catalysts (Pt) and materials (Nafion)	Lower due to cheaper catalysts and materials
Durability	Highly stable in acidic environments	Lower stability in alkaline environments
Technological maturity	Well established	Emerging

12.3.4 Future and Challenges of IEMs

High-temperature operation of PEMFCs in the range of 100–200 °C offers benefits in electrode kinetics, heat management, and better tolerance to fuel impurities, particularly CO in reformate. These improvements could potentially lead to higher overall system efficiencies. However, these gains have yet to be realized, as the gold-standard perfluorinated sulfonic acid (PFSA) membranes lose function rapidly at 100 °C and above if hydration drops below ~100%, and begin to creep in this temperature range, resulting in localized thinning and overall lower system lifetimes. As a result, new anhydrous proton conductors, such as protic organic ionic plastic crystals (POIPCs) and protic ionic liquids, are actively studied for the development of suitable PEMs [58–60].

Anion-exchange membrane fuel cells (AEMFCs) have attracted great interest as a low-cost fuel cell technology for clean energy conversion and utilization for the future. AEMFCs have been considered the most promising succedaneum to PEMFCs for addressing the cost issues associated with PEMFCs due to utilizing non-platinum group metals as electrocatalysts under alkaline conditions (such as Ag, Ni, and Co). Years of study have allowed AEMFCs to achieve unprecedented progress recently, specifically in terms of power density and durability. Researchers have proposed many strategies to boost the performance of AEMs and AEMFCs, including (1) the development of highly stable cationic groups and aryl ether-free polymer backbones, (2) the optimization of microphase morphologies of AEMs, (3) the introduction of crosslinking strategies and composite methods, and (4) optimization of the water management of AEMFCs [61].

12.4 Applications of Ion Exchange Membranes for Electrochemical Water Splitting

Hydrogen is a promising option for sustainable energy storage due to its high energy density, safety for long-term storage, suitability for long-distance transportation, and compatibility with clean energy cycles. Water splitting, or electrolysis, enables the storage of electricity as hydrogen, while fuel cells convert this hydrogen back to electricity by combining it with oxygen to produce water (see Figure 12.10).

Electrochemical water splitting splits water (H_2O) into hydrogen (H_2) and oxygen (O_2) using electrical energy. This process produces green hydrogen, a clean fuel useful for energy storage, fuel cells, and chemical synthesis. Key components of this technology include IEMs and catalysts, which facilitate ion transport and redox reactions while keeping hydrogen and oxygen separated.

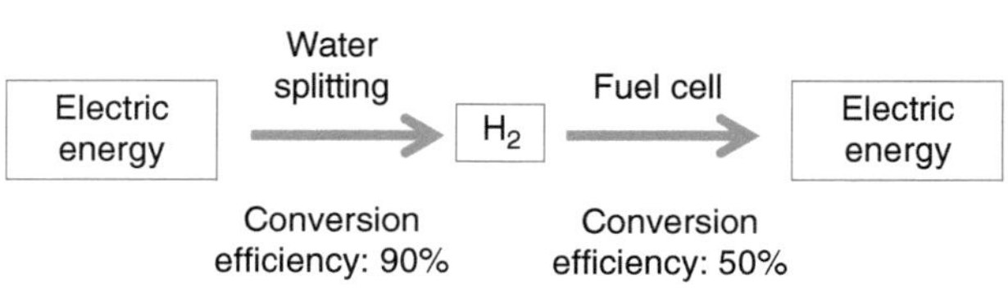

Figure 12.10 Hydrogen as an energy carrier or an energy storage system. *Source:* Y.S. Kang.

IEMs of both PEMs and AEMs are crucial in water electrolyzers, main membrane-based water-splitting technologies. These membranes enable selective ion transport, improve efficiency, and ensure safety by preventing the recombination of hydrogen and oxygen gases. Another significant advancement is the use of BPMs, which consist of cation and anion-exchange layers that promote water dissociation into protons (H^+) and hydroxide ions (OH^-) at their interface of bipolar junction. BPMs enable water splitting at lower voltages, improving ion transport, controlling pH gradients, and reducing overall energy consumption.

This section introduces three types of IEMs (PEMs, AEMs, and BPMs), and highlights their roles in enhancing water-splitting efficiency.

12.4.1 Basic Electrochemistry and Configuration of Water-Splitting Electrolyzers

12.4.1.1 Basic Electrochemistry of Water Splitting with CEMs and AEMs

In a typical water-splitting process, two half-reactions occur. At the anode (oxidation), water is dissociated into oxygen, protons (H^+) (Equation (12.17)), or hydroxide ions (OH^-) is oxidized to generate O_2 and H_2O (Equation (12.18)), depending on pH of the electrolyte.

$$2H_2O \rightarrow O_2 + 4H^+ + 4e^- \quad \text{at pH} = 0 \quad E^0 = 1.23\ \text{V} \tag{12.17}$$

$$4OH^- \rightarrow O_2 + 2H_2O + 4e^- \quad \text{at pH} = 14 \quad E^0 = 0.40\ \text{V} \tag{12.18}$$

At the cathode (reduction), protons (Equation (12.19)) or water (Equation (12.20)) react with the electrons from the external circuit to produce hydrogen gas.

$$4H^+ + 4e^- \rightarrow 2H_2 \quad \text{at pH} = 0 \quad E^0 = 0\ \text{V} \tag{12.19}$$

$$4H_2O + 4e^- \rightarrow 2H_2 + 4OH^- \quad \text{at pH} = 14 \quad E^0 = -0.83\ \text{V} \tag{12.20}$$

$$\text{Overall: } 2H_2O \rightarrow 2H_2 + O_2 \quad E^0 = -1.23\ \text{V} \tag{12.21}$$

Overall, this process requires energy input, usually in the form of electricity by supplying electrons for electrochemical reactions and takes place in an electrolyzer

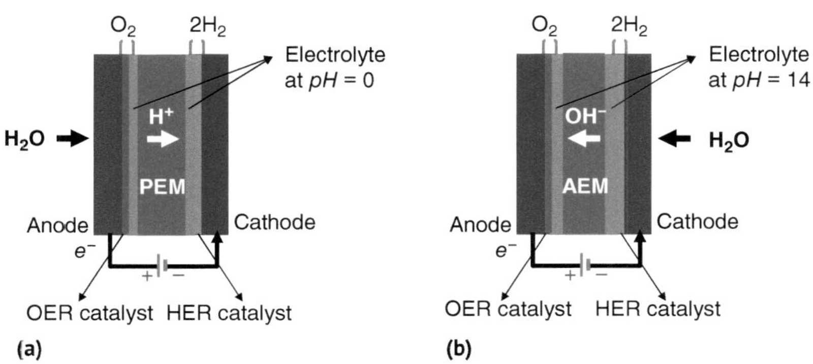

Figure 12.11 Schematic configurations of water-splitting electrolyzers with monopolar (a) proton-exchange membrane (PEM); and (b) anion-exchange membrane (AEM). OER is for the oxygen evolution reaction while HER is for hydrogen evolution reaction. *Source*: Y.S. Kang.

that consists of electrodes and an ion conducting membrane with catalysts, as illustrated in Figure 12.11a and b.

12.4.1.2 Basic Electrochemistry of Water Splitting with BPMs

Recently, BPMs have been paid significant attention to water splitting-applications, mostly because of their significantly improved water dissociation capability by the high electric field generated at the SCR, as described in Section 10.4 of Chapter 10. A BPM is composed of two distinct layers of cation-exchange layer (CEL) and anion-exchange layer (AEL). CEL allows for the selective transport of cations, typically protons (H^+), while blocking anions as shown in Figure 12.12a and b. Conversely, AEL allows for the selective transport of anions, such as hydroxide ions (OH^-), while blocking cations.

In BPM electrolyzers, the electrochemical reactions involve three steps as follows. The first step is a water dissociation reaction into H^+ and OH^- at the bipolar junction of a SCR:

$$4H_2O = 4H^+ + 4OH^- \qquad E^0 = -0.83\ V \tag{12.22}$$

Then, as a second step, H^+ and OH^- pass through CEL and AEL, respectively, to reach cathode and anode. Finally, as a third step, OER at anode and hydrogen evolution reaction (HER) at cathode take place:

$$4OH^- = O_2 + 2H_2O + 4e^- \quad \text{at pH} = 14 \quad E^0 = 0.40\ V \tag{12.23}$$

$$4H^+ + 4\,e^- = 2H_2 \quad \text{at pH} = 0 \quad E^0 = 0\ V \tag{12.24}$$

$$\text{Overall water splitting: } 2H_2O = 2H_2 + O_2 \qquad E^0 = -1.23\ V \tag{12.25}$$

In this case, an additional water dissociation step is required, compared to the two-step processes with monopolar membranes, which consumes a dominant amount of electricity $\left(E^0 = -0.83\ V\right)$. Therefore, a key factor is how to dissociate

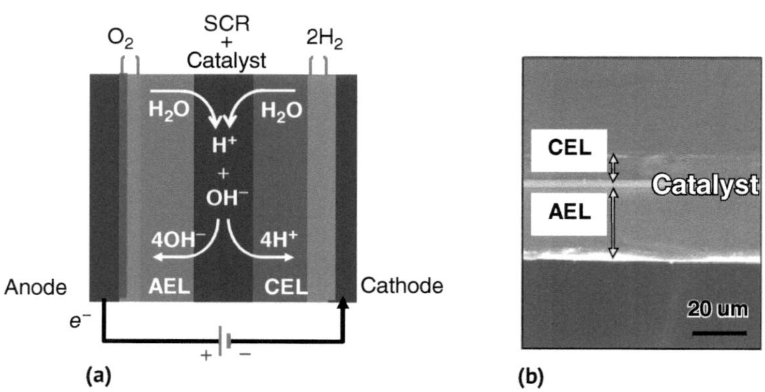

Figure 12.12 (a) Schematic configuration of water-splitting device with bipolar membrane (BPM). The space charge region (SCR, black colored) at the bipolar junction between anion-exchange layer (AEL) and cation-exchange layer (CEL) commonly contains catalysts for enhancing water dissociation reaction. *Source:* Y.S. Kang. (b) Cross-sectional image of a BPM. Adapted from [83].

water into H^+ and OH^- efficiently by BPMs. In addition, OER and HER take place in completely different pH conditions.

Importantly, the two electrode reactions occur under distinctly different pH conditions: OER at pH = 14 and HER at pH = 0. A BPM maintains this pH difference across its junction. When such electrolytes are applied, ΔpH (=14) across the space charge region of the BPM generates an electric field as high as 10^8 V/m as described in Section 10.4 of Chapter 10. This strong electric field effectively lowers the actual energy needed for water dissociation (Eq. 12.22). Additionally, the pH gradient allows each electrode reaction to occur under optimal conditions, acidic for HER and alkaline for OER, enhancing reaction kinetics and reducing overpotential. As a result, the actual operational voltage for water splitting can be lower than 1.23 V, improving energy efficiency. This ability to facilitate water splitting at lower practical voltages makes BPMs highly advantageous for hydrogen production.

12.4.2 Roles and Requirements of Ion Exchange Membranes

In electrochemical water-splitting electrolyzers with monopolar membranes such as CEMs and AEMs, ionic conductivity is crucial alongside mechanical strength and chemical stability. AEMs must withstand the highly alkaline conditions of pH~14.

Since ion transport is closely linked to water transport, both the amount and movement of water within membranes significantly affect cell performance. However, minimizing the transport of fuels and oxidants is essential for improving both efficiency and safety.

In electrolyzers with BPMs, selective ion transport plays an additional critical role in enhancing water-splitting performance. At the bipolar junction, water dissociates into H^+ and OH^-, which then migrate across the CEL and AEL, respectively. Efficient selective ion transport is essential for this process. The strong electric field developed at the bipolar junction facilitates water dissociation, which can be raised by increasing the electric potential difference and/or reducing the junction thickness. Additionally, water dissociation catalysts are typically incorporated at the junction. However, this can lead to delamination of the CEL and AEL, compromising BPM integrity, which must be prevented to ensure long-term stability.

IEMs generally feature a microphase-separated structure with hydrophilic ionic channels dispersed in a hydrophobic matrix. Since both ion and neutral molecule transport occur through these channels, achieving fast and selective ion transport is vital while minimizing transport of product neutral molecules of O_2 and H_2 and maintaining mechanical strength required. These can be achieved by maximizing the number of channels while keeping their size small. Therefore, new chemical structures and morphologies must be developed to optimize performance.

12.4.3 Water Splitting with CEMs and AEMs

In electrochemical water-splitting electrolyzers, IEMs, including CEMs and AEMs, play a crucial role by conducting ions between the electrodes while preventing the mixing of the hydrogen and oxygen gases. IEMs improve the efficiency and safety of water splitting by maintaining the separation of the half-reactions and ensuring selective ion transport.

12.4.3.1 Proton Exchange Membranes (PEMs)

PEMs conduct protons (H^+) from the anode to the cathode in acidic electrolyzers. They are typically made from materials like Nafion. PEM electrolyzers operate in acidic environments and require expensive noble metal catalysts, such as platinum and iridium, to catalyze the reactions. PEMs offer high proton conductivity and excellent gas separation, making PEM electrolyzers efficient and scalable for hydrogen production.

Their advantages include high efficiency, compact design, and fast start-up times, while they need the high cost of PEM materials and noble metal catalysts, along with the acidic environment that can limit material choices for electrodes and components.

While highly efficient and commercially available PEMs, the main challenges for PEMs lie in reducing costs, particularly through the development of less expensive catalysts and membranes. Research into alternative proton conducting membranes with lower cost and higher durability is also ongoing.

12.4.3.2 Anion Exchange Membranes (AEMs)

AEMs conduct hydroxide ions (OH^-) from the cathode to the anode in alkaline electrolyzers. These membranes are based on polymer materials frequently functionalized with quaternary ammonium or other groups that conduct anions like OH^-.

AEM electrolyzers operate in alkaline conditions and allow the use of non-noble, more cost-effective catalysts, such as nickel or cobalt, compared to the expensive platinum group metals required in PEM systems. Lower cost due to cheaper catalysts and less expensive membrane materials, making AEM systems attractive for large-scale hydrogen production. AEMs show lower ion conductivity and stability compared to PEMs, leading to performance and durability challenges that are still being addressed in research.

Figure 12.13 illustrates the complex chemical structures of commercial AEMs designed to ensure stability in such environments. Notably, the Aemion membrane

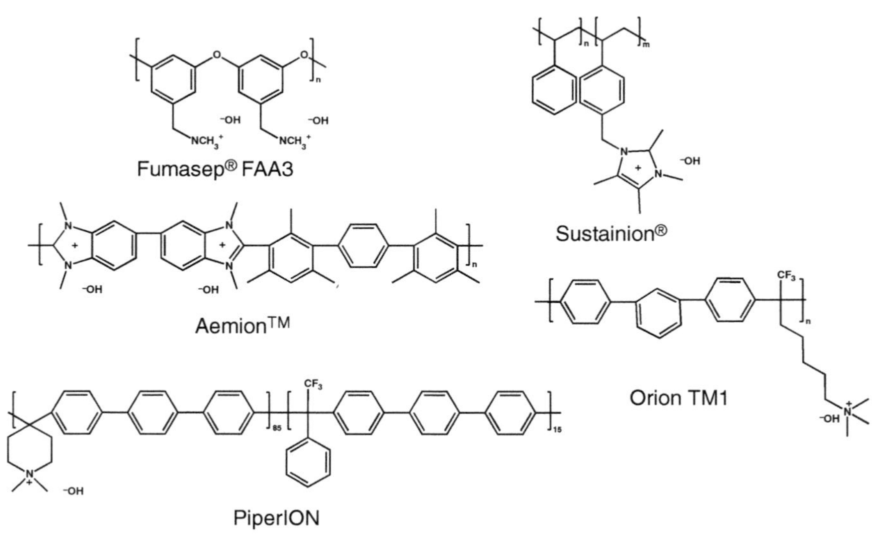

Figure 12.13 Chemical structures of commercial anion-exchange membranes (AEMs) for water splitting. *Source:* Y.S. Kang.

demonstrated durability, operating for 10,000 hours at voltages below 1.8 V under 1 A/cm² [62].

In short summary, PEMs are widely used in commercial PEM electrolyzers due to their high efficiency, but their cost remains a barrier. AEMs offer a promising low-cost alternative, though they face challenges in performance and durability. Both types of IEMs are crucial for the future of hydrogen production, with ongoing research aimed at optimizing their performance and cost-efficiency.

12.4.4 Water Splitting with Bipolar Membranes

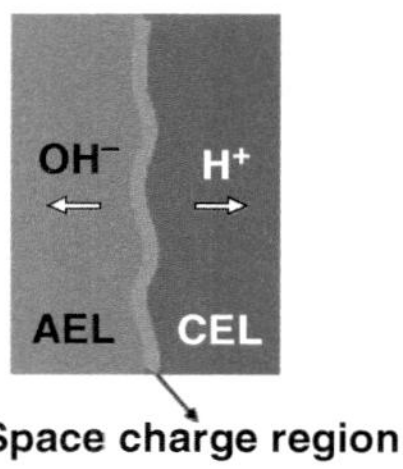

BPMs possess unique properties that distinguish them from traditional monopolar IEMs, including CEMs and AEMs. One notable property is that extremely high electric field is generated at bipolar junction in SCR of BPM, which has been successfully utilized for water-splitting electrolyzers to largely improve the energy efficiency.

Water-splitting performance is commonly illustrated in *J–V* curves, as shown in Figure 12.14. As can be seen, the current density, a measure of hydrogen production rate, increases with applied potential very differently for electrolyzers with monopolar CEMs and AEMs (Figure 12.14a) and with BPMs (Figure 12.14b). The current density increases nearly linearly with increasing potential, known as Ohm's law, for monopolar membranes, while their relationship for BPM is not linear anymore, but exponential. Therefore, the discrepancy between monopolar and BPMs becomes larger with increasing applied potential, demonstrating much higher energy efficiency of electrolyzers with BPM.

The marked improvement in energy efficiency for dissociation rate of water into H^+ and OH^- through BPMs is generally believed to be due to the huge electric field up to 10^8 V/m, called electric field-enhanced (EFE) water dissociation, arising through the bipolar junction at the SCR between CEL and AEL, as described in Section of 10.4 in Chapter 10.

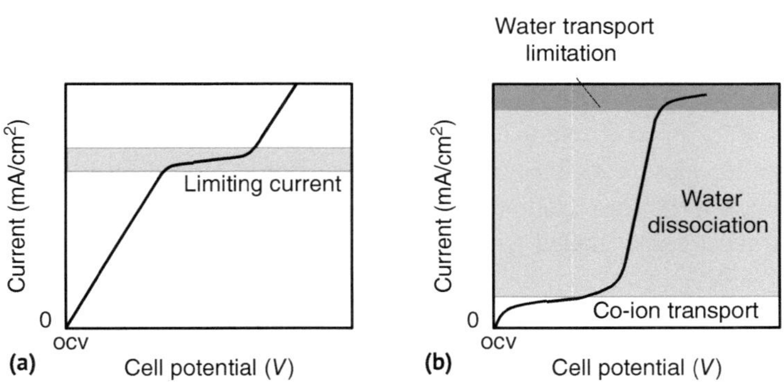

Figure 12.14 Schematic *J–V* curves for water-splitting electrolyzers with (a) monopolar membrane; and (b) bipolar membrane. Replotted from [63, 64].

12.4.5 Mathematical Models for Electric Field-Enhanced Water Dissociation

There have been three mathematical models to explain such high rate of the water dissociation into H^+ and OH^- ($H_2O \rightarrow H^+ + OH^-$). The second Wien effect (SWE) model, the chemical reaction (CHR) model by protonation–deprotonation reactions, and the combined model of the two will be summarized as follows [63, 65].

12.4.5.1 Second Wien Effects

The SWE model for electrolyzers with BPMs considers that water in the SCR at the bipolar junction behaves as a weak electrolyte, and uses Onsager's theory for the increase in the dissociation rate constant k_d of a weak electrolyte in the presence of an external electrical potential [66]. The final result of this model can be written as [65]:

$$\frac{k_{d,\mathrm{SWE}}}{k_d} = 1 + b + \frac{b^2}{3} + \frac{b^3}{18} + \frac{b^4}{180} + \frac{b^5}{2{,}700} + \frac{b^6}{56{,}700} \cdots \tag{12.26}$$

where $k_{d,\mathrm{SWE}}$ and k_d note the dissociation rate constant of water with and without an external electrical field ξ (V/m) and b is defined as:

$$b = \frac{0.096}{\varepsilon_r T^2}\xi \tag{12.27}$$

with the dielectric constant ε_r. It is clear that ε_r is not that of the bulk water but is for the state of the water molecules in the bipolar junction, which is different to that of pure water in the absence of a locally high external electric field. One possible choice for ε_r could be some average value between the dielectric constants of pure water ($\varepsilon_r = 79$) and the organic polymer membrane ($\varepsilon_r = 2$).

When $b > 3$, a condition usually realized at the SCR in BPMs, Equation (12.26) simplifies to [65]:

$$\frac{k_{d,\mathrm{SWE}}}{k_d} = \left(\frac{2}{\pi}\right)^{1/2}(8b)^{-3/4}\exp\left[(8b)^{1/2}\right] \tag{12.28}$$

The high electric field ξ needed for the onset of the water dissociation, known as electric field enhanced EFE) water dissociation, is attained at the bipolar junction, which is already described in Section 10.4 of Chapter 10.

Introducing the typical values for the fixed charge concentrations of cations and anions, $C_{FC} = C_{FA} = 1\ M$, $V = -1$ V and $\varepsilon_r = 20$, we obtain the maximum electric field $\xi_m = 10^8$ V/m according to the abrupt bipolar model described in Section 10.4.1 of Chapter 10. Substituting this value for ξ in Equation (12.28) yields $k_{d,\mathrm{SWE}}/k_d = 10^6$. This increase in the dissociation rate constant of water could explain the anomalously high-water dissociation rate observed in the *J*–*V* curves of electrolyzers with BPMs [67–69].

12.4.5.2 Chemical Reaction (CHR) Model

The CHR model, working via protonation–deprotonation mechanism, states a possibility that H^+ and OH^- ions are generated in proton-transfer reactions between water and fixed charged groups, playing as a catalyst for the water dissociation [70]. When the SWE model predicts similar water dissociating properties for the CEL and AEL junction, the protonation–deprotonation mechanism considers the water dissociation to be dependent on the catalytic activity of the fixed groups. This theory is in line with studies on catalytic activity of various ionic groups [71], the observed larger water dissociation effect for AEL, suggesting that the H^+ and OH^- ions may be generated from protonation–deprotonation reactions between some fixed charge groups (e.g. tertiary amino groups) and the water molecules, and the improvement of water dissociation after introducing a catalyst to the bipolar junction [72, 73].

As a result, the water dissociation mechanism in BPM can be described as a two-stage protonation–deprotonation reaction [74]. In the case of a weak base, B, the reactions can be written as:

$$B + H_2O = BH^+ + OH^- \tag{12.29}$$

$$BH^+ + H_2O = B + H_3O^+ \tag{12.30}$$

where BH^+ refers to the catalytic active center (usually a fixed charge group) for the proton-transfer reactions.

A similar equation could also be written for a cation exchange layer by substituting the fixed charge groups A^- and weak acid AH for B and BH^+ in Equations (12.29) and (12.30) [65]:

$$AH + H_2O = A^- + H_3O^+ \tag{12.31}$$

$$A^- + H_2O = AH + OH^- \tag{12.32}$$

where A^- is the catalytic center in the membrane. Reactions (12.29–12.32) can be summarized as:

$$2H_2O = H_3O^+ + OH^- \tag{12.33}$$

Steps for Equations (12.30) and (12.32) are the rate-limiting steps, as basic groups tend to remain in the protonated form (since the protonation step is much faster than the deprotonation step), while acidic groups remain in the deprotonated form [75].

We will refer to the general mechanism in Equations (12.29) and (12.30) as the CHR model, though it should be emphasized that: (1) this mechanism is not operative in all the existing fixed charge groups but appears to be particularly important only for certain functional groups; and (2) the mechanism is not restricted to the conventional fixed charge groups: catalytic protonation–deprotonation reactions between other active centers in the membrane (e. g. some metallic hydroxides [76], heavy-metal impurities [77], metal complexes [78], and water molecules) are also possible.

It has been assumed that the high electric field in the SCR of the BPM is the mechanism responsible for the increase in the effective dissociation rate constant of water produced by the proton-transfer reactions, and proposed the phenomenological equation [65, 79].

$$k_{d,\mathrm{CHR}} = k^0_{d,\mathrm{CHR}} \exp[\alpha F\xi/RT] \tag{12.34}$$

where $k_{d,\mathrm{CHR}}$ and $k^0_{d,\mathrm{CHR}}$ are the effective dissociation rate for water, which results from the protonation–deprotonation reactions of Equations (12.29) and (12.30) with and without an external electric field, ξ, respectively [80], and α is a microscopic parameter having the dimensions of length. Introducing the typical values $\alpha = 10^{-10}$ m, $\xi = 10^8$ V/m in Equation (12.34), we obtain $k_{d,\mathrm{CHR}}/k^0_{d,\mathrm{CHR}} = 20$ for the increase of the effective dissociation rate constant, a result which could explain the anomalously high water dissociation rate observed in the *J*–*V* curves of BPMs [65, 74].

It has been observed experimentally that the EFE water dissociation increases with temperature. This effect can be incorporated in the SWE and CHR models by assuming an Arrhenius relationship for the rate constants involved in each particular model.

$$k_d = k_d^0 \exp[-E_a/RT] \tag{12.35}$$

where k_d^0 is the pre-exponential (frequency) factor and E_a is the activation energy of the process, which takes different values in the SWE and CHR models. It should also be mentioned that Equation (12.35) gives a qualitative picture of the EFE water dissociation, which is in agreement with the experimental observations: the effective activation energy, E'_a, of the water dissociation reaction is decreased by the application of an electric field ξ from a given value E_a to a new value $E'_a = E_a - \alpha F \xi$ (see Equations (12.34) and (12.35)).

12.4.5.3 Combined Model of SWE and CHR Models

Strathmann et al. developed a model by combining the SWE and CHR models [75]. According to this model, under low electric field strength, only salt ions transport through BPM. With increasing electric fields, the salt ion flux into the junction cannot compensate for the salt ion flux out of the junction required to maintain the current, and the junction becomes depleted of ions. As a result, a limiting current density is reached, whose value depends both on BPM permselectivity and ion diffusion coefficients. When exceeding the limiting current density, water dissociation occurs due to the increased resistance of the depleted junction and the consecutively increased electric field across the junction, generating protons and hydroxide ions to further withstand the current. The dissociated water in the junction is replenished by a diffusive flux of water from the outer solution. The water dissociation is enhanced by protonation–deprotonation and accelerated by the electric field as in the SWE, while water recombination rate is not influenced by the electric field [63].

The water dissociation can be described by a model, which combines the calculation of the reaction zone thickness, the protonation and deprotonation of weakly basic or acidic groups, and the influence of the electric field on the dissociation rate constants, i.e. the SWE. The results indicate that the water splitting in BPMs takes place in a very small region of the monopolar ion exchange layer where uncompensated fixed charges exist. The protonation and deprotonation can take place at basic and at acidic groups. The *pK* value of these groups has to be in a range between 4 and 10. The thickness of the transition region between the ion exchange layers, which includes the reaction zone, is a few nanometers [75].

As a short summary, it has been briefly presented the three mechanisms employed to describe the anomalously high water dissociation rate observed in bipolar IEMs. Those mechanisms emphasize the effect of the high electric field in the membrane SCR on the water dissociation process. The experimental fact that EFE water dissociation occurs mainly in membranes having certain functional groups and/or active centers, such as heavy metals, appears to give more credit to the CHR model [81, 82].

12.4.6 Future and Challenges for Electrochemical Water Splitting

The most distinctive feature of BPMs in water-splitting electrolyzers is their ability to generate an extremely high electric field at the bipolar junction between the CEL and AEL. This high field significantly enhances water dissociation rates, requiring 0.828 V of the total 1.23 V. Additionally, water dissociation catalysts can be incorporated at the bipolar junction to further accelerate reaction rates without compromising electric field generation.

For example, hematite nanoparticles ($\alpha - Fe_2O_3$, 5–6 nm in diameter) dispersed on a graphene oxide (GO) layer act as effective water dissociation catalysts. These nanoparticles promote the formation of ice-like water structures and partially dissociated bound water (PDBW) (Figure 12.15a and b), as confirmed by ^{1}H-NMR spectroscopy, leading to a significant improvement in the water dissociation rate [83]. BPMs utilizing such catalysts reduce energy consumption for water splitting up to 42% compared to electrolyzers with conventional monopolar membranes, offering a major advantage.

BPMs can also establish and maintain a pH gradient across the membrane by conducting H^+ ions to one side and OH^- ions to the other. This capability creates highly acidic and highly alkaline environments on opposite sides, supporting processes that require specific chemical conditions, such as an acidic environment at the anode and an alkaline environment at the cathode. These conditions enable the

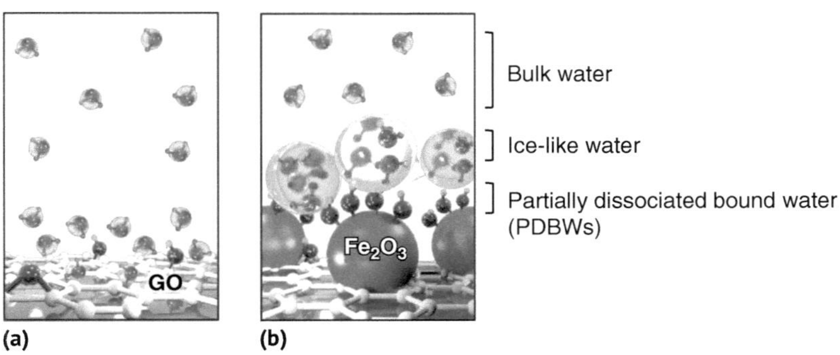

Figure 12.15 Schematic structures of bulk water and interfacial water, including ice-like water and partially dissociated water (PDBW) on surfaces of (a) graphene oxide (GO); and (b) Fe_2O_3@GO catalysts. *Source*: [83]/John Wiley & Sons.

development of catalysts optimized for specific pH environments, enhancing overall system performance.

Despite their advantages, BPMs face challenges such as long-term degradation due to mechanical stress or exposure to harsh chemical environments (e.g. pH = 14), which can reduce performance over time. Additionally, manufacturing high-quality BPMs with precise control over the cation–anion interface is technically demanding. Poor interface quality may result in lower ion conductivity compared to single-ion membranes, impacting overall efficiency. Efficient water management is also crucial to prevent issues like membrane drying or flooding, which can negatively affect performance.

In summary, BPMs represent a significant advancement in electrochemical water-splitting technology. They enhance water dissociation, improve energy efficiency, enable pH gradient control, and facilitate selective ion transport. These features make BPMs particularly suited for hydrogen production, where minimizing energy consumption and maximizing gas purity are critical. While challenges in durability and manufacturing persist, ongoing research and development continue to drive innovation in BPM technology, paving the way for more efficient and sustainable hydrogen production.

References

1 S. Wu, C. Wang, S. Li, J. Weng, Materials Chemistry Frontiers 8(18) (2024) 2924–2943.

2 S.N. Patel, ACS Macro Letters 10(1) (2021) 141–153.

3 F. Wu, K. Zhang, Y. Liu, H. Gao, Y. Bai, X. Wang, C. Wu, Energy Storage Materials 33 (2020) 26–54.

4 H. Wang, L. Sheng, G. Yasin, L. Wang, H. Xu, X. He, Energy Storage Materials 33 (2020) 188–215.

5 J.-M. Tarascon, M. Armand, Nature 414 (2001) 359–367.

6 A. Manthiram, X. Yu, S. Wang, Nature Reviews Materials 2(4) (2017) 1–16.

7 J.B. Goodenough, K.S. Park, Journal of the American Chemical Society 135(4) (2013) 1167–1176.

8 Y. Shan, L. Li, X. Yang, ACS Applied Energy Materials 4(5) (2021) 5101–5112.

9 P. Ding, Z. Lin, X. Guo, L. Wu, Y. Wang, H. Guo, L. Li, H. Yu, Materials Today 51 (2021) 449–474.

10 J. Qiu, X. Liu, R. Chen, Q. Li, Y. Wang, P. Chen, L. Gan, S.-J. Lee, D. Nordlund, Y. Liu, X. Yu, X. Bai, H. Li, L. Chen, Advanced Functional Materials 30(22) (2020) 1909392.

11 J. Li, S. Dong, C. Wang, Z. Hu, Z. Zhang, H. Zhang, G. Cui, Journal of Materials Chemistry A 6(25) (2018) 11846–11852.

12 C. Sångeland, G. Hernández, D. Brandell, R. Younesi, M. Hahlin, J. Mindemark, ACS Applied Materials & Interfaces 14(25) (2022) 28716–28728.

13 Y. An, X. Han, Y. Liu, A. Azhar, J. Na, A.K. Nanjundan, S. Wang, J. Yu, Y. Yamauchi, Small 18(3) (2022) 2103617.

14 M. Geng, G. Su, S. Huang, S. Wang, M. Xiao, D. Han, Y. Meng, Materials Chemistry Frontiers 7(23) (2023) 5963–5988.

15 S.T. Oyakhire, S.F. Bent, Energy Advances 3(1) (2024) 108–122.

16 Q. Liu, B. Cai, S. Li, Q. Yu, F. Lv, F. Kang, Q. Wang, B. Li, Journal of Materials Chemistry A 8(15) (2020) 7197–7204.

17 X. Lin, J. Yu, M.B. Effat, G. Zhou, M.J. Robson, S.C.T. Kwok, H. Li, S. Zhan, Y. Shang, F. Ciucci, Advanced Functional Materials 31(17) (2021) 2010261.

18 H. Miyashiro, Y. Kobayashi, S. Seki, Y. Mita, A. Usami, M. Nakayama, M. Wakihara, Chemistry of Materials 17(23) (2005) 5603–5605.

19 Z. Li, A. Li, H. Zhang, R. Lin, T. Jin, Q. Cheng, X. Xiao, W.-K. Lee, M. Ge, H. Zhang, A. Zangiabadi, I. Waluyo, A. Hunt, H. Zhai, J.J. Borovilas, P. Wang, X.-Q. Yang, X. Chuan, Y. Yang, Nano Energy 72 (2020) 104655.

20 J. Ma, Z. Liu, B. Chen, L. Wang, L. Yue, H. Liu, J. Zhang, Z. Liu, G. Cui, Journal of The Electrochemical Society 164(14) (2017) A3454.

21 L.-P. Wang, X.-D. Zhang, T.-S. Wang, Y.-X. Yin, J.-L. Shi, C.-R. Wang, Y.-G. Guo, Advanced Energy Materials 8(24) (2018) 1801528.

22 J. Lu, J. Zhou, R. Chen, F. Fang, K. Nie, W. Qi, J.-N. Zhang, R. Yang, X. Yu, H. Li, L. Chen, X. Huang, Energy Storage Materials 32 (2020) 191–198.

23 W. Zhou, Z. Wang, Y. Pu, Y. Li, S. Xin, X. Li, J. Chen, J.B. Goodenough, Advanced Materials 31(4) (2019) e1805574.

24 H. Duan, M. Fan, W.P. Chen, J.Y. Li, P.F. Wang, W.P. Wang, J.L. Shi, Y.X. Yin, L.J. Wan, Y.G. Guo, Advance Materials 31(12) (2019) e1807789.

25 P. Barai, K. Higa, V. Srinivasan, Physical Chemistry Chemical Physics 19(31) (2017) 20493–20505.

26 J. Pan, P. Zhao, N. Wang, F. Huang, S. Dou, Energy & Environmental Science 15(7) (2022) 2753–2775.

27 Z. Yao, H. Wu, Y. Li, J. Wang, J. Zhang, M. Zhang, Y. Guo, P. Wang, Energy & Environmental Science 8(11) (2015) 3192–3197.

28 D.E. Fenton, J.M. Parker, P.V. Wright, Polymer 14(11) (1973) 589.

29 M.B. Armand, J.M. Chabagno, M. Duclot, Fast Ion Transport in Solids: Electrodes and Electrolytes, Elsevier North Holland, Inc, New York, NY, 1979.

30 A.F. Nogueira, J.R. Durrant, M.A. De Paoli, Advanced Materials 13(11) (2001) 826–830.

31 A. Nogueira, C. Longo, M.-A. De Paoli, Coordination Chemistry Reviews 248(13–14) (2004) 1455–1468.

32 M.-S. Kang, J.H. Kim, Y.J. Kim, J. Won, N.-G. Park, Y.S. Kang, Chemical Communications (7) (2005) 889.

33 J.H. Kim, M.S. Kang, Y.J. Kim, J. Won, N.G. Park, Y.S. Kang, Chemical Communications (14) (2004) 1662–1663.

34 Y.J. Kim, J.H. Kim, M.-S. Kang, M.J. Lee, J. Won, J.C. Lee, Y.S. Kang, Advanced Materials 16(19) (2004) 1753–1757.

35 T.Y. Kim, W. Wei, T.K. Lee, B.S. Kim, S.C. Park, S. Lee, E.H. Suh, J. Jang, J. Bisquert, Y.S. Kang, ACS Applied Materials & Interfaces 10(3) (2018) 2537–2545.

36 D. Song, Y.-S. Choi, B.S. Kim, H.S. Kim, Y.S. Kang, Materials Letters 246 (2019) 137–140.

37 M.-S. Kang, J.H. Kim, J. Won, Y.S. Kang, Journal of Photochemistry and Photobiology A: Chemistry 183(1) (2006) 15–21.

38 M.-S. Kang, J.H. Kim, J. Won, Y.S. Kang, The Journal of Physical Chemistry 111 (2007) 5222–5228.

39 D. Song, W. Cho, J.H. Lee, Y.S. Kang, The Journal of Physical Chemistry Letters 5(7) (2014) 1249–1258.

40 Y.-G. Lee, S. Park, W. Cho, T. Son, P. Sudhagar, J.H. Jung, S. Wooh, K. Char, Y.S. Kang, The Journal of Physical Chemistry C 116(11) (2012) 6770–6777.

41 J.Y. Kim, T.H. Kim, D.Y. Kim, N.-G. Park, K.-D. Ahn, Journal of Power Sources 175(1) (2008) 692–697.

42 W.S. Chi, J.K. Koh, S.H. Ahn, J.-S. Shin, H. Ahn, D.Y. Ryu, J.H. Kim, Electrochemistry Communications 13(12) (2011) 1349–1352.

43 S.H. Ahn, D.J. Kim, W.S. Chi, J.H. Kim, Advanced Functional Materials 24(32) (2014) 5037–5044.

44 J. Wu, S. Hao, Z. Lan, J. Lin, M. Huang, Y. Huang, P. Li, S. Yin, T. Sato, Journal of the American Chemical Society 130 (2008) 11568–11569.

45 M.S. Su'ait, A. Ahmad, K.H. Badri, N.S. Mohamed, M.Y.A. Rahman, C.L.A. Ricardo, P. Scardi, International Journal of Hydrogen Energy 39(6) (2014) 3005–3017.

46 F. Bella, N.N. Mobarak, F.N. Jumaah, A. Ahmad, Electrochimica Acta 151 (2015) 306–311.

47 S. Rudhziah, A. Ahmad, I. Ahmad, N.S. Mohamed, Electrochimica Acta 175 (2015) 162–168.

48 P. Wang, S.M. Zakeeruddin, I. Exnar, M. Gratzel, Chemical Communications (24) (2002) 2972–2973.

49 K.-M. Lee, V. Suryanarayanan, K.-C. Ho, Journal of Power Sources 185(2) (2008) 1605–1612.

50 M.M. Noor, M.H. Buraidah, M.A. Careem, S.R. Majid, A.K. Arof, Electrochimica Acta 121 (2014) 159–167.

51 R. O'hayre, S.-W. Cha, W. Colella, F.B. Prinz, Fuel Cell Fundamentals, John Wiley & Sons, 2016.

52 H. Qayyum, C.-J. Tseng, T.-W. Huang, S.-y. Chen, Catalysts 6(11) (2016).

53 H.M. Kim, C. Hu, H.H. Wang, J.H. Park, N. Chen, Y.M. Lee, Journal of Membrane Science 644 (2022) 120109.

54 D.R. Dekel, Journal of Power Sources 375 (2018) 158–169.

55 J. Zhang, W. Zhu, T. Huang, C. Zheng, Y. Pei, G. Shen, Z. Nie, D. Xiao, Y. Yin, M.D. Guiver, Advanced Science 8(15) (2021) e2100284.

56 F. Xiao, Y.C. Wang, Z.P. Wu, G. Chen, F. Yang, S. Zhu, K. Siddharth, Z. Kong, A. Lu, J.C. Li, C.J. Zhong, Z.Y. Zhou, M. Shao, Advanced Materials 33(50) (2021) e2006292.

57 J. Wu, X.Z. Yuan, J.J. Martin, H. Wang, J. Zhang, J. Shen, S. Wu, W. Merida, Journal of Power Sources 184(1) (2008) 104–119.

58 J. Luo, A.H. Jensen, N.R. Brooks, J. Sniekers, M. Knipper, D. Aili, Q. Li, B. Vanroy, M. Wübbenhorst, F. Yan, L. Van Meervelt, Z. Shao, J. Fang, Z.-H. Luo, D.E. De Vos, K. Binnemans, J. Fransaer, Energy & Environmental Science 8(4) (2015) 1276–1291.

59 J. Rao, R. Vijayaraghavan, Y. Zhou, P.C. Howlett, D.R. MacFarlane, M. Forsyth, H. Zhu, Physical Chemistry Chemical Physics 20(6) (2018) 4579–4586.

60 H. Zhu, X. Wang, R. Vijayaraghava, Y. Zhou, D.R. MacFarlane, M. Forsyth, The Journal of Physical Chemistry Letters 9(14) (2018) 3904–3909.

61 N. Chen, Y.M. Lee, Progress in Polymer Science 113 (2021) 101345.

62 D. Henkensmeier, M. Najibah, C. Harms, J. Žitka, J. Hnát, K. Bouzek, Journal of Electrochemical Energy Conversion and Storage 18(2) (2021) 024001.

63 R. Pärnamäe, S. Mareev, V. Nikonenko, S. Melnikov, N. Sheldeshov, V. Zabolotskii, H.V.M. Hamelers, M. Tedesco, Journal of Membrane Science 617 (2021) 118538.

64 S.K. Patel, W. Pan, Y.U. Shin, J. Kamcev, M. Elimelech, Environmental Science & Technology 57(11) (2023) 4578–4590.

65 S. Mafé, P. Ramírez, Acta Polymerica 48(7) (1997) 234–250.

66 L. Onsager, The Journal of Chemical Physics 2(9) (1934) 599–615.

67 R. Simons, G. Khanarian, Journal of Membrane Biology 38 (1978) 11–30.

68 S. Mafe, J.A. Manzanares, P. Ramrez, Physical Review A 42(10) (1990) 6245–6248.

69 A. Tanioka, K. Shimizu, K. Miyasaka, H.J. Zimmer, N. Minoura, Polymer 37(10) (1996) 1883–1889.

70 R. Simons, Nature 280 (1979) 824–826.

71 G. Cheng, Y. Zhao, W. Li, J. Zhang, X. Wang, C. Dong, Journal of Membrane Science 589 (2019) 117243.

72 M.-S. Kang, Y.-J. Choi, H.-J. Lee, S.-H. Moon, Journal of Colloid and Interface Science 273(2) (2004) 523–532.

73 M.B. McDonald, M.S. Freund, ACS Applied Materials & Interfaces 6(16) (2014) 13790–13797.

74 R. Simons, Electrochimica Acta 30(3) (1985) 275–282.

75 H. Strathmann, J.J. Krol, H.-J. Rapp, G. Eigenberger, Journal of Membrane Science 125 (1997) 123–142.

76 New Developments in Ion Exchange: Materials, Fundamentals and Applications - Proceedings of the International Conference on Ion Exchange, ICIE '91, Tokyo, Japan, 2–4 October 1991, Elsevier Science Ltd, Tokyo, Japan, 1991.

77 H. Hurwitz, R. Dibiani, R. El Moussaoui, O. Van Volden, M. Maeck, Conference Proceedings of Euromembrane 95, Bath, UK, 1995, pp. 89–94.

78 V. Zabolotskii, V. Ganych, N. Sheldeshov, Soviet Electrochemistry 27(10) (1991) 1098–1102.

79 S.F. Timashev, T.J. Kemp, Physical Chemistry of Membrane Processes, Ellis Horwood, 1991.

80 N.Y. Pivovarov, V.P. Greben, S.A. Ponomarenko, Russian Journal of Electrochemistry 30(10) (1994) 1113–1117.

81 V.I. Zabolotskii, N.V. Shel'deshov, N.P. Gnusin, Russian Chemical Reviews 57(8) (1988) 801–808.

82 V.I. Koval'chuk, Khimiya i Tekhnologiya Vody 15(7–8) (1993) 483–501.

83 B.S. Kim, S.C. Park, D.H. Kim, G.H. Moon, J.G. Oh, J. Jang, M.S. Kang, K.B. Yoon, Y.S. Kang, Small 16(41) (2020) e2002641.

Index

Molecule and Ion Transport through Polymer Membranes, First Edition. Yong Soo Kang.